The New Foundations of Evolution

JAN SAPP

The New Foundations
of Evolution
On the Tree of Life

OXFORD
UNIVERSITY PRESS

2009

OXFORD
UNIVERSITY PRESS

Oxford University Press, Inc., publishes works that further
Oxford University's objective of excellence
in research, scholarship, and education.

Oxford   New York
Auckland   Cape Town   Dar es Salaam   Hong Kong   Karachi
Kuala Lumpur   Madrid   Melbourne   Mexico City   Nairobi
New Delhi   Shanghai   Taipei   Toronto

With offices in
Argentina   Austria   Brazil   Chile   Czech Republic   France   Greece
Guatemala   Hungary   Italy   Japan   Poland   Portugal   Singapore
South Korea   Switzerland   Thailand   Turkey   Ukraine   Vietnam

Published by Oxford University Press, Inc.
198 Madison Avenue, New York, New York 10016

www.oup.com

Oxford is a registered trademark of Oxford University Press.

Library of Congress Cataloging-in-Publication Data
Sapp, Jan.
The new foundations of evolution : on the tree of life / Jan Sapp.
   p. cm.
Includes bibliographical references and index.
ISBN 978-0-19-538849-7; 978-0-19-538850-3 (pbk.)
1. Microorganisms—Evolution.   2. Microbial genetics.   3. Evolution (Biology).
4. Biology—Classsification.   I. Title.
QR13.S27 2009
571.2'9—dc22   2008050915

9 8 7 6 5 4 3 2 1
Printed in the United States of America
on acid-free paper

In memory of my father, Melan P. Sapp

FOREWORD

THE WORD "UNIQUE" best describes the book *The New Foundations of Evolution: On the Tree of Life* and its author, the historian Jan Sapp. Historians of biology typically focus on evolution à la Darwin. Yet, there is much to be understood about the evolutionary process that never met the Darwinian eye, and much more work to be done and biological systems to visit before science can say it "understands" evolution.

I have struggled long and hard to convince biologists that biology owes both science and mankind a genuinely scientific study of the evolutionary process, and that the place to start is not with the birds, beetles, and the bees all over again, where conventional evolutionary language shackles your thoughts before you begin. One starts with the microbial world; starts within the cell, not without; focuses on the origin and evolution of the cell's universal molecular componentry, not the adaptive embellishments. And one does not stick the label "made by natural selection" on anything. Here, in this new venue, is where we can begin to trace organisms back to their roots and begin to talk about evolution in a new, non-anthropomorphic language.

And finally! Along comes a book with an eye-popping title—*The New Foundations of Evolution: On the Tree of Life*, and it is about the microbial world. It is a book I never thought I'd see written by a historian. It says to historian and scientist alike: "Yes, there is evolution after Darwin; and here is what it's going to look like!"

It is impossible to understand the microbial world in any depth without considering the constant evolutionary current that flows through it. To account for the intricate and fascinating molecular structure within microbial cells or the organization of these cells into delicately fabricated microbial communities—so intimately interlinked with their environments—is a weaving of ecology, evolution, and organism, the likes of which are not seen in the larger world "above."

Dr. Sapp's book recounts the story of a basically isolated scientific field struggling to define its venue, find itself, and take its proper place among the other biological disciplines. It is a story of how molecular evolutionists working in the microbial world were able to discover the large-scale structure of the tree of life, and in the process questioned some of the major evolutionary understandings, such as the doctrine of common descent, the notion that evolution occurs only in very small random steps, and the idea that the organisms cannot "learn" from other organisms or share inventions with them. And it is a story of the discovery that there are not two primary lineages of living organisms on this planet, the eukaryotes (animals, plants, fungi, and "protists") and the microscopic prokaryotes, as everyone thought there were, but actually three. The so-called prokaryotes are not all related to one another, but comprise two great classes of (micro)organisms, which are less related to each other than we are to plants. These are the Archaea and the Bacteria, and between them they comprise the bulk of the biomass on this planet and by far the greatest cellular diversity.

Dr. Sapp is as unique among historians of biology as his work is among theirs. His is not a recounting of biology and evolution past, of problems solved and tucked away. His is a story of bringing evolution and biology together, of a new science of biology in the making. Thus, he finds his history on the unpaved trails of contemporary scientific exploration rather than safely recording his travels along the scientific superhighways of the past.

Carl R. Woese

PREFACE

THIS BOOK IS ABOUT the search for the foundations of evolution on this planet, the primary lineages of life, and the most profound differences in life's forms, represented in its highest taxa—the domains and kingdoms. It is a story about a revolution in the way in which biologists explore life's long history on Earth, understand its evolutionary processes, and portray its variety. It is about life's smallest entities, deepest diversity, and largest cellular biomass: the microbiosphere. To come to grips with microbial evolution is to reconsider much of classical biology's understanding of the processes of evolution, its imagery, methods, and doctrines.

Evolution is typically described as "the origin of species," as first summarized in Charles Darwin's legendary work of 150 years ago and articulated in the twentieth century. The problems, protests, and confusions that had sidelined Darwin's theory for many decades were resolved ecumenically in the 1930s and 1940s by what became known as the "modern synthesis"—a fusion of Mendelian genetics, population thinking, and natural history with Darwin's theory of natural selection. Gene mutation and recombination between individuals of a species were the fuel for evolution by natural selection.

The evolutionary synthesis of the last century was forged in terms of a two-kingdom world of animals and plants, whose histories cover at most 25% of the span of evolutionary time on Earth. Focused on the origin of species, that perspective had no concern for the primary groupings of life, the all-embracing kingdoms. In effect, it was a sterile conception of evolution—a world without a microbial foundation.

Evolution is not primarily about the origin of species. That formulation of the problem does not offer useful explanation for the evolution that occurred before organisms as we know them appeared or for understanding the evolutionary process at a deep level, problems such as how life as we know it emerged, and how cellular organization evolved, the genetic code developed, and genomes formed. The origin of species perspective does not really help to rationalize much of the great genetic and biochemical diversity on Earth in the microbial world, where three domains and most of the kingdoms of life are distinguished. Bacterial evolution is not a study in the origin of "species," a doubtful concept at best in that sphere.

Classical evolutionists did, of course, see animals and plants as having evolved somehow from the loosely conceived "lower" or "primitive" organisms, which microbes were taken to be. But that microbial world lay far beyond their purview and interest. Microbiology was largely consigned to pathology, agriculture, and industry before the Second World War, and when microbes emerged at the center of biology afterward, only a very few of them were chosen for study as laboratory domesticates, which were taken as representatives of all. Their study was motivated by their utility for molecular biology and biochemistry.

The elucidation of the structure of DNA, how it is replicated, and how it encodes the genetic information for the synthesis of proteins came to define the biology of the twentieth century. Molecular biology moved with breakneck technological speed and great promise for medicine and agriculture. To the extent that molecular biology showed interest in evolution, it was in guiding the process to the benefit of humans in the future. The interest was generally not there when it came to studying evolution's past to try to understand the process more deeply. Leading microbiologists declared, in effect, that the bulk of organismal evolution simply cannot be known. Leading molecular biologists agreed that the same was true of the evolution of the cell and its parts.

The emergence of molecular phylogenetics, beginning with the comparative study of the amino acid sequences of proteins in the 1960s, broadened to include RNA and individual genes in the 1970s, and then finally whole genomes in the 1990s. All this brought revolutionary change to biology. Startling new and fundamental concepts began to enter evolutionary biology when sequencing was developed for the ancient informational molecules, the "ultimate molecular chronometers" that formed the basis of the genetic system in all organisms. Evolutionary "signatures" from the deep evolutionary past began to emerge from the information encoded in the RNAs of ribosomes—those structures responsible for translating the sequence in the DNA gene into that of its corresponding protein. The information in that ribosomal RNA was held to reveal the universal relationships among all organisms.

There were two separate aspects to classical evolutionary biology. One was concerned with processes of evolutionary change: studies of the change in gene frequencies within populations over time and context. The other was concerned with genealogy or phylogeny: the evolutionary history of a group as it unfolded over time. The group may comprise one or a few species, or it may encompass a larger number of forms, for example, the modern and ancient horse. Organisms were classified and arranged in a hierarchy of groups within a group. Phylogenetic relations were portrayed as a bifurcating tree. Comparative morphology and/or the fossil record were the ways to know the history of a group.

Phylogenetics was reborn as a meticulous experimental science, and investigations of life's history were extended by some three billion years in the process. The new evolutionary biology owed little to the classical period. Its methods were its own, and they unearthed astonishing modes of evolutionary change, unforeseen primordial paths, and an unexpected wealth of genetic diversity.

The new era in microbial phylogenetics began in 1977 with the announcement of the discovery of a form of (microscopic) life, first called the "archaebacteria," phylogenetically as distinct from ordinary bacteria as the latter were from plants and animals. That was soon followed by the publication of the outlines of a universal tree of life, a construction that biologists had long abandoned as beyond scientific possibility. These developments were embedded in novel conceptions of life's emergence: that life as we know it today did not stem from one universal common "ancestral form" (i.e., cell), but had emerged three times, from some primitive communal ancestral state, giving rise to three primordial lineages, or urkingdoms, later referred to as three "domains" of life: the Archaea, the Bacteria, and the Eucarya.

Molecular phylogeneticists further posited that these primordial lineages did not evolve gradually, in the genealogical manner of plants and animals fueled by gene recombination and mutation within species. Transfers of genes, of gene clusters, and of whole genomes between distantly related taxonomic groups are fundamental processes in microbial evolution. In the microbial world, evolution is not therefore solely a matter of changes in the gene frequency in given populations. Populations evolve as do individual microorganisms. The inheritance of acquired genes and genomes is fundamental to microbial evolution. The mixing of genes between widely different taxa can result in rather sudden changes, equivalent to hybridization. In the microbial world, evolution by saltation seems the rule.

Evolutionary innovation through the transfer of genes among distantly related groups is not limited to the bacterial world. It applies also to the eukaryotic protists, microbes whose cells, like our own, possess membrane-bound nuclei and divide by mitosis. The eukaryotic cell also did not evolve gradually in the manner proposed by Darwinian evolutionists. Molecular phylogeneticists have provided rigorous experimental evidence that mitochondria (the respiratory organelles of cells) and the chloroplasts of photosynthetic organisms originated as symbiotic bacteria that had entered a host cell hundreds of millions of years ago. Their gradual integration within the genetic system of their host has meant that many of their genes have been transferred to the cell nucleus, while others needed for their independent life have been lost.

The acquisition of mitochondria and chloroplasts represents merely the tip of the iceberg of symbiosis in evolution. Examples of the inheritance of acquired bacterial symbionts are legion in protists, as well as in multicellular eukaryotes.

Phylogeneticists have shown that genes from bacterial symbionts have also been transferred to the nuclear genomes of their hosts. Whether the nucleus and the outlying cytoplasm of eukaryotic cells are also vestiges of an ancient symbiosis or fusion of some kind is well considered today.

Molecular phylogenetics has revitalized evolutionary biology. The concept of three domains and the fundamental roles of lateral gene transfer and symbiosis in their evolution provide the structure for microbial evolutionary theory today. As this new aspect of evolutionary biology confronted the classical, it was met with fierce debates over its techniques for constructing an outline of a universal evolutionary tree, its evidence for three domains of life, the scope and significance of symbiosis in eukaryotic evolution, and lateral gene transfer among bacteria.

Few (if any) of the evolutionary concepts in this book appear in prior historical accounts of microbiology, shaped as those books were by microbiology's relationship with medicine, agriculture, and industry. Investigations of the evolutionary history and diversity of microbes have largely developed in conflict with those interests. The relationships of microbes to each other and to other organisms were debated and the nature and meaning of kingdoms contested long before the rise of molecular phylogenetics. To appreciate the issues is to understand the methods and principles of classification. Are kingdoms real? Or are they, and other groupings, subjective human constructions of convenience?

We begin our journey in the seventeenth century, when microscopic creatures were first discovered and taken to be little animals. That perspective changed in the eighteenth century when the microbial world was typically understood to lie at the boundaries of the two great kingdoms of plants and animals, sharing properties of both. Microscopic forms were grouped together with other ambiguous or chimeric creatures such as sponges and sea anemones. They were referred to collectively as animal-plants: the zoophyta. By the early part of the following century, there were three interpretations of them. Some saw in the zoophyta evidence that there was no essential difference between plants and animals. Others called for a separate kingdom for such organisms. Still others, equipped with improved microscopes, were intent on establishing their "true" animal or plant natures.

The depiction of microbes as whole animals with parts equivalent to the organs and appendages of a dog or a cat lasted until the middle of the nineteenth century. With the rise of cell theory, the concept of the animal and plant was radically changed. Microbes were understood to be homologous not to whole animals or plants but to the cells of which animals and plants were composed. Microscopic life was considered collectively to include the first animals and first plants. Several proposals for a kingdom of their own were made on that basis in the 1860s: Protozoa, Primegum, Primalia, and Protista. According to the last, not all microorganisms were homologous to the cells of animal and plants;

there were still simpler forms, the "monera," that had evolved at the dawn of life on Earth. That proposal was embedded in phylogenetic speculation.

Darwin, and Jean-Baptiste Lamarck before him, had argued that to classify so as to reveal evolutionary relationships among the great groups of animals, one had to distinguish adaptive characteristics, which resulted from the organism's relationship to its environment, from "essential" organismal characteristics, which were not. "Essential" characters were useful in a genealogical classification; "adaptive" characteristics were genealogically unreliable. Darwin said that the grouping together of all life on Earth was "a grand fact." But there were no methods then available for proving it.

To trace genealogies, one had to distinguish between homologous characters, due to shared ancestry (e.g., the wing of a bat and the arm of a primate), and those that were analogous (e.g., the wing a bat and that of a bird). Whether taxonomy of animals and plants could and should be based on such evolutionary considerations was debated from the beginning of phylogenetic speculations in the nineteenth century. Leading evolutionists, from the nineteenth century to the middle of the twentieth century, had argued that phylogeny and taxonomy should be kept apart, distinguished as hypothesis is from observation. Phylogeny at best could only be deduced from taxonomic classifications.

Nowhere were such discussions of taxonomy more heated than in the classification of the smallest of microbes, the "bacteria." From a medical perspective, they were "germs"; to the botanists interested in the deeper question of bacteria's place in nature, they were the "fission fungi" or "fission plants." The concept of bacteria as a class of plants persisted into the twentieth century; even today, physicians will speak of "gut flora." When bacteriology emerged in the 1860s, coupled with germ theory, it was not certain if they could be partitioned into taxa of any kind or if they had any evolutionary histories to discover. That issue was effectively resolved in the 1880s with the development of techniques for isolating and characterizing microbes in pure cultures. Because bacteriology was primarily an applied science, the study of bacterial diversity was greatly restricted over the next 100 years, focusing almost exclusively on those microbes that directly or indirectly affected the health of humans. Then again, bacteria live in such a complex, co-dependent way in the environment that the great majority could not be cultured in isolation of one another and so characterized.

How bacteria could be distinguished from other microbes was uncertain. A clear biological definition of bacteria was lacking. A kingdom called "Monera" was proposed in the 1930s and 1940s on the grounds that bacteria did not have nuclei. But bacteria were seen at the very limits of the resolution of the light microscope, and their anatomy was debated throughout most of the twentieth century. Whether the organisms grouped together as bacteria stemmed from one, several, or many lineages was unknown. Life itself could have emerged once, a few times, or indeed many times, in which case the "bacteria" could include all manner of organisms. Their commonality could be deduced only from their genealogy or evolutionary history, which was totally unknown and unknowable before the development of molecular phylogenetics.

Various schemes of bacterial classification would be presented, some based on morphological characteristics, and others on physiological ones. All were sooner or later rejected. A committee on bacterial classification formed in the second decade of the twentieth century decided that a system of classification based on as many kinds of characteristics as possible—nutrition requirements, pathogenicity, biochemistry, morphology, and so one—would best approximate a natural, hierarchical, evolutionary order of things. This pluralist approach became the basis of *Bergey's Manual*, the internationally recognized authoritative handbook of bacterial taxonomy. Critics complained that such an unruly approach, which did not distinguish between homologous and analogous traits, but yet that classified hierarchically, group within group, was merely a facade that gave the illusion of a natural phylogenetic classification. These critics sought a classification that distinguished between homologous and analogous traits, before they, too, abandoned the effort in the middle of the twentieth century.

All classification schemes were based on assumptions regarding the origins of life and ever increasing complexity. It was widely believed that bacteria did not possess species as such, and many assumed they even lacked genes; it was commonly believed that bacterial heredity and adaptation involved "Lamarckian" mechanisms. Autotrophic bacteria, those that could acquire energy through inorganic compounds, were commonly considered to have been the first organisms. But that paradigm shifted in the 1940s to one in which the first modern organisms were heterotrophs feeding in a rich primordial soup of organic compounds.

The kingdom Monera was strengthened by the early 1960s when, based on developments in electron microscopy and bacterial genetics, microbiologists agreed that the bacterium lacked a membrane-bound nucleus, organelles, and sex comparable to other organisms and did not divide in the complex manner of other microbes. Bacteria were defined as "prokaryotes," in counterdistinction to cells that possessed true nuclei, the "eukaryotes." Monera or Prokaryotae was typically considered with four other kingdoms: Protista, Fungi, Plants, and Animals. The common origin of prokaryotes as a group of organisms that preceded and gave rise to eukaryotes was assumed by the principle of the "unity of life."

Molecular sequencing inverted the relationship between phylogenetics and taxonomy. Genealogies were not to be deduced from classification—taxa were to be deduced from molecular genealogies. Still, it was uncertain if molecular phylogenetic methods could be applied to the bacterial world. Bacterial geneticists had shown that genes could be transferred laterally among bacterial taxa. To trace the genealogy of a gene was not necessarily to trace the genealogical history of the organism. Lateral gene transfer was recognized to underlie the antibiotics resistance crises in modern medicine and to be at the basis of emergent biotechnology. Its prevalence in nature was unknown, but some bacteriologists of the early 1970s speculated that it could completely scramble the molecular phylogenetic record. Rather than representing a bifurcating tree, and a hierarchy of group within group, bacterial evolution could be highly reticulated.

The RNA of ribosomes (rRNA) emerged as the molecular tool of choice in the 1970s for measuring genealogical relationships among bacteria, and for the construction of a universal phylogenetic tree. It was held to be at the genetic core of all organisms, an "essential" characteristic of all organisms, and far removed from the interactions of the organism and its environment. The rRNA results contradicted many of the existing major bacterial classification schemes. Some major bacterial taxa were split up as unrelated; others were combined. But none compared to the new grouping called the archaebacteria which, according to their rRNA "signatures," had no specific relationship with typical bacteria.

The archaebacteria was an oddball group comprising a diversity of physiological and morphological types. They tended to live in "extreme" environments. The first to be identified were the "methanogens," which lived in anaerobic environments such as rumens, swamps, and hot springs. Next to join the group were salt-loving "extreme halophiles," known for rotting salted fish. Then came a thermophilic microbe, called *Thermoplasma*, that strangely grew in smoldering coal mine refuse piles. No one even remotely suspected that these organisms were related; their odd properties were taken as adaptations to the "extreme" habitats in which the organisms lived. Under the archaebacterial concept, their odd traits were not independent adaptations of unrelated organisms but exactly the opposite: the conserved common characteristics of an ancient lineage, "a third form of life."

Microbiologists, especially those in the United States, were incredulous regarding both the construction of a universal phylogeny and the claim for a "third form of life" based on rRNA. But in Germany, new vibrant research programs on the archaebacteria were started in the 1970s, which greatly expanded knowledge of their biochemistry, molecular biology, and natural history. By the end of the decade, the organisms grouped together as the archaebacteria were shown to share several remarkable characteristics. The chemical structure of their lipid cell membranes was unique, strikingly different from those of typical bacteria; their cell walls lacked peptidoglycan, a presumed defining feature of the prokaryote, and their enzymes responsible for transcribing DNA to RNA were unique. Archaebacterial research focused on understanding the relationships among the three urkingdoms. Efforts to untangle those relationships would persist over the next three decades.

The rRNA-based phylogenetics was applied equally to eukaryotic cell evolution beginning in the 1970s. Scientific speculations about the symbiotic origin of mitochondria and bacteria were transformed into verifiable hypotheses with rRNA phylogenetics. A new era in microbial ecology also began in the middle of the next decade, when it was realized that nucleic acid probes based on rRNA sequences could be used to identify microbes directly from the environment. Microbes no longer needed to be isolated in pure cultures to be identified. The new environmental phylogenetics was to reveal a diversity of microbial life hitherto unimagined. The archaebacteria turned out not to be confined to "extreme" environments. They were discovered in a great variety of "normal" habitats, and they were found to be among the most abundant kinds of organisms on Earth.

The turmoil in microbiology persisted into the 1990s. A fierce dispute broke out when the "archaebacteria" were renamed "Archaea" in a formal taxonomic proposal of three primary "domains" as a replacement for the prokaryote–eukaryote dichotomy. Debates polarized those who preferred to group together the archaea and the bacteria as prokaryotes, and those who argued that the prokaryote concept prevented a proper understanding of the nature of the primary organismal forms and misrepresented the course of evolution.

Microbial genomics gave rise to further complications in unraveling the deep relations among the three domains. Lateral gene transfer among the Bacteria and the Archaea was discovered to be far more extensive than hitherto imagined. The trafficking of genes among taxa is so pervasive over evolutionary time that to follow most genes is to trace the history of a "worldwide web." The nuclear genomes of eukaryotes also were discovered to contain ancient bacterial and archaeal genes in addition to eukaryote-specific ones. How that chimera was formed is unresolved.

ACKNOWLEDGMENTS

THIS BOOK COULD NOT HAVE been written without the help of many friends and colleagues. I thank Carl Woese for many discussions over the years and for his critical readings of early draft chapters, George Fox for his reading of chapters 10–22, and Ford Doolittle for his reading of chapters 21 and 22. For unpublished documents and permission to quote from letters, I thank George Fox, John Fuerst, Roger Garrett, Ingelore Holtz (for Wolfram Zillig's letters), Otto Kandler, Joshua Lederberg, Norm Pace, Mark Wheelis, Carl Woese, and Emile Zuckerkandl. Perspectives in this book were greatly enriched by interviews and discussions with Edward DeLong, Ford Doolittle, Peter Gogarten, Michael Gray, Radhey Gupta, Otto Kandler, Trudy Kandler, Charles Kurland, James Lake, Gary Olsen, Lynn Margulis, William Martin, Miklós Müller, Norman Pace, Karl Schleifer, Mitchell Sogin, Erko Stackebrandt, and Karl Stetter. Ianina Altshuler and Stephanie Chen expertly prepared the figures. I thank Peter Prescott for his editorial expertise. I also thank Carole McKinnon, Will Sapp, and Elliot Sapp. I would like to express my gratitude to the biology department at York University. A special thanks also to the Smithsonian Tropical Research Institute. I am most grateful to the Social Sciences and Humanities Research Council of Canada for support of this work.

CONTENTS

Foreword by Carl R. Woese vii

ONE   Animal, Vegetable, or Mineral? 3

TWO   Microbes First 17

THREE   The Germ of Phylogeny 28

FOUR   Creatures Void of Form 45

FIVE   About Chaos 57

SIX   Kingdoms at Biology's Borders 71

SEVEN   The Prokaryote and the Eukaryote 85

EIGHT   On the Unity of Life 100

NINE   Symbiotic Complexity 115

TEN   The Morning of Molecular Phylogenetics 127

ELEVEN   Roots in the Genetic Code 145

TWELVE   A Third Form of Life 162

THIRTEEN   A Kingdom on a Molecule 177

FOURTEEN   Against Adaptationism 187

FIFTEEN   In the Capital of the New Kingdom 199

SIXTEEN   Out of Eden 216

SEVENTEEN   Big Tree 226

EIGHTEEN   The Dawn Cell Controversy 243

NINETEEN   Three Domains 257

TWENTY   Disputed Territories 267

TWENTY-ONE   Grappling with a Worldwide Web 282

TWENTY-TWO   Entangled Roots and Braided Lives 300

Concluding Remarks 314

Notes 319

Index 405

The New Foundations of Evolution

| Animal, Vegetable, or Mineral?

*I have oft-times been besought, by divers gentlemen, to set down on paper
what I have beheld through my newly invented* Microscopia: *but I have
generally declined; first, because I have no style, or pen, wherewith to express
my thoughts properly; secondly, because I have not been brought up to languages or arts, but only to business, and in the third place, because I do not
gladly suffer contradiction or censure from others.*

—Antony Leeuwenhoek to the Secretary of
the Royal Society of London, August 15, 1673

Little Animals

Antony van Leeuwenhoek (1632–1723) described the invisible world of little
animals in more than 200 letters he sent to the Royal Society of London beginning in 1674.[1] He was not an educated man. His father was a maker of the
wicker baskets used to transport the fine wares produced in Delft. Antony had
spent six years in Amsterdam working in a linen-drapery shop as an appren-
tice to a cloth merchant; he became bookkeeper and cashier before returning
to his native Delft when he was 22 years old. In 1666, the city appointed him
Chamberlain to the Council Chamber of the Sheriffs of Delft, a position that
offered him a permanent source of income and allowed him to concentrate
on his microscopic observations. His marriage in 1671 to Cornelia Swalmius
(daughter of a merchant who dealt in serge) brought him into association with
a more intellectual group.[2] He was a friend of Johannes Vermeer, and it has
been suggested that he influenced Vermeer's use of lenses, the use of the *camera*

obscura, in the creation of his paintings.[3] Some suggest that he is the figure in two of Vermeer's famed paintings, *The Geographer* and *The Astronomer*. He was Delft's proud citizen, visited by dignitaries, distinguished scientists, and nobility.[4]

The key to Leeuwenhoek's discoveries was in the art of the technology: his skill at making microscopes, his subtle use of light, his delicate touch, and his keen eyesight. There is no agreement about exactly who first invented the compound microscope, but similar devices were made in Holland in the early seventeenth century. The compound microscope employed two separate lenses that could be moved relative to each other by means of a sliding tube, allowing the observer to change magnification, to zoom in and out. That device was further developed by Robert Hooke (1635–1703) in London, who attached an eyepiece from a telescope to the viewing lens, thus adding a third lens. Hooke drew stunning illustrations of his microscopic observations in his famed book *Micrographia* of 1665.[5] These early compound microscopes magnified objects no more than about 20 or 30 times.

The microscopes Leeuwenhoek made were simpler and better.[6] Actually, they were just very powerful magnifying glasses. Magnifying glasses had been used to count the threads in cloth, and Leeuwenhoek had graded the density of warp and weft in the textiles in Amsterdam. Intrigued by Hooke's descriptions, he taught himself new methods for grinding and polishing extremely small lenses and experimented with various methods and combinations, changing the forms of the magnifying apparatus, and he tried glass, rock crystal, and even diamonds for his lenses. The strongest of his lenses is said to have given magnifications up to 270 times.[7] Because of its very short focal length, the microscope had to be held a fraction of an inch away from both the observed specimen and the observer's eye. Leeuwenhoek's skill at grinding lenses, combined with his acute eyesight and great care in adjusting the lighting, enabled him to describe a great diversity of "animalcules"—which microscopists 250 years later identified to be forms of yeast, protists, and bacteria, as well as hydra and rotifers.

Microscopists of the twentieth century marveled at his detailed description of protozoa in rainwater and his discovery of bacteria. Leeuwenhoek first observed bacteria in April 1676, accidentally, when trying to discover what it was that made pepper so hot:

> I did now place anew abut 1/3 ounce of whole pepper in water, and set it in my closet, with no other design than to soften the pepper, that I could better study it.... The pepper having lain about three weeks in the water,...I saw therein, with great wonder, incredibly many very little animalcules, of divers sorts.... The fourth sort of little animals, which drifted among the three sorts aforesaid, were incredibly small; nay, so small, in my sight, that I judged that even if 100 of these very

wee animals lay stretched out one against another, they could not reach to the length of a grain of course sand.[8]

The first drawings of bacteria (figure 1.1) accompanied his description of the "animalcules" he had found in the tarter of his teeth and that of two women and an old man as he described them in a letter of September 17, 1683. He took measurements of everything he observed, selecting objects to compare—a hair, a grain of sand. Mites on the rind of cheese were among the smallest creatures visible to the naked eye. The size of some of the animalcules he observed compared with the mite as the bee with the horse. The smallest animalcules were "quite a thousand times thinner than the hair off one's head."[9] "Ten hundred thousand of these living creatures could scarce equal the bulk of a course sand-grain."[10]

Leeuwenhoek also described spermatozoa, "a huge number of little snakes or eels." He made no mention of the medical doctrines of contagion of the time, but others related the discovery of the "little animals" in support of hypothetical infectious entities. Various concepts of the contagion held the air to be infested with invisible wormlike entities that could penetrate into the blood through the mouth, nose, or skin; sometimes the contagion was conceived of as insectlike and to possess wings.[11]

Although animalcules were often associated with the theory of contagions, no real use was made of Leeuwenhoek's observations over the next century and a half.[12] His discoveries excited great interest, but no one seriously extended his observations in his own lifetime. When Hooke wrote on "the fate of Microscopes" in 1692, he noted that they were "becoming almost out of Use and Repute so that Mr. Leeuwenhoek seems to be the principal Person left that cultivates those Enquiries. Which is not for Want of considerable Materials to be discover'd but for Want of the inquisitive Genius of the present Age."[13] Scant research was done in the study of microbial life during the next century.[14]

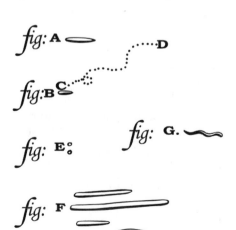

Figure 1.1
Leeuwenhoek's figures of "animalcules" from the human mouth (letter 39, September 17, 1683) in C. Dobell, *Antony Leeuwenhoek and His "Little Animals"* (New York: Russel and Russel, 1958), facing page 239.

Enlightenment Naming

The eighteenth century was an age of natural history. It was marked by the great voyages of James Cook and Joseph Banks, and of Louis Antoine de Bougainville's circumnavigation of the globe. It was a great period of collecting, cataloging, and classifying in newly founded botanical gardens and museums. It was a century in which natural history cabinets and herbaria were owned by kings and wealthy citizens.[15] A new interest in nature entered the writings of philosophers. Inspired by the revolution in physics begun with Newton, explanations of nature in "the age of reason" were to be sought in observation and experimentation. Accurate systems of naming were crucial.

"We think only through the medium of words.—Languages are true analytical methods," Étienne Condillac said. "The art of reasoning is nothing more than a language well arranged."[16] Antoine Lavoisier took heed. "We cannot improve the language of any science without at the same time improving the science itself," he wrote in his treatise on the *Elements of Chemistry* of 1789; "neither can we, on the other hand, improve a science, without improving the language or nomenclature which belongs to it."[17] No matter how certain the facts and correct the ideas, he continued, "we can only communicate false impressions to others, while we want words by which these may be properly expressed."[18] In naming, "the word should elicit the idea, and the idea should depict the fact": "these," he said, "are three imprints of the same seal."[19] While forming a nomenclature, and improving the language of chemistry, Lavoisier explained, his work "transformed itself by degrees, without my being able to prevent it, into a treatise on the *Elements of Chemistry*."

Carolus Linnaeus (1707–1778) in Uppsala revolutionized naming of the living world, when, in his great work *Systema Naturae*, he arranged all "natural bodies" into a nested hierarchy of groupings.[20] Kingdoms of life were divided into classes and then into orders, which were divided into genera, and then into species and varieties. He identified species according to what became a standard nomenclature, a genus name followed by a species name, such as *Homo sapiens*. In describing plants, one could not just compare one plant to another, but parts of a plant with parts of another. And when studying its parts, one had to understand what was of generality and what was not. He created a method for botanists to know plants quickly and with certainty based on what he called a "sexual system" using the number, shape, proportion, and organization of pistils and stamens. In practice, naming was as important in the economy of botanical knowledge as was money in the exchange of goods. Linnaeus said, "Names have the same value on the marketplace of botany as coins have in public affairs, which are daily accepted as certain values by others, without metallurgical examination."[21]

There was nothing like his system before. From Aristotle to the Renaissance, the basis for knowing life had hardly changed.[22] In the Aristotelian order of things life was arranged on an ascending scale, a divinely planned great chain of being. An object's "place" depended on the relative quality and proportion of "spirit" or "soul" it contained—the less "spirit" and the more "matter," the lower down it stood. The Aristotelian order of things was based on the four elements: fire, water, earth, and air. Everything was arranged with a linear ascending order reflecting hidden qualities with various kinds of souls God placed in them. The moister, warmer creatures had a higher type of soul than those that were dryer and colder, which possessed less vital heat. At the bottom were stones and metals, higher up were plants, and then animals, humans, angels, and God. It was a mystical world in which nature was understood as a system of resemblances, a network of relationships to be discerned through analogies. To characterize form was to divine hidden signs in nature that God had put on the surface to know of his intentions. If, for example, a plant resembles the eye, it is a sign that it should be used for treating diseases of the eye.

In classifying in a hierarchical manner of group within group, it has been said that Linnaeus was following the ancient Greek models of logical division. From the time of the ancient Greek philosophers onward, that method was associated with the neoplatonist school of Porphyry and with a pattern of classification later referred to as the "tree of Porphyry."[23] But the progression in the hierarchy of beings of the previous order of things occurred by imperceptible degrees; there were no discrete categories as there were for the Linnaeans, who partitioned nature into genera and species—true forms created by God. The genus was the cornerstone of his classification. "It is not the character which makes the genus," as he famously put it, "but the genus which gives the character."[24] Linnaeus said in 1736 that "there are as many species as the Infinite Being produced diverse forms in the beginning." But several years later he modified his view, at least for plants, and suggested that some new species could arise through hybridization from those original types "created by the Lord."

Animal-Plants

Linnaeus arranged all natural objects within three kingdoms, which defined the great boundaries within the empire of nature, the Imperium Naturae: the Stones Kingdom (Regnum Lapideum) later called Mineralia, composed of unorganized, nonliving bodies; the Vegetable Kingdom (Regnum Vegetabile), organized living bodies without sentientia, that is, without the ability to feel and think; and the Animal Kingdom (Regnum Animale), organized living bodies with the ability to feel and move spontaneously. But there were also creatures that seemed to straddle the boundaries of kingdoms. Who could say that a sponge, medusa, or coral was a plant or an animal?

Creatures possessing the properties of both plant and animal had been spoken about in antiquity. "Indeed, for my own part," Pliny the Elder commented, "I am strongly of opinion that there is sense existing in those bodies which have the nature of neither animals nor vegetables, but a third which partakes of them both:—sea-nettles and sponges, I mean."[25] They were conceived of as animal-plants. Andreas Cesalpino in Florence referred to them as zoophytes in 1583. "Just as zoophytes resemble both animals and plants," he said, "so do mushrooms belong both to plants and inanimate objects."[26]

"Natural bodies are all those which left the hand of the Creator to compose our earth," Linnaeus said; "they are constituted in three kingdoms at the limit of which blend the zoophytes."[27] In the first edition of *Systema Naturae* of 1735, he grouped all known microscopic organisms in his sixth class, Vermes in the order Zoophyta, which also contained such marine organisms as the starfish and the medusa. In the 10th edition of 1758, he placed a few small *Foraminifera* possessing jellylike bodies and calcium carbonate shells with the mollusks; he classified the ciliated infusoria, *Vorticellidae*, as polyps; and corralled all other animalcules into one single genus, *Volvox*, containing two species.[28] In the 12th edition of *Systema Naturae* (1766–1768), he assigned all the animalcules then known to three genera, *Volvox*, *Furia*, and *Chaos*, composed of six species. He placed all the "infusoria" described in "the books of micrographers" in a single species called *Chaos infusorium*.[29]

Classification in the eighteenth century was based on describing what the unaided eye discerned in an organism and rejecting everything that was not obvious. Linnaeus was not partial to microscopic observation, and he doubted its value. He questioned the existence of spermatozoa, and he and his students also questioned whether all the smallest of animalcules (what biologists would later refer to as protists and bacteria) might not actually be stages in the development of fungi.[30] At the end of this classification of the *Chaos* genera, he referred to the nature of *Spermatici vermiculi* of Leeuwenhoek and the *Siphilitidis virus humidum* as problems for the future.[31]

In France, an alternative school of natural history with a philosophy that rivaled that of Linnaeus was led by Georges-Louis Leclerc Buffon (1707–1788), director of the Jardin du Roi. While the Linnaeans emphasized sharp discontinuities in nature, partitioning it into discrete compartments by one or a few essential characteristics, Buffon and his followers stressed the continuity of life, and classification based on several characteristics, not just morphological but also habits, temperament, and instinct.[32] Whereas Linnaeus searched for the essence of natural kinds, Buffon denied their reality. His was a nominalist conception of classification: genera were not realities to be discovered in nature, but rather human inventions, imposed on nature, which, he said, descended "by almost imperceptible degrees from the most perfect creature to the least formed material, from the best organized animal to the most unrefined mineral."[33] "There are really only individuals in nature," he wrote, "and genera, orders and classes exist only in our imagination."[34]

The same was true for the two great kingdoms of life; they, too, were not real in Buffon's philosophy. In the introduction of volume 4 of his monumental *Histoire naturelle*, he said "that there is no absolute, essential, and general difference between the animal and the vegetable, but that Nature descends by degrees and by imperceptible nuances, from that which we call the most perfect and that which is less perfect and from that to the plant."[35] The existence of organisms with the properties of both animal and plants, zoophytes, was evidence of such continuity, and in his view proved that kingdoms were not real. "But as I have already said more than once," he insisted, "lines of separation do not exist in Nature, there are beings that are neither animal or vegetable nor mineral."[36]

Some scholars depict Buffon as an evolutionist; others emphasize that transformations for him were actually limited, representing what he called a "denaturation" of the species type. There were certain species such as the horse, the zebra, and the ass that belonged to the same family with a "main stem" from which "collateral branches" seemed to radiate. But there was nothing uniting these families of species with one another. There was no progression of forms, in time, from the simple to complex in his thinking. His groupings were without ancestral histories.[37] Despite his nominalism in regard to higher levels of classification, Buffon held an essentialist conception of species. They were realities of nature. "The imprint of each species is a type," he said, "whose principal features are engraved in indelible and permanent characters for ever more."[38] At the origin of the living world, he imagined that there were about 40 distinct types from which new forms had sprung.

While Buffon took the existence of animal-plants as evidence of the nonreality of the two living kingdoms, others saw in those organisms an organization that merited a kingdom of its own. "Polyps, nematodes, animals of infusions, do they not have an organization quite different from that of most animals so as to have a different name," Buffon's younger colleague Louis Jean-Marie Daubenton commented in 1795. "Mushrooms, molds, lichens are they true plants? I could report many other observations here that tend to prove that there are a great number of other organized beings that are neither true plants nor true animals."[39] He said little more. A name for the third living kingdom was offered later by a follower of Jean Baptiste Lamarck (1774–1829).

The Lamarckian Transformation

Lamarck developed a true transformationist theory, and he constructed the first large-scale classification based on genealogy. "Nature contains only individuals which succeed one another by reproduction and spring from one another," he wrote in 1809, "but the species among them have only a relative constancy and are only invariable temporarily."[40] Three convergent interests led him to evolution: his thinking on what constituted the essence of life in the simpler organisms (caloric heat and electricity), his view of the "natural" way to arrange taxa, and his geological thinking—of gradual change over long periods of time. Unlike his contemporaries, he believed Earth to be very old, almost incalculable, involving thousands or even millions of centuries.

Few biologists have been more misunderstood and mythologized than Lamarck.[41] He is typically treated as the antihero, who originated a mechanism of evolutionary change based on the inheritance of acquired characteristics: that the characteristics you (and any organism) acquire in your lifetime may be passed on to subsequent generations—not unlike property and wealth. This concept is typically contrasted to that of Darwin's two-step process of variation and natural selection, or "survival of the fittest." Yet, attributing the theory of evolution by the inheritance of acquired characteristics to Lamarck and contrasting it to Darwin's is erroneous for three reasons. First, the concept of the inheritance of acquired characteristics did not originate with Lamarck. It can be traced back to Hippocratic writers and was common in folklore and in the writings of philosophers and naturalists of many countries.[42] Second, although it is generally overlooked in the Darwinian hagiography of biology textbooks today, Darwin also upheld its importance in evolution, and increasingly so in various editions of *The Origin of Species*. "I hardly know why I am a little sorry, but my present work is leading me to believe rather more direct [*sic*] in the actions of physical conditions," he wrote Joseph Hooker in 1862; "I presume I regret it, because it lessens the glory of natural selection, and is so confoundedly doubtful."[43] Six years later, Darwin proposed a hereditary mechanism that would account for the inheritance of acquired characteristics based on hypothetical self-reproducing "pangenes" that would circulate from body to sex cells.[44] Third, and most crucial, the inheritance of acquired characteristics was only one aspect of Lamarck's theory of evolution, and not nearly the most important part of it. To understand the whole, we need only consider the way he ordered life.

Series versus Reticulation

Lamarck worked as a botanist at the Jardin du Roi before the French Revolution, but afterward when it was renamed Muséum d'Histoire Naturelle, he was given a new position as zoologist assigned to the insects, worms, and microscopic animals. His new duties consisted of giving courses and classifying the large collection of these "invertebrates," as he named them. His aim was a classification based on affinities that reflected the course of evolution—from simple to most complex organization. "The aim of a general arrangement of animals is not only to possess a convenient list for consulting," he wrote in his most famous book, *Philosophie Zoologique* (1809), "but particularly to have an order in that list which represents as nearly as possible the actual order followed by nature in the production of animals; an order conspicuously indicated by the affinities which she has set between them."[45] Darwin would argue the same, 50 years later (see chapter 3).

Some writers view Lamarck's hierarchical classification from simple to complex as representing the philosophical principles of the *scala naturae* or "great chain of being."[46] But there was far more to it. There were two aspects to the great chain of being concept. One was what Arthur Lovejoy called "the principle of plentitude," attributed to Platonic idealism in which all material phenomena are imperfect representations of the fixed world of true ideas.[47] Accordingly, there would be no gaps in the material world; everything that could exist would exist somewhere. Buffon was a great advocate of the principle of plentitude, though he was not an advocate of the great chain of being. The great chain of being's best-known advocate in the eighteenth century was the Swiss naturalist Charles Bonnet, whose scale began with inanimate entities at the bottom and ascended to man, and then angels and archangels in the spiritual realm.

Lamarck wrote nothing of a spiritual realm, and his methods for ordering life in an evolutionary system from simple to complex were shared by Darwin, among others. To be sure, not all of those eighteenth- and nineteenth-century naturalists who classified creatures into groups did so in a hierarchical manner.[48] Some arranged life forms in a circular array with groups forming a network along the circumference. Lamarck argued against those who, he said, "have imagined that the affinities among living beings may be represented something after the manner of the different points of a compass. They regard the small well-marked series, called natural families, as being arranged in the form of a reticulation."[49] That idea was "clearly a mistake, and is certain to be dispelled when we have a deeper and wider knowledge of organisation."[50] He insisted that "the list of living bodies should form a series, at least as regards the main groups; and not a branching network" (figure 1.2).[51]

Lamarck observed that the mistake of those reticulated classifications resulted from a failure to distinguish between two kinds of characteristics— those that were shaped by the environment that do not reveal "the growing complexity of organization," and "a system of essential organs" whose organizational features represent the course of evolution from simple to complex. He reasoned that gradations in the scale, recognizable by natural affinities, were "only perceptible in the main groups of the general series, and not in the species or even in the genera." The diverse environmental conditions in which the various races of animals and plants existed had "no relation to the increasing complexity of organisation." They produced "anomalies or deviations in the external shape and characters which could not have been brought about solely by the growing complexity of organisation."[52]

Making this distinction between essential and nonessential characteristics was crucial for showing the increase in complexity, but was revealed only in the larger taxonomic groups, not in that of species, nor always of

SHOWING THE ORIGIN OF VARIOUS ANIMALS.

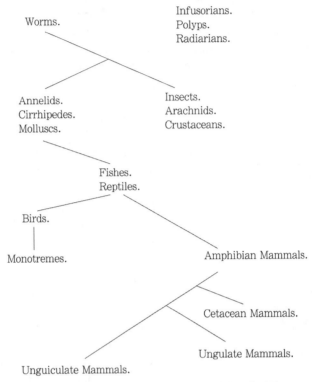

Figure 1.2 From J.B. Lamarck, *Zoological Philosophy* (Chicago: University of Chicago Press, 1984), 179.

genera: "The series from simple to complex can then only be made out among the larger groups; since each of these groups, constituting the classes and bigger families, comprises beings whose organization is dependent on some special system of essential organs."[53]

Lamarck offered two matching mechanisms for these different evolutionary trends. The lateral patterns, those deviations from the progressive series, would be due to the influence of particular environmental circumstances. The inheritance of acquired characteristics would account for a species' external morphological characters as well as instincts and habits. He said that every species "has derived from the action of the environment in which it has long been placed the *habits* which we find in it."[54] The general trend in evolution of "increasing complexity of organization" was primarily due to an unknown inner force in nature which he referred to as "the power of life," a "prime and predominating cause" in life. By this he did not mean an unknowable supernatural vital force. "Life and organisation are products of nature" he said, "and at the same time results of the powers conferred upon nature by the Supreme Author of all things and of the laws by which she herself is constituted: this can no longer be called into question. Life and organisation are thus purely natural phenomena."[55]

Vertebrates and Invertebrates

Lamarck developed his classification of the invertebrates in seven volumes of his *Histoire Naturelle des Animaux sans Vertèbres* (1815–1821). His views on transformationism and principles of classification were elaborated in his best-known work, *Philosophie zoologique* of 1809. Therein he grouped all animals into 14 classes, arranging them into series of increasing complexity; starting with the simplest animals, the infusorians, and working up to mammals, he divided animals into two large divisions: the vertebrates, which comprised four classes—fishes, reptiles, birds, and mammals—and the invertebrates, which comprised 10 classes in descending order: mollusks, cirrhipedes, annelids, crustaceans, arachnids, insects, worms, radiarians, polyps, and infusorians.[56]

Aristotle had also divided the animals into two main groups to distinguish those that possessed blood and those that did not. Naturalists called them red-blooded and white-blooded animals. But attributing "blood" to polyps or to radiolarians, Lamarck observed, made no more sense than attributing it to a plant.[57] In his view, Linnaeus's class of worms was no better. Though "a man of high genius and one of the greatest of naturalists" who taught "the necessity for great accuracy," Lamarck said, "it could not be denied that Linnaeus's class of worms was nothing but a sort of chaos in which the most disparate objects are included."[58]

At first, Lamarck placed the infusoria with the polyps, but by 1807, after "a careful examination of these imperfect animals," he assigned them to a class of their own defined as "amorphous animals, reproducing by fission or budding, with bodies gelatinous, transparent, homogeneous, contractile and microscopic; no radiating tentacles, or rotary appendage; no special organ, even for digestion."[59] To be sure, the infusorians resembled plants in that they did not carry out digestion when feeding, but "the infusorians are irritable and contractile and perform sudden movements which they can repeat several times running."[60] This, he said, indicated their animal nature. He further separated the infusorians into two orders: one for the "naked infusorians," which "are destitute of external appendages," the Monas, Volvox, Proteus (amoeba), Vibrio, Bursaria, and Colpoda; the other for the "appendiculate infusorians," which "have projecting parts, like hair, kinds of horns or a tail," which included the Cercaria (Trematodes), Trichocerca, and Trichoda.[61]

The infusoria were the most primitive of organisms. But they were *not* the oldest in Lamarck's conception. To understand why is to recognize his conception of spontaneous generation and the cycles of life. While higher organisms had the power of reproducing themselves, Lamarck noted that it was not possible to demonstrate the same for infusorians, especially those he referred to as the Monas. Nor was it possible to demonstrate that the simplest plants, such as the *Byssus* of the first family of algae, have all sprung from individuals similar to themselves. Indeed, he suspected that the "extremely small and transparent animals and plants of gelatinous or mucilaginous substance" were unable to reproduce themselves. It was "far more probable" that their new individuals arose by spontaneous generation.[62] He was quite certain that spontaneous generation occurred "at the beginning of the animal and vegetable scales, and perhaps also at the beginning of certain branches of those scales." He saw these transitions to be part of "the immense cycle of changes" in which some bodies "are passing from the inert or passive condition to that which permits of the presences of life in them, while the rest are passing back from the living state to the crude and lifeless state."[63]

Lamarck's conceptions were proposed at the threshold of a new tradition in comparative anatomy and paleontology. Animals became the main objects of comparative anatomy because the interdependent correlation of parts was most obvious in them. Given a tooth of an extinct animal, such as a dinosaur, one could make numerous predictions about the anatomy of that animal. Some comparative anatomists were evolutionists; some were not, including Lamarck's adversary at the Muséum National d'Histoire Naturelle, Georges Cuvier (1769–1832), who rejected the linear order of increased complexity as well as transformationism.[64] For Cuvier, the history of nature was not gradual and continuous from beginning to the present. There were no traces of a single line of descent in the animal kingdom, and no series of changes by which each species might have been gradually transformed from another. In 1812, he declared that there were four distinct and completely unrelated divisions (*embranchements*) of animals, and that there was no evolutionary connection between them: vertebrates (fish, reptiles, birds, and mammals), mollusks (snails, squids, and octopuses), articulates (annelids and arthropods), and radiates (starfish, jellyfish, anemones, corals, and hydras).

While Cuvier's scheme broke the unity of life, other anatomists in France led by Étienne Geoffroy Saint-Hilaire (1772–1844) aimed to reunite it by searching for a common plan underlying animal organization.[65] Following Lamarck, Geoffroy maintained that while the environment affected some structures, the basic plan was always conserved: that fundamental organization transcended adaptive function. The "transcendental anatomists" argued that the way to find the plan was to ignore differences and search for similarities in the relative positions of the parts.

A Kingdom for Beings with Two Souls

Lamarck had argued that it was difficult to classify plants based on affinities because much less was known of their organization.[66] It was more difficult still for the infusoria. They were typically cataloged as a class of worms, as zoophytes but in which, Cuvier said, "all is reduced to a homogeneous pulp."[67] Despite Lamarck's arguments for the true animal nature of some of the "invertebrates," "zoophyte" remained the common word for creatures such as hydras, sponges, and the obscure ones referred to collectively as the infusoria. Naturalists made combinations with the rock kingdom for beings such as hard corals. Some referred to them as "lithozoa"; others, as "lithophyta." While naturalists argued for the true plant or animal nature of such "ambiguous" creatures, some saw that aim to be futile and simply wrong-headed.

In 1824, traveler and naturalist Jean Baptiste Bory de Saint Vincent (1780–1846) proposed the kingdom Psychodiaire (beings with two souls) for those that are "at the same time Animals, Plants or Minerals, and which can not therefore be put exclusively in one or the other three kingdoms."[68] Those beings that Linnaeus had called zoophytes, he said, had only "thrown confusion on the two empires and tortured the minds of naturalists who attached too much importance on distinguishing plants from animals." While botanists considered some to be plants, zoologists saw in them animal characteristics such as a nervous system where none actually existed. It was as if the zoophyta "were supposed to exist between two bands of colors in a rainbow."[69] Indeed, that is precisely how Buffon had interpreted them: they were gradations in a continuum.

Bory de Saint Vincent argued that the three-kingdom model of Linnaeus was derived from ancient commonsense views of three forms of existence: "an inanimate state, vegetation, and life properly said."[70] How was one to classify those zoophytes that, because of their form, color, or intimate organization, were impossible to distinguish from plants, and that however moved spontaneously by instinct?[71] Lamarck had begun to clear up the confusion. And although those beings had been previously looked down upon, Bory de Saint Vincent said, Lamarck showed that they held a very important rank in nature as "rudimentary productions of the organizing power of nature."[72] Lamarck referred to most of the organisms in the Psychodiaire kingdom as *animaux apathiques* in that they lacked organs of respiration, locomotion, circulation, and nervous apparatus.

Bory de Saint-Vincent called these apathetic beings "psychodes," as beings that were "above plants" in possessing a "degree of animality": they possessed "instinct," but not "intellect."[73] He divided the kingdom into three classes: Phytozoaires (most of the beings traditionally defined as zoophytes), Lithozoaires (rock-animals, such as hard coral, which had previously been called lithophytes by Cuvier and others), and Ichnozoaires (Greek *ichno*, footprint, track, or trace). These comprised two families: one for the polyps that lived independently, such as *Difflugia* (later referred to as a protozoan or protist). The other comprised the polyps that lived in masses such as the bryozoans and soft coral.[74]

There were five kingdoms of "natural bodies" in Bory de Saint Vincent's proposed order of things. The "Etherial" kingdom comprised "invisible *molecules*" lacking appreciable form: "imponderable fluids such as light, fire, electricity and perhaps magnetic fluid."[75] That kingdom was followed by minerals composed of *molecules* of forms, which could be more or less easily determined by our senses, such as salts, rocks, and mineral substances. Then came the vegetable kingdom composed of *individuals* lacking in sensation, without consciousness, and entirely lacking in the faculty of locomotion (those that botanists regarded as plants minus some of the cryptograms). The fourth kingdom, Psychodiaire, contained all the "apathetic *individuals*" lacking in consciousness and that "develop and grow in the manner of plants and minerals, up to the moment where animated propagules disperse the species" (*anthrodiées*, sponges and most polyps).[76] The kingdom of animals, "each *individual*, sensitive, having the consciousness of its being and possessing the faculty of locomotion, choosing, in order to live, a site suitable for its species."[77]

The Psychodiaire may have represented the first time in history that a third kingdom of beings was formally named, and it drew few adherents. Debates continued over the animal and/or plant status of the creatures that composed the zoophyta. Bory de Saint-Vincent's scheme represented the end of an era, not the beginning of a new one. When, three decades later, a few other naturalists proposed a third kingdom of life, it would be on very different organizational grounds: that of the cell theory. The place of the simplest of the ambiguous creatures grouped as the Psychodiaire would change. The third kingdom was not placed between plants and animals, but before them.

| # Microbes First

There are, very probably, three kingdoms or great primary groups of organized beings, as distinct from each other as any subordinate groups and as readily defined by valid and recognizable characters.

—Thomas Wilson and John Cassin, "On a Third Kingdom of Organized Beings" (1863)

THERE WERE THREE proposals in the early 1860s for a third kingdom of life. They were set against the background of new explorations in microbial natural history and of a new conception of the organisms that emerged with the cell theory. Surveys of microscopic life expanded during the first half of the nineteenth century. Microscopic technology improved, allowing a great plethora of hitherto invisible organisms to be described and more details of their internal structure to be seen and compared.

Infusorial Expeditions

If Antony Leeuwenhoek was the "Columbus" of this new world, then it was said that Christian Gottfried Ehrenberg (1795–1876) was its "Humboldt," for he "thoroughly investigated these hidden provinces to the outermost boundaries."[1] Examining water, soil, sediment, and chalks, he described thousands of new species of infusoria, euglena, *Paramecia*, diatoms, and radiolarians, as well as many small fossils. He had studied medicine at the University of Berlin, completing his degree in 1818 with a thesis on the lower fungi, which he studied using a wooden microscope he made for himself.[2] He described some 250 species of fungi from the Berlin district, of which 62 were new to science. He also demonstrated sexuality in some of the lower fungi, which was something of a sensation because it contradicted the widely held view that fungi arose abiotically.

Between 1820 and 1825 Ehrenberg and friend Wilhelm Hemprich made extensive natural history expeditions in the Middle East. During the first three years, they collected animals and plants as well as infusoria in Egypt and Nubia, the Sinai, and Lebanon.[3] Four years later, Ehrenberg set out on another expedition through eastern Russia to the Chinese frontier led by Alexander von Humboldt. After his return, he concentrated on the classification of "microscobes," an effort that culminated with his famed book, *Die Infusionstierchen als vollkomme Organismen* (The Infusoria as Whole Organisms) of 1838.

Ehrenberg's discovery of fossilized remains of infusoria, and his discovery that calcium carbonate–secreting creatures actually made up the chalks in Egypt, Syria, and Sicily, attracted the interests of geologists and paleontologists. "For these organisms," he said, proved "that the dawn of the organic nature coexistent with us reaches further back in the history of the earth than had hitherto been suspected."[4] Microscopic organisms, he declared, "are very inferior in individual energy to lions and elephants, but in their united influences they are far more important than all the animals."[5] His book *Mikrogeologie* of 1854 contained illustrations of the greatest variety of microfossils ever assembled within the covers of a single work.[6]

In Ehrenberg's day, two problems that had obstructed microscopy since the time of Leeuwenhoek had been resolved. First, the early lenses had the defect of surrounding objects in the field of view with distracting fringes of color. These were known as chromatic aberrations, and they limited magnification.[7] That problem was overcome when Dutch designers in the late eighteenth century combined a convex lens made of "crown" glass with a concave lens made of "flint" glass. Stability and focus precision were also increased when brass instead of wood was used for microscope construction. Still, the curvature in the lens introduced another defect: spherical aberration, which often completely negated the benefits of the achromatic lens. That problem was solved in 1824 when the lenses of flint glass were spaced at specific distances from those of crown glass. The refractive problems of one then offset those of the other. That innovation opened the way to the construction of high-power microscopes that could magnify 1,000× with clarity, providing a brilliant and clear window into the microscopic realm.

Ehrenberg used magnification no greater than 300×, but the famous pictures of the microscopic sea creatures that he drew, engraved with subtle coloring and fine detail, captured the imagination of scientists, artists, and architects as well as the general public.[8] His descriptions of the infusoria incited lively discussion among botanists, zoologists, and paleontologists. They pondered the true nature of the infusoria and how they ought to be classified. What one "saw" depended on more than the power of the microscope—the viewed object was shaped and interpreted by theory. Ehrenberg was a follower of *Naturphilosophie*, prevalent in German natural history of the eighteenth and early nineteenth centuries, the movement led by Wolfgang Goethe, Friedrich Schelling, and Lorenz Oken, who were heavily influenced by idealist philosophy of Plato, Kant, and Hegel.[9] *Naturphilosophers* saw in nature an ideal succession of progressive perfection driven by a divine spirit imposing its will and purpose on the overall structural plan or unity of nature.[10] These thinkers searched for a common structural organization, a common divine plan underlying life, a *bauplan*, or "archetype."[11]

Despite his laborious efforts, many of Ehrenberg's species were difficult to confirm. And some of his descriptions seemed to be wholly imaginary, nothing more so than his claim that such infusoria as rotifers, desmids, and diatoms and bacteria possessed a complex organization on par with that of any "higher animal."[12] The infusoria were "whole organisms" organized in the same way as higher animals, as Ehrenberg saw them; they were "Polygastria," possessing multiple stomachs, digestive glands, and sexual organs.[13] Vacuoles of ciliates, for example, were stomachs, and the cilia that cover them were similar to the hair of higher animals; he also claimed that he could identify their sex organs and their eggs. There were 22 families of Polygastria, each containing several genera. His family Vibrionia, for example, encompassed five genera: *Bacterium*, rod-like forms; *Vibrio*, undulating flexible forms; *Spirochaeta*, spirally wound flexible filaments; *Spirillum*, spirally wound inflexible filaments, and *Spirodiscus*, compressed inflexible spiral filaments.

Followers of Naturphilosophie initially embraced Ehrenberg's theory, and he himself, it was said, "after spending some years in vainly defending his cause, withdrew entirely from all research work."[14] In his *Histoire naturelle des infusoires* of 1841, Félix Dujardin (1801–1860) was among the first to oppose Ehrenberg's concepts. He insisted to the other extreme that the infusoria actually possessed no permanent organs, but rather consisted of a homogeneous mass that he called "sarcode" from which vacuoles, granules, and the nucleus were secreted.[15] He characterized the family Vibrionia, comprising the genera *Bacterium*, *Vibrio*, and *Spirillum*, as "extremely thin filamentous animals without any appreciable organization, and without visible organs of locomotion."[16] Dujardin's book sat at the threshold of a revolutionary understanding of organismic organization, of an apparent common underlying structure of all life, embodied in cell theory.

Reformed by Cell Theory

The concept of beings radically changed in the middle of the nineteenth century. A plant or an animal was no longer conceived of as a singular entity, but comprised thousands of millions of mutually interdependent living units living in "cell-commonwealth." This was no mere analogy. "Each cell leads a double life," wrote Matthias Schleiden in 1838, "an independent one, pertaining to its own development alone; and another incidental, in so far as it has become an integral part of a plant."[17] A plant or an animal represented a colony of these smaller individuals; it was a "cellular state." The cell was perceived as the universal unit of both structure and function. All distinctive vital processes—metabolism, growth and reproduction, sexual phenomena, and heredity—were ultimately due to activities taking place in cells. As Rudolf Virchow phrased it in 1858, "Every animal appears as a sum of vital units, each of which bears in itself the complete characteristics of life."[18] All cells arise by division of pre-existing cells—*omni cellula e cellula*, he wrote in 1855. Cells were "elementary organisms," as Ernst Brücke called them in 1861.[19]

Far from being the hollow chamber that the word denotes, in fact the cells of plants and animals were found to be highly organized, and they were structurally and functionally the same as an amoeba or a ciliate. The most conspicuous parts were the spherical body that in 1831 Robert Brown dubbed the nucleus, and its surrounding material, which Rudolf Kölliker named "cytoplasm" in 1862. All cells of plants and animals were thought to have this basic division of labor. There was a fundamental common plan of organization underlying diversity because every organism is, or at some time had been, a cell. The only real difference between the cell of an animal and infusoria was that in the latter case the whole body could, for whatever reason, reach no higher degree of complexity than the single cell. This view of life framed the way that organisms were conceived.

All higher organization was supposed to evolve through the principle of "physiological division of labor" and ultimately to result in the mutuality of the constituents. This was held to be as true for the development of complex organisms as it was for human societies. The usual conception of this division of labor was, as Herbert Spencer phrased it in 1893: "An *exchange of services*—an arrangement under which, while one part devotes itself to one kind of action and yields benefits to all the rest, all the rest, jointly and severally performing their special actions, yield benefits to it in exchange."[20] Organisms were accordingly conceived as cooperative assemblages with parts integrated into organs that live for and by one another. With the exception of the theory of evolution itself, no other biological generalization, it was said, accomplished more for the unification of biological knowledge than did the cell theory.[21]

Protozoa

The conception of the infusoria as homologous, not to the entire plant or animal, but to their cells was at the heart of a new kingdom, the Protozoa, proposed by Richard Owen (1804–1892) in his book *Paleontology* of 1860.[22] The term "protozoa" had been used earlier with different connation. Georg Goldfuss had coined the term in 1817 for a class of invertebrates that included the Infusoria, Lithozoa, Phytozoa, and Medusinae.[23] His protozoa were effectively the same as Bory de Saint Vincent's Psychodiaire. Karl von Siebold subsequently redefined the word in his *Manual of Comparative Anatomy* of 1845 when he restricted it to the flagellata, rhizopoda, ciliata, and sporozoa.[24] He conceived of the protozoa as simple cells, each with a nucleus and vacuoles, that reproduced by division without any special sexual organs. Owen's concept of the protozoa was similar to Siebold's class, though of a different grade.

Owen was one of the most distinguished comparative anatomists of the nineteenth century.[25] He rose to fame as "the British Cuvier," naming and describing a vast number of living and fossil vertebrates His best-known taxonomic act followed his studies of reptile-like fossil bones of *Iguanodon*, *Megalosaurus*, and *Hylaeosaurus* found in southern England—he named the creatures the Dinosauria ("terrible lizard") in 1842. He also described the anatomy of a newly discovered species of ape reported in 1847—the gorilla—and in 1863 he described the first specimen of an unusual Jurassic fossil from Germany, the famous bird *Archaeopteryx lithographica*, which he bought for the British Museum.

Owen's comparative anatomy was deeply influenced by idealist thought of the early nineteenth century. Like Ehrenberg, he searched for a common structural plan, a *bauplan* or "archetype" for the main taxa, such as the essence of the vertebrate form, when stripped of its adaptations. This approach led him to formulate the concepts of "homology" and "analogy." The former referred to "the same organ in different animals under every variety of form and function." Structures that are as different as a bat's wing, a cat's paw, and a human hand show a common plan of structure—a very similar arrangement of bones and muscles; they are homologous. Flippers of dolphins and penguins serve a similar purpose, but their underlying structures are different; they are analogous. Analogous structures would have no significance for taxonomy.[26]

The diverse Infusoria fossils that Ehrenberg discovered in chalks and limestone held great interest for Owen. They pushed the dawn of life farther back into Earth's history. They held the secret of life's origin, which Owen placed in the bottoms of lakes and oceans: "If it be ever permitted to man to penetrate the mystery which enshrouds the origin of organic force in the widespread mud-beds of fresh and salt waters," he wrote, "it will be, most probably, by experiment and observation on the atoms which manifest the simplest conditions of life."[27]

Scholars have attended to Owen's comparative anatomy of large animals, but they have universally overlooked his discussions of microscopic life and the problem of defining kingdoms in light of them. Like Buffon in France, some naturalists in Britain maintain that the distinction between plants and animals was not real but only of human invention. The botanist John Lindley at University College London commented in his *Natural System of Botany* in

1831: "Plants are not separable from animals by any absolute character, the simplest individuals of either kingdom not being distinguishable by our senses."[28] Dutch zoologist Jan van der Hoeven spoke to the same issue in his *Manual of Zoology*, translated into English in 1856:

> At first sight, it seems easy to distinguish an animal from a plant, and even the most unskilled person thinks he has a clear notion of the difference. Yet it is just his want of knowledge that causes the difference to appear so prominent, whilst he overlooks the intermediate links, and thinks, for instance, of a dog and a pear tree.[29]

Owen disagreed, and he sought criteria to define the kingdoms as natural kinds. In his Hunterian Lecture "On the Structure and Habits of Extinct Vertebrate Animals" of 1855, he observed:

> Nothing seems easier than to distinguish a plant from an animal, and in common practice, as regards the more obvious members of both kingdoms, no distinction is easier; yet as the knowledge of their nature has advanced, the difficulty of defining them has increased, and seems now to be insuperable.[30]

He began his inquiry with the definition that Linnaeus gave 91 years earlier: "Minerals are unorganized; vegetables are organized and live; Animals are organised, live, feel, and move spontaneously."[31] Movement, as the mark of an animal, was belied by some of Ehrenberg's Polygastria, which possessed locomotion like an animal but released oxygen like a plant. Friedrich Wöhler had reported in 1843 that some of the free and locomotive Polygastria eliminate pure oxygen; and on the other hand, mushrooms and sponges exhale carbon dioxide. "Chemical antagonism," Owen said, "fails as a boundary line where we must require it—viz., as we approach the confines of the two kingdoms." He noted further that the "green coloured matter called 'chlorophyll' which is common in most plants, exist in the *Polygastria*, in the green *Planariae*, and the fresh-water *polype*."[32]

Owen pointed to cell structure as the common plan of all organisms, and then to those creatures that lacked structures of either plant or animal: "When a certain number of characters concur in the same organism," he said,

> its title to be regarded as a "plant" or an "animal" may be readily and indubitably recognized; but there are very numerous living beings, especially those that retain the form of nucleated cells, which manifest the common organic characters, but without the distinctive superadditions of either kingdom. Such organisms are the *Diatomaceae, Desmideae, Protococci, Volvocinae, Vibriones, Astasiaeae, Thalassicolae,* and *Spongiae,* all of which retain the character of the organized fundamental cell with comparatively little change or superaddition.[33]

When Owen proposed the kingdom Protozoa five years later, he again defined it in terms of what its members lacked. He began with the "organism" and then offered new descriptions for plants and animals based on morphology, physiology, chemistry, locomotion, and underlying cellular structure:

> Organisms, or living things, are those which possess such an internal cellular or cellulo-vascular structure as can receive fluid matter from without, alter its nature, and add it to the alternative structure....
>
> When the organism is rooted, has neither mouth nor stomach, exhales oxygen, and has tissues composed of "cellulose" or of binary or ternary compounds, it is called a "plant." When the organism can also move, when it receives the nutritive matter by a mouth, inhales oxygen, and exhales carbonic acid, and develops tissues, the proximate principles of which are quaternary compounds of carbon, hydrogen, oxygen, and nitrogen, it is called an "animal." But the two divisions of organisms called "plants" and "animals" are specialised members of the great natural group of living things; and there are numerous organisms, mostly of minute size and retaining the form of nucleated cells, which manifest the common organic characters, but without the distinctive super-additions of true plants or animals. Such organisms are called "Protozoa," and include the sponges or *Amorphozoa*, the *Foraminifera* or Rhizopods, the *Polycystineæ*, the *Diatomaceæ, Desmidiæ, Gregarinæ*, and most of the so-called *Polygastria* of Ehrenberg, or *infusorial* animalcules of older authors.[34]

Without further ado, Owen divided the Protozoa into three classes: (1) the Amorphozoa for sponges; (2) the Rhizopoda (root-feet) for microscopically minute organisms "of a simple gelatinous structure, commonly protected by a shell" (e.g., *Difflugia* and the *Foraminifera*); and (3) the Infusoria, which he referred to simply as "the '*Polygastria*' of Ehrenberg."[35]

Owen's kingdom protozoa differed from Bory de Saint Vincent's Psychodiaire in several respects. It was based on cell theory as the common organismal plan, and its members were not considered to be a chimeric blend of the properties of plants and animals. Still, breaking the plant-animal dichotomy with a third kingdom was not the obvious solution to all who considered Owen's protozoa. Some continued to search within it to distinguish true plants and animals.

Primigenum

When the ink in Owen's *Paleontology* was barely dry, botanist John Hogg (1800–1869) criticized and rejected Owen's kingdom Protozoa and proposed the kingdom Primigenum instead. It would contain the same creatures as Owen's Protozoa, but Hogg considered their nature differently.[36] The word "protozoa," he said, implied their animal nature. Although the suffix "-zoa" could also be translated as an adjective "living," it was generally accepted to mean animal.[37] And in Hogg's view, many of the organisms in that group were actually plants.[38] He told of how he had examined the "simpler living Beings of the Creation" for some 20 years, and like so many others, he had difficulty in discerning whether the infusorians that Ehrenberg had termed Polygastria belonged to the vegetable or animal kingdom. But he had no intention of abandoning the effort to do so. "Although strictly speaking there may be no actual distinction between these two kingdoms, and that *life* in the lowest animal and that in the simplest plant may be the *same*," he said, "still the naturalist must endeavor to draw a line of demarcation between these two great provinces, for the sake of the arrangement and classification of the infinitely numerous living beings or organisms existing in the word."[39]

Some of Ehrenberg's Polygastria were "known to exhale oxygen after the manner of true vegetables."[40] And although Ehrenberg "viewed" *Desmids* and *Diatoms* as true animals, Hogg suspected that they, too, would turn out to be plants.[41] He insisted that it was not true that all fungi and all sponges exhale carbon dioxide, and therefore were animals. He had shown that some sponges contained chlorophyll and released oxygen.[42] He noted that Louis Agassiz, von Siebold, and van der Hoeven also believed sponges were plants.[43]

The Primigenum kingdom, Hogg argued, would thus comprise both proto-plants and proto-animals. It would include

all the lower creatures, or the primary organic beings,—"Protoctista"—*first created beings*;—both *Protophyta*, or those considered now by many as, lower or primary beings having more the nature of plants; and *Protozoa*, or such are esteemed as lower or primary beings, having rather the nature of animals.[44]

Owen's definitions of the three remaining kingdoms, while the best yet, Hogg said, were nonetheless flawed. He doubted that defining animal tissue in terms of the chemical elements hydrogen, carbon, nitrogen and oxygen would be useful.[45] The principal characteristics of an animal, in Hogg's view, were the muscular and nervous systems. Accordingly, he proposed new definitions:

Minerals are bodies, hard, aggregative, simple or component, having bulk, weight, and of regular form; but inorganic, inanimate, indestructible by death, insentient, and illocomotive.

Vegetables are beings, organic, living, nourishable, stomachless, generative, destructible by death, possessing some sensibility; sometimes motive, and sometimes locomotive in their young or seed state; but inanimate, insentient, immuscular, nerveless, and mostly fixed by their roots.

Animals are beings, organic, living, nourishable, having a stomach, generative, destructible by death, motive, animate, sentient, muscular, nervous, and mostly spontaneously locomotive, but sometimes fixed by their bases.[46]

Darwinian Misconceptions

Proposals for the Protozoa and the Primigenum kingdoms were made the year following the publication of Darwin's *Origin of Species*, and one might suppose—indeed, it has been supposed—that those kingdoms were informed by that work.[47] This might *appear* to be so because Darwin's classification was based on evolutionary relationships. But this is misleading on several counts. First, classification based on evolutionary relationships was the basis of Lamarck's classification long before Darwin's. Second, the roots of the kingdom problem were much deeper; Owen, Hogg, and others before them had discussed these issues long before the appearance of Darwin's *Origin*. Darwin's book centered on the origin of species; he wrote nothing on the origin of the major kingdoms and, strikingly, nothing on the infusoria—nothing on the deeper foundations of diversity. None of the proposals for a new kingdom mentioned Darwin. Third, neither Owen nor Hogg was a Darwinian.

Owen's relationship to Darwin was one of antipathy. The fossil record for Owen constituted the unfolding of a divine plan.[48] In *Paleontology*, Owen wrote of a "great First Cause" with "wisdom and power" to produce all the diversity and "perfect adaptations" exhibited in the history of Earth.[49] Unlike Lamarck and Darwin, Owen conceived nature to involve "the beneficence and intelligence of the Creative Power":

> Everywhere in organic nature, we see the means not only subservient to an end, but that end accomplished by the simplest means. Hence we are compelled to regard the Great Cause of all, not like certain philosophic ancients, as a uniform and quiescent mind, as an all-pervading *anima mundi*, but as an active and anticipating intelligence.[50]

Similarly, Hogg's view of kingdoms carried no Darwinian signature. He said nothing of transformationism based on natural causes. Some writers at the end of the twentieth century have commented that "what is significant in Hogg's depiction is that plants and animals share a common ancestry from the protoctista."[51] Actually, what is significant is that he explicitly did not conceive of the Protoctista as an evolutionarily coherent group with shared properties arising from common descent. A second misinterpretation is coupled with the first. Hogg, it is said, had depicted the Protoctista "as a group of organisms having the common characters of both plants and animals."[52] In fact, he argued to the contrary. The notion that there was something other than a plant or an animal was questioned throughout his paper. His kingdom Primigenum represented a sundry group of organisms, most of which, he suspected, would be discovered one day to be either plants or animals.

Hogg was reluctant to even assign a fourth kingdom of nature. He was emphatic about this:

> And although I at present do not feel quite convinced of the immediate necessity of doing so, or that it will ever remain…impossible for man to determine whether a certain minute organism be an animal or a plant, I here suggest a fourth or additional kingdom, under the title the *Primigenal* kingdom.[53]

Protoctista were lower creatures "which are of a doubtful nature, and can in some instances only be considered as having become blended or mingled together."[54] The kingdoms for Hogg represented states of "perfection." The word "protoctista" literally meant first created beings, and Hogg repeatedly referred to the Protoctista as the "simpler beings of the Creation," which he compared to the "highest or more perfect state of plants" and to the "more perfect condition of animality."[55]

Primalia

Virtually unnoticed by biologists then and now, there was another proposal in 1863 for a third kingdom of organisms based on evolutionary relationships and grades or levels of organization. The ornithologist John Cassin (1813–1869) and his Philadelphian patron, Thomas B. Wilson, a wealthy physician, proposed the kingdom Primalia, as "a natural and primary division," comprising five subkingdoms: Algae, Lichens, Fungi, Spongiae, and Conjugata:[56]

> All the groups properly of [the Primalia] kingdom are, in our opinion, readily demonstrable as having a greater degree of relationship to each other than to any groups whatever in the other two kingdoms. This circumstance is held, very properly, as of the first importance in all classifications.[57]

They contrasted their views to the common assumptions of life's development either into "two series or great classes of existences—animal and vegetable" or into one series only, "the chain of being, from the lowest vegetable to the highest animal."[58] The transformations in life, they said, "though evidently progressive" occurred "under circumstances coincident with and dependent upon the laws or conditions of existence of organic life in any geological period."[59]

Wilson and Cassin's kingdoms represented "three very distinct grades or specializations of development."[60] The Primalia was characterized by the "possession of an organization exclusively providing for Nutrition and Reproduction, and these functions only." In that group, there were

> no other organs than those performing the function of Reproduction, and the structure is exclusively cellular without vascularity; or, perhaps

it may be more properly stated to consist of mere unicellular aggregation. The possession of organs for, and the first development of the function of Reproduction is the specialization of this kingdom.[61]

The kingdom Vegetabilia was "marked by the high development of the organs performing the functions of Nutrition and the superposition or superaddition of organs providing for the cooperative or identical functions of Respiration and Circulation, and these only." Animalia was characterized by the high development of the nervous system and its sphere of functions: "the possession of organs for Nutrition and Reproduction, Circulation and Respiration, and for Sentiency, Voluntary motion, and all other functions and relations of the Nervous system."[62]

These discussions of the primary groupings of life were far removed from the main thrust of evolutionary concerns in the second half of the nineteenth century. But there was still another proposal for a third kingdom in the 1860s that was planted deeper than the others. It included organisms hypothesized to be of simpler organization than any of those included in the kingdoms Protozoa, Primigenum, or Primalia.

THREE | # The Germ of Phylogeny

I endeavored in my Generalle Morphologie *(1866) to draw the attention of biologists to these simplest and lowest organisms which have no visible organization or composition from different organs. I therefore proposed to give them the general title of monera. The more I have studied these structureless beings—cells without nuclei!...the more I have felt their importance in solving the greatest questions of biology—the problem of the origin of life, the nature of life, and so on.*

—Ernst Haeckel, *The Wonders of Life* (1904)

IN 1866, ERNST HAECKEL (1834–1919) proposed a new kingdom based on the concept of phylogeny, which included precellular organisms he called monera. He called the kingdom Protista, and he elaborated principles of phylogeny in his landmark two-volume *Generelle Morphologie*. That book, he said, was the "first attempt to introduce the Descent Theory into the systematic classification of animals and plants, and to found a 'natural system' on the basis of genealogy; that is, to construct hypothetical pedigrees for the various species of organisms."[1] The book contained tables with the genealogical histories of plants, animals, and protists, as well as the first outline of a phylogenetic tree of life. Haeckel coined the terms "ecology," "ontogeny," "phylum," and "phylogeny" (Greek *phylum*, tribe, and *genesis*, origin).

Generelle Morphologie was both impressive and exasperating. Some evolutionists considered it to be "one of the greatest books ever written."[2] But it was long and difficult reading. Haeckel wrote *The History of Creation* as a popular version. It was an enormous success and went through six editions between

1868 and 1892.[3] His book *The Riddle of the Universe* of 1899 was even more successful. It was among the most spectacular accomplishments in the history of printing. It sold 100,000 copies in its first year, went through 10 editions by 1919, was translated into 25 languages, and by 1933 had sold almost half a million copies in Germany alone. Little wonder that some suggest that Haeckel was as important as Darwin for the advancement of evolutionary theory.[4] He had studied medicine at Würzburg and Berlin before turning to natural history at the University of Jena in 1859, where he later was appointed professor of zoology. His thesis was on the planktonic radiolarians—a group of one-celled marine organisms that secrete siliceous skeletons. He named nearly 150 new species of them during a trip to the Mediterranean.[5]

Descent with Modification

Haeckel had been a pious protestant, a lover of *Naturphilosophie*, the idealist philosophy of Goethe, Hegel, and Schelling, who believed that an unknowable spirit or creative organizing force gave nature purpose.[6] He turned to philosophical materialism and evolutionism after he read Darwin's *Origin*. There were hints of transformationism in the writings of Goethe and Owen, but in Haeckel's view, Lamarck had offered the only systematic treatment before Darwin. "We find in Lamarck a preponderant inclination to *deduction*, and to forming a complete, monistic scheme of nature; in Darwin we have a prudent concern to establish the different parts of the theory of selection as firmly as possible on the basis of observation and experiment."[7]

Darwin (1809–1882) and Alfred Russel Wallace (1823–1913) postulated that evolutionary change occurred primarily by a struggle for existence, giving rise to a "natural selection" of the most fit. Evolution, for the most part, was a process resulting from chance and necessity—from the production of variation and the preservation of favorable variations and rejection of injurious ones. Life was contingent, for if the conditions of life were different, then life would be different. It was a tinkering process: life *is* because it can be, with no direction, design, or purpose. Though Darwin would not convince his contemporaries of his explanation for the process of evolution, he did convince most of his contemporaries of evolution itself, and he elaborated on what that would mean for classification.

As it was for Lamarck, so for Darwin—"descent with modification" could explain the affinities that naturalists had observed. The similarities between species would reveal their common descent, whereas differences would reveal species' modifications. "From the first dawn of life," he said, "all organic beings are found to resemble each other in descending degrees, so that they can be classified in groups under groups. This classification is evidently not arbitrary like the grouping of the stars in constellations."[8] Evolution would aid classification, and inversely, classification should express evolutionary relatedness. Thus, the course of evolution was represented as a hierarchical arrangement from a process of descent with modification:

> All of the foregoing rules and aids and difficulties in classification are explained, if I do not greatly deceive myself, on the view that the natural system is founded on descent with modification; that the characters which naturalists consider as showing true affinity between two or more species, are those which have been inherited from a common parent, and, in so far, all true classification is genealogical; that community of descent is the hidden bond which naturalists have been unconsciously seeking, and not some unknown plan of creation, or the enunciation of general propositions, and the mere putting together and separating objects more or less alike.[9]

Darwin's Simile

There was only one diagram in the *Origin*, that of hypothetical trees representing species divergence (figure 3.1). For Darwin, the tree of life was a simile for relatedness, reflecting the process of evolution:

> The affinities of all of the beings of the same class have sometimes been represented by a great tree. I believe this simile largely speaks the truth. The green and budding twigs may represent existing species; and those produced during former years may represent the long succession of extinct species.... The limbs, divided into great branches, and these into lesser and lesser branches, were themselves once, when the tree was young, budding twigs, and this connection of the former and present buds by ramifying branches may well represent the classification of all extinct and living species in groups subordinate to groups.[10]

The branching order of diversity arose from specialization—which Darwin saw as a kind of ecological "division of labor."[11] Any locality could support more life if it is occupied by diverse organisms partitioning the resources than if it is occupied by similar organisms requiring the same resources. Divergence is advantageous because organisms avoid competition that way. Ever-increasing specialization of niche within larger niche would create the hierarchical order of taxa within taxa.[12] In the course of evolutionary time, a small number of similar organisms would therefore produce a large number of descendants that diverge from the original type. "This grand fact of the grouping of all organic beings," Darwin wrote, "seems to me utterly inexplicable on the theory of creation."[13]

The "grouping of all organic beings" and showing affinities between species through characteristics "inherited from a common parent" were certainly not a "grand fact" of empirical science. To establish lines of descent, it was crucial

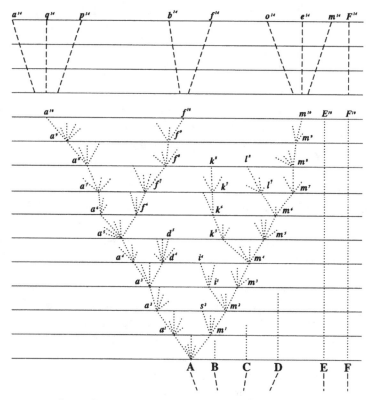

Figure 3.1 "In each genus, the species, which are already extremely different in character, will generally tend to produce the greatest number of modified descendants; for these will have the best chance of filling new and widely different places in the polity of nature." In this figure, an extreme species (A) has diverged and given rise to new varieties and species. The other species of the original genus (marked by capital letters), he said, "may for a long period continue transmitting unaltered descendants." Charles Darwin, *On the Origin of Species*, facsimile of 1859 edition (Cambridge: Harvard University Press, 1964), 121.

to be able to distinguish "adaptive or analogous" characteristics from "homologous characters."[14] The similar characteristics of a whale to a fish were of no importance for classification; they were analogous. A wing of a bat and a bird also did not have common ancestry, but the hand of a man and fin of porpoise were homologous and had taxonomic significance for evolutionists. Without genealogies, Darwin could not distinguish between homology and analogy and rigorously test the reality of the tree of life. As he wrote to Thomas Huxley (1825–1895) in 1857:

But as we have no written pedigrees, you will, perhaps, say this will not help much; but I think it ultimately will, for it will clear away an immense amount of rubbish about the value of characters & will make the difference between analogy & homology, clear. The time will come I believe, though I shall not live to see it, when we shall have very fairly true genealogical trees of each great kingdom of nature.[15]

In the *Origin*, Darwin pointed to the essential principles for such great genealogical trees. Genealogical trees of great size and scope required comparisons of what he called "essential characters," highly preserved ancient features far removed from the everyday life of the animal or plant.

It might have been thought (and was in ancient times thought) that those parts of the structure which determined the habits of life, and general place of each being in the economy of nature, would be of high importance in classification. Nothing can be more false.... It may even be given as a general rule that the less any part of the organisation is concerned with special habits, the more important it becomes for classification.[16]

Recall that Lamarck had made essentially the same argument 50 years earlier—that one had to distinguish characteristics resulting from the organism's relationship to its environment, from the "system of essential organs" in order to arrange organisms in a series of "growing complexity," as he called it. Evolution from simple to complex for Lamarck was due to an unknown "power of life." Darwin denied that there was a separate mechanism to account for the progressive evolution of complexity, and he mistook Lamarck's writings about adaptations as due to "the willing of animals," but his conclusions, he said, were similar: "Heaven forfend me from Lamarck nonsense of a 'tendency to progression,'" he wrote to Joseph Hooker in 1844, "'adaptations from the slow willing of animals' &c, but the conclusions I am led to are not widely different from his—though the means of change are wholly so I think I have found out (here's presumption!) the simple way by which species become exquisitely adapted to various ends."[17]

Darwin pointed to embryology as providing the way to document evolutionary progression and affinities: the course of ontological development would reflect the course of evolution. Those organizational characteristics through which one could follow descent were those that had become ever integrated within the organism as evolution proceeded.[18] He said that "embryological characters are the most valuable of all."[19] That there was some sort of parallel between the stages in the course of individual development and the natural order of life forms from lowest to highest can be traced to the ancient Greeks.[20] Aristotle had classified beings into those with a nutritive soul (plants), a nutritive and sensitive soul (animals), and those that also had a rational soul (man). And he imagined that during the course of development these three kinds of souls came into operation in succession. Some anatomists of the early nineteenth century believed that the development of a higher animal actually passed through adult stages that lay below it. So, for example, birds and mammals pass through a fish stage, evidenced by gill slits in the embryo. For the *Naturphilosophen*, the parallelism between the stages of ontogeny and the stages in a great chain of being was evidence of the unity of nature and of God's divine plan.[21]

This concept of a transcendental recapitulation of *adult* types in the course of development was modified into a recapitulation of *embryonic* types by Karl von Baer in 1828, advancing his biogenetic law according to which development progressed from the general or primitive to the specific or advanced, and that this developmental order reflected the natural order between kinds of organisms.[22] The first features to appear were those of the phylum, and these were followed by the class, order, family, genus, and species. The human embryo was first a single cell, then a colony like a sponge, and when a liver was added it reached the level of organization of a mollusk. For von Baer, this was only a comparative argument, not an evolutionary one: developmental progression of the embryo simply paralleled the taxonomic order of life from simple to complex.

Darwin saw it differently: the course of development revealed the course of evolution. "This process, whilst it leaves the embryo almost unaltered, continually adds, in the course of successive generations, more and more differences to the adult." So, he reasoned, "the embryo comes to be left as a sort of picture, preserved by nature, of the ancient and less modified condition of each animal."[23] Embryology also provided some of the strongest evidence for the transmutation of species. If each species had been created independently by divine inspiration, Darwin argued, one would expect that the route from egg to infant would be direct. But there were extraordinary detours.[24] Embryos of land-living vertebrates go through a stage that has gill slits. Embryonic baleen whales develop teeth, and higher vertebrates have a notochord, a flexible rod that develops along the back of the embryo, a characteristic of chordates such as tunicates and very primitive filter-feeding fish, the lancets.

Comparative embryology was considered to be the best means for uncovering relationships and ancestry in the nineteenth century. Haeckel championed the field. The course of development from embryo to infant documented the path of evolution. All organisms are historical records because each preserved the forms of ancestors as key stages in the embryonic growth of the embryo. "The history of individual development, or Ontogeny," he said "is a short and quick recapitulation of palaeontological development, or Phylogeny."[25] Recapitulation theory had to be toned down considerably by the end of the nineteenth century, as embryologists came to recognize that not all embryological features reflect ancestral patterns, nor did they reflect adult types. Not all embryonic characters were conserved over evolutionary time. Development also evolves.[26]

Haeckel was notorious for his unbridled phylogenetic speculations.[27] Darwin was anxious about it from the moment he read the proofs of *Morphologie Generelle*. "Your chapters on affinities and genealogy of the animal kingdom strike me as admirable and full of original thought," he wrote Haeckel. "Your boldness, however, sometimes makes me tremble, but as Huxley remarked, someone must be bold enough to make a beginning in drawing up tables of descent."[28] Audacious they were, and Haeckel's conjectures got everyone going on phylogeny. As one of his contemporaries commented in 1873, "It matters nothing that he has repeatedly been obliged to correct himself, or that others have frequently corrected him; the influence of these pedigrees on the progress of zoology of Descent is manifest to all who survey the field of science."[29] No one was more deeply moved by Haeckel's program than Huxley, who turned to investigate the relations between fossil hominids and living apes and the descent of birds from dinosaurs, and he expounded on the fossil horse sequence as proof of evolution.[30] "In Professor Haeckel's speculation on Phylogeny, or the genealogy of animal forms, there is much that is profoundly interesting," Huxley commented in his review of *The History of Creation* in 1869. "Whether one agrees or disagrees with him, one feels that he has forced the mind into lines of thought in which it is more profitable to go wrong than to stand still."[31]

On Life's First Breath

Haeckel's conjectures went from one end of the phyletic spectrum to the other. To bridge the evolutionary gap between ape and humans, he postulated that fossil evidence of human evolution would be found in the Dutch East Indies, and he even gave a name for the speechless ape-man, *Pithecanthropus alalus*. At the other end, he speculated about a more fundamental missing link at the origin of life on Earth, life-forms he called the monera that lacked the common structure of the cells of all other organisms. Monera would bridge the gap between life and non-life, a bridge that not all evolutionists seemed willing to cross.

There were degrees of evolutionism in the nineteenth century. Some naturalists were wary about including humans; others about the origin of life by natural means. Darwin concluded in *The Origin of Species* that "Light will be thrown on the origin of man and his history."[32] He pursued the evolution of humans in his *Descent of Man* (1871). But Darwin said little about the origins of life and virtually nothing about microbes. His views of natural history, like most other naturalists, were of life's splendid diversity as readily seen all around us. He paid little attention to developments in cell theory or the diversity of the "invisible world."

In *Origin*, Darwin mentioned the origin of life in the briefest of terms when commenting on the unity of life, that "all living beings have much in common in regard to their chemical composition, their germinal vesicles, their cellular structure, and their laws of growth and reproduction."[33] By analogy, he said

"all plants and animals have descended from one common prototype" and that
"probably all organic beings which have ever lived on this earth have descended
from some one primordial form, into which life was first breathed."[34] Or perhaps it was a few primordial forms, as he suggested in the last sentence of the
book: "There is grandeur in this view of life, with its several powers, having
been originally breathed into a few forms or into one."[35]

The sentiment in that last sentence, ringing as it did with the biblical Book
of Genesis, seemed to contradict his whole argument that evolution operated by
natural means. Haeckel noted that Darwin simply did not take the problem of
the origin of life seriously:

> The greatest fault of the Darwinian theory is that it yields no clues on the
> origin of the primitive organisms from which all the others have grad-
> ually descended—very probably a simple cell. When Darwin assumes a
> special creative act for this first species, he is not consistent at any rate
> and, I think, it is not intended to be taken seriously.[36]

Darwin seemed to have been of two minds about this. Privately, he said that
he was repentant for having used that phrasing, writing to Joseph Hooker in
1863, "I have long regretted that I truckled to public opinion, and used the
Pentateuchal term of creation, by which I really meant 'appeared' by some
wholly unknown process."[37] Yet, strikingly, in later editions of *Origin*, he modified its last sentence to "originally breathed by the Creator into a few forms or
into one."[38]

By introducing contingency in nature, the theory of natural selection was
to be a replacement for design and the purposefulness of supernatural intelligence as explanations for adaptations and the origin of species. But God
was displaced, not replaced. Darwin acknowledged that there may well be a
"First Cause having an intelligent mind in some degree analogous to that of
man....I cannot pretend to throw the least light on such abstruse problems.
The mystery of the beginning of all things is insoluble by us, and I for one
must be content to remain an Agnostic."[39] He did not consider a natural
origin of life in any of his publications. But biologists a century later often
quoted from a letter he wrote to Hooker in 1870 about a hypothetical "warm
little pond":

> It is often said that all the conditions for the first production of a living
> organism are now present, which could ever have been present. But if
> (and oh! what a big if!) we could conceive in some warm little pond, with
> all sorts of ammonia and phosphoric salts, light, heat, electricity, &c.,
> present, that a proteine [*sic*] compound was chemically formed ready
> to undergo still more complex changes, at the present day such matter
> would be instantly devoured or absorbed, which would not have been
> the case before living creatures were formed.[40]

Monera and the Origins of Life

Historians' writings on Haeckel have focused on his recapitulation theory of animal development, his monist philosophy about the inseparability of spirit and matter, and his "social Darwinist" writings and their political ramifications.[41] Yet, Haeckel considered his theory of the monera in the kingdom Protista to be of the utmost importance for solving the greatest of all problems of biology: the nature and origin of life. One hundred years after the terms "protista" and "monera" were coined, biology textbooks began to present them as kingdoms. "Monera" in that more recent scheme referred to the bacteria and blue-green algae (cyanobacteria) or what biologists today simply call "prokaryotes"—organisms lacking a nucleus (see chapters 6–8). Although the term "monera" is attributed to Haeckel, his concept differed in three fundamental ways: (1) monera included other organisms in addition to bacteria and blue-green algae; (2) Haeckel was uncertain whether the monera had a single origin or originated independently several times; and (3) he supposed that monera lacked all traces of the hereditary substance of other organisms.

To make room for a natural cause for the origins of life, Haeckel pointed to the misinterpretations of experiments that supposedly disproved spontaneous generation. The classical episode in the controversy is well known.[42] Félix Pouchet reported that microbes are spontaneously generated after sterile air is passed through mercury and introduced into a flask containing hay infusions. Louis Pasteur conducted a counterexperiment. He made long-necked swan-shaped flasks containing infusions of various organic substances. Unboiled flasks became infected with microbes, but the flasks that he boiled for a few minutes remained sterile even when air was free to pass through it because germ-carrying dust particles passing down the long necks would adhere to the sides before reaching the water.

A deeply religious Catholic, Pasteur understood that spontaneous generation, in his time or in the remote past, was an argument for materialism and atheism. "What a victory for materialism," he decried in 1864,

> if it could be affirmed that it rests on the established fact that matter organizes itself, takes on life itself; matter which has in it already all known forces!...Ah! If we could add to it this other force which is called life...what would be more natural than to deify such matter? Of what good would it then be to have recourse to the idea of a primordial creation, before which mystery it is necessary to bow?[43]

Spontaneous generation continued to be debated in the context of materialism versus belief in a divine primordial creation.[44]

In Haeckel's view, the great popularity of Pasteur's experiments had sown intellectual confusion. Those experiments actually said nothing about the origin of life from inorganic materials in the remote past or even today.[45] They proved only that in certain artificial conditions, infusoria are not formed in decomposing organic compounds.[46] The conclusions drawn from those experiments were sterile. How could one construct the laboratory conditions for generating life from inorganic precursors when no one knew what the material basis of life was exactly? Huxley had called it protoplasm in 1868. "Protoplasm, simple or nucleated, is the formal basis of all life. It is the clay of the potter: which, bake it and paint it as he will, remains clay, separated by artifice, and not by nature, from the commonest brick or sun-dried clod."[47]

All forms of protoplasm were composed of four elements, carbon, hydrogen, oxygen, and nitrogen, in very complex union. "To this complex combination," Huxley wrote, "the nature of which has never been determined with exactness, the name of Protein has been applied." Carbonic acid, water, and certain nitrogenous substances were unto themselves lifeless, but when brought together under certain conditions, they would give rise to protoplasm exhibiting the phenomena of life. The matter of life differed from inorganic matter only in the manner in which its atoms are aggregated. "All vital action," Huxley said, "is the result of the molecular forces of the protoplasm which it displays." Haeckel imagined "the plasma-molecule" albumin or protein to be "extremely large, and made up of more than a thousand atoms, and that the arrangement and connection of the atoms in the molecule are very complicated and unstable."[48]

For Haeckel, a theory of the origin of life from abiotic beginnings due to natural causation was not to push God away from earthly affairs, but to draw God deeper in it. Godliness in the unity of nature had been the foundation of those *Naturphilosophen* before him who with Goethe rejected a Christian God outside of nature. Haeckel's monist philosophy on the inseparability of spirit and matter was nonteleological, involved no "immaterial natural forces."[49] "While Monism establishes the unity of the whole of nature," he said, "at the same time it demonstrates that only one God exists, and that this God reveals himself in a sum total of natural phenomena."[50]

Monist philosophy demanded the removal of explanatory boundaries between life and non-life, for to "reject abiogenesis is to dismiss Monism from consideration."[51] Thus, he postulated, the existence of "precellular" organisms: monera that formed the trunk of the tree of life (figure 3.2). They were

> the simplest conceivable organisms... most of which are minute, microscopic, and formless bodies, consisting of a homogeneous substance, or an albuminous or mucous, soft mass, and which, though they are not composed of diverse organs, are yet endowed with all the vital qualities of an organism.[52]

Haeckel considered them to be more closely related to inorganic crystals than to nucleated cells.

> The difference between the monera I have described and any higher organism is, I think, greater in every respect than the difference between

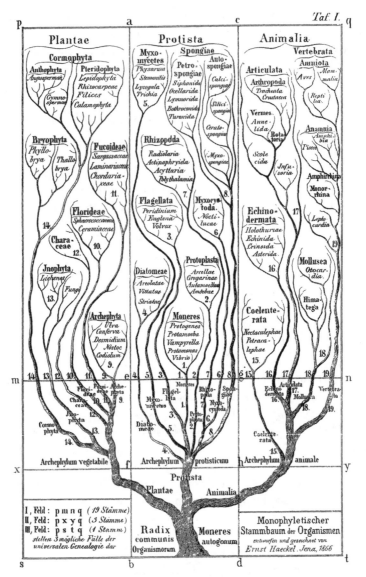

Figure 3.2 Haeckel's three kingdoms of life. From *Generelle Morphologie der Organismen*, Vol. 2 (Berlin: Reimer, 1866).

the organic monera and the inorganic crystals. Nay, even the difference between unnucleated monera (as cytodes) and the real nucleated cells may fairly be regarded as greater still.

Monera lacked the primary division of labor of "real cells" of nucleus, and the outer cell body, the cytoplasm. In those true cells, he said, "the nucleus discharges the functions of reproduction and heredity; the cytoplasm of the cell body accomplishes the metabolism, nutrition, and adaptation. Here we have, therefore, the first, oldest, and most important process of division of labor in the elementary organism."[53]

Protista

Haeckel's Protista differed radically from Richard Owen's Protozoa and John Hogg's Primigenum. Most members of the Protista were not related to either plants or animals. He partitioned the kingdom into several phylogenetic sub-groups. The protophyta, the primordial "one celled plants" (freshwater algae), possessed "synthetic or reducing plasma."[54] Protozoa, the primordial, or so-called "one-celled animals" (infusoria, especially the ciliates and the rhizopods), possessed analytical or oxidizing plasma. But "the large majority of all Protista, Haeckel said, "have independent roots which stand in no direct phylogenetic relation either to the vegetable or to the animal kingdom."[55] The organisms in this "neutral" phylum of Protista had morphologies and physiologies that were so peculiar that "an endless dispute arises over whether they were 'primaeval plants' or 'primaeval animals.'"[56] They included the Flagellata, the Lobosa, the Gregarinae, the Diatomeae, and the Mycetozoa.[57]

Protists in turn were derived from the monera, of which there were also two types. (1) The "phytomonera," blue-green algae (*Cyanophyceae* or *Chromacea*), were the most ancient.[58] They arose from inorganic materials under the influence of sunlight. (2) The "zoomomera" evolved later. "These 'rapacious animal-monera' found it more convenient to procure their nourishment directly from their vegetable sisters than to form the plasma synthetically themselves."[59] The zoomonera were the most common kind of monera yet described. Haeckel claimed to have discovered some of them, which he named *Protamoeba*, *Protogenes*, and *Protomyxa*.[60]

The bacteria, in his scheme, were "a special group" of zoomonera. They were the smallest of the monera, and they were known agents of putrefaction and fermentation and of such dangerous diseases as cholera, tuberculosis, and leprosy. "The entire minute body of the Bacteria," he said, "consists of a homogeneous little lump of plasma, like all other Monera. As there is no cell-kernel, they cannot be termed cells; but are rather simple cytodes. They increase simply by division."[61]

Did the monera descend from one stalk or from several? Haeckel explained that

the unitary or *monophyletic* hypothesis of descent will endeavour to trace the first origin of all individual groups of organisms, as well as their totality, to a single common species of Moneron which originated by spontaneous generation. The multiple, or *polyphyletic*, hypothesis of descent, on the other hand, will assume that several different species of Monera have arisen by spontaneous generation, and that these gave rise to several different main classes (tribes, or phyla).[62]

In his *History of Creation*, he suggested a multiple origin of life and the development of nuclei in several distinct lines. But the question was empirically irresolvable. Because monera lacked morphological complexity, genealogies could only be constructed and the differences among them distinguished by compositional analysis of their albumen. "But these subtle and complicated differences of mixture of the infinitely manifold combinations of albumen are not appreciable by the crude and imperfect means of human observation."[63]

There Must Be a Moneran Somewhere

Haeckel's phylogenetic speculations were considered inspirational to some and unscientific and valueless to many others. Indeed, many taxonomists advocated abandoning phylogenetics as outside the boundaries of empirical objective science. Huxley himself warned that phylogeny as speculation should be kept separate from taxonomy based on verifiable fact. "But while the logical value of phylogeny must be fully admitted," he wrote in 1874,

> it is to be recollected that, in the present state of science, absolutely nothing is positively known respecting the phylogeny of any of the larger groups of animals. Valuable and important as phylogenetic speculations are, as guides to, and suggestions of, investigations, they are pure hypotheses incapable of any objective test; and there is no little danger of introducing confusion into science by mixing up such hypotheses with Taxonomy, which should be a precise and logical arrangement of verifiable facts.[64]

Such discussions about trying to keep phylogenetic speculation separate from taxonomy persisted throughout most of the twentieth century (see chapter 6).

Taxonomists of the nineteenth century argued that Haeckel's kingdom Protista only doubled the problem of defining kingdoms. Its effect was to replace a single indefinite boundary between plants and animals with two indefinite boundaries.[65] Others added that one-celled microorganisms were far too diverse to be grouped into one kingdom based on phylogeny. Moreover, if life had arisen many times with each phylum representing a completely independent order of life, as Haeckel considered, then the kingdoms of plants and animals would be inherently artificial, and separating them from the Protista would be meaningless.[66]

No aspect of Haeckel's speculations drew more criticisms than his concept of the monera. At first it seemed to be confirmed from Huxley's study of sticky mud dredged from the North Atlantic depths in 1857 during the *Challenger* expedition. Huxley had noted the presence of "a multitude of very curious rounded bodies." At first he suspected they were not organic. But in 1868, when he reexamined that mud with a more powerful microscope, he saw "innumerable lumps of a transparent, gelatinous substance" in which were embedded "granules, coccoliths and foreign bodies." He identified the "Urschleim" with the monera and gave it the name of *Bathybius haeckellii*.[67] He thought that large swaths of the ocean floor were covered with a layer of the primordial protoplasmic slime.[68]

It turned out that *Bathybius* was an artifact, probably a precipitate induced by the application of chemical preservatives to organic substances in the seawater. Many of the organisms that Haeckel had claimed to be primitive anucleated organisms were either nonexistent or, with improvements in microscopic staining, were shown to possess a nucleus like other organisms.

By the end of the nineteenth century, the monera were on the skids. Haeckel observed in 1904 that those who were opposed to the whole concept of evolution happily denied the existence of such forms.[69] By then, it seemed to him that he would have the last laugh. He pointed to the bacteria and the blue-green algae, the *Chromacae*. He had mentioned them only briefly, on two pages of the last and sixth edition of *History of Creation* of 1889.[70] But he played them up in his *Wonders of Life*, asserting that they were as different from nucleated cells as "a hydra was from a vertebrate" or "a simple alga from a palm."[71] "The much-discussed question of the bathybius is superfluous as far as our monera theory [is] concerned," he declared, "since we have now a better knowledge of the much more important monera-forms of the chromacea and bacteria."[72]

The *Chromacae* he insisted were "the oldest phyletically, and the most primitive of all organisms known to us." Their very simple forms "correspond exactly to all the theoretical claims which monistic biology can make as to the transition from the inorganic to the organic."[73] They were classified as a class or family of algae in most botany books. But Haeckel asserted that the only real point of comparison between them and plants was the chloroplasts inside plant cells that reproduced themselves just like the free living chromacae[74] (on the symbiotic origin of chloroplasts, see chapters 9, 18, 23). He insisted that the "whole vital activity of the simplest monera, especially the chromacae is confined to their metabolism, and is therefore a purely chemical process, that may be compared to the catalysis of inorganic compounds."[75]

In 1904, Haeckel considered three conceptions of the origin of life: (1) that all organisms descended from "one single root" in primordial times; (2) that life forms were frequently repeated "under like conditions—in the same form in primordial times," and (3) that life has been repeated several times, down to our present day. "As to locality," he said, "the sea-shore probably affords the most favorable conditions; as for instance, on the surface of fine moist sand the molecular forces of matter in all its conditions—gaseous, fluid, viscous, and solid—find the best conditions for acting on each other."[76] One could conceive of the monera as having "remained unchanged or having made little advance in organization since the beginning of life—more than a hundred million years; or else the phylogenetic process of their transformation has been frequently repeated in the course of this period."[77]

Bacteria received only the briefest mention in the texts of those nineteenth and early twentieth-century microscopists who focused on the development and heredity of animals and plants.[78] And as Haeckel saw the situation in 1904, "cells without nuclei" were simply downplayed, ignored or denied by microscopists. The universal presence of nucleus-like granules was no small matter. By the late nineteenth century, microscopic anatomists had come to conclude that there was a deep chemical unity of the living world in that all cells possess nucleus and cytoplasm.

The nucleus had an abundance of a substance called "nuclein" which was made up of a protein base, rich in nitrogen, and a complex organic acid containing phosphorus named "nucleic acid" in 1889.[79] A number of ribbons, bands, and threads could be observed in the nucleus during cell division. Since they could be stained, Walther Flemming called them "chromatin" (*chroma*, colored, and *tin*, thread). The cell cytoplasm (sometimes referred to as the "cell body") was believed to contain no true nuclein or chromatin, but it was rich in protein.

Did bacteria and blue-green algae possess a nucleus like the cells of plants and animals? Much depended on the definition of the term "nucleus," whether or not one defined it *chemically* or *morphologically*. Chemically, the nucleus was defined in terms of chromatin. Otto Bütschli reported granules distributed in bacteria that stained intensely with hematoxylin and other "nuclear dyes" used in studies of animal and plant cells, and many microscopists claimed that these bacterial granules had the power of division like the chromatin granules of "higher forms." "For these reasons," Edmund Wilson commented in 1900,

> most observers...regard them as true chromatin granules which represent a scattered or distributed nucleus not differentiated as a definite morphological body. If this identification is correct, such forms probably give us the most primitive condition of the nuclear substance, which only in higher forms is collected into a distinct mass enclosed by a membrane.[80]

Other microbiologists insisted that the concept of the nucleus could not be reduced to chemistry: it was essentially a structural conception. Edward Minchin commented in 1915, "A nucleus is not merely an aggregation of chromatin; it is not simply a central core of some chemical substance or material differing in nature from the remainder of the protoplasm. A pound of chromatin would not make a nucleus." The concepts of nucleus and chromatin differ as do those of "table" and wood."[81] The true nucleus of metazoan cells was a membrane-bound cell organ of considerable complexity that had been elaborated progressively in the course of the evolution of the cell; it was considered to be as much an organ of the cell as the brain was an organ of the human body.[82]

The cells of animals, plants and infusoria divided in a complex manner, a process that Flemming called "mitosis": the chromatin threads, at first long and slender, grow shorter, thicker, and straighter to form "chromosomes" (colored bodies). Then, the nuclear membrane disappears, the chromosomes move to the center of the cell, and each one splits lengthwise into two "daughter" chromosomes that move apart toward the two poles of the cell, where they in turn form two daughter nuclei. The cell body then divides by constriction to form two daughter cells. In view of such an elaborate mechanism for transferring chromosomes from one cell generation to the next, the nucleus was typically regarded as the primary factor in development and heredity.[83]

In a morphological sense, there *seemed* to be little question for many microscopists that a nucleus was lacking in bacteria. It was less obvious in the blue-green algae. Those organisms seemed to possess a central body at the time of cell division, which in the view of some microscopists perhaps even divided by mitosis like other cells.[84] Whether bacteria and blue-green algae possess something like a nucleus and even whether they divide in the manner of the cells of plants and animals remained controversial throughout the first half of the twentieth century. There was a range of conflicting viewpoints, from one extreme—that bacteria lack anything resembling a nucleus—to the other: that the bacterium as a whole was a nucleus.[85]

Whether bacteria possessed a morphological nucleus was one issue, but Haeckel's conception of bacteria as precellular entities that lacked chromatin and whose metabolism "compared to the catalysis of inorganic compounds" was rejected outright by leading microscopists. "The Monera as defined by Haeckel must be rejected and struck out of the systematic roll as a nonexistent and fictitious class of organisms," Minchin said.[86] Wilson agreed that "there seems to be no present justification for admitting the existence of 'Monera' in Haeckel's sense."[87]

The breach between life and nonlife had only been enlarged by microscopic studies. It was naive to believe that a cell could be generated from disorganized material as some of the champions of materialism had hoped. As Wilson put it, "The study of the cell has on the whole seemed to widen rather than narrow the enormous gap that separates even the lowest forms of life from the inorganic world." It was time now to put aside the "ultimate problems" of ontogeny and phylogeny. "The magnitude of the problem of development, whether ontogenetic or phylogenetic, has been underestimated," he said; "the progress of science is retarded rather than advanced by a premature attack on its ultimate problems."[88]

Haeckel's phylogenetic speculations had little impact on empirical investigations of microbes. During the 1870s, when botanists distinguished bacteria as the simplest creatures, their aim was modest: to establish that bacteria possessed different *forms* that persisted as such and could be ordered hierarchically into taxonomic groups, as had plants and animals since the time of Linnaeus.

| # Creatures Void of Form

Are the bacteria animals or plants? A review of the literature shows that the bacteria were earlier considered to be animals, but now most researchers consider them to be plants.

—Ferdinand Cohn, "Untersuchungen über Bacterien" (1872)

Just Germs

Microorganisms had been studied predominantly from a medical perspective in the nineteenth century. The concept of germs as agents of killer diseases against which one could take action was the wonder of the 1880s and 1890s (table 4.1).[1] The demonstration that invisible organisms were inducers of putrefaction and fermentation is also one of the earmarks of modern biology. Life does not result from decayed organic matter—it is its cause. This was the conclusion of the debates and experiments over spontaneous generation, brought to a head in the early 1860s.[2] Led by Louis Pasteur (1822–1895) and Robert Koch (1843–1910), new methods were developed for detecting the fearful enemies and for learning how they were transported, how they multiplied, and how they could be arrested. Public health laboratories expanded and multiplied as pathologists developed vaccines, antitoxins, and antisera. In 1888, the Pasteur Institute was founded, and an international network of 40 other Pasteur institutes was subsequently established. In 1891, the German government furnished Koch with a research institute. That same year, the British Institute for Preventive Medicine was founded in London, which in 1903 changed its name to the Lister Institute to honor its champion, Joseph Lister (1827–1912).[3]

Professors of pathology in the universities taught practical classes in bacteriology to medical and veterinary students and candidates for diplomas in

Table 4.1   Discoveries of Microbes Causing Deadly Diseases

Year	Disease
1875	Amebic dysentery
1876	Anthrax
1879	Gonorrhea
1880	Typhoid fever, leprosy, and malaria
1882	Tuberculosis
1883	Cholera
1884	Diphtheria, tetanus, and pneumonia
1887	Epidemic meningitis
1892	Gangrene
1894	Plague and botulism
1895	Syphilis
1898	Bacillary dysentery

public health, sanitation, and food inspection. Bacteriology was also vital for such dairy products as cheese and yogurt and in industrial fermentation, primarily brewing.[4] Bacteriologists were admired for their detailed studies of the physiological properties of bacteria and for ascertaining their great importance for explaining disease, and they were also admired for their refined methods of culture, preparation, staining, and observations with ever improving microscopes. But what bacteria were, exactly, remained uncertain.

Medical researchers typically called them "germs" as Lister had first referred to them in 1874.[5] "Microbe" was introduced two years later.[6] Some thought they were little animals, *Microzoa*; others thought they were little plants, *Microphytes*. Pasteur spoke casually of "*végétaux cryptogames microscopiques*," sometimes of "*animalcules*," of "virus" (the Latin word for poison), "*Champignons*," or "*Infusoires*," or of "*Bactéries, Vibrioniens, Monades*." The word "bacteria" (from the Greek meaning little rod or staff), the plural of Christian Gottfried Ehrenberg's Bacterium, had also begun to be used to refer to the smallest of the germs, "all those minute, rounded, ellipsoid, rod-shaped, thread-like or spiral forms."[7] When perceived as mortal enemies in medical science, bacteria were typically named according to the disease they caused (e.g., the tubercle bacillus), and when they were perceived as little workers in industry, they were named according to the products they made (e.g., the lactobacillus).

Little Plants

In the universities, bacteria were studied by botanists who thought of them as plants, a class of lower fungi—"one-celled fungi, reproducing by simple fission only."[8] They were "fission fungi," or *Schizomycetes*, as Carl von Nägeli named them in 1857 at the University of Munich. The *Schizomycetes* embraced all lower forms of plant life in which chlorophyll was absent and that could utilize as food only living or dead organic matter: *Bacterium*, *Vibrio*, *Spirillum*, *Sarcina*, mother of vinegar, yeast fungus, and the small fungus-like organism associated with silkworm disease.[9]

Nägeli and other botanists did not seek to classify the bacteria with the proper compound Latin names for genus and species accorded to plants and animals. This was not because it was technically difficult to do so, nor was it for lack of interest. It was because they believed that there were no species or genera of bacteria to classify. Bacteria showed diversity only in size and a few basic forms: spherical (or cocci), rod shaped (or bacillus), or spiral. It was thought possible that the morphological differences one observed in a bacterial culture reflected different aspects of the life history of one and the same organism.[10] In other words, that bacteria could change with environmental circumstances from cocci to bacilli to spirals. They were pleomorphic (from the Greek meaning "doctrine of many shapes"). If so, there would not be specific germs for specific diseases, any more than there would be evolutionary groups to be classified. "For ten years I must have examined thousands of fission forms and with the exception of Sarcina," Nägeli wrote in 1877, "I cannot say that I see any necessity for the differentiation of even two forms."[11]

Ferdinand Cohn's paper of 1872, "Research on the Bacteria," marked the turning point in bacteriology. The use of the term "bacteria" seems to begin with him. William Bulloch commented in 1938 that Cohn

> was successful in disentangling almost everything that was correct and important out of a mass of confused statements on what at that time was a most difficult subject to study. His work was entirely modern in its character and expression, and its perusal makes one feel like passing from ancient history to modern times.[12]

Cohn had studied botany at the University of Breslau and then began his doctorate at the University of Berlin in 1846, where Ehrenberg introduced him to the study of the microbial world.

Cohn (1828–1898) is legendary in the history of medicine for the help he afforded in 1876 to Koch, then an unknown country doctor, who wrote to him about his discovery of the life history of the anthrax bacillus. Thirteen years earlier, Casimir Davaine found the anthrax bacillus in the blood of dying sheep and in the blood of healthy sheep that he had inoculated.[13] Carrying the experiments further, Koch cultivated successive generations of the bacteria in artificial media, observed their growth into long filaments, and discovered the formation within them of oval, translucent bodies—dormant spores. Such spores could resist adverse conditions, such as lack of oxygen, desiccation, heat, and cold, that would normally kill bacteria. But when conditions were favorable, they would germinate and again give rise to bacilli. That discovery could explain the recurrence of anthrax in pastures long unused for grazing.

Cohn immediately organized a meeting with other colleagues to witness Koch's demonstration, and he published Koch's epochal paper on the etiology of the anthrax bacillus in the journal *Beiträge zur Biologie der Pflanzen*, which he had established two years earlier.[14] Cohn had independently discovered heat-resistant spores of bacteria that same year, relating his finding to the problem of spontaneous generation. Hay infusions at 80°C did not kill bacteria, which then were generated anew from organic matter; the bacteria lay dormant in spores. His studies on the heat resistance of spores, it was said, "contributed perhaps more than the studies of Pasteur to the final overthrow of the doctrine of spontaneous generation."[15]

Cohn's research and writings wandered far from medical interests. He popularized concepts of bacteria both as agents of disease and as the font of all life on Earth. He explained that they were omnipresent in air, soil, and water, and that they attach themselves to the surface of all solid bodies and develop in masses where putrefaction is present. In the Franco-Prussian War, he said, bacteria killed more victims by blood poisoning through wounds than did the enemy's cannon balls.[16] And while emphasizing how mercilessly diphtheria destroyed so many promising lives, he noted that most bacteria did not cause disease; on the contrary, they were essential and beneficial to all life on Earth, vital to the cycling of elements, breaking down organic material and making nutrients available to other creatures:

> The whole arrangement of nature is based on this—that the body in which life has been extinguished succumbs to dissolution, in order that its material may become again serviceable to new life. If the amount of material which can be moulded into living beings is limited on the earth, the same particles of material must ever be converted from dead into living bodies in an eternal circle; if the wandering of the soul be a myth, the wandering of matter is a scientific fact. If there were no bacteria, the material embodied in animals and plants of one generation would after their death remain bound, as are the chemical combinations in the rocks; new life could not develop, because there would be a lack of body material. Since bacteria cause dead bodies to come to the earth in rapid putrefaction, they alone cause the springing forth of new life, and therefore make the continuance of living creatures possible.[17]

Cohn did not distinguish bacteria based on their internal anatomy, whether or not they possessed a nucleus; he said that little or nothing could be distinguished in terms of their internal parts.

So far as microscopy could tell, bacteria formed the boundary line between life and nonlife. But Cohn's views of life's origin on Earth contrasted sharply with those of Ernst Haeckel; bacteria and blue-green algae did not arise by spontaneous generation today or in the remote past from inorganic matter on Earth. Cohn favored the idea of Scottish naturalist Wyville Thomson, director of the famed *Challenger* expedition, according to which life was conveyed to our world from another.[18] Perhaps some germ could have survived the heat while entering our atmosphere, Cohn wrote in 1872, adding that "the commencement of life may have descended from Heaven upon this lifeless earth; as according to the myth, the living spark was brought down by Prometheus from Olympia."[19]

Panspermia—that life came to Earth from interstellar space—was well considered in the late nineteenth century and promoted notably by the Swedish chemist Svante Arrhenius in the early twentieth century.[20] Life forms could have come on meteorites, as Thomson suggested, or they could have been transported here as bacterial dust that had floated in space for eons, as Cohn suspected. Bacteria were thought to be killed at a temperature of about 60°C. As soon as the primitive sea temperature cooled, "the stray life form would find in the sea, richly saturated with salt, conditions for unlimited multiplication." Bacteria were known to divide every hour in optimal conditions. Left unchecked by food supply or competition, in three days, one organism would give rise to 48 trillion.[21] The whole ocean might be filled with such forms in a few days.

> From these first living germs, in which the peculiarities of the animal and vegetable kingdoms are not yet separated, the laws of development, the battle for existence, natural increase, geographic distribution, and many other known and unknown forces might have produced the different forms of the animal and plant world, which inhabited the earth in the past, as they do in the present.[22]

Cohn was convinced that bacteria belonged to the plant kingdom. To be sure, they had some animal qualities. Under certain conditions bacteria were extremely active. They swarmed in a drop of water, moving among each other in all directions similar to that of "a swarm of gnats or an ant-hill." They presented an attractive spectacle, he said, as they "swim rapidly forward, then, without turning about, retreat; or even describe circular lines. Sometimes they moved with a rocket-like spring, at another they turn upon themselves like a top; or they remain motionless for a long time, in order as quick as lightning to be up and away." The longer, fiber-like bacteria bent their bodies in swimming, "sometimes sluggishly, sometimes with address and agility, as if they were troubled to find their way through some impediment, like a fish that seeks its way among aquatic plants."[23] Still, Cohn suspected that their movements were not voluntary like those of animals. "The appearance of volition is only a delusion," he said. "No mind energies, as they lie in our conception of volition, and which in fact govern the least of the actions of the higher animals, are at play in the bacteria."[24]

Similar movements were known among microscopic plants, the diatom and the fresh water algae *Oscillariae*. Bacteria were motionless when temperature was unfavorable, nourishment scarce and oxygen lacking. In that state, he said, they could not be "entirely distinguished from common plant cells; and certain kinds of spherical bacteria (micrococcus) appeared never to move."[25] Bacteria had been classified as fungi because they lacked chlorophyll, and like fungi, no bacteria were then known to make their own food and live on organic matter. On the other hand, blue-green algae seemed to reproduce by fission. They were referred to as simple plants, but Cohn saw them to be closely related to the bacteria.[26] In 1875, he united all those forms that multiplied by simple fission, regardless of whether they possessed chlorophyll, calling them fission plants: "Perhaps the designation of *Schizophytae* may recommend itself for this first and simplest division of living beings, which appears to me to be naturally segregated from the higher plant groups, even though its distinguishing characters are negative rather than positive."[27] Few botanists followed his lead; bacteria were generally regarded as fungi, and the *Schizomycetes* and the blue-green algae as organisms apart.

No principles had been established for classifying bacteria. Sometimes names were given without consideration to others who had already named them, and sometimes they were distinguished by insignificant details; a new name merely reflected differences in the aims of observers who emphasized certain features.[28] Technical limitations were also imposed by design and magnification of microscopes. Cohn's father had bought him the finest microscope of the times, made by Plossl of Vienna. The tube was 24 inches long, impossible to use from a sitting position.[29] Classifying the bacteria thus strained back and eyes. The size difference in bacteria was comparable to that of whales and mice. Most bacteria were less than 1/1,000th of a millimeter—one micrometer (1 μm)—in diameter. Even with the strong immersion lens available in the 1870s, they remained at the limits of resolution. "The strongest of our magnifying lenses, the immersion system of Hartnack, gives a magnifying power of from 3000 to 4000 diameters; and could we view a man under such a lens, he would appear as large as Mont Blanc, or even Chimborazo. But even under this colossal amplification," Cohn wrote, "the smallest bacteria do not appear larger than the points and commas of good print.... These smallest bacteria may be compared with man about as a grain of sand to Mont Blanc"[30] (figure 4.1).

There were few characteristics for classification. At least, as Cohn saw them, the life history of a bacterium seemed to be limited to alternate processes of elongation and transverse fission. Bacteria could be visually distinguished only by their size, whether they clumped into groups, and their three basic forms—bacillus, coccoid, or spiral.[31] The lack of sexuality in bacteria excluded species and genera to be discerned on that basis.[32] But that was not a useful criterion to paleontologists when classifying fossils, either; they classified based on morphology alone. Bacteria exhibited great physiological diversity in fermentation and disease. But Cohn suspected physiological differences alone would indicate varieties or races, whereas with improvements in microscopic technique, any true species detected on physiological grounds would display a characteristic morphological difference.[33] He thus named four tribes and six genera, based exclusively upon gross shape and size:

Tribe I Sphaerobacteria
    Genus 1. *Micrococcus*
Tribe II Microbacteria
    Genus 2. *Bacterium*
Tribe III Desmobacteria
    Genus 3. *Bacillus*
    Genus 4. *Vibrio*
Tribe IV Spirobacteria
    Genus 5. *Spirillum*
    Genus 6. *Spirochaeta*[34]

Cohn proposed these genera in direct conflict with the concept of pleomorphism of Nägeli and many others who insisted throughout the 1880s that ordering the fission-fungi into genera and species was simply valueless.[35] Hans Buchner insisted that *Bacillus subtilis* could be transformed into *Bacillus anthracis*, and conversely, by shaking at different temperatures.[36] Theodore Billroth insisted that the staphylococci, diplococci, and streptococci were but stages in the development of a single type, which he called *Coccobacteria septica*.[37]

Pure Cultures

The only way to know for sure that different forms were not different stages in one organism was to grow bacteria in pure cultures, that is, from the stock of one form only. Pure-culture methods could show whether bacteria grown on one medium maintained their form and function when transplanted to another under different conditions.[38] In the culturing techniques of Pasteur, bacteria were grown in a liquid broth. Making pure cultures required a medium that was solid, transparent, and sterile. Koch and his colleagues worked out what would become the standard methods. At first, they used gelatin for culturing some organisms, but gelatin liquefied at temperatures of body heat that pathogens required to grow.

That problem was overcome in 1882 with the introduction of agar-agar, a gelling agent used for centuries in Asia for cooking, especially in the making of jam. Fanny Angelina Hesse, wife and assistant of one of Koch's coworkers, Walther Hesse, had learned of it as a youth from her Dutch neighbor in New York City, who had emigrated from Java.[39] In 1887, another of Koch's assistants, Richard Petri, modified the method by pouring the melted medium into

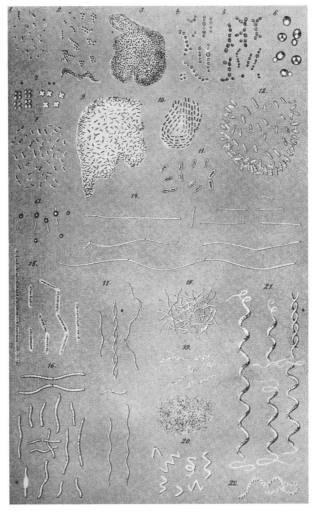

Figure 4.1 Cohn figure and description from Ferdinand Cohn, *Bacteria: The Smallest of Living Organisms* (1872), translated by Charles S. Dolley (1881) (Baltimore: Johns Hopkins Press, 1939), 38, 39.

a covered glass dish, which permitted repeated examination of the plate without risk of aerial contamination.[40]

Koch published the new pure culture techniques for isolating bacteria in 1883. He isolated the tubercle bacillus; his colleagues Friedrich Loeffer and Georg Gaffky isolated the diphtheria and typhoid bacilli, respectively, in 1884. The concept of bacterial specificity was established: typhoid germs descended from typhoid germs, and tubercle bacilli from tubercle bacilli. The term "pleomorphism" continued, but in altered form, in the early twentieth century when it was used to refer to slight changes in bacterial forms seen under different conditions of nutrition.

FIG. 1.—*Micrococcus prodigiosus* (*Monas prodigiosa,* Ehr.). Spherical bacteria of the red pigment, aggregated in pairs and in fours; the other pigment bacteria are not distinguishable with the microscope from this one.

FIG. 2.—*Micrococcus vaccinæ.* Spherical bacteria, from pock-lymph in a state of growth, aggregated in short four to eight-jointed straight or bent chains, and forming also irregular cell-mass.

FIG. 3.—Zoöglœa-form of micrococcus, pellicles or mucous strata characterized by granule-like closely set spherules.

FIG. 4.—Rosary-chain (Torula-form) of *Micrococcus ureæ,* from the urine.

FIG. 5.—Rosary-chain and yeast-like cell-masses from the white deposit of a solution of sugar of milk which had become sour.

FIG. 6.—*Saccharomyces glutinis* (*Cryptococcus glutinis,* Fersen.), a pollulating yeast which forms beautiful rose-colored patches on cooked potatoes.

FIG. 7.—Sarcina spec,* from the blood of a healthy man,** from the surface of a hen's egg grown over with *Micrococcus luteus,* forming yellow patches.

FIG. 8.—*Bacterium termo,* free motile form.

FIG. 9.—Zoöglœa form of *Bacterium termo.*

FIG. 10.—Bacterium-pellicle, formed by rod-shaped bacteria arranged one against the other in a linear fashion, from the surface of sour beer.

FIG. 11.—*Bacterium lineola,* free motile form.

FIG. 12.—Zoöglœa form of *B. lineola.*

FIG. 13.—Motile filamentous bacteria, with a spherical or elliptical highly refringent " head," perhaps developed from gonidia.

FIG. 14.—*Bacillus subtilis,* short cylinders and longer, very flexible motile filaments, some of which are in process of division.

FIG. 15.—*Bacillus ulna,* single segments and longer threads, some breaking up into segments.

FIG. 16.—*Vibrio rugula,* single, or in process of division.

FIG 17.—*Vibrio serpens,* longer or shorter threads, some dividing into bits, at * two threads entwined.

FIG 18.—" Swarm " of *V. serpens,* the threads felted.

FIG. 19.—*Spirillum tenue,* single and felted into " swarms."

FIG. 20.—*Spirillum undula.*

FIG. 21.—*Spirillum* volutans,* two spirals twisted around one another.

FIG. 22.—*Spirochæte plicatilis.*

All the figures were drawn by Dr. Ferdinand Cohn with the immersion lens No. IX. of Hartnack Ocular III., representing a magnifying power of 650 diameters.

Figure 4.1 *Continued.*

With ever improving microscopic techniques, more morphological differences were disclosed in groups previously regarded as homogeneous: these included the formation of bacterial aggregates such as chains among the spherical bacteria, the ability to produce spores among the rod-shaped bacteria, and the arrangement of organs of locomotion (the flagellum) among the motile bacteria. New diagnostic staining techniques were used to distinguish bacteria that were morphologically similar but physiologically dramatically different. The colon bacillus *Klebsiella pneumoniae* and the typhoid germ, for example, were morphologically indistinguishable, one being a normal intestinal resident, and the other a cause of severe disease. In 1884, Danish bacteriologist Hans Christian Gram (1853–1938) developed a procedure to distinguish pneumococci from *Klebsiella.* When stained with crystal violet, some bacteria are dyed a dark blue or violet color (Gram positive), whereas others are not affected by the stain (Gram negative).

The Birth of Microbial Ecology

New culturing techniques based on well-defined nutritional requirements were used to isolate and identify various types of microbes from soil and water. Food for one type of microbe was less so or entirely unsuitable for another, so by manipulating the contents of the medium one could select for certain microbes and study their metabolic activities. This was called the principle of elective or enrichment culture, and it yielded a rich harvest of microbial diversity. Sergie Winogradsky (1856–1953) and Martins Beijerinck (1851–1931) led explorations of the microbial world using this method, moving bacteriology far beyond the boundaries of medicine and the malignant conception of bacteria it inspired. Their research was embedded in concepts of cycles, exchanges of compounds between species, and symbiosis. Though overshadowed by the germ theory of disease, the late nineteenth century was a time in which the classical studies of symbiosis emerged. These were based on investigations of the exchange of organic compounds between species, the dual nature of lichens as an emergent property resulting from fungi and algae, the important role of root fungi (mycorrhiza) in providing minerals and nitrogenous compounds from the soil to forest trees, and nitrogen-fixing bacteria in the root nodules of legumes.[41]

Winogradsky discovered chemosynthetic bacteria, those that do not require organic food but use inorganic chemicals as an energy source, without the aid of sunlight. His research focused on sulfur-oxidizing species that were capable of metabolizing sulfur and its compounds, iron bacteria that derive their energy by oxidizing dissolved ferrous iron, and nitrifying soil bacteria that convert soil ammonia to nitrates usable by plants.[42] Winogradsky was a student of Andrei Famintsyn at the University of St. Petersburg, known for his work on lichens and his theorizing about cycles of life, exchanges of organic matter, and transformation of energy. Famintsyn was also well known for his views of chloroplast origin by symbiosis (see chapter 9).[43] Between 1885 and 1888, Winogradsky worked in Strasbourg with Anton de Bary, one of the discoverers of the dual nature of lichens, who is credited with coining the term "symbiosis" in 1879.[44] He turned to the study of sulfur bacteria in an attempt to resolve debates between those who believed that bacteria could be characterized by morphological types and the pleomorphists who believed that bacteria comprised only a very few species. Winogradsky, like de Bary, took the side of the first group.[45] In studying bacterial morphology, he also became interested in their physiology.

The prevailing belief was that the sulfur bacteria were responsible for the formation of hydrogen sulfide and its appearance in sulfur springs. Winogradsky inverted the causality when he applied the elective culture technique in 1888. In this work, he showed that bacteria led to the disappearance of hydrogen sulfide, not its formation. That year he completed his studies of the iron bacteria, another group of organisms that also had served as a basis for various pleomorphic conceptions.[46] When he subsequently turned to study the nitrifying bacteria, his biographer noted he was "free from the prejudices concerning bacterial nutrition that dominated many of the bacteriologists trained primarily in medical schools."[47]

Leeuwenhoek had trained no school of microbiologists. But microbiology was said to have been born again in Delft under Beijerinck in the nineteenth century. The tradition later dubbed the "Delft School" was maintained by Albert Jan Kluyver and his students, including C.B. van Niel, who in 1928 moved to the United States, where he developed a great following (see chapters 7 and 8).[48] Beijerinck completed his doctorate in biology at the University of Leiden in 1877. He worked as a microbiologist at the Netherland's Yeast and Alcohol Manufactory in Delft beginning in 1885. Ten years later, Dutch government officials created a special position for him at the Delft Polytechnic School.[49] Beijerinck is perhaps best known for his discovery in 1898 of the existence of a new kind of infectious agent, which he called a "filterable virus."[50]

The word "virus" had been used for centuries to denote a poison or contagion of infectious disease, and it was still used as a general term after pathogenic bacteria were discovered. Infectious fluids were often tested for the presence of pathogenic bacteria by passing them through filters known to retain bacteria. If the filtration rendered the fluids innocuous, a bacterial agent was indicated. Beijerinck studied a disease that stunts the growth of tobacco plants and mottles their leaves. He pressed out the juice of infected tobacco leaves and found that it was able to infect healthy plants even after it was passed it through a filter that typically removed bacteria. The causative agent could not be a toxin because he could infect a healthy plant and from that plant infect another healthy plant. He concluded that it was a "contagium vivum fluidum"—a contagious living fluid. Subsequent researchers developed a particulate notion of such viruses as submicroscopic entities. If and how such filterable viruses could be distinguished from the smallest of bacteria remained a subject of debate until the rise of electron microscopy and molecular biology (see chapter 7).

Beijerinck's isolation of nitrogen-fixing root-nodule bacteria in pure cultures in 1888 exemplified what he called "the ecological approach."[51] These bacteria were capable of taking nitrogen gas (N_2) from the air and reducing it to ammonia (NH_3). Isolating an organism capable of fixing nitrogen from the air simply required making a medium free of organic nitrogen; any microbe growing in the medium had to be able to derive its nitrogen by performing aerobic nitrogen fixation. Using the elective culture technique, he and his students isolated many other microbes, and by the early twentieth century, that exploration of the microbial world had revealed a great metabolic diversity as reflected in the diverse nutritional requirements of various microbial types. "This approach can be concisely stated as the study of microbial ecology, that is, of the relation between environmental conditions and the special forms of life corresponding to them," Beijerinck commented in 1905 when he was awarded the Leeuwenhoek Medal:

> It is my conviction that, in our present state of understanding, this is the most necessary and fruitful direction to guide us in organizing our knowledge of that part of nature which deals with the lowest limits of the organic world, and which constantly keeps before our minds the profound problem of the origin of life itself.[52]

Studies in microbial ecology were certainly not predominant in the field of bacteriology. The vast majority of microbiologists were concerned with practical problems. Industrial microbiologists that searched for specific microbes that were able to decompose certain organic materials, cause food spoilage, and produce specific chemicals also isolated and defined physiological groups and their nutritional physiology.[53] But for those evolutionary-minded bacteriologists, no discoveries attracted more attention than those of the chemosynthetic forms that derived their energy by using oxygen to feed on inorganic compounds: nitrifiers in the soil that live on ammonium and carbon dioxide, bacteria that oxidize methane and convert it to carbon dioxide and hydrogen, bacteria that convert ferrous to ferric oxide or hydrogen sulfide to sulfur dioxide, bacteria that oxidize carbon monoxide or hydrogen, and the sulfur-metabolizing bacteria. These kinds of organisms incited a radically new systemic approach to classifying bacteria when they were considered to be the progenitors of all life on Earth.

FIVE

About Chaos

The present chaotic condition of bacterial classification is in large measure due to the carrying over into bacteriology of the conception—valid enough among the higher plants—that classification must be based on visible structure or morphology.

—C.-E.A. Winslow et al., "The Families and
Genera of the Bacteria" (1917)

A review of the literature will show that the most popular term that has been used to describe systematic bacteriology is "chaos"; and this irrespective of the period of history under consideration.

—Robert Breed, "The Present Status of
Systematic Bacteriology" (1928)

BACTERIOLOGY DEVELOPED PRIMARILY as an applied science in relation to pathology, public hygiene, agriculture, veterinary science, sanitary engineering, and industry. In 1899, the Society of American Bacteriologists was founded, and according to T.W. Sedgwick, its first president, bacteriology was hardly heard of in the United States before 1885.[1] But by 1916 when the *Journal of Bacteriology* was launched, Sedgwick could boast that bacteriology was one of the most "esteemed" and "cultivated" few fields of biology. "The extent, variety, magnitude and interest of the microbial World," he said, was

> second only in importance and impressiveness to that other and distant world which has been revealed by the telescope and the methods of

astronomy.... But the revelations of the microscope and the lessons of bacteriology have so direct, so intimate, and so fateful association with almost every aspect of the conduct of our daily and personal life—with food and drink, with health and disease, with life and death even—that they gain in intimacy what they lose in grandeur.[2]

Bacteriology was indeed a utilitarian field composed of specialized practitioners who used similar methods of microscopy and culturing, but who were trained in the various practical arts, reflected in the titles of the papers introducing the *Journal of Bacteriology*: "Bacteriology of Food," "Bacteriology of Soils," "Bacteriology of Water and Sewage," "Dairy Bacteriology," "Disinfection Immunology," "Laboratory Technique," "Medical Bacteriology," "Physiology of Bacteria," and "Plant Pathology."

The Society of American Bacteriologists aimed to bring together workers to consider their problems "in the light of the underlying unifying principles of bacteriology as a member of the group of the biologic sciences."[3] A common vocabulary was required to identify the different types, and a common scheme was required to organize them. Although a common nomenclature for animals and plants was developed since the time of Linnaeus, not so for bacteria. Various taxonomic schemes had been proposed by individual bacteriologists based on a number of conflicting criteria; none was accepted.

In 1915, the Society of American Bacteriologists appointed a committee to assess the various schemes and work out a taxonomic framework that might be followed and developed by all bacteriologists. It was a landmark committee—nothing like it had ever been formed in biology before. The methods and scheme the committee presented would be followed throughout most of the twentieth century—The manual covered what the bacteria were; how to delimit and define them as a group was not an immediate concern. They were taken to be a class of plants: the fission-fungi—*Schizomycetes*, as Carl von Nägeli had named them in the middle of the nineteenth century. What was important was to find a common ground for classifying their diversity.

More Than a Storage System

Biological classification was doubly important. On the one hand, it was to characterize and sort organisms in such a way that they could be readily found: an efficient storage and retrieval system. On the other hand, it aimed at arranging forms into hierarchical groupings of species within genera, genera within families, and families within orders, respecting evolutionary relationships. It was far from clear that one system of classification for bacteria could serve both purposes.

Pathologists, hygienists, brewers, and chemists regarded the organism simply as an object to be named for convenience, because it brought about certain changes in tissues, water, and other media with which they were concerned.[4]

Their way of identifying bacteria was accordingly far removed from the principles for a natural classification. Characters that might be of paramount importance for pathology or industry might be most trivial and arbitrary from an evolutionary perspective.

Two bacteria may be morphologically and culturally identical, but one of them might be pathogenic and the other not, or one might differ in the kind of immune reaction it elicits. Those traits might not be of sufficient importance to warrant creating a new species, but they were of considerable importance in the diagnosis and treatment of disease. The industrial bacteriologists manufacturing vinegar might classify bacteria on the basis of such properties as the ability to grow on media with high alcohol concentrations, or to produce certain specified concentrations of acetic acid. It was thus useful to put pathogenic bacteria and the vinegar bacteria into groups. But those groups held no biological meaning; they merely reflected the interests of the bacteriologist who studied them.

The thorny issue for those who sought a classification based on evolutionary relationships was in deciding what kinds of characters should and could be used for classification. In arranging forms in a hierarchical evolutionary system, paleontologists and morphologists distinguished what characteristics were fundamental and what were relatively trivial. The former were those features common to higher taxonomic groups; the latter would be all the minor points of difference, the more recently acquired traits. Distinguishing homologous characteristics from analogous characteristics was difficult enough for animals, which possessed complex morphologies and for which there was a fossil record. But that distinction was virtually impossible to make among bacteria, which lacked complexity of form and a fossil record.

Several schemes had been proposed since the late nineteenth century in which bacteria were arranged into orders, families, and genera on a morphological basis: the shape of the bacterial cell, whether the bacteria in question formed spores, whether they possessed flagella and were motile, the plane of cell division, the ability to form colonies, whether the cells were connected, and whether they were branching.[5] Classification based on ever increasing morphological complexity was considered by some well into the middle of the twentieth century.

There were also many physiological traits by which bacteriologists could and did distinguish bacterial cultures: staining reactions, ability to develop in certain culture media, relation to temperature and oxygen, pigment production, pathogenicity, and a great diversity of biochemical powers. These traits were of great practical significance, but those who sought a natural classification were wary of using them. They argued, just as Ferdinand Cohn had, that traits of taxonomic value should be reflected in form.

In 1909, Sigurd Orla-Jensen (1870–1949) in Copenhagen made a radical turn when he proposed a phylogenetic system based principally on physiological characters.[6] His system was rooted in assumptions about the origins of life and on increasing physiological complexity reflecting nutrient requirements and habitat. The first organisms on Earth would have developed in an environment where there was neither light nor organic matter. Autotrophs, organisms that live on inorganic compounds, would have been the first organisms, the progenitors of life on Earth. Among them were those that oxidize methane (CH_4) and convert it to carbon dioxide (CO_2) and hydrogen (H_2), those that oxidize carbon monoxide, and those that oxidize hydrogen, as well as the nitrifiers in the soil, capable of living on ammonium and carbon dioxide as their only sources of energy.

Bacteria that derived their energy from carbohydrates and their nitrogen from inorganic nitrogen compounds came next, followed by organisms requiring nitrogen in organic form. The ability to utilize such substances as milk sugar and urea would have come only in relatively recent geological times; human pathogens would be among the youngest of bacterial types. Orla-Jensen's system was revolutionary in not giving primacy to morphology in classification.

The Commonsense Revolution

The Committee on Characterization and Classification of Bacterial Types formed by the Society of American Bacteriologists was chaired by Charles-Edward Winslow (1877–1957), professor of public health at the Yale School of Medicine and first editor-in-chief of the *Journal of Bacteriology*. He had called for the formation of the committee in his address as president of the Society of American Bacteriologists the previous year.[7] While recognizing that microbiology was, above all, an applied science, he explained that as a biological science bacteriology needed a sound characterization and classification of the bacteria and "a clear conception of the relation between these kinds."[8]

Winslow's aim was to determine bacterial relationships using as many characteristics as they exhibited—both physiological and morphological. Classifications based on morphology had failed. And that failure had "led many bacteriologists to abandon any attempt at a natural classification and to seek refuge in frankly arbitrary schematic groupings."[9] He pointed to the elaborate system of classification proposed by Walter Migula, who compiled and integrated most of the new acquired morphological knowledge in his two-volume *System der Bakterien* of 1897–1900.[10] His scheme was not only of limited practical use but also deeply flawed. He used the presence or absence of flagella as the basis of two genera, the *Bacillus* and the *Bacterium*. In doing so, he had grouped together widely different types, only because they did *not* have a certain characteristic. "These genera based on motility," Winslow declared, were "on par with a division of animals into those with wings and those without, which would place bats and birds and flying fish and bees in one group and cats and ordinary fishes and worker ants in another."[11] Even if one accepted the motility genera,

Migula's scheme left more than one-third of all known bacteria in one unwieldy genus, *Bacillus*.

In Winslow's view, Orla-Jensen's scheme was groundbreaking. It was high time that physiological traits were used in classification because those traits best reflected the nature of bacterial diversity. Imitating botanical and zoological classifications based on morphology had a paralytic affect on classification and distorted a true understanding of the bacteria. Whereas plants and animals had developed complex structural modifications to obtain food materials of certain limited kinds, the bacteria had acquired the power of assimilating simple and abundant foods of various sorts. Evolution had developed gross structure in one case without altering metabolism, and it had produced a diverse metabolism in the other case without altering gross structure. As Winslow put it "There is as wide a difference in metabolism between pneumococci and the nitrifying bacteria as there is in structure between a liverwort and an oak."[12]

The problem with using physiological traits was that so many of them seemed to be directly adaptive according to environmental circumstances. Similar adaptive characters were likely to arise in different groups under the influence of similar environmental conditions and therefore would not reflect to true phylogenetic relationships. The same physiological trait may have been acquired relatively recently in one case, and long ago in another.[13] It was well recognized that the physiology of bacteria could change in response to the environment, for example, in the case of the increase in virulence of a pathogen on passage through susceptible animals or the converse process of attenuation by exposing the pathogen to unfavorable environmental conditions, and using the weakened form to produce a vaccine.[14] Some environmentally induced changes were temporary; others were known to be inherited.[15] Bacteria seemed to evolve in a "Lamarckian" manner of environmentally induced adaptive hereditary changes.

To avoid "confusing independently acquired adaptive characters with those which indicate real community of descent," Winslow reasoned, one could tie physiological traits with other characters. "When we find a number of different characters, which have no necessary connection, correlated together, the presumption is warranted that common descent is the connecting link which has united them."[16] He called this "the general principles of statistical classification."[17] In this "common sense interpretation of all the characters of the organisms," groups would be defined based on the most important and greatest number of general statements that can be made, both morphologically and physiologically. "There is no fundamental distinction between morphological and physiological properties," Winslow declared, in what would become an oft-repeated statement of the 1920s, "since all are at bottom due to chemical differences in germ plasm, whether they happen to manifest themselves in the size and arrangement of parts or in the ability to utilize a certain food stuff."[18] There was a mass of taxonomic work that needed to be assessed, all of which was potentially valuable to the pluralist methodology.

When the Committee on Characterization and Classification of Bacterial Types published a preliminary report in 1917, it thus recommended using both morphological and physiological characters. Orla-Jensen's scheme, though "revolutionary and illuminating," was flawed because he rode "his physiological principle much harder than even Migula did his morphological one, and makes no attempt at a judicial, consideration of the grouping of common characters which should mark natural biological subdivisions."[19]

The committee claimed that its approach was not based on theoretical preconceptions, like the others. It was a strictly empirical approach that included "what is valid and discard what is arbitrary in the older classifications—with no idealistic conceptions, either morphological or physiological in mind—but with the sole purpose of recognizing and defining the principal groups of bacteria which exhibit circumstantial evidence of common evolutionary relationship."[20] The committee thus divided the bacteria (Class Schizomycetes) into four orders: the Myxobacteriales (those with "an amoeboid stage and rather complex cyst production"), Thiobacteriales (purple-sulfur bacteria), Chlamydobacteriales ("cells united in elongated filaments, and lacking sulfur"), and Eubacteriales (true bacteria "least differentiated and least specialized"). They defined families and genera within them using a broad range of characteristics.

After its outline of bacterial classification was scrutinized by members of the Society of American Bacteriologists, the committee hoped that a revised classification would be adopted and form the starting point for progress toward a "natural system."[21]

> It is easy to continue in the old slipshod ways and there will be a powerful force of inertia to overcome. If, however, this Society can actually bring into use a system of classification which approximately represents natural bacterial relationships and terminology which is uniform and in accord with the rules of priority, a great service will have been rendered to the science of bacteriology.[22]

Some bacteriologists objected to the whole idea that such a scheme should be recommended by a committee and endorsed by the society.[23] Others wanted only to ensure that certain evolutionary assumptions were revealed and not concealed in a broad system of classification endorsed by the society.[24]

On Autotrophic Origins

Despite the committee's claims of being atheoretic, its scheme included two fundamental assumptions regarding the origin of life: (1) that life had a common ancestry, and (2) that the common ancestor was an autotroph, capable of living on inorganic substances. The family *Nitrobacteriaceae* was of the primordial forms. Orla-Jensen had called it *Oxydobacteriaceae*. In renaming it *Nitrobacteriaceae*, the committee said, "Their power to live without complex organic substances would have made it possible, as Jensen points out, for them to flourish at a very early period in the world's history, and their simple structure is in harmony with the view that they represent the ancestral type of all other bacteria."[25]

That autotrophic bacteria were the common ancestors of all life on Earth was the favored view among origin of life theorists of the first decades of the twentieth century. Paleontologist Henry Fairfield Osborn, at Columbia University and curator at the American Museum of Natural History, depicted them as the universal ancestor in his book on *The Origins and Evolution of Life* of 1916.[26] In his view, the earliest forms of life would have been the "*heat-loving* and *light-avoiding*" nitrifying bacteria.[27] A similar view was shared by Osborn's colleague I.J. Kligler at the American Museum of Natural History (see figure 5.1).[28] He considered the possibility of *several independent multiple origins*, but he favored methane bacteria as the universal ancestral form. The subsequent evolution of bacteria would be in the direction of mobilizing more and more enzymes to enable the organism to utilize more complex nitrogen and energy-yielding substances:

> It is not at all improbable that the four groups of oxidizers—the carbon, sulphur, iron and nitrogen oxidizers, respectively arose at about the same time, independently of one another. Nevertheless it seems fairly certain from the important part played by carbon compounds in the vital activities of our common bacteria, especially as a source of energy, that the carbon oxidizers are the forerunners of the bacteria of today. Starting, therefore, with methane, the simplest carbon compound, at the base line, the oxidizers of CO would follow, and from them would arise in succession those organisms capable of utilizing CO_2, formic acid, acetic acid, alcohol, etc.[29]

In Kligler's scheme, as it was in that of Orla-Jensen, Osborn, and the society's committee, life emerged when Earth was both dark and devoid of organic compounds. Therefore, the universal ancestor was a nonphotosynthetic autotroph. Recall that Ernst Haeckel had speculated that blue-green algae were the most primitive of all. The time had long past since one considered that such organisms might arise by spontaneous generation today, but that such phototrophic organisms might be the primordial forms gained plausibility in light of the planetesimal theory of Earth's origin put forward by the Committee on Characterization and Classification of Bacterial Types in 1916. He postulated that Earth was made from smaller objects that gradually built the planets by accretion. Accordingly, sunlight would have reached Earth's surface even in the earliest times, thus supporting the possibility of a primordial form that was photosynthetic.[30]

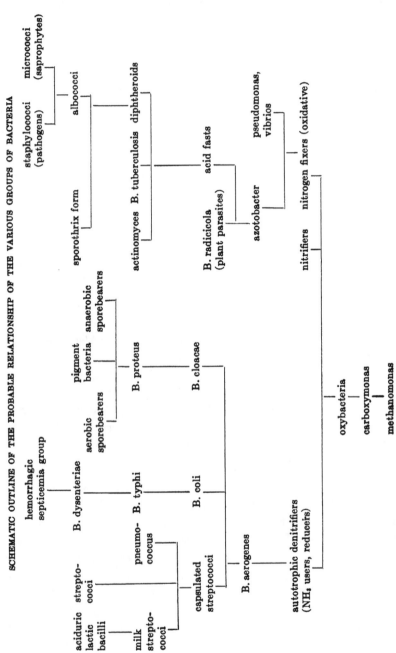

Figure 5.1 Schematic Outline of the Probable Relationship of the Various Groups of Bacteria.

But there were problems with the photosynthetic-origins proposal, at least as it concerned blue-green algae. Three years earlier, biochemists Benjamin Moore and T.A. Weber in Liverpool reported that the chlorophyll of blue-green algae was far too complex to be thought of as the first stage in the evolution of organic from inorganic matter.[31] Moore considered another possibility: that the earliest bacteria were actually heterotrophs, not autotrophs, and that the organic food matter they consumed had been formed by inorganic means by the agency of inorganic colloids acting as transformers or catalysts for radiant solar energy. He first proposed this theory at a meeting on the origin of life held by the British Association, in Dundee, Scotland, in 1912, and then amplified it his book *The Origin and Nature of Life* published the same year.[32]

Robert Breed (1877–1956) and coworkers at the New York Agricultural Experiment Station considered still another possibility when commenting on the committee's classification scheme—that bacteria were not the most primitive forms of life at all. Organic carbon compounds at the dawn of life on Earth could have been produced by life forms of a simpler nature than bacteria that were either extinct today or had escaped detection.[33] Breed also cautioned bacteriologists about assuming that all the bacteria could be grouped together under a common ancestry and descended from a universal ancestor that was still living today. To accept the committee's proposed family, *Nitrobacteriaceae*, they observed, was "really to endorse the theory that its members are modern representatives of primordial bacteria."[34]

When considering the relationships of bacteria, Breed and colleagues warned, one had to keep in mind—just as one did when dealing with macrobiological forms—that the species alive today "were only the ends of evolutionary lines, and that one modern form must not be considered the ancestor of another."[35] It was probably true that there has been a greater persistence of ancient types among bacteria than in any group of animals or higher plants. Breed suspected that this was "because the environment of many bacteria—salt water, fresh water and soil—has presented fairly uniform conditions throughout long geologic periods." But there was no reason to assume that the most primitive types of bacteria were still living; it was more likely that the primordial types were all extinct. Breed also considered the shape of the cell when considering the most ancient. The most primitive would have been a nonmotile spherical organism.[36]

Big Fleas Have Little Fleas

In 1915, Frederick W. Twort (1877–1950), working at the Brown Institute for Animals of the University of London, described the possible lysis of bacterial colonies by ultramicroscopic viruses. He suspected that the virus was "an acute infectious disease of micrococci," perhaps "a minute bacterium," but he did not pursue the matter further.[37] Two years later, while investigating an outbreak of dysentery among soldiers during the First World War, Félix d'Herelle (1873–1949) at the Pasteur Institute in Paris reported that an "invisible microbe" acted against the bacteria (*Shigella*) that caused dysentery.[38] He called it "bacterio-phage" (bacteria eater). He conceived of the bacteriophages as organisms, and he investigated and theorized on their importance in medicine and evolution. He speculated that such viruses would have appeared at the dawn of life before bacteria.[39] The phages today were not ancient forms because of their long adaptations with their hosts. The most primitive of phages, or "a protobe," probably had coalesced to form the first bacteria, which D'Herelle suspected would have been the sulfur-using bacteria.[40]

Few agreed with d'Herelle that phages were organisms that arose before bacteria and gave rise to them. Most considered them to be agents produced by bacteria. Phages could mutate and perpetuated such changes, but they could not reproduce outside their "host." In 1922, the geneticist H.J. Muller (1890–1967) identified the "d'Herelle bodies," not with organisms, but with the gene. Classical geneticists had located genes on chromosomes and associated them with various hereditary differences between individuals. But the physical nature of the gene, or even whether it was a material thing, was unknown. Still, given the self-propagating properties of phages and their ability to mutate like the gene, Muller prophesied their study would open up

> an utterly new angle from which to attack the gene problem. . . . It would be very rash to call these bodies genes, and yet at present we must confess that there is no distinction known between the genes and them. Hence we cannot categorically deny that perhaps we may be able to grind genes in a mortar and cook them in a beaker after all.[41]

J.B.S. Haldane (1892–1964) adopted a similar perspective in 1928 when he considered phage in the origins of life. The phage was part of the organism just as was the gene. He said that "the bacteriophage is such a part which has got loose." He quoted the poem:

Big fleas have little fleas
Upon their backs to bite 'em;
The little ones have lesser ones;
And so ad infinitum.

The phage was "a cog, as it were, in the wheel of a lifecycle of bacteria."[42] Haldane's theory of the origins of life is considered later with that of Aleksandr Oparin (see chapters 7 and 17). Conceptions of the phage as a living microbe persisted until the 1950s; while some microbiologists grouped them together with the bacteria, others assigned them to a kingdom of their own (see chapter 8).

Bergey's Manual

In its final report of 1920, the Committee on Characterization and Classification of Bacterial Types retained the family *Nitrobacteriaceae* when presenting its scheme as "the most reasonable outline for true biological relations among the bacteria, which can be drawn up in the state of present knowledge."[43] That outline and the methods it recommended formed the basis of what became the most important handbook of bacterial taxonomy in the twentieth century, *Bergey's Manual of Determinative Bacteriology*, first published in 1923. Upon the final report of the committee, the Society of American Bacteriologists appointed a board that included Breed and was chaired by David Bergey (1860–1937), professor of hygiene at the University of Pennsylvania. He had published a well-known *Handbook of Practical Hygiene* in 1899 and *The Principles of Hygiene* in 1901.[44]

As indicated in its title, *Bergey's Manual* did not make a claim that its classification was phylogenetic; it claimed only a "determinative bacteriology," a reliable system of identification, not an "evolutionary bacteriology." Other manuals tended to have an emphasis on medical bacteriology, but the board of *Bergey's Manual* obtained descriptions of bacteria by workers in the broadest range of specialties who distinguished bacteria using whatever traits that suited their own interests: morphological traits, biochemical activity, pathogenicity, and serum reactions.

The first edition of *Bergey's Manual* was presented as "a progress report leading to a more satisfactory classification in the future" rather than a definitive classification. It aimed to "stimulate efforts to perfect the classification of bacteria."[45] The manual was sponsored by the Society of American Bacteriologists but not endorsed by it. The first four editions carried the rider: "In publishing this Manual the Society of American Bacteriologists disclaims any responsibility for the system of classification followed. The classification given has not been formally approved by the Society, and is in no sense *official* or *standard*."[46]

By consulting widely with specialists, *Bergey's Manual* quickly became the internationally recognized and authoritative handbook on bacterial taxonomy. It grew through many editions, becoming increasingly more voluminous and complex. By its fourth edition in 1939, it had doubled in size from 500 to 1,000 pages. The classification changed from edition to edition as new facts were continuously learned about bacteria, and new species were discovered that often shed new light on possible relationships not known before. Groups were split up, and species were shifted from one genus to another.

Though it was a great success, *Bergey's Manual* received criticism from various points of view. Some bacteriologists complained that they wanted a fixed utilitarian classification, like a key, not a system that was persistently changing.[47] Others insisted that the concept of species simply could not be applied to bacteria because there were no known interbreeding groups: bacteria seemed to reproduce solely by fission, and the differences between types were very subtle.[48] It was argued that because of the speed of their reproduction, one could not expect a sufficient fixity of characteristics to make it possible to recognize species.[49] And although the old debates over pleomorphism were long past, there were still controversial reports of a phenomenon called "cyclogeny"—that some bacteria possessed complex life cycles, sometimes comparable to the "higher fungi," stages of which might be confused as different species.[50] Still other bacteriologists demanded a natural classification based on principles of homology and morphology in line with classification of "higher" plants and animals. They insisted that bacterial taxonomy simply could not be advanced using the pluralistic approach of *Bergey's Manual*; classifying on the basis of physiology, pathology, utility, and morphology resulted in a complete lack of consistency and homology among the groups.[51]

Those who defended *Bergey's Manual* argued that stability in classification and nomenclature could not be achieved for some time. About 1,000 species were described before the Second World War, and that number was considered to be a very small proportion of the actual number because bacteriologists had studied only those habitats that were of some practical importance.[52] Breed, an author of the first five editions of *Bergey's Manual*, addressed those critics who continued to use the word "chaos" to describe the state of bacterial taxonomy:

> It had its beginning with Linné (1774), who proposed the term *Chaos* as a generic term to be applied to those species of microscopic life that were so poorly known that their natural relationships could not be expressed clearly. From the time of Linné to that of the latest writer in the field of systematic bacteriology, the word has been overworked. We would challenge the accuracy of this term as applied to systematic bacteriology.
>
> Can we prove our point when we accuse nature of being chaotic?...Is it not that we have groped about in an orderly world shouting "chaos" when we ourselves were but ignorant and blind to the orderliness about us? It is our knowledge of the natural relationships and evolutionary development of this great group of living things that is chaotic, not nature.[53]

Bacteriologists were simply overly cautious about basing taxonomy on relationships, in Breed's view. The reports of bacteria with complex life cycles were doubtful. But even if they were proven to be true, and such stages in life cycles were sometimes misinterpreted as distinct species, bacterial taxonomy would not end. None of these issues was unique to bacteriology. The larva stages of tapeworms, for example, had also been considered different organisms. The sedentary polyps of hydra and jellyfish had been thought to be different species, but they turned out to be alternate generations of free-swimming medusa.[54] Those who studied black bears of North America, Breed observed, "have never agreed whether these animals should be placed in one, or in three or more species. The vines known generally as the blackberries of North America puzzle the best of our systematists."[55]

Breed dismissed "the specter" of the old debate in regard to morphological versus physiological characters. "Why should we care whether one or the other type of character is more useful in defining a genus, or whether we should use one type of character in defining a genus and another in defining a species?"[56] The distinction between morphological and physiological was a false dichotomy, as he saw it; there was no fundamental difference between them since both kinds of traits resulted "from chemical differences in the germplasm of the organism whether they happen to manifest themselves in the size and arrangement of parts or in the ability to utilize certain foodstuffs."[57] Such debates would continue through the first half of the twentieth century (see chapter 7).

There was no manual comparable to *Bergey's* for the classification of other microorganisms. But those who studied the evolution of the more complex protists also advocated a classification based on natural relationships. In his address to the British Association for the Advancement of Science in 1915, Edward Minchin explained that the study of Protista was in its infancy; groups were recognized and given designations but knowledge of the affinities and relationships of the groups were highly speculative.[58] Still, he argued there were several methods for determining natural relationships. First of all, there was the comparative method, whereby different types of cell structure could be compared with one another in order to determine what parts were invariable and essential and what were sporadic in occurrence and of secondary importance. Second, there was the developmental method, the study of the mode and sequence of the formation of the parts of the cell as they come into existence during the life history of the organism. Then, Minchin prophesied there would be the possibility of classifying based on chemical activities of organisms. That method, he observed, had already been applied to bacteria, but not yet to the more complex protozoa.[59]

Discussions of bacterial classification also filtered into animal taxonomy. Zoologists had come to learn that the rate of evolutionary change in different lines of descent could vary dramatically. Traits thought to be due to common ancestry were actually the result of convergent evolution resulting from adaptations to similar environments. One could classify in two ways: by genealogy, or by "grades" which considered the amount of similarity regardless of common ancestry. But "equal insistence on both makes your system look more like a cross-word puzzle than either a tree or a key," paleontologist Francis Bather at the British Natural History Museum observed in his anniversary address of the Geological Society of London in 1927.[60] If one classified by lineage, he said, "the last word is with the experimental geneticists; and one cannot deny that to-day the *infima* species, the ultimate element, of biological classification is the gene."[61]

At the genetic level, there was no distinction between morphology and physiology. Genes were responsible for physiological differences of multicellular organisms just as they were "in the minutest disease-germs, such as the streptococci associated with scarlet fever, the bacilli of dysentery, of paratyphoid, and other diseases." Such "physiological species," Bather insisted, "surely have a chemical difference, transmitted by the chromosomes."[62]

Yet, it was not at all certain if bacteria possessed genes. Classical genetics was based on the study of multicellular animals and plants, the fruit fly *Drosophila*, guinea pigs, and corn. A few protists and fungi were domesticated for genetic use during the 1930s. But genetic methods and theory was not extended to bacteria until after the Second World War (see chapter 7). While protists were observed to possess chromosomes and nuclei that divided by mitosis, debates persisted over whether or not the same was true of the bacteria.

SIX

Kingdoms at Biology's Borders

The living things are not all plants or animals. Nature has been more resourceful, more thorough in trying out the possibilities. Another kingdom, that of the bacteria, using the word in an inexact sense, is likewise world-wide in distribution, probably most numerous in individuals and very important in its human relations. And, besides these three major kingdoms, there are a number of minor kingdoms, not unsuccessful, but much less successful lines of evolution from the primitive beginnings of life.

—E.B. Copeland, "What Is a Plant?" (1927)

"Pourquoi ne pas avoir le courage de dire: le Règne Bactérien?"

—André Prévot, *Manuel de Classification et de Determination de Bactéries Anaérobies* (1940)

THERE HAD BEEN FIVE PROPOSALS for a third kingdom of organisms by the 1870s. Few biologists accepted any of them. Those early discussions fell into oblivion, as nature was partitioned according to biology's overriding institutional dichotomy: zoology and botany. Protista and protozoa were divided among botanists and zoologists as "lower plants" and "lower animals." The green flagellates from which plants were thought to have descended, along with the bacteria, were in the domain of botanists, and the colorless "protozoa," that of zoologists.[1]

The Fears of Secession

There seemed to be a fierce antipathy to breaking the plant-animal dichotomy—even among those taxonomists who recognized that many of the organisms they dealt with had no relationship to others of the kingdom they surveyed. The most common guide to plant taxonomy was *Syllabus der Pflanzenfamilien* first published in 1892 by Adolf Engler at the University of Berlin. In the 10th edition of 1924, he recognized that 10 out of 13 divisions or phyla in the plant kingdom actually had "no direct connection with higher plants": Schizophyta (blue-green algae), Myxomycetes (slime molds), diatoms, various flagellates, and red algae. Still, Engler maintained that the guiding principle to his system of classification was the principle of phylogeny.[2]

Perhaps it is not so surprising that when renewed proposals were published to recognize kingdoms other than plants and animals, they came from biologists outside the institutional mainstreams.[3] Edwin Copeland (1873–1964) had considered a kingdom for the bacteria, along with several "minor kingdoms" of other microbes, in his teaching at the University of the Philippines since 1914.[4] He did not publish his views until 1927, at which time he had accepted a position as associate curator of the Herbarium at the University of California–Berkeley. His was a somewhat obscure missive in *Science* based mainly on long quotations from an unidentified "old botany text" that treated the schizophytes as a distinct kingdom. "If two trunks grow from a common root," the unidentified author wrote, "who shall decide to which trunk the root belongs?"[5] Copeland said that a "plant kingdom comprised of all the organisms listed in the texts of botanical taxonomy is no more 'natural' than a kingdom of the stones."[6]

Copeland focused on the bacteria but suggested that there might be other kingdoms, too. He reassured botanists that to grant bacteria a kingdom of their own would not reduce botany's holdings: "These various creatures do not disappear from the course in botany just because they are not plants." Botany was "still the most convenient place to study them." Convenience aside, he asserted, "there is no other one thing so important in systematic biology as the fact that the grouping of organisms reflects and expresses their true relationships. It is inconsistent and unreasonable to begin the course in botany by doing violence to this basic principle."[7]

Copeland's note received little attention until 1938, at which time his son, Herbert Copeland (1902–1968), an instructor at Sacramento Junior College, published a classification system for all organisms in *Quarterly Review of Biology*. He had learned botany and taxonomy from his father, and he taught an "elementary" course addressing biology as a whole and the problems of matching bacteria, algae, fungi, and protozoa to the fixed definitions of plants and animals. "The scientists who find themselves under pressure to devise a more satisfactory system of kingdoms," he observed

are those charged with elementary instruction in biology or in one of its main branches, as botany or zoology. . . . The one who taught me elementary botany made clear to his freshman students the principles of classification. . . . When it became my turn to undertake elementary instruction, my efforts to recognize a series of natural kingdoms led me to distinguish four of them, called Monera, Protista, Plantae, and Animalia.[8]

Copeland reflected on this system for 12 years before publishing it. He gave Monera kingdom status on the assumptions "that they are the comparatively little modified descendants of whatever single form of life first appeared on earth, and that they are sharply distinguished from other organisms by the absence of nuclei."[9] By nucleus he meant that "part of the protoplast set apart (at least when it is not dividing) by a membrane. Its most definite characteristic is the process, mitosis or karyokinesis, by which it divided into two."[10]

Copeland knew well that there was no unanimity that bacteria and blue-green algae lacked nuclei.[11] There was a range of virtually every conceivable viewpoint. Discussion of the question of whether they possessed nuclei or not had not advanced since the nineteenth century. When in 1911 Cambridge microbiologist Clifford Dobell reviewed in detail the work of 49 authors, he concluded that bacteria definitely possess nuclei, and often a complex life cycle. "There is no evidence that enucleated Bacteria exist. The Bacteria are in no way a group of simple organisms, but rather a group displaying a high degree of morphological differentiation coupled in many cases with a life-cycle of considerable complexity."[12] While some saw in the bacteria evidence of a primitive nucleus composed of chromatin, others denied all traces of a nucleus with hereditary determinants comparable to other life forms.

Still others maintained that blue-green algae were an intermediary form between the bacteria, and truly nucleated forms, which as Benjamin Franklin Lutman commented in *Microbiology* of 1929, "may undergo a simple form of mitotic division."[13] Lutman's book focused on molds, yeast and bacteria as a group of colorless plants. He said that the "many common physiological characteristics of the colorless plant give such a group a unity in spite of the diverse morphology and distant relationship of its components."[14] Blue-green algae were also excluded in medical perspectives of bacteria. The following definition from *Pathogenic Microorganisms* in 1929 was typical:

Bacteria may be defined as extremely minute, simple, unicellular microorganisms, which reproduce themselves under suitable conditions with exceeding rapidity, usually by transverse division, and grow without aid of chlorophyll. They have no morphological nucleus, but contain nuclear material which is generally diffused throughout the cell body in the form of larger or smaller granules.[15]

From the perspective of general biology, Copeland postulated a series of four kingdoms, from the most simple to the most complex, each lacking characteristics of the next:

1. Monera, Haeckel. Organisms without nuclei, cells solitary or physiologically dependent. Groups included, bacteria and blue-green algae. Ancestral form, the original form of life; it is believed to be most nearly represented among living organisms by the nitrifying bacteria. Nomenclatorial type, *Bacillus subtilis*.

2. Protista, Haeckel. Organisms, largely unicellular with nuclei; typically with permanent nuclear membranes, centrosomes, and intra nuclear spindles, though all of these may be lost in evolution; lacking the combinations of characteristics to be listed as characteristics of plants and animals. Groups included, Flagellata (construed as excluding Volvocales), Rhizopoda, Sporozoa, Infusoria, diatoms, red algae, brown algae, and Fungi. Ancestral form, the first nucleated organism; this is presumably most nearly represented among living forms by the Chrysomonadida. Nomenclatorial type, *Amoeba Proteus*.

3. Plantae, Linneaus. Organisms (with few and derivative exceptions) having plastids containing the four pigments chlorophyll A, chlorophyll B, carotin, and xanthophyll, and producing true starch and cellulose. The primitive plants were unicellular and motile and have a nucleus as in Protista. Higher plants are non-motile, of elaborate structure, have no centrosomes nor intranuclear spindles. Groups included, Chlorophyceae an Embryophyta. Ancestral group, Volvocales.

4. Animalia, Linnaeus. Organisms which are multicellular, typically diploid and holozoic passing through blastula and gastrula stages in development. Centrosomes are present; spindles are generally formed outside the nuclear membrane, and enter the nucleus only as the membrane dissolves. Groups included, the Metazoa as usually construed (except possibly the Porifera, which might fall into Protista). Ancestral group, Porifera; or if that be excluded, Coelenterata.[16]

Monera for Expedience

Copeland recognized that the Protista and Monera kingdoms might not be natural groupings. He thought the Protista could actually be divided into several kingdoms, but not knowing how to do so, he thought it expedient to put them into one kingdom (see figure 6.1). Still, he suspected that the nucleated cell originated from nonnucleated life only once, as evidenced by "the uniformity of the nucleus, in its structure and in its behavior, in mitosis, in sexual reproduction, and as the vehicle of Mendelian heredity, wherever it occurs."[17] There was no comparable evidence that monera had a common origin; they

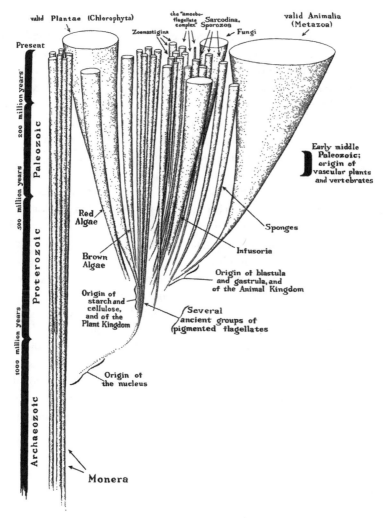

Figure 6.1 "Diagram of the General Phylogeny of Organisms," by Herbert Copeland, "The Kingdoms of Organisms," *Quarterly Review of Biology* 13 (1938): 383–420, at 410. With permission.

could have evolved multiple times. Even if "life originated more than once," he reasoned,

> it might be expedient to make an exception to the rule by gathering into one group all of the original forms and their comparatively little modified descendants. It is possible that this is done in establishing the kingdom Monera; one prefers to suppose that this group, like any other satisfactory taxonomic group, is natural.[18]

Copeland turned to arguments in regard to the unity of life to support a common origin, just as Darwin had in 1859. "All life resides in mixtures of essentially the same materials; all life obtains the energy for its immediate operations by processes of oxidation; all life...exhibits cellular organization."[19] The first life forms would have spread over the world and so change the planet as to prevent its repeated origin in the fashion that Haeckel had considered. The curtain was drawn. The door to further entry closed. He thus divided life on earth into two eras: the Proterozoic of the past billion years, and the Archeozoic representing all preceding time. Great deposits of elemental carbon in the Archeozoic provided definite proof that life was in existence at least 1.5 billion years ago, but there was no indication of what life forms were in existence then.[20]

The first organisms were autrotrophs, in his scheme. Blue-green algae were ruled out as the universal common ancestor because photosynthesis was too complex a system and involving two different forms of chlorophyll, as well as other pigments, yellow, brown, red, or blue. Filterable viruses were ruled out because they were not complex enough: none was known to possess any capacity for making organic compounds from inorganic ones.

Like others, Copeland pointed to the autotrophic bacteria (see chapter 5). The nitrifying bacteria, which oxidize ammonia and nitrites and use the energy released to make organic compounds from carbon dioxide, were the most primitive forms, in his view. They were important for agriculture and, though of great economic importance, were little studied. Close to them were the sulfur-oxidizing bacteria along with several genera of other obscure organisms that oxidize such substances as hydrogen, carbon monoxide, methane, alcohol, and acetic acid. The pigmented sulfur bacteria exhibited the property of swimming toward light and seemed to possess a primitive form of photosynthesis; the blue-green algae descended from them.[21]

The Map Is Not the Territory

Copeland elaborated on the basic postulates of his theory of classification in 1947. In his view, the patterns of life possessed an objective reality to be discovered from genealogies and common descent. Within interbreeding groups, life was a reticulated network. But where there were sterility barriers, from the species outward, it was a branching tree-like system. Evolutionary lineages did not form a network in which streams branch out and then reconnect as did for example the arteries of the human body. The natural system, he said, "consists of lines which divide, diverge and redivide, and are incapable of anastomosis."[22] A natural classification was something to be sought but perhaps never fully attained in full detail because so much of it had perished, and what remained possessed "an intricacy which has defied extensive detailed knowledge."[23] The taxonomic system he said "is not and can not become identical with the natural system; the two are related as an artist's representation, of a tree for example, is related to an actual tree."[24]

There were three fundamental conventions for constructing a natural taxonomic classification as Copeland conceived of it. The first concerned the rules for grouping nature's diversity into delimited fragments. Common descent was a requirement of any "natural group."[25] Characters or combinations of characters that appeared to have come into existence only once were indicators of natural groups. Because the plant and animal kingdoms possessed such characters they were undoubtedly natural groups.[26] The Protista was a natural group; its members include the original form of nucleated life and all of its descendants except those two specialized secondary developments, the familiar kingdoms of plants and animals.[27] Still, he emphasized, "The most profound of all distinctions among organisms is that which separates those without nuclei— bacteria and blue-green algae—from those which possess them."[28]

The second convention concerned the established categories of taxonomy. Taxonomists could give broader limits to any group by taking account a more distant common ancestor, or narrower limits by considering a more recent one. Therefore, the delimitation of any particular group, named and assigned to a taxonomic category, was "always artificial, arbitrary, and decided by convenience." But, one could not cross phylogenetically divergent lines. In taxonomic convention groups were assigned to seven grades or categories: kingdom, phylum or division, class, order, family, genus, species. None of them in Copeland's view was an objective category determined by nature. They were conventions, a practice that he traced to the writings of Darwin's contemporary Louis Agassiz (1807–1873), founder of the natural history museum of Harvard.[29] Agassiz perceived each of the seven categories to be realities, facts of nature; they were natural kinds to be discovered, not invented categories. "We know that he was mistaken," Copeland said. All taxonomic groups "are parts of a branching system considered separately for purposes of human thought."[30]

The third convention pertained to rules for assigning a name to a group. One was a useful invention called the type method, whereby a group is named with a specific example representing the group. The type meant the exemplar of the group whose members possess many but not all group characteristics.[31] As Copeland put it, it was a "standard" and allowed for diversity within the polythetic group.[32] The second rule was that of priority: recognizing the oldest valid name and only that name. On that basis, in 1947 Copeland abandoned the terms "protista" and "monera" because he had come to learn of Richard Owen's and John Hogg's proposals (see chapter 2). He rejected "protozoa" because it had been applied to a class before Owen made it a kingdom, and he adopted Hogg's term "Protoctista."[33] He rejected the term "monera" because the standard example of it for Haeckel in 1866 was Protoamoeba, which, as it turned out, was not an organism at all but rather a broken-off fragment of an amoeba. Actually, by 1904, Haeckel's standard examples for monera were bacteria and blue-green algae (see chapter 3). Nonetheless, Copeland adopted the term "Mychota" for that kingdom, as coined by Günther Enderlein in 1925.[34]

A Synthesis of Sorts

Discussions of kingdoms were as far removed from the concerns of most biologists as were microbes themselves. Copeland's four-kingdom proposal appeared at a time when biologists' thinking was converging on an outlook much like that which Darwin had originally proposed. Following the publication of *Origin*, many biologists rejected natural selection as the primary mechanism of evolution.[35] Some argued that the inheritance of acquired characteristics was the most important mechanism; others insisted that new species arose rather suddenly by large leaps, not gradually from minute variations within species as Darwin had proposed; still others adopted a plurality of evolutionary mechanisms including a directive tendency, an evolutionary trend in lineages: orthogenesis. During the Second World War, there was a conscientious effort to show how the conclusions from various biological specialties might come together over Darwin and establish a set of general evolutionary principles.

T.H. Huxley's grandson Julian Huxley (1887–1975) called it "the modern synthesis."[36] The evolutionary synthesis has been depicted then and now as an attempt at the "unification of biology," a time to bridge the gaps among disparate disciplines in the life sciences: ecologists, zoologists, paleontologists, zoological and botanical taxonomists in museums and herbariums, embryologists, and geneticists who developed the chromosome theory and who studied the genetics of populations.[37] In its conceptual foundations, it was about animals and plants. Evolution would be gradual, and evolutionary change would be based on Mendelian gene mutations and recombination together with the natural selection of the fittest. The first generation of Mendelian geneticists had been non-Darwinian saltationists who opposed gradual evolution of species by natural selection. But by the 1920s, most Mendelian gene differences were shown to be small and subtle.

Saltational leaps in the emergence of new species were ruled out, as was the inheritance of acquired characteristics. The mathematical underpinnings of the merger of selection theory and Mendelian genetics with the statistical studies of variations within populations was led by R.A. Fisher, Sewall Wright, and J.B.S. Haldane.[38] Selection acted on variations within populations, gene pools, which evolve. "Evolution is a change in the genetic constitution of populations," Theodosius Dobzhansky (1900–1975) explained in *Genetics and the Origin of Species*.[39]

Microbes were not included in that synthesis.[40] Bacterial evolution seemed to lie beyond neo-Darwinian conceptions. Bacterial adaptations to their physical environment, their resistance to lethal viruses, and the transformations resulting from the acquisition of what some considered to be infectious genes from the environment that then exist in symbiosis within their hosts, were viewed to belong to a neo-Lamarckian realm. As Stanford bacteriologist

W.H. Manwaring commented in 1934:

> About the only conventional law of genetics and organic evolution that is not definitely challenged by current bacteriologists is the nineteenth century denial of the possibility of spontaneous generation of a bacterial cell. Even this is questioned by certain recent theorists in their hypothetical transformation of certain normal enzymes into "pathogenic genes" or "filterable viruses", and in their apparently successful synthesis of "Twort genes" by the chemical oxidation of certain heat-sterilized organic products.
>
> Whether or not future refinements in immuno-chemical technique can or will bridge the gap between the apparent Lamarckian world of bacteriology and the presumptive Darwinian world of higher biological science is beyond current prophecy.[41]

Haeckel's conception of bacteria as lacking genelike entities also persisted well into the twentieth century. In *The Evolution of Genetic Systems* of 1939, C.D. Darlington (1903–1981) referred to "asexual bacteria without gene recombination" and "genes which are still undifferentiated in viruses and bacteria."[42] In *Evolution: The Modern Synthesis*, Julian Huxley summarized what many believed about the gene-less bacteria in 1942:

> Bacteria (and *a fortiori* viruses if they can be considered to be true organisms), in spite of occasional reports of a sexual cycle, appear to be not only wholly asexual but pre-mitotic. Their hereditary constitution is not differentiated into specialized parts with different functions. They have no genes in the sense of accurately quantized portions of hereditary substances; and therefore they have no need for accurate division of the genetic system which is accomplished by mitosis. The entire organism appears to function as soma and germplasm, and evolution must be a matter of alteration in the reaction-system as a whole. That occasional "mutations" occur we know, but there is no ground for supposing that they are similar in nature to those of higher organisms, nor since they are usually reversible according to conditions, that they play the same part in evolution. We must, in fact, expect that the processes of variation and evolution in bacteria are quite different from the corresponding processes in multicellular organisms. But their secret has not yet been unraveled.[43]

The New Systematics

The "modern synthesis" was spearheaded by naturalists-systematists who aimed at revitalizing systematics. In 1938, when Copeland proposed four kingdoms of organisms, Huxley, then secretary of the Zoological Society of London, was in the process of organizing a volume with 22 contributors drawn from botany, zoology, embryology, paleontology, and ecology, as well as museum and herbarium taxonomists. No microbiologist was included. He called it *The New Systematics*.[44] Systematics, he said, was perceived by experimentalists as "a rather narrow branch of biology on the whole empirical and lacking in unifying principles, indispensable as a basis for all biological workers, but without much general interest or application to other branches of their science."[45] His aim was to shake off the dust of the stark museum and make systematics one of the focal points of biology, by applying the fresh findings of experimentalists and exploring unifying principles.

The New Systematics along with Harvard ornithologist Ernst Mayr's (1904–2005) *Systematics and the Origin of Species* of 1942,[46] included discussions of the philosophical and methodological principles of taxonomy. What is meant by a "natural classification"? Are such taxonomic categories as species, genus, and family artificial creations of the systematist, merely matters of convenience, or are they real natural groupings with an objective reality? Could and should taxonomy be based on phylogeny? Can the course of evolution be visualized as a tree? To what extent is the course of evolution reticulated?

ARE SPECIES REAL?

The problem of conceptualizing species was characterized by the dispute between Darwin and Agassiz, a lifelong opponent of the theory of evolution. For Agassiz, species were real, natural types around which individual variations occurred; they were the "thought of God."[47] He criticized Darwin's *Origin* in part because of Darwin's seemingly nominalist conception of species. For Darwin, like Lamarck before him, species were not real—if real meant eternal or immutable, or if the boundaries between them had to be sharp. There was a great amount of variation within species, and this variation was the fuel of evolution. Darwin did not develop a clear conception of species. Sometimes he considered the inability to interbreed as the decisive criterion, as did Buffon and Lamarck; other times he considered differentiating morphological characters.[48] "In determining whether a form should be ranked as a species or a variety," Darwin said, "the opinion of naturalists having sound judgment and wide experience seems the only guide to follow."[49]

Those who forged the neo-Darwinian synthesis disagreed. Species was a real category of nature. "There is the question," Mayr wrote, "whether the higher categories represent phylogenetic groups or not. Do genera and families have an objective reality or are they artificial creations of the taxonomist?"[50] Whether one put 50 species into one genus, or split them into two groups of the most closely related ones, each group, a separate genus, might be a matter of convenience. The same might be true for higher categories of family, class, phylum, or perhaps even kingdoms if they had considered them. But the reality of species was a cornerstone of "the new systematics."[51] Huxley explained that although there had been a time in which biologists denied

> any greater objective validity to the species-category than to categories of higher order such as genus, family or class, there seems now to be a general recognition among those who have concerned themselves with taxonomic facts, whether from the standpoint of the museum systematist, the ecologist and physiologist, or the geneticist, that the species are in some valid sense natural groups.[52]

Referring to John Stuart Mill's concept of "natural kinds," William Turrill wrote in 1925:

> Kinds differ one from another in an indefinite number, "an unknown multitude" of properties and characters. We select a set of characters to discriminate each Kind from all other Kinds. Our selection of these characters is arbitrary and matter of convenience, but separate Kinds really exist.[53]

Species were the nexus of the evolutionary process; speciation was real. Like no other stage of divergence, H.J. Muller commented, it "involves the entrance of a qualitatively different factor, having a direct influence upon the process of divergence itself."[54] Dobzhansky's *Genetics and the Origin of Species* of 1937 included a chapter titled "Species as Natural Units." In it, he defined the species as "that stage of the evolutionary process at which the once actually or potentially interbreeding array of forms becomes segregated into two or more separate arrays which are physiologically incapable of interbreeding."[55] Mayr adopted Dobzhansky's definition and proposed three years later, "Species are groups of actually or potentially interbreeding natural populations which are reproductively isolated from other such groups."[56] He dubbed this "the biological species concept."

Delimiting species by means of interbreeding and sexual isolation could not be applied to so-called asexual organisms.[57] Dobzhansky was convinced that bacteria evolved without speciation.[58] And it was also obvious to bacteriologists of the 1920s and 1930s that "bacterial species" was an artificial division that did not actually exist.[59] Even when genetic recombination by conjugation was demonstrated in bacteria, it was considered to be a rare event, and leading bacteriologist preferred to speak of "biotypes" instead of species.[60]

Many plants also violated the "biological species concept" because they reproduced asexually. Speciation in plants could also occur by polyploidy, where more than two sets of chromosomes are formed in offspring derived from seeds in which a halving of chromosome number (haploid) by meiosis did not occur. The diploid seeds give rise to a tetraploid plant. When the newly arisen tetraploid plant tries to breed with its ancestral species, triploid offspring are formed. These are sterile because they cannot form gametes with a balanced assortment of chromosomes. However, the tetraploid plants can breed with each other. A new species could be formed in one generation.

New plant species could be formed sympatrically by hybridization.[61] Geneticists had also shown that what taxonomists regard as perfectly good species and genera of plants may be crossed and leave fertile offspring. Hybridization could confuse a phylogenetic classification of plants. If hybridization were a significant process, then phylogenetic interpretation would be more difficult in plants; the pattern of the phylogenetic "tree" would, in many places, be hopelessly obscured by anastomosis of the branches. The scope and significance of hybridization had been one of the most controversial topics in the field of plant evolutionary biology since the 1920s.

George Ledyard Stebbins's (1906–2000) *Variation and Evolution in Plants* of 1950 helped to complete "the evolutionary synthesis" from the macrobiological point of view.[62] He recognized hybridization to be a significant mechanism of evolution in numerous groups of higher plants:

> In fact, the accumulating evidence may make possible the generalization that nearly all of the plant genera which are "critical" or intrinsically difficult of classification owe their difficulty largely to either the direct effects of interspecies hybridization or the end results of hybridization accompanied by polyploidy, apomixis or both.[63]

Still, while experimentalists assigned such an important role to hybridization, systematists assigned little importance to it.

Animal systematists also dismissed hybridization as insignificant.[64] "As far as animals are concerned," Mayr said, "the possibility of 'reticulated' evolution ... may be largely disregarded. Reticulated evolution is possible only where species, genera, and families can hybridize successfully, and this occurs only exceptionally in animals."[65] Both Mayr and Dobzhansky recognized that a species concept could not be based on sterility alone—because two species of animals may mate and produce fertile offspring in artificial conditions such as a zoo—but in natural conditions this would not occur because of courtship patterns and other animal behaviors.[66] Once speciation occurred in animals in the wild, only further branching was possible. And that bifurcating, tree-like pattern would reflect the evolutionary process. Speciation for Mayr was based on geographic isolation, which he called allopatric speciation.

TAXONOMY AND PHYLOGENY

The first three decades of the twentieth century marked a tremendous period of ordering the expanding known diversity of macrobiological life. About one million species of animals alone were thought to exist. Organisms were classified into a hierarchical array of group within group as if to reflect a phylogenetic process of divergence. But there was certainly no consensus among the contributors of *The*

New Systematics that taxonomy reflected phylogeny. "Fundamentally, the problem of systematics," Huxley wrote, "is that of detecting evolution at work."[67] But, just as his grandfather had argued six decades earlier, there was "no little danger of introducing confusion into science by mixing up such hypotheses with taxonomy" (see chapter 3).[68]

The methodological relations between taxonomy and phylogeny were indeed as muddled as the relations between fact and theory. On the one hand, Julian Huxley maintained, taxonomy should be based on comparisons of the largest possible number of characters, and phylogeny could only be subsequently deduced by the characters useful in taxonomic evaluation. On the other hand, he conceded that "a phylogenetic interpretation may sometimes suggest an improved taxonomy."[69] In Huxley's view, there were two areas of taxonomy that simply could not have a phylogenetic basis. One was in plants, in which lineages would be obscured by hybridization, resulting in a reticulated descent; the other was in "the early evolutionary stages" where there was a lack of fossil material.[70]

Few would disagree that, in principle, classification should double as an easy means of identification and as a summary of existing knowledge of phylogeny.[71] But there was a range of views about the actual situation. Paleontologists said that their classification was a phylogenetic one, "in which fossils are arranged as nearly as possible in accordance with the supposed course of their evolution."[72] But at the best of times, their results were hypothetical and subjective; experts seldom agreed on the details of phylogeny.[73] William Thomas Calman at the British Museum was more confident. He argued that taxonomists had classified about three-quarters of a million animal species, and while unknown species were brought in every day, seldom did they come across a species for which there was not already a place in the existing taxonomic order of things. No reorganization of the puzzle was necessary. "As a result of this experience, we come to have confidence in the Natural System of classification that is perhaps not always shared by our colleagues of the laboratory. The *Systema Naturae* becomes for us an objective reality, not merely a convenient filing device."[74] Most taxonomists, he said,

> would find it difficult to divest themselves entirely of their evolutionary preconceptions, and nearly all would agree that it is not only legitimate but necessary to be guided by ideas of probable phylogeny when the mere balancing of resemblances and differences leaves the position of an organism uncertain.[75]

Botanists were polarized in their views of the relations of phylogeny and taxonomy. At one extreme, Thomas Sprague at the Royal Botanical Garden said: "Taxonomy may be defined as scientific classification of the different kinds of living organisms according to their proved or inferred phylogenetic relationships."[76] At the other extreme were the views of his colleague at Kew,

John Gilmour, who insisted that speculation about phylogeny ought to be abandoned altogether in taxonomy.[77]

None of the contributors to *The New Systematics* mentioned the microbial world where the proposed new kingdoms lay hidden. Discussion of kingdoms was exceptional in biology. In 1939, botanist Henry Conrad at Grinnell College in Iowa considered a third kingdom for protists.[78] But, as André Prévot at the Pasteur Institute in Paris noted when he proposed "the bacterial kingdom," few bacteriologists were willing to speak of it. The main reason was that bacteria lacked definition. Should blue-green algae be considered as part of the group? How were bacteria to be distinguished from protists, on the one hand, and from viruses, on the other? "It is somewhat difficult to define the bacteria," Arthur Henrici wrote in *The Biology of Bacteria* in 1939, "They have been defined as one-celled Fungi reproducing by simple fission only; this is probably the most widely accepted definition."[79] Prévot's "bacterial kingdom" did not include the blue-green algae.[80] Still, the kingdom Monera, as Copeland defined it in 1938, was recognized by a few leading microbiologists in the 1940s.[81]

| # The Prokaryote and the Eukaryote

It is a waste of time to attempt a natural system of classification for bacteria, . . . bacteriologists should concentrate instead on the more humble practical task of devising determinative keys to provide the easiest possible identification of species and genera. This opinion, based on a clear recognition and acceptance of our ignorance concerning bacterial evolution, probably represents the soundest approach to bacterial classification.

—Roger Stanier, Michael Doudoroff, and
Edward Adelberg, *The Microbial World* (1957)

THE DECADES FOLLOWING the Second World War were heady times for microbiology. Microbes surfaced at the center of biology, as a new generation of scientists converged on *Escherichia coli* and a few other select microorganisms to investigate the chemical basis of the gene and how it functions in the cell.[1] Methodological debates about bacterial classification continued, and so too did the issue of how bacteria could be distinguished from other microbes. Members of what came to be called the "Delft School" led those discussions: Albert Jan Kluyver (1888–1956), his former student Cornelis van Niel (1897–1985), and van Niel's former student Roger Stanier (1916–1982).[2]

Best known then and now for his broad survey of microbial metabolism, Kluyver is considered one of the founders of comparative biochemistry.[3] In 1921, he succeeded Martins Beijerinck in the general and applied microbiology chair at the Delft Institute of Technology. Van Niel completed his Ph.D. in 1928 before moving to Stanford University, where between 1930 and 1962 he taught a whole new generation the principles of microbiology in his famed summer course at the Hopkins Marine Station on Monterey Bay.[4] Stanier was van Niel's student between 1938 and 1942. Together with Michael Doudoroff and Edward Adelberg, Stanier wrote *The Microbial World*, a book that appeared in five editions, synthesizing the work in microbiology, biochemistry, molecular biology, genetics, and cell biology.

In 1962, Stanier and van Niel would close the controversy once and for all regarding the anatomy of bacteria and blue-green algae when they introduced the central concepts that came to define microbiology and cell biology: the distinction between the prokaryote and the eukaryote. Their views regarding the cellular organization and classification of bacteria had taken sharp turns over the previous two decades. At first, they advocated a phylogenetic classification based largely on morphology and homology, and they endorsed Copeland's concept of the Monera. But by the mid-1950s, they would deny the possibility of a natural classification and renounce the kingdom Monera, never to support it again.

Realists versus Idealists

Bergey's Manual used a great diversity of characteristics to differentiate bacteria into a hierarchical classification of orders, families, genera, and species. Recall from chapter 5 that Winslow and the Committee on Characterization and Classification of Bacterial Types for the Society of American Bacteriologists sought an empirical approach with no "idealistic" preconceptions regarding the course of evolution.[5] Bacteria were to be grouped together using whatever traits were available on the assumption that naturalness of a group would be decided by the greatest number of characters its members have in common. Although it became the most widely used source on bacterial taxonomy, *Bergey's Manual* received severe criticisms from such leading bacteriologists as Sergie Winogradsky and André Prévot as well as from the Delft School, which pursued a system of classification based on principles of homology and the weighting of characters.[6]

The open conflict with the authors of *Bergey's Manual* continued throughout the 1930s and 1940s. Debates were framed between evolutionary hypothesis and morphology, on the one hand, and empiricism and eclecticism, on the other. "Anyone who has had the opportunity to peruse the book," Kluyver and van Niel wrote in 1936, "will have been struck by the fact that morphological, physiological, nomenclature, utilitarian, cultural and pathogenic properties have been used in the building up of the system in the most arbitrary way. The result is a complete lack of homology in the various groups."[7] Robert Breed replied in the fifth edition of *Bergey's Manual* in 1939, noting how Kluyver and van Niel's classification scheme was "drawn up to express the ideas held by their authors of the natural relationships of the various groups of bacteria":

> It is surprising that there has been so little appreciation of the fact that both idealism and realism are needed for the best development of systematic bacteriology. Idealists have been very impatient of the imperfections in Manual keys, descriptions and classification outlines; but few of these critics have done much toward working out the monographs that are needed before the Manual can be improved. Realistic workers have been impatient with idealists who have introduced many unnecessary names and unjustified speculations regarding relationships between the various groups of bacteria.[8]

In 1941 when Stanier and van Niel responded to these remarks in a paper titled "The Main Outlines of Bacterial Classification," they ignored Breed's statement that both realism and idealism were necessary, and painted *Bergey's Manual* as being against phylogeny. "In most biological fields," they wrote,

> it is considered a truism to state that the only satisfactory basis for the construction of a rational system of classification is the phylogenetic one. Nevertheless, "realistic" bacteriologists show a curious aversion to the attempted use of phylogeny in bacterial systematics. This is well illustrated, for example, by the statement of Breed (1939).[9]

Although "the true course of evolution can never be known and that any phylogenetic system has to be based to some extent on hypothesis," they said, "there is good reason to prefer an admittedly imperfect natural system to a purely empirical one."[10] A phylogenetic system, they reasoned, could be altered and improved as new facts come to light, and its very weaknesses would point to the kind of experimental work necessary for improvement. A determinative classification had no predictive ability; it did not allow microbiologists to predict those properties of newly discovered organisms that might be closely related from the properties of previously described organisms. A classification scheme lacking an evolutionary framework would not permit studies of the origin and evolution of such functions as drug resistance or photosynthesis. Stanier and van Niel put the issue in a nutshell:

> An empirical system was largely unmodifiable because the differential characters employed are arbitrarily chosen and usually cannot be altered to any great extent without disrupting the whole system. Its sole ostensible advantage is its greater immediate practical utility; but if the differential characters used are not mutually exclusive (and such mutual exclusiveness may be difficult to attain when the criteria employed are purely arbitrary) even this advantage disappears. The wide separation of closely related groups caused by the use of arbitrary differential characters naturally enough shocks "idealists," but when these characters

make it impossible to tell with certainty in what order a given organism belongs, an empirical system loses its value even for "realists."[11]

The weakness of "the empirical" approach was more than apparent in the vague way in which *Bergey's Manual*'s defined bacteria simply as "relatively primitive in organization."[12] The manual said nothing about the absence of a "true nucleus," or about the absence of sexual reproduction. And blue-green algae were not included.[13] Instead, bacteria were classified as a class of fungi, the fission-fungi—Schizomycetes, as they had been named in 1857—even though the claim that chlorophyll is absent was invalidated by all the purple bacteria. "A more inadequate definition than that given by Bergey would be hard to conceive," Stanier and van Niel scoffed.[14] Following Copeland, they assigned bacteria and blue-green algae to the kingdom Monera: "The common features of true bacteria and blue-green algae may be summarized as follows:

1. Absence of true nuclei.
2. Absence of sexual reproduction.
3. Absence of plastids."[15]

Heterotroph and Sphere First

With the exception of such energy-providing processes as photosynthesis, Stanier and van Niel maintained that morphological characters should generally take priority over physiological characters when defining taxa above the species. The main problem of drawing up a phyletic system on physiological basis, they said, was "the necessity of making a large number of highly speculative assumptions as to what constitute primitive and advanced metabolic types."[16] They pointed to Sigurd Orla-Jensen's scheme as illustrative. Recall that he had assumed that chemosynthetic bacteria were the most primitive group, as did many other microbiologists, including the authors of *Bergey's Manual* (see chapters 5 and 6). Although autotrophs-first was the favored view during the first three decades of the century, a shift in preference to heterotrophs-first occurred after 1938 when Aleksandr Oparin's book *Origin of Life* of 1924 was translated into English.[17]

Oparin postulated that life did not originate in an environment devoid of organic matter, but rather that many organic compounds had emerged from strictly chemical reactions before organisms appeared. Benjamin Moore argued similarly in 1912 (see chapter 5). Autotrophs-first models had assumed an oxidizing atmosphere in which aerobic chemotrophs would thrive. But J.B.S. Haldane suggested in 1929 that if the primitive atmosphere was reducing (oxygen-poor), and if there was an appropriate supply of energy, such as lightning or ultraviolet light, then a great variety of organic compounds might be synthesized abiotically.[18] Without microbes to decompose them, they could have persisted for long periods of time, and as a consequence, more and more complex molecules or molecular aggregates could have arisen through chemical interactions. With the emergence of such complexes, a fortuitous combination of circumstances might have yielded systems with properties of self-propagation. The synthetic ability of those entities would emerge gradually, step by step. Consequently, the earliest living forms would have been heterotrophs; the development of autotrophism was a later adaptation when organic materials had become scarce because of the overpopulation of heterotrophs (see chapter 17).[19]

As Stanier and van Niel saw it, autotrophy was too complex to be primordial. The ability of an organism to synthesize all its cellular constituents using carbon dioxide, for example, as the only carbon source, they said, would require a highly developed enzymatic system, and it was hard to imagine how such an apparatus could have originated by any mechanism in an inorganic world. Of course, there was the hypothesis of panspermia, which postulated transportation of extraterrestrial life by one means or another (see chapter 4). But, in their view, that seemed unlikely because such an organism would have to withstand the heat of atmospheric entry, and during interstellar transit the germs would be exposed to ultraviolet irradiation of such intensity that it seemed inconceivable that any germ could reach Earth without being killed. Even then, the problem of the origin of life would not be solved; it would merely be shifted from one part of the universe to another.[20]

Stanier and van Niel argued that it would be foolhardy to construct a classification system based on the increased metabolic complexity because many of the differences in power of synthesis may be due to losses. It was "rather naive," they said, to believe that one can discern the trend of physiological evolution.[21] The real evolutionary trend in their view was toward morphological complexity beginning with a simple sphere, the original shape of all bacteria: "the hypothetical coccus type."[22] Breed had made the same suggestion in 1918, and van Niel developed the concept with Kluyver in 1936 (see figure 7.1).[23] Evolution proceeded toward increasingly intricate form, toward more complex life cycles,

> in the direction from unicellularity to multicellularity. The highest developmental stage in the group of spherical organisms is in all probability displayed by the cocci able to form endospores.... Endospore forming rods with pritrichous flagella present a higher stage of development in these groups.... Further development of these universally immotile bacteria can have given rise to the mycobacteria which apparently form the connecting link with the simpler actinomycetes.[24]

In their scheme, morphological properties would be used for the demarcation of larger taxonomic units: tribes, families, and so on, whereas both morphological and biochemical characters would be used to define genera.[25] An

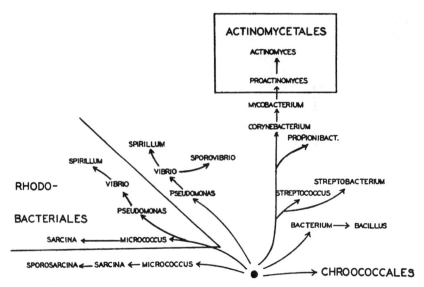

Figure 7.1 "Phylogeny of the Eubacteriales and Related Groups, According to the Scheme of Kluyver and van Niel with Slight Modifications." R.Y. Stanier and C.B. van Niel, "The Main Outlines of Bacterial Classification," *Journal of Bacteriology* 42 (1941): 437–466, at 446. With permission.

exception to that rule was made for the blue-green algae because of their photosynthesis by chlorophyll a and b.[26] Stanier and van Niel insisted that there were enough discernable morphological differences for the job:

> Clearly paramount is the structure of the individual vegetative cell, including such points as the nature of the cell wall, the presence and location of chromatin material, the functional structures (e.g., locomotion), the method of cell division, and the shape of the cell. A closely allied character is the type of organization of cells into larger structures. In addition, the nature and structure of reproductive or resting cells or cell masses deserve due consideration.[27]

Abandoning Phylogenetics

Despite his public optimism, van Niel was actually far from confident about approximating a natural bacterial classification. He conveyed his despair in a letter to Stanier in 1941:

> Many, many years ago I often went around with a sense of futility of all our (my) efforts. It made me sick to go around in the laboratory (this was in Delft) and talk and think about names and relations of microorganisms; about the fate of substrates and hydrogen atoms, about—well about

everything. During those periods I would go home after a day at the lab, and wish that I might be employed somewhere as a high-school teacher. Not primarily because I liked that better. But simply because it would give me some assurances that what I was doing was considered worth-while.[28]

When van Niel addressed bacterial classification and natural relationships of bacteria at a famous Cold Spring Harbor Symposium in 1946, his support for the phylogenetic outline he, Kluyver, and Stanier had proposed had weakened, mainly because of "the paucity of characteristics, especially those of a developmental nature."[29] Nonetheless, he avowed, "the search for a basis upon which a 'natural system' can be constructed must continue."[30]

But all hope for a natural classification of bacteria would perish a few years later. In 1952 Winogradsky asserted that phylogenetic classification was simply "impossible to apply to bacteria."[31] Van Niel followed suit in 1955: "The sooner bacteriologists recognize that comparative morphology was inappropriate for microbial classification the better."[32] Both of them reminded bacteriologists that the hierarchical arrangement in *Bergey's Manual* of species, genera, tribes, families, and orders was only a facade. To avoid the delusion that it represented a natural ordering, they suggested using the term "biotypes" instead of "species" and using common names such as "sulfur bacteria," "photosynthetic bacteria," and "nitrogen-fixing bacteria" instead of Latin names with their phylogenetic implications.[33]

When the first edition of *The Microbial World* was published in 1957, Stanier, Doudoroff, and Adelberg followed van Niel in denouncing a natural bacterial classification. They said that the system for classification for plants and animals was natural because clues to evolutionary relationships could be inferred from the fossil record, comparative morphology, and embryology. But bacteria offered few points of comparison because of the simplicity of their morphology and development. "Hence the construction of the broad outlines of a natural system of bacterial classification involves much guesswork and affords the possibility for endless unprofitable disputes between the holders of different views."[34]

Renouncing Kingdom Monera

Van Niel and Stanier also renounced the kingdom Monera in the mid-1950s on the same grounds on which they had endorsed it. By that time, bacteria were reconceived as possessing the three characteristics that they and others had previously claimed that they lacked: nuclei, plastid-like entities, and sexual reproduction. New evidence based on improvised staining techniques seemed to indicate that bacteria did indeed have a primitive nucleus. All earlier claims of a bacterial nucleus had been based on using the basic aniline dyes that stain the chromatin of other organisms. All the nucleic acid in bacteria stained, giving some the impression that the entire bacterial cell was equivalent to the nucleus.

But in the early 1940s, C.F. Robinow at Cambridge University developed new methods for staining bacterial cells during spore germination and reported the existence of bacterial "chromosomes" and a form of mitosis.[35] "Lack of nuclei, if universal, would indeed place bacteria in an isolated position as to their taxonomy," he wrote in 1945. "However, it now seems that the cytology of bacterial *spores* may become a factor in the breaking down of this isolation and in the tracing of affinities to other groups of microorganisms."[36] René Dubos commented that year "that the blue-green algae possess a structure—the central body—which is rich in desoxyribonucleic acid, and which has been considered as a primitive nucleus sharply differentiated from the protoplasm."[37] Ten years later, van Niel agreed "that bacteria contain discrete structures that might be considered, on the behavior and chemical nature as nuclei."[38]

Whether anything like mitosis occurred in bacteria had long been disputed.[39] Electron microscopy developed during the 1940s: using a beam of electrons instead of a beam of light increased magnification to 200,000×.[40] The first electron microscopic photographs of bacteria seemed to confirm the presence of intracellular round bodies called "nucleoids," "chromatinic bodies," or "chromosomes" that, for some observers, appeared to pair or constrict and indicate a probable division.[41] "The presence of nuclei in bacteria was established with certainty only about 15 years ago," Stanier and colleagues wrote in 1957. "It has been shown that all bacteria contain discrete, intracellular bodies with the chemical properties expected of nuclei, which appear to divide in a co-ordination with the division of the cell."[42]

There was also new evidence suggesting that both blue-green algae and bacteria possessed plastid-like structures. Stanier himself provided some of it. After completing his Ph.D. in 1942, he returned to his native Canada to take part in the development of large-scale industrial production of penicillin during the Second World War. In 1947, he accepted a position at Berkeley, where for the next 24 years he worked on the taxonomy, metabolism, and physiology of bacteria and blue-green algae. In 1952, he and his collaborators reported the presence of uniform spherical particles containing photosynthetic pigment in the bacterium *Rhodospirillum rubrum*.[43] Each possesses thousands of specialized structures harboring photosynthetic pigments; they referred to those structures as "chromatophores," a word often used to include chloroplasts of plants.[44] But those photosynthetic purple bacteria did not produce oxygen as did blue-green algae, which carried out photosynthesis in the manner of plants.[45] Although it had been generally assumed that the pigmentation of blue-green algae were uniformly distributed throughout the cytoplasm, there had been occasional reports that blue-green algae contained grana or chloroplasts. The reality of those bodies seemed to be confirmed in 1952.[46]

Finally, there was evidence of genetic recombination and sexuality in bacteria. During the first three decades of the century, geneticists had mapped genes on chromosomes and correlated various alleles with diverse phenotypic traits of plants and animals. But what the gene was and how it affected the characteristic said to be under its control was unknown. The gap between gene and character began to be filled in during the 1940s when biochemists and geneticists teamed up to explored how genes affect metabolic reactions by means of enzymes. The methods of the new biochemical genetics required that both the sex life and growth of an organism be brought under control. Microorganisms were most suitable. Microbiologists had studied their growth and life cycles and in many cases succeeded in identifying their nutritional requirements, their ability to use certain compounds as a carbon source, and their sensitivity to antimicrobial substances. The geneticist would isolate mutant strains that were found to be unable to grow, or that grew poorly on a well-defined medium, and the biochemist sought the reason for this inability. Mutants would be selected that were unable to synthesize known metabolites such as vitamins and amino acids.

Sexual genetic recombination had been demonstrated in various microorganisms by the 1940s: the bread mold *Neurospora*, the green alga *Chlamydomonas*, the ciliate *Paramecium*, and baker's yeast, *Saccharomyces cerevisiae*.[47] All became model genetic organisms. Experiments using bacteria would provide some of the first evidence that DNA was the basis of the gene. That conclusion began to emerge with research on *Pneumococcus* transformations published in 1944 by Oswald Avery (1877–1955), Colin M. MacLeod (1909–1972), and Maclyn McCarty (1911–2005) at the Rockefeller Institute. *Pneumococcus* could absorb and incorporate the DNA released by dead bacteria (transformation).[48] Two years later, Joshua Lederberg (1925–2008) and Edward Tatum (1918–1994) demonstrated genetic recombination in *Escherichia coli*.[49] Lederberg referred to it as sex and called it "conjugation."

There were two competing models for bacterial recombination in the early 1950s. Lederberg and collaborators assumed the traditional Mendelian-chromosome model of genetic recombination between strains resulting from cell fusion chromosome pairing crossing over, reduction and zygote formation as in other organisms. Such sexual fusion was considered to be rare: one in a million. The alternative model was proposed by William Hayes (1918–1994) at the Royal Postgraduate Medical School in London who maintained that (1) there was only a partial transfer of DNA during bacterial conjugation, which was (2) uni-directional from donor to recipient, and that (3) recombination was mediated by an infectious virus-like genetic particle: an infectious "fertility factor F" determined sexual compatibility.[50] Those cells possessing the specific fertility, F+ cells, act as donors; F− cells act as gene receivers and "only part of the genetic material of the F+ parent is effectively transferred to the zygote." By the middle of the decade, the hypothesis of one-way partial chromosome transfer was demonstrated beyond reasonable doubt.

Still, as Stanier, Doudoroff, and Adelberg saw it in 1957, sex in bacteria was similar to that in some protozoa.[51] In *Paramecium*, for example, two cells pair and form cytoplasmic connections between them. Each partner retains one haploid nucleus and transmits a second to the other partner. Nuclear fusion with the formation of a diploid nucleus takes place within each cell, following which the two cells separate. This resembled the process in bacteria, except that in bacteria one cell acted strictly as a genetic donor, transferring only a limited portion of its genetic material to the receptor.

These data about bacterial "sex," nucleus, and organelles, in the view of Stanier and colleagues, invalidated the kingdom Monera. "It is clear that the criteria for a kingdom of organisms without nuclei do not apply to the bacteria and blue-green algae," van Niel wrote in 1955.[52] Both van Niel and Stanier would from then on maintain that bacteria belonged within Haeckel's kingdom Protista. *The Microbial World* of 1957 spoke of "lower" and "higher" protists.[53] The cells of the "higher" forms were identical to those of plants and animals. Bacteria and blue-green algae were "lower" protists characterized by "a relatively simple cell structure."[54] In defining the bacteria, Stanier and colleagues excluded the blue-green algae:

> Probably the best positive definition of the bacteria that can be made at present would run as follows: The bacteria are a morphologically varied collection of small microorganisms with a primitive cellular organization, like that of the blue-green algae. Most of them are nonphotosynthetic. The photosynthetic representatives differ from blue-green algae physiologically, for they carry out a special kind of photosynthesis in which oxygen is never involved.[55]

Bacteriology's Scandal

Stanier and van Niel changed their views radically five years later when they aimed to resolve the issues about the anatomy of the bacterium once and for all in a famous paper titled "The Concept of a Bacterium" in 1962. "Any good biologist," they wrote, "finds it intellectually distressing to devote his life to the study of a group that cannot be readily and satisfactorily defined in biological terms; and the abiding intellectual scandal of bacteriology has been the absence of a clear concept of a bacterium."[56] Though they had "become sceptical about the value of developing formal taxonomic systems for bacteria, the problem of defining these organisms as a group in terms of their biological organization is clearly still of great importance, and remains to be solved."[57] Microbiologists had little difficulty distinguishing a bacterium from another kind of microorganism, they said, disputes occurred at the borders where the largest of the bacteria might be confused with other protistan groups, and the smallest of bacteria confused with the largest of viruses.[58] Their purpose was to provide a definition that would "permit a clear separation of the bacteria *sensu lato* both from viruses and other protists."[59]

The meaning of the word "virus" had changed over centuries. In the sixteenth century it referred to poison or venom; in the eighteenth it referred to a poisonous substance produced by the body resulting from disease in the sense of contagious pus, and in the nineteenth century viruses were considered as microbes—germs. As Pasteur declared in 1890, "*Tout virus est un microbe.*"[60] Pathogens were isolated using filters permeable to toxins but impermeable to bacteria. Infectious agents that could pass through a bacterial filter were called "filterable viruses" (see chapter 4).

But it was unclear whether there was any organizational difference between viruses and the smallest of bacteria. Small obligate parasitic bacteria of the rickettsia type were often considered to be transitional. Stanier and his colleagues wrote in *The Microbial World* of 1957, "In fact there is no sharp line of distinction between the largest animal viruses and the rickettsiae."[61] That year the authors of *Bergey's Manual* discussed the idea of a kingdom composed of bacteria, blue-green algae, and viruses, the "*Protophyta.*"[62] Eight years earlier, Theodore Jahn and Frances Jahn proposed that viruses and bacteriophage be assigned a kingdom of their own, the "Archetista," "as submicroscopic organisms" living in hosts.[63]

Viruses had been conceptualized as "naked genes" by Muller in 1922 (see chapter 5).[64] And during the 1940s and 1950s, the genetic study of bacteria and their viruses was developed most prominently in the United States by Max Delbrück, Alfred Hershey, Salvador Luria, Joshua Lederberg, and Norton Zinder, and in France by André Lwoff, Elie Wollman, François Jacob, and Jacques Monod at the Pasteur Institute. In the early 1950s, Lwoff synthesized the work on the phenomenon of lysogeny, whereby the genome of a bacteriophage is integrated into the "chromosome" of its host. It remains there unless the bacterium is exposed to certain adverse stimuli. Then the prophage genome is excised from the bacterial "chromosome," the virus reproduces, and it kills its host.[65] Viruses could also act as vehicles to transfer genes between bacteria. In 1952, Lederberg and Zinder called this process "transduction" (see chapter 9).[66]

In a paper titled "The Concept of Virus" in 1957, Lwoff (1902–1994) articulated the difference between the virus and the bacterium based on electron microscopy and chemistry, and without regard to phylogeny. He made the distinction hard and unambiguous: the virus was not a cell; it contained only one kind of nucleic acid, either RNA or DNA, enclosed in a coat of protein; it possessed few if any enzymes; and it did not reproduce by division like a cell.[67] Its replication occurred only within a susceptible cell, which always contains both DNA and RNA, an array of different proteins endowed with enzymatic functions that are mainly concerned with the generation of ATP and the synthesis of varied organic constituents of the cell from chemical compounds present in the environment. "Viruses should be treated as viruses," he concluded, "because viruses are viruses."[68] Lwoff found it difficult to conceive of any biological entities that could be transitional between a virus and a cell. (Only the word "virulence" for toxicity would be retained for both bacterial and viral diseases.) Stanier and van Niel agreed that it was "indeed difficult to visualize any kind of intermediate organization."[69]

When Stanier was on sabbatical leave at the Pasteur Institute in Paris in 1961, Lwoff recommended to him the terms "procaryote" and "eukaryote," which his former mentor Edouard Chatton (1883–1947) had coined.[70] Notwithstanding the myths perpetuated by the end of the century (see chapter 21), Chatton said very little about the distinction. He first used the terms in a paper in 1925, "*Pansporella perplex*: Reflections on the Biology and Phylogeny of the Protozoa." At the center of that work is the life history of the amoeba *Pansporella*, a parasite in the intestines of *Daphnia*, which he discovered as a student in 1906 and later studied at the Pasteur Institute before the First World War. As the subtitle suggests, he made some general comments on the phylogeny of what he considered to be primitive flagellated protozoan, Protomastigotes. He used the terms without definition in two diagrams (figure 7.2) in which he referred to Cyanophyceae (blue-green algae) Bacteriacae, and Spirochaetaceae as "procaryotes" (Greek *pro*, before, and *karyon*, nucleus or kernel) in contradistinction to protozoa or protists, which he called "eucaryotes" (Greek *eu*, true).[71]

Chatton used the terms again in a rare publication in 1938 in which he reviewed his most important scientific achievements as part of the requirements for his appointment to a chair in marine biology in the Faculté des Sciences in Paris. He again mentioned them only in passing when discussing the nature of the first protozoa:[72]

> Protozoologists agree today in considering the flagellated autotrophs as the most primitive of the Protozoa possessing a true nucleus, Eucaryotes (a group which also includes the plants and the Metazoans) because they alone have the power to completely synthesize their protoplasm from a mineral milieu. Heterotrophic organisms are therefore dependent on them for their existence as well as on chemotrophic Procaryotes and autotrophs (nitrifying and sulfurous bacteria, Cyanophyceae).[73]

Stanier first used the terms "procaryotique" and "eucaryotique" in a paper published in the *Annales d'Institut Pasteur* in 1961.[74] In it, he referred to Lwoff's arguments about the difference between a virus and a cell, and wrote that bacteria and blue-green algae shared a prokaryotic structure, characterized mainly by lack of a nuclear membrane. Stanier and van Niel's subsequent more detailed treatment, "The Concept of a Bacterium," was a sister paper to Lwoff's "The Concept of Virus" published four years earlier.[75] In effect, they retracted everything they had said in the 1950s about the presence of the nucleus, cytoplasmic organelles, and sex in bacteria.[76]

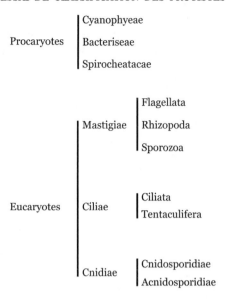

		Cyanophyeae
Procaryotes		Bacteriseae
		Spirocheatacae

			Flagellata
	Mastigiae		Rhizopoda
			Sporozoa
Eucaryotes	Ciliae		Ciliata
			Tentaculifera
	Cnidiae		Cnidosporidiae
			Acnidosporidiae

Essai sur la Phylogénie des Protistes

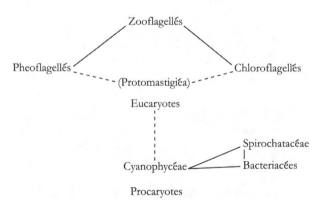

Figure 7.2 Based on images from E. Chatton, "Pansporella Perplexa: Reflexions sur la Biologie et la Phylogénie des Protozoaires," *Annales des Sciences Naturelles, Zoologie.* 10e serie, 7 (1925): 1–84. Redrawn by Ianina Altshuler and Stephanie Chen.

It was now clear from the continued studies by Robinow that there was nothing in bacteria or in blue-green algae that could be equated with a mitotic mechanism in eukaryotic cells, and all the DNA of the bacterial nucleus was associated with a single structural element, a circular chromosome.[77] Electron microscopic studies revealed the fine structure of chloroplasts and mitochondria. No similar structures were possessed by bacteria or blue-green algae.[78] Although Stanier had previously compared sex of bacteria to that of ciliated protozoa, he and van Niel now stated that gene transfer and genetic recombination in bacteria was not "homologous with the sexual process as it exists in eukaryotic organisms; it does not give rise to reciprocal recombinants." Not only was conjugation unidirectional, but the donor did not transfer its full complement of genes to the recipient.[79]

A definition of a bacterium was possible only if one included the blue-green algae. "It is now clear" they wrote,

> that among organisms there are two different organizational patterns of cells, which Chatton (1937) [*sic*] called, with singular prescience, the eukaryotic and procaryotic type. The distinctive property of bacteria and blue-green algae is the prokaryotic nature of their cells. It is on this basis that they can be clearly segregated from all other protists (namely, other algae, protozoa, and fungi), which have eucaryotic cells.[80]

Eukaryotic cells had a membrane-bound nucleus that divided by mitosis, a cyto-skeleton, an intricate system of internal membranes, mitochondria that perform respiration, and in the case of plants, chloroplasts. Prokaryotic cells lacked all of these structures, and their nuclei reproduced by fission:

> The principal distinguishing features of the procaryotic cell are: 1 absence of internal membranes which separate the resting nucleus from the cytoplasm, and isolate the enzymatic machinery of photosynthesis and of respiration in specific organelles; 2 nuclear division by fission, not by mitosis, a character possibly related to the presence of a single structure which carries all the genetic information of the cell; and 3 the presence of a cell wall which contains a specific mucopeptide as its strengthening element.[81]

When Stanier, Doudoroff, and Adelberg reintroduced the terms in the Second edition of *The Microbial World* in 1963, they attributed the distinction wholly to the triumph of electron microscopy.[82]

The words prokaryotic and eukaryotic spread fast and far. They were adopted without hesitation, and they the came to signify a great organizational schism in evolution on Earth. "In fact," Stanier, Doudoroff, and Adelberg wrote in the second edition of *The Microbial World* 1963, "this basic divergence in cellular structure, *which separates the bacteria and blue-green algae from all other cellular organisms*, probably represents the greatest single evolutionary discontinuity to be found in the present-day world."[83] Just as there would be no transitional forms between viruses and bacteria, they said, there would be none between prokaryote and eukaryote. Evolutionary diversity was constrained by these cellular forms. Only the eukaryotic cell contained the potentialities for highly differentiated multicellular organisms.

EIGHT

On the Unity of Life

All these organisms share the distinctive structural properties associated with the procaryotic cell ... and we can therefore safely infer a common origin for the whole group in the remote evolutionary past.

— Roger Stanier, Michael Doudoroff, and
Edward Adelberg, *The Microbial World*, 2nd ed. (1963)

Monophyly is a principal value of systematics, but like other values is not absolute and will not always be followed to the sacrifice of other objectives.

—Robert Whittaker, "New Concepts of
Kingdoms of Organisms" (1969)

THE PROKARYOTIC–EUKARYOTIC DISTINCTION was warmly welcomed by biologists. It seemed to dispel all the confusion about bacterial cell structure that had persisted for 100 years.[1] It also had the less conspicuous but equally important effect of closing discussion about the common ancestry of the bacteria, depicted as forms that preceded and gave rise to eukaryotes. To appreciate this significance is to recognize the extent to which microbiologists had previously supposed that bacteria constituted a polyphyletic group of distantly related organisms.

A Group Negatively Defined

In the absence of a phylogenetic understanding, bacteria could have originated once or many times, as discussed since the time of Haeckel. It was widely assumed among Stanier and van Niel's contemporaries that bacteria actually comprised several distinct lineages. The similarities between them were thought to be due to convergence, most likely resulting from the loss of characteristics—not to common ancestry. Adolf Pascher in Prague noted in 1931 that "very diverse structures are still accepted as schizomycetes: true bacteria, apochromatic blue-green algae, probably very reduced fungi, and possibly organisms of yet another origin."[2] The blue-green was a case in point. They differed from bacteria in both structural complexity and diversity. Some blue-green algae possessed an elaborate multicellular branching and possessed fine hairlike structures called trichomes. No such structures were known for any bacteria. Blue-green algae also differed in the way they moved—by a gliding motion; no blue-green algae possessed flagella, but many groups of bacteria did. And, blue-green algae did not produce spores, but some groups of bacteria did.

Copeland was uncertain if the kingdom Monera was a natural monophyletic grouping when he had proposed it in 1938. Although he was unsure whether there was a common origin for all extant life forms, he liked to think so (see chapter 6). René Dubos (1901–1982) was confident that the kingdom Monera was not a monophyletic group of related forms.[3] In *The Bacterial Cell* (1945), he commented that the false assumption about their common ancestry was derived from false beliefs about their simplicity:

> Some investigators have looked upon bacteria as a primitive homogenous group from which higher types have arisen. It appears more likely, however, that these microorganisms constitute a heterogeneous group of unrelated forms. Even among the Eubacteriales—the so-called true bacteria—one finds strange bedfellows, such as small Gram-negative autotrophic organisms, the Gram-positive proteolytic spore formers, the acid-fast bacilli which differ so profoundly from each other in metabolism, structure, and even mode of division as to have little in common except microscopic dimensions. One may indeed wonder whether the apparent unity of the group is not due to a narrow range of cellular size which determines, by a sort of convergent evolution, a number of physical and chemical characteristics.[4]

Well known for his research in isolating antibacterial substances from certain soil microbes, Dubos hoped that comparative serological reactions—that is, the reaction of bacteria to antibodies, contained in specifically prepared sera— might be used to help establish phylogenetic relationships among them.

Blue-green algae were excluded from the "bacterial kingdom" that André Prévot proposed in 1940: "The absence of a definitive nucleus, of indisputable sexuality, and of animal or plant respiratory pigment indicate that the bacteria constitute a singular group, clearly outside the animal kingdom and the plant kingdom."[5] Ernst Pringsheim (1881–1970) was critical of the unnatural state of bacterial classification ever since the appearance of the first edition of *Bergey's Manual* in 1923.[6] He addressed that issue in an extensive review in 1949. Their small size, the multitude of similar forms, lack of sexuality, and morphological characteristics were all familiar reasons why there was not a natural classification of bacteria. He also noted that because microbiology was tied so closely to medicine and industry, microbial diversity was neglected in favor of investigations of a few organisms of practical importance.[7]

Pringsheim denounced the kingdom Monera on two grounds: no bacteria possessed the structural complexity of the blue-green algae, and the kingdom was defined negatively.[8] Only the least specialized of the blue-green algae could possibly be "confused" with bacteria, and their similarities to bacteria, he said, most likely "originated by convergence and afford no evidence of descent from a common ancestry."[9] Pringsheim suggested that new "methods of extracting specific proteins and other compounds of high molecular weight" might eventually offer a means for revealing true relationships.[10] He noted how, in 1941, Stanier and van Niel had defined the Monera negatively:

> They believe that Bacteria and blue-green Algae have originated from common ancestors and summarize their common characteristics as follows: (1) absence of true nuclei, (2) absence of sexual reproduction, (3) absence of plastids.... The entirely negative characteristics upon which this group [Monera] is based should be noted, and the possibility of convergent evolution of the two classes be seriously considered.[11]

The "procaryote" as used by Edouard Chatton in his phylogenetic tree of 1925 (see chapter 7) clearly implied common ancestry for the grouping, just as did the kingdom Monera of the 1930s and 1940s. Although Stanier was introduced to the terms while on leave from Berkeley in 1961, ironically, they had been coined at Berkeley independently of Chatton four years earlier. In 1957, Ellsworth Dougherty introduced the adjectives "eukaryotic" and "prokaryotic" as well as nouns "eukaryon" (Greek *eu*, true, *karyon*, nucleus or kernel) for the nucleus of "higher organisms" and "prokaryon" (Greek *pro*, before) for the "moneran nucleus" of bacteria and blue-green algae.[12]

There was an important difference between Chatton's evolutionary considerations and those of Dougherty. Chatton assumed that "procaryotes" preceded and gave rise to eucaryotes. But Dougherty considered two alternatives: (1) that the moneran and protistan lineages may have diverged early from a hypothetical universal common ancestor that he called "Premonera," or (2) that protistans and blue-green algae diverged from a common ancestor after "the true bacteria split off."[13] He had a tentative preference for the latter hypothesis (figure 8.1).

Stanier's own views changed when, in the 1950s, he denounced the kingdom Monera, and then again in the 1960s, when he and van Niel advanced the prokaryotic and eukaryotic cell dichotomy. In the first edition of *The Microbial World* of 1957, he, Doudoroff, and Adelberg asserted that the bacteria

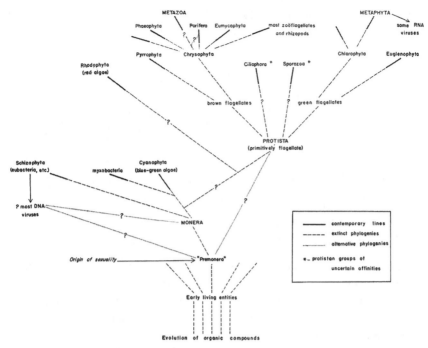

Figure 8.1 "Suggested phylogeny of contemporary living groups (highly speculative!), with the four 'major' groups in capital letters." E.C. Dougherty, "Comparative
Evolution and the Origin of Sexuality," *Systematic Zoology* 4 (1955): 145–169, at 149.

represented an unnatural group of organisms of remote ancestry, all of which
were distinct from the blue-green algae:

> It seems quite certain that the resulting assemblage of organisms that are
> recognized today as "bacteria" does not constitute a natural group, but
> consists rather of several groups that are remotely related. This biological
> heterogeneity of the bacteria really calls for a negative definition rather
> than a positive one: bacteria include all lower protists that do not show
> the characteristic properties of blue-green algae.[14]

Five years later, Stanier and van Niel argued that bacteria and blue-green
algae had a common origin within the prokaryotic form:

> It thus appears that the procaryotic cell has provided a structural frame-
> work for the evolutionary development of a wide variety of microorgan-
> isms.... If we look at the microbial world in its entirety, we can now see
> that evolutionary diversification ... has taken place on two distinct levels
> of cellular organization.[15]

When the second edition of *The Microbial World* appeared in 1963, without
acknowledging any shift in their previous statements regarding the polyphyletic

nature of bacteria, the authors asserted that, based on their prokaryotic cell structure, one could "safely infer a common origin for the whole group in the remote evolutionary past":

> We can also discern four principal sub-groups, blue-green algae, myxobacteria, spirochetes, and eubacteria, which seem to be distinct from one another.... Beyond this point, however, any systematic attempt to construct a detailed scheme of natural relationships becomes the purest speculation, completely unsupported by any sort of evidence. The only possible conclusion is, accordingly, that the ultimate scientific goal of biological classification cannot be achieved in the case of bacteria.[16]

The prokaryotic form ("lower protists"), they said, was as a "stage" in evolution leading to the eukaryotic cell ("higher protists"), the essential structure of which had become fixed, its evolution over.

> It is not too unreasonable to consider that the bacteria and blue-green algae represent *vestiges of a stage in the evolution of the cell* which, once it achieved a eukaryotic structure in the ancestors of the present-day higher protists, did not undergo any further fundamental changes through the entire subsequent course of biological evolution.[17]

Despite, the assertion about safely inferring "a common origin for the whole group," Stanier knew that it was actually an unwarranted assumption. In 1971 he commented: "Indeed, the major contemporary prokaryotic groups could well have diverged at an early stage in cellular evolution, and thus be almost as isolated from one another as they are from eucaryotes as a whole."[18]

An essential similarity in internal structure would indeed indicate common descent, but a negative definition of prokaryote would not. When defined as lacking a membrane-bound nucleus, lacking organelles, and lacking sexuality comparable to "higher" forms of protists, the prokaryote concept might well be similar to Lamarck's grouping "invertebrate," which included such diverse creatures as insects, worms, and infusoria. It would be an illusion, as in the nineteenth-century assumption that rhinos, hippos, and elephants descended from a single large ancestor—it is now known that each of these animals evolved from a separate small ancestor, and the common ancestor of all of them was small and slightly built, with presumably thin skin and fur. Rhinos, hippos, and elephants share derived characters that originated several times by convergence.

Of E. coli and Elephants

Life's common origin was generally assumed implicitly or explicitly by biologists since the emergence of evolutionary theory, cell theory, and germ theory. To be sure, Lamarck had suggested repeating origins of life as did Haeckel. But others, including Darwin, pointed to evidence of a common chemistry and cellular structure underlying life's diversity in support of its essential unity (see chapters 1, 3, and 6). The theory of evolution itself was embedded in the concept of common descent and coupled with life's common chemistry. "Thus it becomes clear," T.H. Huxley commented in 1868, "that all living powers are cognate, and that all living forms are fundamentally of one character. The researches of the chemist have revealed a no less striking uniformity of material composition in living matter."[19] On "The Dynamic Side of Biochemistry," Frederick Hopkins (1861–1947) at Cambridge wrote in 1913, "If there be any basis for our views concerning the specificity of, say, the tissue proteins, they must apply to the amoeba no less than to the higher animal."[20]

Albert Jan Kluyver coined the catch phrase "the unity of biochemistry," in 1926: "From the elephant to butyric acid bacterium—it is all the same!"[21] The similarity of biochemistry, further borne out by the occurrence of the same amino acids, vitamins, and enzymes in diverse forms of life, gave more credence to the concept of one universal common ancestor. "These two aspects of life—its constancy and variability—are reflected in many ways," van Niel wrote in 1949:

> From the point of view of comparative biochemistry, the constancy finds its expression and counterpart in the unity of the fundamental biochemical mechanisms, that is, Kluyver's concept of the "unity of biochemistry." This, to-day, is also the most compelling argument in favor of a monophyletic origin of life.[22]

Molecular biologists also assumed it when inferring the universality of the genetic code. During the early 1960s an outline was sketched by which information encoded in nucleic acid sequences is translated into amino acid sequences that composed proteins. The genetic information in DNA was understood to be a combinative system composed of sequences of four components: the four nitrogenous bases, adenine (A), thymine (T), cytosine (C), and guanine (G). While the diversity of genetic information stored in nucleic acids was based on combinations of four components, the diversity and specificity of proteins were based on combinations of 20 amino acids. Information from nucleic acid to protein was transmitted through three-letter code words; GCG, for example, encoded the amino acid alanine.

Kluyver's expression, "From the elephant to butyric acid bacterium—it is all the same!" was transformed into a new maxim: "If the codes in *Serratia* and *Escherichia coli* and perhaps a few other genera turn out to be the same," Jacques Monod and François Jacob wrote in 1961, "the microbial-chemical-geneticists will be satisfied that it is indeed universal, by virtue of the well-known axiom that anything found to be true of *E. coli* must also be true of Elephants."[23] All organisms lived by the same code etched in the structure of DNA at life's dawning. One universal mother cell, one common ancestor of life appeared to be axiomatic. "If the genetic code is universal," Jacob commented in 1970, "it is probably because every organism that has succeeded in living up till now is descended from one single ancestor."[24] That aspect of evolution was essentially over, "frozen" thousands of millions of years ago, just like the structures of the prokaryote and eukaryote. That the basic molecular genetic mechanisms have been preserved from bacteria to baboons, essentially intact for billions of years of evolution, was one of the greatest discoveries of the twentieth century.[25]

Kingdom or Superkingdom?

Microbiologists eagerly welcomed prokaryotic–eukaryotic cell dichotomy as signifying the structural foundations of all life on Earth. But lacking a phylogeny, Stanier and van Niel did not assign kingdom status to those forms. In 1962, R.G.E. Murray, one of the editors of *Bergey's Manual*, expressed views about bacteria similar to those of Stanier and van Niel's prokaryote concept. But Murray supported Copeland's proposal for a kingdom for the bacteria and the blue-green algae.[26] He did not understand why Stanier and van Niel did not recognize a major difference in cell organization by a fundamental taxonomic separation.[27] Stanier explained to him that he would "certainly not object to setting up a separate kingdom for the prokaryotic microorganisms if such an operation would serve as a handy device for emphasizing the fundamental differences between these types and organisms that possess a eukaryotic cellular organization." But he and van Niel considered

> detailed system building at the microbial level to be an essentially meaningless operation, since there is so very little information that can be drawn on for the purposes of phylogenetic reconstruction. For this reason I prefer to use common names rather than Latin ones for every bacterial group above the level of genus.[28]

In 1968, Murray proposed "Procaryotae" as a kingdom of microbes "characterized by the possession of nucleoplasm devoid of basic protein and not bounded from cytoplasm by a nuclear membrane."[29] He suggested "Eucaryotae" as a kingdom that would include other protists, plants and animals. The next year, Allan Allsopp suggested that Procaryota and Eucaryota might be given a higher status of "superkingdoms since the differences between them transcend those between the often accepted kingdoms, Protista, Plantae and Animalia."[30] Allsopp recognized that there was no definitive evidence that Prokaryota represented a monophyletic grouping. He considered the views of those who suggested that bacteria may have evolved by loss from blue-green algae, and of others who insisted that bacteria were polyphyletic. "It is perhaps surprising that the balance of evidence does not permit an outright rejection of any of these opposed hypotheses."[31]

Robert Buchanan (1883–1973), chairman of *Bergey's* board of editors, had had three concerns about a kingdom for prokaryotes: (1) the almost entirely negative characteristics by which the group was identified, (2) whether or not the blue-green algae should be included, and (3) how viruses could be completely distinguished from such small obligate bacterial parasites such as *Rickettsia* and *Chlamydia*.[32] He consulted with Stanford botanist Peter Raven, who agreed that the prokaryote was defined largely negatively, but considered "viruses as a group outside the usual classifications of living organisms" as "by-products of bacterial reproduction, in which segments of DNA or RNA protected with protein coats, spread from cell to cell, directing the host cell's metabolism to reproduce more of the viral DNA or RNA."[33] Raven consulted Stanier on the relationship between blue-green algae and bacteria. Stanier replied that

> the assignment of the blue-green algae to the "algae" is just an unfortunate historical accident.... As to what one might do about this situation in formal taxonomic terms, I don't really care very much, since taxonomic system-building (especially in the realms of the biological world) isn't an operation that seems very useful.[34]

In the eighth edition of *Bergey's Manual* (1974), the kingdom was called "Procaryotae," and as Murray explained in the introductory chapter, nothing was said about a kingdom for eukaryotes so as not to disturb the rest of biology:

> The assumption of a new Kingdom is both appropriate and helpful to the bacterial taxonomist, but a kingdom including all the eukaryotes would be disturbing to botanists and zoologists causing a realignment of their respective hierarchies. It is probably best to leave matters as they have been expressed above and only recognize, at the moment, the Kingdom *Procaryotae*.[35]

Still, biology as a whole remained largely a two kingdom world. And those authors who did considered kingdoms could not agree on the terms or taxonomic level for the prokaryotes. Some referred to kingdom Monera; others, to kingdom Prokaryotae.[36] Few recognized Prokaryotae and Eukaryotae as superkingdoms.[37] Confusion reigned regarding the highest of taxonomic categories and on what basis they should be drawn.

On Evolution's Direction

Five kingdoms—Plantae, Animalia, Fungi, Protista, and Monera—as proposed by ecologist Robert Whittaker (1920–1980) at Cornell University, was the most widely accepted of the new systems.[38] The novelty of Whittaker's scheme lay in its eclectic mixture of ecology and morphology. The first three kingdoms were assigned, not in accordance with phyletic principles, but in accordance with three main functional "roles" within ecosystems: producers (green plants), which use solar chemical energy to synthesize organic compounds from inorganic and provide all the food energy available to the community; consumers (animals), which harvest the energy they provide; and reducers (fungi), which break down the dead remains of producers and consumers. Plantae, Animalia, and Fungi were "three functional kingdoms of nature," as Eugene Odum called them in his *Fundamentals of Ecology.*[39]

Whittaker's concepts grew from his criticisms of Copeland's kingdom Protoctista (formerly Protista) as described in his book *The Classification of Lower Organisms* of 1956 (see figures 8.2 and 8.3). Recall that Copeland, and others before him, recognized that protists were so diverse as to be divided into several kingdoms.[40] Whittaker agreed that the kingdom Protoctista was an "incoherent grouping," composed of diverse organisms, some of which were plants and some animals, even though unicellular; and it included the fungi, which as he saw it was neither plant nor animal.[41] But he broke from a strictly monophyletic taxonomy, maintaining instead that a broad system of classification should reflect "evolutionary direction," not genealogy. In 1959, he argued against the assumption that fungi were derived from algae, as often assumed,

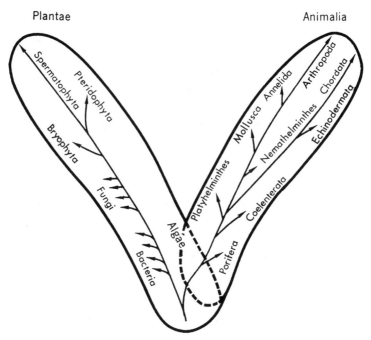

Figure 8.2 "A simplified evolutionary scheme of the two-kingdom system as it might have appeared early in the century." R.H. Whittaker, "New Concepts of Kingdoms of Organisms," *Science* 163 (1969): 150–163, at 151. With permission.

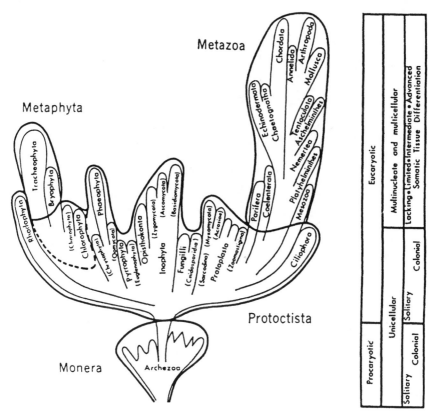

Figure 8.3 "The Copeland system, with relationships of phyla to kingdoms and levels of organization." R.H. Whittaker, "New Concepts of Kingdoms of Organisms," *Science* 163 (1969): 150–163, at 156. With permission.

and he proposed a separate kingdom for the fungi based on their organization as multinucleated organisms lacking chlorophyll.[42]

Theodor Jahn and Francis Jahn had proposed the same taxon 10 years earlier in a six-kingdoms scheme: (1) Archetista, for "submicroscopic organisms" that lived in animals and plants or bacteria and that cause disease, for example, the tobacco mosaic virus and typhoid bacteriophage; (2) Monera, for bacteria and blue-green algae; (3) Protista, for largely unicellular organism with a nucleus, including protozoa, green algae, red algae, and brown algae; (4) Fungi, usually containing multiple nuclei and no chlorophyll; (5) Metaphyta or Embryophyta, multicellular organisms with chlorophyll, usually sessile; and (6) Metazoa, multicellular, typically holozoic and mobile.[43]

Whittaker's views about the bacteria changed repeatedly. In 1957, he placed them with the fungi, where they had been situated for 100 years.[44] Two years later, he assigned them to a subkingdom of protists (figure 8.4).[45] Then in 1969, he promoted them to a kingdom Monera on par with four others.[46] He assumed

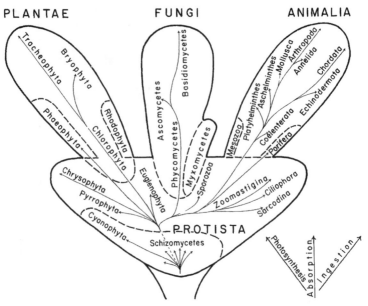

Figure 8.4 "A schematic view of the kingdoms and phyla according the view presented here with evolutionary relations suggested in simplified form." R.H. Whittaker, "On the Broad Classification of Organisms," *Quarterly Review of Biology* 34 (1959): 210–226, at 217. With permission.

that Monera was a monophyletic kingdom, but he was confident that the same was not true for plants and animals. The line between protists and plants and animals, he said, "is primarily by degree of tissue differentiation," and that line was crossed several times (figure 8.5).[47]

Few evolutionary biology texts considered evolution in terms of ever increasing complexity from its earliest beginnings. Verne Grant's (1917–2007) *The Origin of Adaptations* of 1963 was exceptional when it considered a six-kingdom scheme. Grant set the main themes of neo-Darwinian evolutionary biology (population genetics, levels of selection, genetic drift, divergence, and modes of speciation) in a broad cosmic context, as one of four stages in evolution.[48] The first was that of "atomic evolution" in which matter was formed from a tremendous explosion, when some 15 billion (15 × 10⁹) years ago the hydrogen atom underwent nuclear reactions inside stars to form more complex atoms such as helium, carbon, nitrogen, oxygen, sulfur, and lead.

The second phase was that of "chemical evolution," when about 3.5 billion years ago atoms combined into chemical compounds of various degrees of complexity and formed carbon compounds that served as food for primitive life, which occurred about 500 million years later. Life emerged in a nutrient bath from self-reproducing "naked genes" (virus-like entities) and "saprophytic nucleo-proteins."[49] Though at first naked and single, simpler nucleic acids

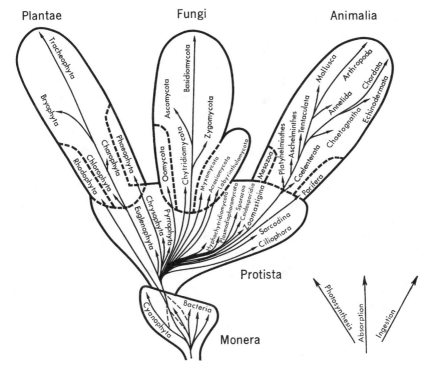

Figure 8.5 "A five-kingdom system based on three levels of organization—the procaryotic (kingdom Monera), eucaryotic unicellular (kingdom Protista), and eucaryotic multi-cellular and multinucleate." R. H. Whittaker, "New Concepts of Kingdoms of Organisms," *Science* 163 (1969): 150–163, at 157. With permission.

chains that preceded DNA would grow, interconnect, and become wrapped up in the cell, chromosome, and the body "to ensure the better survival of the genes."[50]

The beginnings of life initiated the third phase, when forces of "organic evolution" predominated. A fourth phase occurred about a million years ago with the production of the modern human. It marked the beginning of "cultural evolution" and the beginning of the end of organic evolution. "Man's control over nature is bringing the period of organic evolution to a close" Grant said, "although the processes responsible for organic evolution will continue to operate as before, their influence on future developments in man's world will be relatively slight by comparison with the effects of man himself."[51] From this cosmic perspective, he proposed six kingdoms: animals, plants, fungi (for "slime molds and several of the groups of true fungi"), protists (nucleated single-celled organisms), monera (for bacteria, blue-green algae and viruses), and a hypothetical kingdom of "simple precellular organisms" that would be home to "the hypothetical naked genes and the hypothetical saprophytic nucleoprotein particles."[52]

Prokaryotes First?

Though evolutionists since the times of Lamarck and Darwin had assumed that life evolved from microbial ancestors, there was no firm evidence in the fossil record until the 1960s.[53] The known history of animal and plant life in the fossil record was presented as progression from spore-producing to seed-producing to flowering plants, from marine invertebrates to fish to amphibians and then reptiles, birds, and mammals at the beginning of the Cambrian era (550–490 million years ago).[54] The most famous store of Cambrian fossils, discovered in 1909 near Burgess Pass in the Rocky Mountains of British Columbia, held a great diversity of arthropods, the most common animals on Earth; trilobites, now extinct; lobsters, crabs, and shrimp; spiders and scorpions; and insects. In 1946, another great store of Cambrian fossils was found north of Adelaide, Australia, in an abandoned copper, lead, and zinc mine called "Ediacaria." The Ediacaran biota comprised soft-bodied marine creatures, unusual fossils, originally interpreted as jellyfish, strange worms, and frondlike corals. Many were of organisms now extinct. Why the "Cambrian explosion" of diverse organisms occurred is not fully understood, but the extent of the extinction was so great that some evolutionists have considered the subsequent history of animal life to be one of massive removal.[55]

Zoologists have pondered why no new body plans have appeared in the evolution over the past 500 million years.[56] While that aspect of evolution seemed to be over, those who concerned themselves with primordial kingdoms recognized that the real explosion out of which plants and animals emerged had occurred millions of years earlier—with the emergence of the eukaryotic form. But hard fossil evidence for their early evolution was lacking, and paleontologists considered the chances of finding it to be exceedingly small. Indeed, just as microbial taxonomists had maintained that a microbial phylogeny was impossible, so, too, leading paleontologists maintained that microbial fossils would be impossible to discover.[57]

Not only do microbes lack bits of shells, bones, or other hard parts resistant to decay, but even when they were entombed in sediment they would be almost always flattened beyond recognition. Then there were problems of finding ancient rocks that might contain them: the older the rock, the rarer it is. Fossil hunters of the twentieth century had been greatly aided by geological surveys aimed at exploring and identify rock strata. Determining the age of rocks was vital because certain types of economically important rocks were formed abundantly during particular geological times. For example, iron-rich rocks for making steel, and certain types of uranium deposits for nuclear reactors, are plentiful only in Precambrian strata older than two billion years, whereas coal and oil reserves were common in particular periods of the past 550 million years.

Specific radioactive isotopes were needed to date ancient rock. Radioactive carbon-14 (^{14}C) is used for dating fossil humans or human artifacts. However, ^{14}C rapidly decays to a stable isotopic form of nitrogen (^{14}N) after 50,000–60,000 years, so it is useless for dating anything older. To determine the age of ancient rocks, uranium-238 (^{238}U) was used; its rate of decay to ^{206}Pb is extremely slow, with a half-life (the time it takes for half of it to be converted to the new form) of about 4,500 million years.

In the 1950s, there were reports of microbes in fossils dating back to 2,000 million years, but the methods of identifying them were error prone, and Precambrian fossils were regarded with skepticism.[58] The authenticity of such microfossils was verified in the mid-1960s; the fossil "microflora" was reported to be "thread-like bacteria and blue-green algae," the builders of those large "*Cryptozoon*-like structures, stromatolites."[59] Startling evidence of fossils formed some 3.2 billion years ago was reported in 1968.[60] In the mid-1970s, paleobiologists confirmed fossilized eukaryotic microbes that had lived about 1,400 million years ago.[61]

Few questioned that prokaryotes preceded and gave rise to eukaryotes. Kenneth Bisset at the University of Birmingham was exceptional.[62] "Nowadays, I find that biology students at all levels, and even schoolchildren, have been indoctrinated with this prokaryotic idea, and according it the sanctity of holy writ," he commented in 1973.[63] "At the back of such theories as that of the prokaryotic cell lies the concept that bacteria are primitive creatures, mainly because they are small."[64] That was indeed the argument of Gunther Stent's *Molecular Genetics* in 1971: prokaryotes were 1,000–10,000 times smaller, they have 1/1,000th the amount of DNA as a mammalian cell, they lack a nuclear membrane, their DNA is not combined with protein to form structures like eukaryotic chromosomes, they lack mitosis and meiosis, and they lack mitochondria and centrioles. "There can be little doubt," he concluded "that the simpler prokaryotes are the evolutionary antecedents of the more complex eukaryotes."[65]

As Bisset saw it, such views were not only wholly speculative, but probably incorrect: it was possible that bacterial evolution paralleled eukaryotic evolution or even that bacteria had evolved from eukaryotes. He favored the latter view: "The bacteria are not fundamentally different from larger cells, but have simply reduced the size and complexity of their structures to a minimum, for purposes of efficiency in pursuing an opportunistic saprophytic or parasitic existence."[66] The "pro-eukaryotic theory...is wrong," he said, "at least in the form that it is commonly stated, because the nuclear membrane may not, in fact, be missing."[67] The inner layer of the bacterial cell envelope was actually comparable to the nuclear membrane of protists, and therefore proposed that "the structure which now represents the cell membrane of bacteria may have originated in the nuclear membrane of an ancestral form." In short, bacteria may be economized forms of eukaryotes that transformed their nuclear membrane. Like others he pointed to the negative characterization of the prokaryote: "Positive grouping based on negative criteria, are seldom durable in biology."[68]

Stanier also knew that one could not safely assume the pro-eukaryotic view of bacteria, anymore than one could assume a common origin for the group. In 1970 he entertained the notion that prokaryotes may have arisen from eukaryotes. At that time, the notion that chloroplasts and mitochondria of eukaryotes had originated as symbiotic microbes was being reconsidered:

> Is the comparative structural simplicity of prokaryotic organisms really indicative of great evolutionary antiquity? In view of their similarities to mitochondria and chloroplasts, it could be argued that they are relatively late products of cellular evolution, which arose through the occasional escape from eukaryotes of organelles which had acquired sufficient autonomy to face life on their own. This is a far-fetched assumption; but I do not think one can afford to dismiss it out of hand.[69]

The symbiotic origin of mitochondria and chloroplasts was rigorously debated for a decade before the question was resolved with the development of molecular phylogenetic methods (see chapters 9, 17, 22).

| # Symbiotic Complexity

The evolution of the complex cell, with its array of more or less autonomous organelles, from the simpler organization found in Monera is a question that has been neglected. With the demonstration of ultrastructural similarity of a cell organelle and free living organisms, endosymbiosis must again be considered seriously as a possible evolutionary step in the origin of complex cell systems.

—Hans Ris and Walter Plaut, "Ultrastructure of DNA-Containing Areas in the Chloroplasts of *Chlamydomonas*" (1962)

Innovation Sharing

Symbiosis as a means of evolutionary innovation had been discussed since the late nineteenth century.[1] The dual nature of lichens as fungi and algae, nitrogen-fixing bacteria in the root nodules of legumes, fungi (mycorrhiza) in the roots of forest trees and orchids, algae living inside the bodies of protists, hydra, sea anemones, and the flat worm *Convoluta roscoffensis*—all these showed how intimate physiological relationships could be established between distantly related organisms. Symbiosis often entailed the emergence of new structures and sometimes, as in the case of the lichen, to the evolution of whole new organisms.

When Anton de Bary (1831–1888) used the word "symbiosis" in 1878, he defined it as "the living together of unlike-named organisms," and he understood it to be a means of saltational evolutionary change.[2] "Whatever importance one wants to attach to natural selection for the gradual transformation of species," he said, "it is desirable to see yet another field opening itself up for experimentation."[3] When the clear-cut examples of symbiosis were considered together with cytological evidence for various "self-reproducing" entities within all cells, there were proposals that the nucleus, cytoplasm, chloroplasts, mitochondria, and centrioles had also evolved from symbiosis.

That green plants originated from symbiosis was suggested by one of de Bary's former students, Andreas Schimper (1856–1901), when he coined the term "chloroplast" in 1883.[4] Recall that Haeckel suggested in 1904 that chloroplasts had evolved in the remote past from blue-green algae that came to live inside nonpigmented cells (see chapter 2).[5] That idea was widespread.[6] Shôsaburô Watase (1862–1929) extended the reach of symbiosis to account for the origin of all nucleated cells in 1893. In a lecture before the Biology Club of the University of Chicago in 1893, he proposed that the cell cytoplasm was "formed of a group of small, living particles, each with the power to assimilate, to grow and multiply by division," and that each chromosome was "a colony of minute organisms of another kind," endowed with similar attributes of vitality.[7] The deep physiological interdependence between nucleus and cytoplasm and reciprocal interchange of metabolic products could only be understood as symbiosis "in a more restricted sense, *the normal fellowship or the consortial union of two or more organisms of dissimilar origin, each of which acts as the physiological complement to the other in the struggle for existence.*"[8] In 1904, Theodor Boveri (1862–1915) similarly suggested that the nucleated cell arose from a "symbiosis of two kinds of simple plasma-structures—Monera, if we may so call them—in a fashion that a number of smaller forms, the chromosomes, established themselves within a larger one which we now call the cytosome."[9]

Watasé suggested that symbiosis might also account for the origin of centrosomes or centrioles. Those organelles had aroused great interest in the 1890s when they were reported to play a key role in separating daughter chromosomes during cell division in animals.[10] At the onset of cell division, the centrosome seems to divide in two; the pair then separates, and each moves to opposite sides of the nucleus. When the nuclear membrane dissolves, starlike structures ("asters") form around each centrosome, and rays of threads run through the nuclear area so as to constitute a "spindle." One could observe chromosomes split into two and watch each daughter chromosome attach to a spindle and move to opposite poles where the centrosomes lie. Two daughter nuclei are then formed. Centrosomes, asters, and spindle constitute an apparatus for accurately separating the daughter chromosomes and for the division of the cell body. It was also reported that centrosomes could move from the cell center to the cell membrane were they were converted to structures that multiply and function as organs of motility.[11] Those structures at the base of cilia of many protists were called "kinetosomes" or "basal bodies."

Russian botanist Constantin Merezhkowsky (1855–1921) is known today as the early-twentieth-century champion of the concept of "symbiogenesis."[12] Though remembered mainly for his argument that chloroplasts originated as symbiotic blue-green algae, in 1909 he also offered a detailed theory for the origin of nucleus and cytoplasm from two kinds of organisms and two kinds of protoplasm, which he called mycoplasm and amoeboplasm. The chromatin of the nucleus, chloroplasts, and bacteria were supposed to be of the nature of mycoplasm; the cytoplasm, of amoeboplasm. Each kind of protoplasm had an origin in different epochs of Earth's history (see figure 9.1).

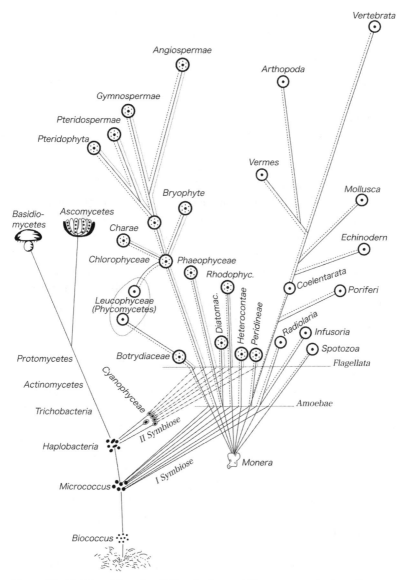

Figure 9.1 C. Merezhkowsky, "Theorie der zwei Plasmaarten als Grundlage der Symbiogenese, einer neuen Lehre von der Entstehung der Organismen," *Biologisches Centralblatt* 30 (1910): 277–303, 321–347, 353–367, at 366.

The earliest forms of life were minute particles of mycoplasm, capable of existing in temperatures close to the boiling point, in the absence of oxygen, and subsisting on inorganic materials. They appeared when Earth's crust had cooled sufficiently for water to be condensed upon it. Organisms consisting of amoeboplasm emerged during a subsequent terrestrial epoch when the waters covering the globe were cooled down to below 50°C and contained dissolved oxygen and abundant organic food. They crept in an amoeboid manner on the floor of the ocean, devouring bacteria. The next great step in evolution occurred when, in some cases, the bacteria they ate resisted digestion. At first the ingested bacteria were scattered in their host, but later they became concentrated at one spot, surrounded by a membrane, thus giving rise to the cell nucleus. In this new kind of organism, the locomotor powers of the host were combined with the great biochemical powers of the symbionts. That was the beginning of the animal kingdom. Meanwhile, the free bacteria gave rise to different kinds of cyanophytes. The plant kingdom began when some of the cyanophytes, red, brown, or green, became symbiotic in nucleated cells.

That mitochondria were also symbionts is typically traced to the writings of Richard Altmann (1852–1900), remembered today for coining the term "nucleic acid." In 1890 he proposed that intracellular granules, which he called "bioblasts," were elementary organisms that reproduced by division and built up the cytoplasm of the cell by secreting fat, glycogen, and pigments, and they could be transformed into, or produce, various rods and fibers.[13] At first, Altmann's granules were rejected as artifacts of the staining techniques he used.[14] Their reality was established beginning in 1898 when Carl Benda introduced crystal violet as a stain. He renamed them "mitochondria" from the Greek *mitos* (thread) and *khondrion* (little granule) because they took on forms as both threads and granules.[15]

During the first decades of the twentieth century, mitochondria were intensely investigated by European cytologists.[16] Mitochondria were generally held to be pleomorphic entities that could transform themselves into diverse cell structures, including Golgi bodies, cilia, and flagella, as well as chloroplasts. Paul Portier (1886–1978) developed the concept of mitochondria as ancient symbionts in his book *Les Symbiotes* of 1918.[17] Ivan Wallin (1883–1969) advanced a similar view in his book *Symbionticism and the Origin of Species* of 1927 when he proposed that acquired mitochondria were the source of new genes.[18] In his view, evolution was governed by three principles: symbiosis was concerned with the origin of species; natural selection, with their survival and extinction; and an unknown principle was responsible for the direction of evolution to ever more complex ends.[19]

The concept of symbiosis was applied to bacteria and their viruses, too. Recall that in 1917, Félix d'Herelle reported on an "invisible microbe" that he named "bacteriophage" that decimated a colony of dysentery bacilli.[20] Two years later, he noticed that not all bacteria were destroyed by bacteriophages.

Sometimes, mixed cultures of phage and bacteria could be subcultured indefinitely, and there were transformations in the morphology and physiological properties of the infected bacteria. D'Herelle referred to these mixed cultures as "microlichens." In 1926, he declared that "symbiosis is in large measure responsible for evolution."[21]

No Room at the Inn

Despite claims for its fundamental role in evolution, microbial symbiosis was generally considered to be a rare, exceptional phenomenon of little evolutionary significance.[22] Several antagonistic forces kept concepts about the integrative power of symbiosis close to the margins of "polite biological society."

First, microbial symbiosis was eclipsed by the study of disease. The notion that bacteria played any beneficial role in the tissues of their hosts was in virtual conflict with the aims and basic tenets of germ theory. Rather than viewing microbes from "the window of medicine," Portier said, he looked at "microbiology from the window of comparative physiology" and envisaged "a new form of bacteriology: physiological and symbiotic bacteriology."[23] Similarly, Wallin commented in 1927, "It is a rather startling proposal that bacteria, the organisms which are popularly associated with disease, may represent the fundamental causative factor in the origin of species."[24]

Second, hereditary symbiosis conflicted with Mendelian geneticists' conceptions of inheritance in terms of the sexual transmission of genes from parents to offspring, the concept of one germplasm, one organism, and its underlying neo-Darwinian tenets. Harvard geneticist E.M. East remarked in 1934:

> There are several types of phenomena where there is direct transfer, from cell to cell, of alien matter capable of producing morphological changes. It is not to be supposed that modern biologists will cite such instances when recognized, as examples of heredity. But since an earlier generation of students used them, before their cause was discovered, to support arguments on the inheritance of acquired characteristics, it is well to be cautious in citing similar, though less obvious, cases as being illustrations of non-Mendelian heredity.[25]

Third, symbiosis conflicted with the basic tenets of the evolutionary synthesis of the 1930s and 1940s based on natural selection acting on gradual transformations resulting from gene mutation and recombination between individuals of a species. As discussed in chapter 6, the tree of life would be a bifurcating one, not a reticulated one, except for hybridization among some plants. The basis of classification was a hierarchical ordering of group within (one) group based on characters inherited from one common parent—a common ancestry.

Fourth, symbiosis was frequently allied with mutualisms and confronted the emphasis on an incessant struggle for existence between species. Known cases of symbiosis were treated as curiosities, "special aspects of life," depicted as "strange bedfellows."[26]

Fifth, symbiotic theories of cell organelles were beyond the range of experimental inquiry. E.B. Wilson's comment of 1925 would be echoed for decades: "To many no doubt, such speculations may appear too fantastic for mention in polite society; nevertheless it is within the range of possibility that they may someday call for some serious consideration."[27]

Infective Heredity

Hereditary symbiosis captured the attention of geneticists after the Second World War with the emergence of microbial genetics, and with new evidence of non-Mendelian inheritance based on cytoplasmic entities.[28] In some cases, such as kappa in *Paramecium*, the cytoplasmic particles normally transmitted sexually could also be transmitted artificially by infection. The evidence for non-Mendelian heredity became caught in cold war rhetoric between communists led by T.D. Lysenko in the Soviet Union, who denied the existence of genes and who advocated the inheritance of acquired characteristics, on the one hand, and Western geneticists who insisted that chromosomal genes in the nucleus were the sole source of hereditary change, and who denied the inheritance of acquired characteristics, on the other.[29]

Leading geneticists trivialized the significance of cytoplasmic heredity; they dismissed the evidence as resulting from infection and therefore of little value to genetics. Others, including Tracy Sonneborn (1905–1981), Cyril Darlington (1903–1981), and Joshua Lederberg, called for a broader concept of genetics that would embrace "infective heredity."[30] Sonneborn argued in 1950 that the concept of heredity had to be expanded to include kappa and other infectious genetic particles, and he considered that mitochondria and chloroplasts, which were not infectious today, may once have been.[31] Darlington prophesied that recognition of cytoplasmic genetic entities would enable geneticists "to see the relations of heredity, development and infection and thus be the means of establishing genetic principles as the central framework of biology."[32]

Lederberg interpreted lysogeny—cases in which the bacterial genome is infected by a virus but the bacterium continues to grow and divide—as "stable symbiotic associations." In 1952, he and Norton Zinder reported that such bacteriophages could also transmit bacterial genes from a host cell to another cell. A small piece of the host chromosome is incorporated into the phage; when the particle infects a new host cell, it injects not only its own DNA but also the DNA from the former host. They called the phenomenon "transduction."[33] In

Lederberg's view, transduction was "functionally and perhaps phylogenetically, a special form of sexuality."[34]

In addition to its main chromosome or genophore, many bacterial species have independent loops of DNA. Lederberg coined the word "plasmids" for extrachromosomal hereditary determinants of bacteria and of plant and animals cells regardless of their origin.[35] He conceptualized a graded series of symbiosis, from cohabitants of a single chromosome, through to plasmids, and to ecological associations of variable stability and specificity. Symbioses obscured biological definition of the individual. Given the lack of phylogenetic methods, Lederberg remained agnostic about the symbiotic origin of chloroplasts and mitochondria: "We should not be too explicit in mistaking possibilities for certainties," he said. "The general criteria that have been used to decide the historical origin of certain plasmids are unverifiable, and such controversies have tended to be sterile."[36]

In 1961, René Dubos emphasized the creative role of microbial infections in bringing about new structures, functions, properties, and products. He pointed out that diphtheria was due to the infection of a bacterium with a toxogenic virus, that symbiotic phage brought about profound changes in the morphology of *Salmonella*, and that lambda phage carrying certain genes from a host cell could confer on a recipient the ability to produce enzymes for utilizing galactose. He added these to the classic examples of symbiosis: nitrogen-fixing bacteria of legumes, the dual nature of the lichen, and crown galls induced by inoculating certain plants with bacteria.

Still, examples of the creative and evolutionary effects of microbial infections could not compare in funding or interest to the war between humankind and microbes. The role of symbiosis and gene transfer between different species in evolution remained largely unexplored by virologists and bacteriologists, who, as Dubos lamented, maintained themselves as "poor cousins in the mansion of pathology."[37] He prophesied that there would "soon develop a new science of cellular organization, indeed perhaps a new biologic philosophy."[38]

The Tipping Point

Symbiotic theories of mitochondria and chloroplasts found a stronger footing when these organelles were shown to possess their own DNA in the early 1960s.[39] The DNA in both organelles was circular, like that of bacteria, and both organelles possessed ribosomes and a protein synthesis apparatus. Genetic research programs on mitochondria and chloroplasts had also emerged.[40] The electron microscope revealed similarities in the fine structure of mitochondria and chloroplasts on the one hand, and bacteria and blue-green algae on the other.[41] Hans Ris (1914–2004) and Walter Plaut at the University of Wisconsin commented in 1962 that the new evidence regarding chloroplasts lent support to the old hypothesis that they "originate from endosymbiotic blue-green algae."[42] The next year, when Sylvan Nass and Margit Nass in Sweden reported evidence of DNA in mitochondria, they commented similarly that a "great deal of modern biochemical and ultrastructural evidence may be interpreted to suggest a phylogenetic relationship between blue-green algae and chloroplasts, and between bacteria and mitochondria."[43]

That issue was discussed in more than 50 papers during the 1960s and 1970s. A symposium on the question was hosted by the International Society for Cell Biology in 1966.[44] This new interest did not signify a simple rediscovery of past ideas; much had changed. Mitochondria were no longer considered the principal basis of cellular differentiation and the source of other organelles. Biochemists of the 1950s had shown that mitochondria were the seat of respiration; they were the energy-generating organelles of aerobic eukaryotic cells.[45]

As it was in the beginning, symbiosis was also extended by some to account for the origin of the nucleated cell. In 1967, Norwegian microbiologist Jostein Goksøyr at the University of Bergen imagined that the nucleated cell may have emerged when anaerobic bacteria were brought into contact without intervening cell walls. The DNA of the two kinds of cells accumulated in the center, a primitive mitotic mechanism developed, and a nuclear membrane formed as the mitotic process evolved further. Aerobic eukaryotes emerged later when blue-green algae built-up oxygen in the primitive atmosphere. To adapt, the anaerobic eukaryotes established an endosymbiotic relationship with the aerobic bacteria. During its further evolution, this protomitochondrion lost much of its autonomy as some of its DNA became incorporated in the nuclear DNA. The aerobic eukaryote would subsequently enter a new symbiotic relationship with a primitive form of blue-green algae, which evolved into the chloroplasts of plant cells.[46]

That same year, Lynn Sagan (Margulis) (b. 1938) published a more elaborate account of the evolution of the eukaryote that included the symbiotic origin of centrioles/kinetosomes, the origin of cell motility and the microtubular system.[47] The flagellum or cilia was conceived of as an important invention that conditioned eukaryotic evolution.[48] Margulis proposed that the flagella had arisen independently in symbiotic events that occurred many times in the course of evolution. And she pointed to spirochetes that attach themselves to the protist *Mixotricha paradoxa* living in the hind gut of a termite.[49] In the scenario she presented in 1970, mitochondria were the first symbionts; then, certain motile spirochete-like organisms became symbiotic to their hosts, forming a motile amoeboid organism containing the flagellar symbiont.[50] The endosymbiont's genes were eventually utilized to form the chromosomal centromeres and centrioles.[51] During the course of the evolution of mitosis, various lineages of protists would subsequently acquire blue-green algae to generate the ancestral algae.[52]

Many biologists supported the concept that plastids and mitochondria originated as symbionts, but few supported the same for centrioles.[53] There were several difficulties with that proposal: (1) the actual existence of DNA within basal bodies and centrioles was questionable; (2) centrioles did not divide as did mitochondria and chloroplasts: new centrioles or basal bodies assembled from the general vicinity of existing ones, and in some organisms they could be formed *de novo*; (3) it was difficult to believe that the eukaryotic cell, with its nucleus and mitotic nuclear division, could have arisen repeatedly from the simple prokaryotic type independently 27 times, as Margulis had suggested.[54] Discussions in the 1970s focused on the idea that chloroplasts and mitochondria evolved as symbionts inside eukaryotic protists. Their multiple origins seemed plausible, especially for chloroplasts.[55]

Anthropomorphisms and Eating Peanuts

That mitochondria and chloroplasts originated as symbiotic bacteria was recognized by many biologists as a revolution in scientific thought. Interpretations of the underlying dynamics of symbiosis and its scope and significance continued to be laden with seemingly inescapable anthropomorphism and politics. "Just as the Copernican revolution demonstrated that man is not the center of the Universe," Seymour Cohen commented in 1970,

> so the investigation of this problem may show that a man (and indeed any higher organism) is merely a social entity, combining within his cell the shared genetic equipment and cooperative metabolic systems of several evolutionary paths. We suspect that governments should be interested in such a possibility, although their responses many not be readily predictable.[56]

Gunther Stent (1924–2008) expressed an opposite interpretation of the politics in the cell:

> Thus a eukaryotic cell may be thought of as an empire directed by a republic of sovereign chromosomes in the nucleus. The chromosomes preside over the outlying cytoplasm in which formerly independent but now subject and degenerate prokaryotes carry out a variety of specialized service functions.[57]

Interpretation of the relationship between host and symbiont in terms of master and slave was common, but in *The Lives of a Cell* of 1974, Lewis Thomas (1913–1993) maintained that conceptualizing symbionts as captured "enslaved creatures" was a one-sided subjective anthropocentric perspective.[58] Far from being a case of one-sided exploitation, he commented, "If you were looking for something like natural law to take the place of the 'social Darwinism' of a century ago, you would have a hard time drawing lessons from the sense of life alluded to by chloroplasts and mitochondria, but there it is."[59]

However, it seemed declarations of a revolution were premature: the evidence that chloroplasts and mitochondria actually arose from symbiosis was far unequivocal in the early 1970s.[60] Some critics associated the symbiotic theory with erroneous arguments that invoked supernatural cause by intelligent design to account for complex organs. In 1974 Thomas Uzell and Christina Spolsky asserted:

> The endosymbiosis hypothesis is retrogressive in the sense that it avoids the difficult thought necessary to understand how mitochondria and chloroplasts have evolved as a result of small evolutionary steps. Darwin's *On the Origin of Species* first provided a convincing evolutionary viewpoint to contrast with the special-creation position. The general principle that organs of great perfection, such as an eye, can evolve provided that each small intermediate step benefits the organisms in which it occurs seems appropriate for the origin of cell organelles as well.[61]

Non-Darwinian it certainly was. Biologists of the 1970s agreed that the symbiotic origin of chloroplasts and mitochondria was plausible, though it was not certain if that conjecture could ever be tested in a scientific manner. Initially, Margulis imagined that biologists might learn to culture chloroplasts, mitochondria, and centrioles. However, it soon became evident that these organelles were highly integrated into the nuclear genetic system: only a small fraction of the genes needed for mitochondrial and chloroplast functions were actually located in the organelles themselves. Most of the organellar proteins were encoded in the DNA of the nucleus. The "definitive proof" of their symbiotic nature or origin, their culturing outside the cell, was an experimental ideal that could not be realized.

There were acute difficulties in verifying any account of the origin of the eukaryotic cell.[62] The events one was trying to reconstruct took place millions of years ago. The fossil record was virtually absent. The clues had to be found by unraveling the nature of contemporary systems. The case for the symbiotic origin of mitochondria and chloroplasts was based on (1) the fact that different degrees of integration of a symbiont into the life of a cell were known to occur, (2) that the plastids and mitochondria are separated by a double membrane from the rest of the cell, (3) that they reproduced by fission, and (4) that their ribosomes were similar in size to prokaryotic ribosomes.[63]

Still, none of the evidence could be taken as direct support for symbiotic origins. Change the assumptions and one could argue that the gene traffic had moved in the opposite direction—not from organelle to nucleus but from nucleus to organelles. Those who favored the notion that organelles arose gradually and endogenously by compartmentalization within cells also offered plausible reasons for why a microbe might sequester DNA and protein synthesis into mitochondria and chloroplasts. Those organelles were concerned with vital functions. The "selective advantages" to the cell to having some of its DNA localized there was obvious, Aharon Gibor explained: "Ontogenetically, the cell is provided with an efficient and rapid mechanism for the synthesis of essential proteins; phylogenetically the multiplicity of templates can function to stabilize very important enzyme systems from the hazards of deleterious mutations."[64] Ruth Sager (1918–1997), who pioneered chloroplast genetics, expressed a similar view in 1966: "The existence of a nonchromosomal genetic system designed to minimize variability leads one to wonder whether [non-chromosomal] genes control particular traits of crucial value to the organism."[65] If organelle genomes had evolved at a slower rate than nuclear genomes, one could easily explain the similarity of organelles and prokaryotes without recourse to symbiosis. The features of mitochondria and chloroplasts that were similar to bacteria and blue-green algae were "retained primitive states."[66]

Blue-green algae had long been suspected to be transitional organisms, the missing link that would to some extent bridge the gap between prokaryote and eukaryote.[67] If blue-green algae were the ancestors from which eukaryotes gradually evolved, the "Uralga," as Richard Klein and Arthur Cronquist named the hypothetical lineage in 1967, then there was no need to invoke the symbiotic origin of chloroplasts.[68] In the view of those authors, "The evidence [was] clear that the photosynthetic bacteria gave rise to the primitive algae, and that these 'Uralgae' were the ancestors of all other life above the bacterial level."[69]

The strength of alternative theories boiled down to a question of parsimony, that is, which one had the fewest assumptions. Known as Ockham's razor, the scientific principle was one of frugality: "Neither more, nor more onerous, causes are to be assumed than are necessary to account for the phenomena."[70] Here, too, each side claimed victory. "In my opinion," Allan Allsopp wrote in 1969,

> all the striking similarities, outlined above, between cell organelles of eucaryotes and complete procaryotic cells cannot be taken as providing direct support for the symbiont hypothesis: they are rather the precise resemblances that might be expected if the eucaryotic pattern of cell structure had evolved by gradual transformation of procaryotes....On the general principle that the simpler hypothesis is always to be preferred in the absence of any evidence to the contrary, the direct transformation of procaryotes seems a more reasonable hypothesis than the symbiotic view.[71]

That organelles had evolved gradually by internal compartmentalization also seemed the simplest solution to Tom Cavalier-Smith in 1975: "Cell compartmentalization explains not only the origins of mitochondria, plastids and nuclei, but also their characteristic properties, more simply than does the symbiosis theory."[72] But Peter Raven had the opposite opinion. In 1970, after listing the names of 26 authors who supported the symbiotic origin of plastids and mitochondria, he concluded that the arguments necessary to defend an endogenous view were "far more complex than those supporting what is now clearly the majority opinion."[73]

From the outset of the debates of the 1960s and 1970s, some had argued that theories of cell origins did not belong to the realm of science at all. In 1967, W.H. Woolhouse suggested that speculations about the origin of cytoplasmic organelles be buried "by way of an epitaph, a parody of Wittgenstein's well-known remark, 'Whereof one cannot know, thereof one should not speak.' "[74] Stanier had a less harsh perspective in 1970:

> Evolutionary speculation constitutes a kind of metascience, which has the same fascination for some biologists that metaphysical speculation possessed for some medieval scholastics. It can be considered a relatively harmless habit, like eating peanuts, unless it assumes the form of an obsession; then it becomes a vice.[75]

Eight years later, when J.T.O. Kirk and R.A.E. Tilney-Basset weighed the alternative theories for chloroplast origins in their well-known book, *The Plastids*, they still had no idea how the issue could be resolved.[76]

Speculations about the origin of organelles, whether gradual or by saltational symbiosis, had produced a confusing mix of evolutionary scenarios based on diverse interpretations of a complex array of facts drawn from genetics, biochemistry, cytology, ecology, geology, and natural history. Backed by possibilities, and plausibility, and even "majority opinion" they amounted to what those critical of such adaptationist accounts would refer to as "just-so stories," as in the title of Rudyard Kipling's fables about how the elephant got its trunk, the camel its hump, and the leopard its spots.[77] While leading biologists insisted that one could not know the course of cell evolution in any scientifically testable way, others turned to comparative analysis of nucleic acids and proteins to do precisely that.

| The Morning of Molecular Phylogenetics

One of the grand biological ideals is to be able to work out the complete, detailed, quantitative phylogenetic tree—the history of the origin of all living species, back to the very beginning. Biologists have had this hope for a long time; biochemistry now has the actual capabilities of accomplishing it.

—Margaret Dayhoff and Richard Eck, "Inferences from
Protein Sequence Studies" (1969)

Yet the introduction of reticulated evolution presents severe conceptual and practical problems, which so far as we know have not been adequately discussed. The occurrence of highly similar proteins or DNA fractions in two different bacterial taxa may indicate recent gene transfer rather than recent common ancestry.

—Dorothy Jones and Peter Sneath, "Genetic Transfer and
Bacterial Taxonomy" (1970)

THAT MOLECULES WOULD be indicators of evolutionary history was an old idea. In 1892 Ernst Haeckel supposed that phylogenetic lineages among monera could be known only by the study of the atomic composition of their albumen (protein). In 1927, Francis Bather wrote of the gene as "the ultimate element of classification" (chapter 5). In 1948, Ernst Pringsheim suggested that "specific proteins and other compounds of high molecular weight" might eventually afford the clue to bacterial phylogeny.[1] At that time it was widely believed that the gene acts through protein enzymes that catalyze chemical reactions in the cell. The discovery of DNA as the basis of the gene, how it is reproduced, and how it affects protein synthesis was the work of molecular biology of the 1950s and 1960s.

Within the Gene

What emerged with molecular biology was an entirely new concept of life, not in terms of formless colloidal gels as it was for biochemists of the early twentieth century, but in terms of the specific structure of informational macromolecules, proteins, and nucleic acids.[2] The DNA molecule was understood to contain a code that was transcribed to RNA and then translated into the specific amino acid sequences that compose each specific protein. It was a digitized system in which unique sequence patterns of the nucleotide bases—adenine (A), guanine (G), cytosine (C), and thymine (T)—serve as the "templates" for the assembly of a specific protein. The gene was understood to be a sentence made up of many three-letter words, each representing one of the 20 kinds of amino acids in the long chains that composed proteins; the triplet AGG, for instance, meant the amino acid arginine, and CCG meant proline.

Molecular biology brought with it great advances in technology: recombinant DNA, transgenic organisms, and the patenting of genes from microbes to humans. And promise of a new age of gene-based medicine was to follow from the fundamental discoveries of the 1950s and 1960s. Although largely shaped by such biomedical engineering pursuits, molecular biology held concepts and techniques of profound evolutionary implication. The molecular conception of life brought with it wholly new kinds of characters through which to follow the dynamic of evolution. Instead of comparative morphology and physiology, one could determine organismal relationships based on differences in the order or sequence of the nucleotide bases within selected genes and on the amino acids of proteins. One would follow the changes *within* a gene and *within* a specific protein.

The process of evolution inside the gene differed from that which operated on the organism's physiological and morphological manifestations As genes mutate, the simplest change would be to replace one nucleotide base for another, for example, replacing the G in the sequence "AAG" to C, yielding "AAC," which in turn could change the amino acid glutamic acid (AAG) to aspartic acid (AAC) at a certain position in some protein. Some of these small changes in amino acids would have little effect, but others could drastically change the encoded protein—small changes can have large effects. Genetic mutations that either have no immediate effect or that improve protein function would accumulate over time. As two species diverge from an ancestor, the sequences of the genes they share also diverge, and as time advances, the genetic divergence will increase. One could therefore reconstruct the evolutionary past of species and make phylogenetic trees by assessing the sequence divergence of genes or of proteins isolated from those organisms.

The molecular evolutionary trait had little resemblance to those visible manifestations in forest, garden, or Petri dish. Each amino acid position in each protein was considered to be a variable trait with 20 potential levels of distinction, just as was each nucleotide of DNA and RNA. Nucleic acids and proteins were not mere indicators of evolutionary history; they contained precisely inscribed historical signals to be discerned by precise analysis and measurement.

Paleogenetics

The use of the amino acid sequence of proteins for taxonomic purposes was foreseen by some of molecular biology's founders a few years before the genetic code (the correspondence of nucleotide triplets to amino acids) was cracked. Francis Crick (1916–2004) prophesied in 1958:

> Biologists should realize that before long we shall have a subject which might be called "protein taxonomy"—the study of amino acid sequences of proteins of an organism and the comparison of them between species. It can be argued that these sequences are the most delicate expression possible of the phenotype of an organism and that vast amounts of evolutionary information may be hidden away within them.[3]

Still, phylogeny was as far away from Crick's own research interests as it was from the minds of most of the others who forged the new field. Crick's paper of 1958 is remembered for giving molecular biology what he called its "central dogma": the "text" of DNA determined protein specificity, and information flowed from nucleic acid to protein, not in the reverse direction, nor from protein to protein. "The gene gives orders. The protein executes them," François Jacob wrote in 1971.[4] That depiction no longer holds today, but it certainly captured the conceptions of the times.[5]

Protein sequencing for phylogenetic purposes rode the back of a great wave of medical interests. Frederick Sanger (b. 1918) at Cambridge University succeeded in developing the techniques for amino acid sequencing. He deployed them to determine the complete sequence of insulin in 1955, showing that proteins have definite structures, which were the key to their function.[6] He was awarded his first Nobel Prize in Chemistry in 1958 (see also chapter 19). In his Nobel lecture, he said, "One may also hope that studies on proteins may reveal changes that take place in disease, and that our efforts may be of more practical use to humanity."[7] Sanger's method for determining amino acid sequences opened up the archives of molecular evolution locked away in proteins.

Linus Pauling (1901–1994) and Emile Zuckerkandl (b. 1922) at the California Institute of Technology called the field "chemical paleogenetics" in 1963. They pioneered the use of amino acid sequences of hemoglobin to infer evolutionary relationships of primates.[8] In a key colloquium, "Evolving Genes and Proteins," held in 1965, they articulated how evolutionary time might be kept in what they called "the molecular clock."[9] Their idea was one of rate constancy at the molecular level—that changes in the amino acid sequence of a protein from different species should be "approximately proportional in number to evolutionary time."[10] In other words, informational macromolecules (nucleic acid and proteins) that they called "semantides" would not evolve in the same (irregular) way that the morphological features of an organism did by adaptation and natural selection.[11] The very same enzyme function could be found in a variety of taxa, despite the fact that the underlying protein sequence was never exactly the same. Many changes simply had no effect on protein function and therefore were of no adaptive value. The evolution of proteins and nucleic acids without selection was called "the neutral theory" by Motoo Kimura (1924–1994) and "non-Darwinian evolution" by Jack King and Thomas Jukes in 1969.[12]

As it turned out, the time kept in those clocks was not linear. Different proteins and nucleic acids changed at different rates, and different parts of the same molecule did too. Still, Zuckerkandl and Pauling crystallized the idea of molecular evolution for many who entered this field. When comparing differences in a protein shared by two different kinds of organisms, they argued, one need not be too concerned with problems of convergence, that is, the independent development of a similar (analogous) structure in different groups. The likelihood that the molecular character was due to convergent evolution was greatly diminished because the space of possible outcomes was so very large. As they put it, "The ease with which variations of a given type of protein can be produced by duplication and mutation of a gene would be so much greater than the ease of convergent evolution from independent starting points."[13] Zuckerkandl was the first editor-in chief of the *Journal of Molecular Evolution* founded in 1971; that journal played a key role in facilitating the development of the new field.

A few choice proteins were initially used to discern evolutionary relationships as biochemists, physicists, and computer experts converged on the new sequencing technology. At the University of California–San Diego in 1964, Russell Doolittle pioneered the use of blood clotting proteins to determine evolutionary relationships among vertebrates.[14] Other proteins would have to be used for a broader based phylogeny. In 1967, Emmanuel Margoliash at the Abbott Laboratories in North Chicago, and Walter Fitch at the University of Wisconsin teamed up to show how the amino acid sequence of cytochrome *c* could be used to infer the evolutionary relationships among 20 eukaryotes ranging from horses, humans, pigs, rabbits, chickens, tuna, to baker's yeast.[15] Cytochrome *c* is the terminal enzyme in the respiratory chain and is located in the inner membrane of mitochondria and of aerobic bacteria. Much of the molecular phylogenetic data in the 1970s was based on cytochrome *c* sequences.[16]

No one championed the use of sequence data and the promise of a universal tree of life more than did Margaret Dayhoff (1925–1983). Remembered today as one of the founders of the field of bioinformatics, she was the first to create an extensive computerized protein sequence and nucleic acid databases for deducing evolutionary histories.[17] In 1959, she joined the newly founded National Biomedical Research Foundation, a nonprofit organization in Silver Spring, Maryland, whose primary aim was to apply technology to medical research. Protein sequence studies promised various medical applications; there was also the engineer's hope of a new "era of control" in which scientists would be able to synthesize proteins to their own designs.[18] But Dayhoff's interests lay in the construction of a phylogenetic tree that would encompass all life on Earth.

Beginning in 1966 Dayhoff, Richard Eck, and coworkers collected all the published protein sequences, transformed those data into a uniform format, and made them available to the scientific community in a small book, *Atlas of Protein Sequence and Structure*. The first atlas contained sequence information on 63 organisms and 13 viruses.[19] Those data included fewer than 50 partial protein sequences with a length of 30 or more amino acid residues. The third volume of the atlas of 1969 included more than 200 protein sequences and six nucleic acid sequences.[20]

Dayhoff and Eck aimed to construct an outline of a universal phylogeny traceable back to the common ancestor of all life on Earth. They had no doubt that "the exact relationship and order of derivation of the living kingdoms and phyla," would be worked out through protein studies.[21] Evolution had a conserved nature. The genetic code was essentially the same in all organisms, as were a number identical compounds, mechanisms, and reaction pathways.[22] The common ancestor of life was therefore already complex enough to possess all of them:

> The inferred biochemical structure of this primitive cell provides evidence that it was itself the product of many evolutionary steps. . . . Implicit in the principles of evolution and in the inferred structure of the proto-organism is a wealth of information about the far more primitive organisms which preceded it.[23]

For Dayhoff and Eck, the origin of life on Earth involved cosmic principles far beyond Darwinian principles. Someday when organic evolution was traced through many steps all the way to chemical simplicity, they prophesied, "it will be seen that the 'origin of life' was not a single unique event, but rather a continuous development of the potentialities inherent in the building blocks and in the energy flow of the universe."[24]

By DNA Composition

The technology was not available for sequencing a gene in the 1960s. But there were two other methods for extracting the taxonomic information therein. One was based on DNA hybridization. It consisted of labeling the DNA from one organism with either heavy or radioactive isotopes, annealing it with ordinary DNA from other organisms, and determining what percentage of the labeled DNA hybridizes (binds). Since hybridization occurs only between regions that have sufficiently similar nucleotide sequences, the results indicated the percentage of the labeled DNA that is similar. This technique was applied to bacterial taxonomy by Brian McCarthy and E.T. Bolton in the Department of Terrestrial Magnetism at Carnegie Institution in Washington, D.C., in 1963.[25]

The second method was based on a quantitative study of the nucleotide composition of DNA. Recall that DNA contains four kinds of bases, adenine (A), thymine (T), guanine (G), and cytosine (C), as its major components. DNA is double stranded, and A bonds with T, and G with C. Therefore, the percentage of G is always the same as that of C, and the percentage of A is the same as T. Since many genes differ from one group of organisms to another, the percentages of A–T and G–C also differ. In practice, this came to be represented as percent G–C. Belgian microbiologist Jozef De Ley (1924–1997) at the University of Ghent pioneered this approach to bacterial taxonomy. "Bacterial classification is confused and biased by the personal opinion of the investigators," De Ley commented in 1968. "Because of the lack of paleontological, embryological and comparative anatomical data, the actual evolutionary tree of bacteria is unknown to us. Percent GC values may give us our first clue."[26]

De Ley calculated that a difference between organisms in percent GC of 30% and higher meant that there were practically no nucleotide sequences in common.[27] Those values often conflicted with bacterial "species" that had been identified by phenotypes. The GC values of what had been called different species were frequently found to be too similar to be distinguished as such. "The only biological unit which has survived the close scrutiny of the modern bacterial taxonomists," De Ley said, "is the genus."[28] In many cases, the molecular conclusions confirmed the phenotypic identifications of genera; agrobacteria, for example, were of one genus; they had 70% DNA homology.[29] But sometimes what had been considered to be two genera was actually one genus, and sometimes the inverse was the case: organisms that had been considered to be of genus were off the scale of relatedness.[30]

By the end of the decade, many kinds of data could be used to classify bacteria: protein sequences, nucleic acid hybridization, and nucleic acid composition in addition to morphological, physiological, and biochemical characters. Roger Stanier, who had been so defeatist about a bacterial phylogeny, had a slight change of heart about the possibility, commenting in 1971 that there were now

> a variety of methods for ascertaining (within certain limits) *relationships* among the bacteria; and that where relationship can be firmly established, it affords a more satisfactory basis for the construction of taxa than does mere resemblance. As the philosopher G. C. Lichtenberg

remarked 200 years ago, there is significant difference between *still* believing something and believing it *again*. It would be obtuse still to believe in the desirability of basing bacterial classification on evolutionary considerations. However, there may be solid grounds for believing it again, in the new intellectual and experimental climate which has been produced by the molecular biological revolution.[31]

Cladistics versus Phenetics

There was no question that the new molecular methods held great promise for taxonomy, but it was still far from certain that these new molecular characteristics actually implied phylogenetic relatedness. Critics insisted that all the uncertainties that had plagued bacterial phylogenetics earlier in the century still applied. There was also no consensus regarding whether bacteria, or any other kind of organisms, *should* be classified phylogenetically. Taxonomists had not established rigorous methods for determining phylogenetic relations (see chapter 6). Monophyletic groups were not always insisted upon, and sometimes they were classified based on judgments about the amount of evolutionary change that had occurred between groups.

Taxonomy of the 1960s and 1970s was embroiled in a controversy between "cladists," who insisted that taxonomy must be based on phylogenies, and "pheneticists," who were adamant that classification be based on overall likeness of organisms regardless of how those traits came to be.[32] The aims of both groups were similar—make classification more scientific, that is, quantitative and verifiable. But they differed in how this could be achieved. Cladists sought to rid taxonomy of its subjectivity by always classifying organisms according to ancestry—that is, based on branching points. They were led by German entomologist Willi Hennig (1913–1976).[33] Cladistics took its name from Hennig's concept of a "clade" (Greek *klados*, branch), which he defined as a group of organisms related by common descent from a unique ancestor that it does not share with any other group. Descent and branching points were the only criteria for classification, they argued, because, as Sergius Kiriakoff put it, "it is epistemologically the truest, and the only one free of arbitrariness. The purely physical succession of parents and children cannot be interpreted in several ways, nor are there any gradations in the process which might allow discrepancies in appreciation."[34]

Pheneticists were led by Robert Sokal, an entomologist at the University of Kansas, and by Peter Sneath, a bacteriologist at Medical Research Council in London. They aimed to replace the subjective judgments of naturalists with a statistically based system of classification that they called "numerical taxonomy." Their book *Principles of Numerical Taxonomy* of 1963 described the new methodology.[35] To preclude subjective judgments of similarity based on the "most important characters," they searched for quantitative measures of an unprecedented large number of characters (often 100–200), assigning to them equal value.[36] Although pheneticists claimed to avoid subjectivity by equally weighting many characters, critics noted that equal weighting *is* weighting. Essentially, it was the same as the "statistical taxonomy" of old (chapter 5) but with computerization.[37] Sneath and Sokal's *Numerical Taxonomy* of 1973 became the standard text and started a rush to employ these methods in classification as computing technology advanced into biology.[38]

The taxonomies of cladists and pheneticists were not always discordant. In fact, Sokal and Sneath believed that "numerical taxonomy will in general give monophyletic taxa."[39] Their results conflicted when the organisms to be classified were not closely related but had evolved similar traits as a result of having to adapt to similar environments, that is, when convergent evolution had occurred.[40] They also clashed when evolutionary rates of change had been very different in the organisms to be classified. For example, birds and reptiles have a common ancestor among the dinosaurs. Should one classify birds as reptiles according to their ancestry? Under the phenetic system, birds are put in Class Aves, while dinosaurs were included in Class Reptilia. A cladistic definition of dinosaurs, on the other hand, is based on the last common ancestor of *Tyrannosaurus* and *Triceratops* and all of its descendants. Because birds are descended from this common ancestor they are classified as dinosaurs. The amount of evolutionary change is irrelevant to the cladist's analysis.

Cladistics was simply an abomination to science for classical taxonomists such as Ernst Mayr because it ignored uneven rates of evolutionary change: "the amount and nature of evolutionary change between branching points."[41] As he saw it, a purely genealogical classification ignored the "most interesting aspect of evolution and phylogeny: namely that of radiative and divergent adaptiogenesis."[42] Putting birds and crocodiles at the same categorical rank "may be logically impeccable, but is simply wrong biologically."[43]

> Even if it should be impossible to compare numerically the (slow) rate of evolution between the stem species and the modern birds, every beginner can see how much more drastically the birds differ from the common ancestor than the crocodilians. To ignore this altogether, because it can not (yet) be measured accurately would seem a poor escape from a difficulty.[44]

Mayr called his own brand of classification "evolutionary taxonomy" and claimed for it higher information content than the pure phylogenetic systematics of the cladists. As he put it, Hennig had "traded biological meaning for a hoped-for logical consistency"; giving biological meaning to classification required subjective judgments.[45] Hennig replied that by considering traits that were nongenealogical, Mayr's approach was pre-Darwinian, and, indeed, Aristotelian.[46]

The phenetic approach was the most widely used approach to microbiology taxonomy in the 1960s and 1970s. And as Sneath saw it, that was not a matter of choice but of necessity. Convergence could be rampant, and both slow and rapid evolutionary change might occur. One could not decide which traits were primitive and which were not.[47] There was no logical reason why any one feature should be given greater weight in classification than any other. The same applied to protists: those who had discussed the phylogeny of protozoa and of algae and fungi emphasized the difficulty of deciding what properties of microorganisms were primitive.

"The history of phylogenetic studies in microorganisms is strewn with wrecks of broken theories," Sneath commented in 1974.[48] He pointed to Orla-Jensen's suggestion that the earliest life forms must have been photosynthetic or chemoautotrophic; autotrophs were primitive, "living fossils." Oparin's theory of the origin of life turned that argument on its head and proposed that the earliest organism were heterotrophs living in a "prebiotic soup" of the primeval oceans. Then there was the model based on increased morphological complexity from the coccoidal type, as proposed and then later rejected by Albert Jan Kluyver, Cornelis van Niel, and Stanier (see chapter 7).

There was no reason to favor any of these models or even to assume that bacteria alive today are similar to ancient ones.[49] Part of the problem, as Sneath saw it, was that scientists were simply ignorant of the changes that might have occurred in the major habitats over enormous periods of time. They knew no more about the difference between an ocean now and one a billion years ago than they did between a Devonian swamp of 400 million years ago and one today. If these environments had been the same for eons, this could mean that present-day microbes had reached the limit of their adaptation many millions of years ago. Bacterial evolution would be essentially over.

De Ley had addressed that question based on the range of percent GC values. Although it had often been tacitly assumed that primitive and modern bacteria were similar and that there had not been much evolutionary divergence since the Precambrian, his molecular biological evidence pointed to the contrary.[50] Bacteria and other microbes were "in essence much more varied than the most widely varied groups of vertebrates."[51] In light of their diversity, he said, it was "quite possible that the present living bacteria are quite different from those living in other geological periods, particularly the Precambrian seas."[52]

Is Lateral Gene Transfer Natural?

There was still another potentially fundamental obstacle in basing bacterial taxonomy on gene phylogenies: lateral gene transfer. In addition to the "vertical" gene transfer that occurs when a parent cell divides into two daughter cells, bacterial geneticists had shown that genes could be transmitted "horizontally" or "laterally" between taxa. That transfer and recombination between strains could occur by three means: (1) the uptake of small bits of foreign bacterial DNA from the environment (transformation), (2) transfer of genes by viral "infections" (transduction), and (3) transfer of genes by direct cell-to-cell contact (conjugation). Thus, a bacterium of one type may acquire genes from an unrelated organism. Therefore, similarities in a gene may not be a measure of genealogical relationship. If organism type A and organism type B carry the same gene, it may not be because both belong to the same taxonomic group because of common descent, but rather because one, or both, of them acquired that gene. Lateral gene transfer could potentially scramble the phylogenetic record.

Though lateral gene transfers were well known in the laboratory, it was not certain how extensive they were in nature. Could gene transfers occur between all kinds of bacteria? Could all genes be transferred laterally, or were those located in cytoplasmic factors or plasmids more commonly transferred? Which mechanisms of lateral transfers were more prevalent? Recall that in 1952 Joshua Lederberg considered the saltational effects of lateral gene transfers by phage when recommending that the definition of heredity be enlarged from sexual inheritance to include what he called "infective heredity." By the end of the decade, René Dubos had also become a strong advocate of the integrative and creative effects of viral infection (chapter 9).

Transformation and transduction experiments by Lederberg and others in the early 1950s showed that gene transfers by these means could occur between different taxonomically defined "species" and genera.[53] Still, the main thrust of bacterial genetics was to bring bacteria into the Darwinian fold. Bacteriologists had long suspected that the environment induced adaptive hereditary changes in bacteria (see chapter 6). Felix d'Herelle had postulated, for example, that resistance to lethal viruses was induced by the virus. But in their famous "fluctuation tests" of 1943 Salvador Luria (1912–1991) and Max Delbrück (1961–1981) aimed to show the opposite, that bacterial mutations arise independently of the action of the environment and then are subject to natural selection.[54] Bacterial resistance to lethal viruses was not the result of the direct action of the virus, but rather, "resistant bacteria arise by mutations of sensitive cells independently of the action of the virus."[55]

In 1948, Milislav Demerec confidently asserted that "bacterial resistance to penicillin and streptomycin is not induced by these compounds but originates spontaneously through genetic changes comparable to gene mutations."[56] The demonstration that bacteria possessed genes like other organisms helped to reinforce the concept that mutations and selection were the primary means of evolutionary innovation. When Joshua and Esther Lederberg at Wisconsin introduced replicating techniques to bacterial genetics in 1952, they did so, they said, to further confirm "previous evidence for the participation of spontaneous mutation and populational selection in the heritable adaptation of bacteria to new environments."[57]

Darwinian selection acting on preexisting variations in a population applied to bacteria as much as it did to plants and animals. Lateral gene transfer between taxa was a decidedly non-Darwinian mechanism of evolution, but leading bacterial geneticists denied its significance. Many seemed to be confident that lateral gene transfer was not pervasive in bacteria and that gene mutation was the major mechanism of evolutionary change, just as it was in other organisms.

Bruce Stocker at the Lister Institute in London was certain that gene transfers between bacterial taxa did not occur naturally. He carried out genetic hybridization experiments between different strains of *Salmonella* using phage transduction. "Our ability to hybridize two bacterial strains in the laboratory has no more (and no fewer) implications for taxonomy than has the artificial hybridization of two higher plants or animals," he wrote in 1955:

> There is, as yet, no evidence that these or other phenomena of bacterial hybridization occur under natural conditions. It is, therefore, still possible to treat bacteria as organisms which in nature multiply only by fission, without sexual process or other mechanism of gene interchange; we may then consider all strains of a valid species, genus or larger taxonomic group as members of a clone, all derived from a single common ancestor. If we assume this, we may regard any good practical classification (that is one based on multiple correlated characters) as being of necessity an approximation to a phylogenetic classification.[58]

Lateral gene transfer by conjugation was indeed believed to be rare, one in a million by Lederberg and Tatum's reckoning in 1946. In 1952, William Hayes showed that transfer of DNA in bacterial conjugation was unidirectional, and dependent on the presence of a specific fertility factor plasmid in the donor cell.[59] Conjugation triggered the transfer of the F factor plasmid, along with any other DNA that might become integrated with that plasmid.[60]

In 1957, Luria and Jeanne Burrows at the University of Illinois showed that the F plasmid could be transferred between *Escherichia coli* and *Shigella*, the bacilli responsible for dysentery. Still, they suspected that such lateral gene transfer across phylogenetic groups was not significant. After all, what counted as a bacterial species, and genera typically had no meaning in terms of relatedness.[61] Luria had long suspected that many of the groups that taxonomists classified as bacterial "species" and even "genera" or "tribes" were separated on the basis of character differences that may be brought about by a single mutational step.[62] "Hybridization capacity, even if present," he and Burrows wrote, "might indeed not be responsible for significant amounts of gene flow among natural bacterial populations, which can propagate indefinitely by vegetative reproduction alone."[63]

Stanier, Doudoroff, and Adelberg adopted a similar view in the first edition of *The Microbial World* of 1957. They considered sex to be infrequent for any microorganism. "Among microorganisms endowed with sexual capabilities," they said, "the occurrence of the sexual act may be an extremely rare event, most reproduction being asexual."[64] In the second edition of 1963, they noted that the transfer of genes between bacteria of soluble DNA had been "demonstrated in only a few types of prokaryotic cells."[65] Although transformation may be rare, not so for transduction: "Since all species of bacteria harbour temperate phages, they said, "transduction may well be a universal mechanism of recombination in bacteria."[66]

The main focus of bacteriology was on its practical uses for humans, not on new evolutionary principles. The use of bacterial viruses in the biomedical industry was in the discussion stages in the late 1960s.[67] Bacterial viruses could be used as models for the treatment of viral infections. Viral-mediated gene transfer might be used to repair genetic errors that result in metabolic diseases. Viruses might have utility in controlling bacterial populations. Lateral gene transfer would indeed have important applications in the engineering of new strains of bacteria when recombinant DNA technology was developed.[68]

Antibiotic-Resistant Outbreaks

While lateral gene transfer held the promise of a new biomedical technology, it also revealed a great danger in the way in which the war against germs was waged by the emerging industrial medical complex. Indeed, lateral gene transfer captured the attention of microbiologists with outbreaks of multiantibiotic-resistant germs. Though leading bacteriologists considered gene mutation and selection to be its basis, it was ultimately shown that the adaptation from sensitivity to resistance occurred rapidly by lateral gene transfer in response to antibiotics. An antibiotic-resistant factor, like the F factor, was transmitted among bacteria by cell contact. In environments where exposure to antibiotics was high, such as hospitals, those plasmids spread quickly.

The production and use of antibiotics was one of the most spectacular frontiers of the biomedical industry. In the 1930s sulfa drugs (sulfamides) were introduced to treat urinary tract infections, ear and eye infections, bronchitis, bacterial meningitis, and pneumonia. Such medicines led to a wave of optimism in the fight against hemolytic streptococcus (the cause of strep throat and scarlet fever) and other microbes. Confidence was boosted when penicillin was introduced. Its use seemed to signal the imminent defeat of staphylococci, which caused a wide variety of diseases and were especially associated with hospital-acquired infection following surgery or other invasive medical procedures. But optimism turned to anxiety in the decade following the Second World War. No sooner were new antibiotics announced than new resistant strains appeared.[69]

Mary Barber (1911–1965) at the Postgraduate Medical School in London and at Hammersmith Hospital led the movement to expose the outbreaks of antibiotic-resistant strains and curtail their spreading when she reported that the widespread use of penicillin and tetracyclines had engendered them.[70] "The rate of increase in this hospital at present is so rapid as to be somewhat alarming," she said in 1947.[71] It was evident that the cavalier use of penicillin and other antibiotics had been a gross tactical error.[72] She called for the "control of the use of antibiotics in hospitals" as "essential both to prevent the emergence of drug-resistant bacteria and also to avoid rendering patients more susceptible to infection by such bacteria, through elimination of the normal flora."[73] Fourteen years later, she reported the existence in hospitals of staphylococci that were resistant to multiple drugs: erythromycin and novo-biocin, as well as tetracycline and penicillin.[74]

Discussions of bacterial antibiotic resistance were framed by the classical opposition that had shaped the development of evolutionary theory regarding plants and animals since the nineteenth century. Did inherited modifications in bacteria arise in a "Lamarckian" fashion by environmentally induced adaptive hereditary changes, as had long been thought, or in a "Darwinian" two-step fashion by random mutation and selection?[75] Barber supposed that multidrug-resistant staphylococci "almost certainly arise by single or multiple step mutations" involving a number of genes.[76]

Tsutomu Watanabe at Keio University in Tokyo explained the story about lateral gene transfer and multiple-antibiotic resistance to English-speaking bacteriologists in 1963. Although sanitary conditions in Japan were considered very good, bacillary dysentery was one of the most important infections, and shortly after the Second World War a high incidence of antibiotic resistance in *Shigella* was reported. In 1957, *Shigella* strains were found that were resistant not only to the antibiotic used to treat it (chloroamphenicol), but also to several other antibiotics that had not been used to treat it.[77]

Two years later, Japanese researchers demonstrated that multiple drug resistance could be easily transferred between *Shigella* and *E. coli* in mixed cultures. Multiple resistances did not arise in a series of discrete steps, each corresponding to a single drug, but usually appeared fully developed as groups of resistances against some or all of the drugs. Multiple drug resistance factors (R factors) were carried and transferred by an "episome." Francois Jacob and Eli Wollman had coined that term for those genetic elements that could replicate autonomously in the cytoplasm or could be integrated into the main bacterial chromosome and replicate with the chromosome.[78] "Multiple drug resistance," Watanabe concluded, "is therefore an example of "infective heredity."[79] Antibiotic resistance could be easily transferred between *E. coli* and *Shigella* in the intestines of humans.[80] He warned that multiple drug resistance "could become a serious world-wide problem."[81]

R factors were subsequently demonstrated to have a very wide host range. And by 1967, there were reports of lateral transfer of drug resistance among staphylococci by transducing phages, as well.[82] Still, the complacency of the medical establishment and the pharmaceutical industry about the threat of antibiotic resistance led to the continued manufacture of large numbers of antibiotics.[83] Before the end of the decade, another antibiotics controversy broke out over the widespread use of antibiotics in animal food. Adding small amounts of antibiotics to the food of farm animals increased their rate of growth and improved the efficiency with which they converted their food into meat and other products.[84] It was uncertain if infective drug resistance in the bacteria of animals could be spread to bacteria in the human gut.[85] A number of committees of inquiry were set up to assess the risks and make recommendations.[86]

Patchwork Evolution

The outbreaks of drug resistance sparked speculations about the evolutionary importance of lateral gene transfer. How many other characteristics could be transmitted laterally? What was the range of such transmission? How would it affect the speed of evolution in bacteria? How would it affect phylogenetics?

In 1970, Norman Anderson at the Oak Ridge National Laboratory in Tennessee made a radical suggestion that the lateral transfer of viruses was a mechanism of saltational evolutionary change in all organisms from the origins of life to the present; it would account for parallel evolution such as eyes of squid and vertebrate, and it would also account for the universality of the genetic code itself.[87] Why was there only one version of the genetic code today? "If the information from the entire biome was read by any and all organisms, only one code could (and would) survive."[88] He also noted the problems such ideas presented for phylogenetics: "The greatest objection to the concepts presented here is that they undermine the foundations of a favourite pastime—the reconstruction of evolutionary relationships by comparing amino acid sequences."[89]

Anderson's speculations regarding lateral gene transfer among all organisms were exceptional, but microbiologists certainly recognized the problem it might pose for bacterial phylogenetics. "Genetic studies on procaryote are complicated by a phenomenon not known to exist among eukaryote," Stanier commented in 1971.

> A bacterium may be a genetic chimera, some of its phenotypic traits being determined by episomes that are transferable among (and expressed in) a considerable range of species, having markedly different chromosomal genomes. It is therefore conceivable that false inferences concerning the relatedness of a series of bacteria could be reached by the study of one or more shared characters determined by episomal genes.[90]

British microbiologists seemed to be especially receptive to the issue: the classical conception of evolution as descent with modification might not be applicable to the bacterial world. The cytoplasmic factors concerning drug resistance were the best studied, but there was no reason that other virus-like cytoplasmic factors (plasmids) carried genetic characteristics that could be laterally transferred like the F factor and the R factor. The mechanism of infection by such factors was known to be highly efficient. In effect, it was the same as "Lamarckian principles" of environmentally induced adaptive hereditary change. As E.S. Anderson at the Public Health Laboratory Service in London commented, "It often leads to infection of 100 percent of cells."[91] He suspected that such transfer factors had more evolutionary importance than hitherto imagined in speeding up the evolutionary process of bacteria. "The evolutionary time-scale may have been telescoped into a shorter span than that envisaged purely in terms of the selection of mutants with survival advantages."[92] The question remained whether such mechanisms operated in all bacteria and, indeed, if such mechanisms were limited to bacteria. If the latter, he concluded, "some rethinking may be called for in relation to evolution in other fields of biology," as well.

When Dorothy Jones and Sneath assessed the phylogenetic implications of lateral gene transfer a few years later, they suspected that plasmid transfer by conjugation would have the widest range.[93] Transformations resulting from the transfer of genes as soluble DNA from a culture medium into a recipient cell seemed to occur primarily between species, not between genera. Transduction was known in only a few groups of bacteria, notably the Enterobacteriaceae, pseudomonads, and *Bacillus*. However, gene transfers through conjugation could cascade across the bacterial kingdom. "The genera could behave, as it were, as stepping stones for the genes. At each step, it might be necessary to achieve integration into the recipient chromosome and this might be the limiting factor."[94]

Because gene transfer among bacteria involved only a small part of the genome, it made transfer across taxonomic boundaries easier. "This in turn," Jones and Sneath reasoned,

> could favor extremely reticulated modes of evolution, with numerous partial fusions of phyletic lines (i.e., involving only a few genes at a time). It may well be that gene exchange is so frequent that the evolutionary pattern in bacteria is much more reticulate than is commonly believed and cannot be satisfactorily represented by the usual cladogenies that repeatedly branch with time, but show no fusions of branches.[95]

At the Royal Postgraduate Medical School in London, where Mary Barber had first sounded the alarm about antibiotic resistance, Robert Hedges offered the most extensive discussion of lateral gene transfer to date. He posited that the structure of the bacterial genome itself had evolved in a way that would favor such gene transfers.[96] To begin with, he pointed to the circular organization of bacterial DNAs: the main chromosomes or "genophore," plasmids, and bacteriophages, compared to the linear organization of the chromosomes of eukaryotes. The simplest explanation for that circularity, he said, was that the genetic elements evolved so as to exploit the evolutionary opportunities presented by genetic interaction. A linear DNA molecule could attach itself to the end of another DNA molecule, but it could only integrate into the interior of such a molecule by replacing a homologous segment. This very limited potential for interaction contrasted with "the almost infinitely wide opportunities" open to circular structures, which can undergo recombination at any point.

Then there was the way in which bacterial genomes were organized into clusters of functionally related genes.[97] Gene clusters based on metabolic functions would also favor lateral gene transfer from one genus to another. If only a small piece of genetic material could be transferred, those genes whose functions were metabolically coordinated should be transferable as a unit. "Genomes of bacteria evolved in a patchwork fashion," Hedges observed; the consequences for phylogenetics would be profound "even if one knew the complete evolutionary history of the amino-acid sequence of a particular protein one could deduce little about the evolutionary history of bacteria."[98]

The concept of species simply did not apply to the bacterial world, in Hedges's view. Within sexually reproducing eukaryotes, extreme reticulation occurred within the boundaries of the species. Certainly there was a doubling of chromosome number through hybridization in some plants. Hybridization occurred among some animals as well. In bacteria, though, genes from a wide range of taxonomic groups could be reassembled to produce a progeny species for which no single parent species could be assigned. As Hedges argued, it was possible that hierarchical ordering of one species within a genus, and one genus within one family based on common descent did not apply.

Certainly, many bacteriologists of the 1950s and early 1960s had also recognized that the concept of species did not apply to bacteria. But this was not because bacteria could exchange genes between unrelated groups, but because laboratory studies indicated that sexual reproduction was a rare event for bacteria—just as it was considered to be for other microorganisms.[99] As Samuel Cowan commented in 1962, before lateral gene transfer received great consideration, "The microbial species does not exist; it is impossible to define except in terms of nomenclatural type; and it is one of the greatest myths of microbiology."[100]

In the face of evidence of lateral gene transfer, Hedges suggested that it was

best to think of bacteria as constituting one gene pool from which any "species" may draw genes as these are required. The most recent "demand" has been for genes conferring resistance to the various antibiotics. On this view, whilst the "phylogenetic tree" is a reasonable representation of the evolution of eukaryotic species, a reticulated network would be required to represent the evolution of the genome of a bacterial species.[101]

As for taxonomy, any classification "depicting well-defined species or groups of related species would be inherently unnatural." The "real situation," he concluded, "is best represented by a picture of a continuous gene pool in which the 'species' merely represent sets of genetic information which happen to be adaptive." This way of seeing things, he said, was in line with numerical taxonomy.[102]

The Core Concept

Those microbiologists who used molecular methods to classify bacteria typically ignored the issue of lateral gene transfer. Some also rejected the principles of numerical taxonomists, according to which all traits were considered to be equal. Some aimed to distinguish the most conserved parts of the genome, those sequences that were similar among widely divergent taxonomic groups. Just as Lamarck and Darwin had done, they distinguished characteristics of adaptive value to the organism from those that were essential and of universal phylogenetic significance.

Already in the 1960s, a consensus was growing regarding what those most unchanging parts of the genome were. "Genes coding for components of the translation mechanism are likely candidates," David Dubnau and collaborators at the Albert Einstein College of Medicine in New York argued in 1965.[103] After all, the genetic code, being universal, would have evolved at the dawn of life, together with certain components of the translation mechanism, and once the mechanism for translating nucleic acid to amino acid was optimized, there would be selection pressure against any major changes to it. "Any major change in this apparatus would tend to be highly disadvantageous, since the selection pressures for an efficient error-free mechanism of translation must be very great."[104]

Dubnau and coworkers focused on two kinds of RNAs: the RNAs that formed part of the structure of ribosomes, the structures at which the information in nucleic acids was translated into the amino acid sequences of proteins; and the transfer RNAs, the molecules that carry amino acids to the growing polypeptide chain in the ribosome. They conducted molecular hybridization experiments to determine the relationships between eight members of the genus *Bacillus* and concluded that "many properties of ribosomes and sRNA [soluble RNA (see chapter 11)] seem to be phylogenetically invariant in the bacteria."[105] Based on the great amount of hybridization that occurred, they said that the ribosomal RNAs and transfer RNAs were at "the conserved 'core' of genetic material."[106]

That concept of a highly conserved genetic core was shared by Roy Doi and Richard Igarashi at Syracuse University. When, in 1965, they conducted DNA–RNA hybridization experiments on several species of *Bacillus*, their results also indicated that there was "a small number of identical sequences shared by all *Bacillus* species" and that the ribosomal RNA genes "appeared to be more highly conserved relative to other genetic sequences."[107] Moore and McCarthy at the University of Washington concluded similarly in 1967 when they carried out DNA–RNA hybridization experiments with various species of Enterobacteriaceae: "The relative extent of hybrid formation is always greater for ribosomal RNA."[108]

None of it could be trusted, in Sneath's view.[109] One could no more weight such molecular characteristics than one could phenotypic characters. Lateral gene transfer closely mimicked evolutionary convergence. The effects of the two were similar; the only difference was the cause.[110] "Concepts such as that of a conserved core of genetic material resistant to evolutionary change," he commented in 1974, "depend on whether this material is evolutionary conservative or has been transferred by recent gene exchange. Arguments based on the conservatism of functional classes of genes (e.g., ribosomal genes) are likely to prove as unsafe in micro-organisms as in higher organisms."[111] Nothing was certain about bacteria evolution or about molecular phylogenetic method:

> We are faced with four possibilities: a) bacteria have not altered greatly for long periods; b) some parts of the genome are extraordinarily stable (conservative); c) present-day bacteria are descended from a recent ancestor and have not diverged much; d) recent transfer of genes accounts for the recognizable homologies, whereas other older genes are no longer recognizable as homologous at all. There is little critical evidence yet to support any one of these, and all may be partly true.[112]

Still, Sneath was hopeful that a phylogenetic classification might be accomplished in the future with the development of new methods of reading sequences—perhaps by electron microscopes, he thought—combined with the statistical methods of numerical taxonomy.[113] "Even the severe problem of unraveling reticulated evolution may be feasible with abundant data."[114] In the meantime, a new approach to microbial phylogeny was emerging based on comparing partial *sequences* of ribosomal RNA. These studies would revolutionize the field of microbiology taxonomy and challenge some of its most venerable evolutionary assumptions.

| # Roots in the Genetic Code

In a restricted sense, it can be said that the genetic code has been solved. That is to say, we can construct...a table...describing the mapping relationship between the primary structure of any gene and that of its corresponding protein.... The relationships in the table are formal; they convey nothing about the actual manner in which the mapping, the translation, occurs. Neither have they provided definite clues as to what interactions, principles, etc., might underlie these relationships.

—Carl Woese, "The Problem of Evolving a
Genetic Code" (1970)

A NEW APPROACH to microbial classification emerged in the 1970s, far removed from the many-characters method of numerical taxonomy and from the molecular methods of the 1960s based on amino acid sequence, GC values, and nucleic acid hybridization. The research program was based on comparing short sequences within one molecule of ribosomal RNA. It was led by Carl Woese (b. 1928) and collaborators at the University of Illinois, whose methods and concepts revitalized microbial phylogenetics and challenged fundamental assumptions about the primary course of evolution. The methods and concepts they developed revitalized microbial evolutionary biology and microbial taxonomy; they also challenged core assumptions about the course of evolution.

Within the theoretical framework Woese's group constructed, prokaryotes were not a genealogically coherent group that gave rise to eukaryotes. Instead, there were three fundamental forms of life, each representing a distinct primeval lineage and possessing a unique cellular organization. Nor did these lineages take the form of a genealogical tree branching from a common prokaryotic ancestor. Rather, all three lineages emerged from simpler more primitive ancestors, the progenotes, hypothetical ancient life forms in the throes of the evolving relationships between nucleic acid and protein. And in direct conflict with the consensual conception of the origin of life according to which aboriginal organisms were heterotrophs feeding in a rich primordial soup of organic compounds, these phylogeneticists argued that the first organisms were autotrophs that synthesized their own organic compounds.

Code Cracking

Woese had not been brought up within the tradition of microbiology and was oblivious to its tumultuous taxonomic discussions. The roots of his program were embedded in concepts about the origin and evolution of the genetic code. He was a graduate student of the distinguished physicist Ernest Pollard, the founder of biophysics at Yale University. He completed his doctoral research on the inactivation of viruses by ionizing radiation and heat, in 1953, the year in which James Watson (b. 1928) and Francis Crick reported the double-helical structure of DNA. He subsequently entered medical school at the University of Rochester—"for two years and two days," as he recalled.[1] At the beginning of the third year, students were assigned to wards; Woese was assigned to pediatrics. After two days in the pediatrics ward, he quit. Through the kindness of Pollard, he returned to Yale as a postdoctoral fellow; his research focused on the radiation resistance of bacterial spores and the reemergence of ribosomes during spore germination.[2] Five years later, he signed on at General Electric's Knowles Laboratory in Schenectady, New York.[3]

In the fall of 1960, while setting up his lab at GE, Woese turned to one of the most important problems in the molecular biology of the day: cracking the genetic code.[4] DNA was understood to be language or code that used a four-letter alphabet of nucleic acid bases, A (adenine), G (guanine), C (cytosine), and T (thymine), which somehow specifies the amino acid sequence of a particular protein. There were 20 or so amino acids from which proteins were assembled. The question was how a message written in a 20-letter alphabet of amino acids could be encoded in a four-letter alphabet of nucleic acid bases.

It was known as the "coding problem," and in the 1950s it was addressed theoretically as a formal problem, concerned neither with chemical steps nor with the arrangements of atoms in molecules.[5] Was the code based on groups of two, three, or four nucleotides? A one-to-one correspondence between DNA bases and amino acids would not suffice; that obviously would allow the encoding of only four amino acids. Therefore, at least three bases or base pairs were required to specify one amino acid. A triplet code would allow $4 \times 4 \times 4 = 64$ different trinucleotide combinations, more than enough.

Physicist-cosmologist George Gamow (1904–1968), champion of the big bang theory of the universe, started the theoretical work to solve the genetic code in 1954.[6] He suspected that there was a fundamental law of nature that underlay the correspondence between nucleic acid triplet and amino acid, comparable to the complementarity of purine and pyrimidine bases of the DNA double helix: G with C, A with T. As he envisaged it, the genetic information would be "translated" into the specific amino acid sequence of proteins directly from the DNA template. Specific amino acids would fit into what he saw to be diamond-shaped cavities in the surface of the double-stranded DNA structure.[7]

As it turned out, there were exactly 20 such diamond-shaped configurations, corresponding exactly to 20 amino acids, "the magic twenty," as they came to be called.[8] In a purely informational sense, Gamow's model was based on the notion that the code was triplet, but overlapping. Every nucleotide base would therefore form part of three triplets, or, put differently, every nucleotide triplet shared two bases with its adjacent triplet. That diamond code was also "degenerate"—that is, several sets of three letters stood for a particular amino acid. The code contained synonyms.

Though it would not stand up to empirical scrutiny, Gamow's model was met with infectious enthusiasm among prominent physicists who were challenged by the mathematical properties of the code. An overlapping code such as his was highly restrictive; there would always be some forbidden sequences of amino acids. For example, in a fully overlapping code the sequence of, say, ACG could be followed only by four other coding sequences: CGA, CGC, CGG, or CGU. However, as the peptide sequence data accumulated, it became increasingly clear that no such neighbor restrictions existed. Surveys of amino acid sequences published in 1957 by Sydney Brenner (b. 1927), who coined the word "codon" for each nucleotide triplet, ruled out any fully overlapping code.[9] Three years later, Crick and his colleagues provided evidence (from their "frameshift" experiments) supporting a nonoverlapping triplet code.[10]

Cracking the code as a cryptographic problem ultimately meant assigning each amino acid to its respective nucleic acid codon(s).[11] Decoding had captivated Woese; he published a series of theoretical papers on the code in *Nature* in 1961 and 1962. By that time, it was widely assumed that the genetic information from DNA was transcribed to RNA, the close chemical kin to DNA—one of the principal differences between them was that in RNA, uracil (U) replaced thymine (T) as one of the bases. But it was not certain which RNAs, exactly, served as the DNA messenger. Bodies then called microsomes (ribosomes) were known to contain RNA; there were thousands of them in the cell. Biochemists also identified what they called soluble RNA (so named because its relatively small size allowed it to be chemically separated in bulk from other RNAs).

Woese derived a correspondence table based on comparing the GC/AU composition of those soluble RNAs to the GC/AT ratio of DNA. Based on comparisons in three different kinds of bacteria, those RNAs did not have a base ratio even approaching that of DNA.[12] He subsequently tested the hypothesis of Martynas Yčas that, in viruses, one nucleotide, rather than a triplet, encoded one amino acid. Woese reinterpreted Yčas's data and suggested triplet codes for 18 amino acids. Among them, he successfully predicted that CCC encodes proline.[13] By the end of 1961, the theoretical approach to the code was replaced by an experimental one, and the cryptographic aspect of the code began to be resolved.[14]

The beginning of the end of the theoretical phase occurred when Marshall Nirenberg (b. 1927) and Heinrich Matthaei (b. 1929) at the National Institutes of Health in Bethesda, Maryland, announced that the synthetic RNA poly-U translates into polyphenylalanine.[15] It was called "the U-3 incident"—in reference to the U-2 crisis that occurred the year before when an American U-2 spy plane was shot down over the Soviet Union. Nirenberg and Matthaei's experiments involved preparing a cell-free extract from *E. coli* that seemed to synthesize protein/peptides when they added an external source of RNA. At first they used inputs such as viral RNA to stimulate their system, but ultimately they used the simple synthetic RNA UUU as an input, and to their surprise, even it had the capacity to stimulate peptide formation. To sort this out, they refined their experiment by testing individually labeled amino acids one by one. When they added the synthetic UUU RNA to the extract containing the radioactively labeled amino acid phenylalanine, the resulting "protein" was composed solely of phenylalanine.

The Genetic Code

		U	C	A	G	
U		UUU UUC] Phe	UCU UCC UCA UCG] Ser	UAU UAC] Tyr	UGU UGC] Cys	U C
		UUA UUG] Leu		UAA UAG] Stop	UGA] Stop UGG] Trp	A G
C		CUU CUC CUA CUG] Leu	CCU CCC CCA CCG] Pro	CAU CAC] His	CGU CGC CGA CGG] Arg	U C
				CAA CAG] Gln		A G
A		AUU AUC] Ile AUA]	ACU ACC ACA ACG] Thr	AAU AAC] Asn	AGU AGC] Ser	U C
		AUG] Met		AAA AAG] Lys	AGA AGG] Arg	A G
G		GUU GUC GUA GUG] Val	GCU GCC GCA GCG] Ala	GAU GAC] Asp	GGU GGC GGA GGG] Gly	U C
				GAA GAG] Glu		A G

Figure 11.1 Table of the genetic code.

Over the next few years, many similar experiments were done using other synthetic RNA inputs (random sequences of simple polymers such as UUC, UUG, and UUA). In 1964, Nirenberg announced that he and Philip Leder had devised a new and more powerful decoding technique. Within a year, the cryptographic problem of the code was essentially solved; all 64 possible triplet codons had been assigned to their corresponding amino acids (figure 11.1).[16]

On the Origin of Translation

Woese was a visiting researcher for four months at the Pasteur Institute in Paris in 1962. At that time, molecular biologists were beginning to form a consensual outline of *how* the information in nucleic acids was translated into the amino acid sequences of proteins. Led by Jacques Monod (1910–1976) and François Jacob (b. 1920), the Pasteur Institute emerged as a major center for research on the molecular mechanisms of gene expression and gene regulation. In 1961, Monod and Jacob published their famous "operon theory" of gene regulation in bacteria.[17] That year, Brenner, Jacob, and Mathew Meselson made another fundamental contribution to understanding protein synthesis by distinguishing what they had called "messenger RNA" (mRNA) that carried the encoded information of DNA to the site of protein synthesis in the cytoplasm.[18]

By 1962, a new outline of protein synthesis was formulated, with names depicting the special functions of the different types of RNAs in protein synthesis. Each DNA gene is *transcribed* onto mRNA molecules, which carry information for the synthesis of a specific protein from DNA to ribosomes. The mRNA temporarily associates with a ribosome, where the nucleic acid message is *translated* into the specific amino acid sequence of a protein.[19] Translation occurs when other RNAs called *transfer* RNAs (tRNAs) each carrying a specific amino acid, line up on its corresponding codon on the mRNA template; the amino acid polymers are processed to form specific proteins by the ribosome.[20]

Woese had begun to think of the genetic code in evolutionary terms: how codon assignments might have originated, and how the translation from nucleic acid to amino acid sequence might have evolved.[21] Chemists, led by Stanley Miller's experiments of 1953, showed how certain amino acids might have been synthesized in a presumed abiotic soup billions of years ago, as postulated by Oparin and Haldane.[22] That was important, but it hardly settled the issue of the origins of life. How the amino acid sequences of proteins and the corresponding sequences of DNA and RNA had evolved and come together in the appropriate way to make a functional protein was much more difficult to fathom. Why did three nucleic acid units (trinucleotides) "mean" one specific amino acid and not another?"[23]

In 1962, Sol Spiegelman (1914–1983) passed through Paris looking for a molecular biologist to fill a vacant position at the University of Illinois in

Champaign-Urbana. Spiegelman had studied gene–enzyme relations in yeast during the 1940s, at which time he argued that genes produced partial replicas of themselves called "plasmagenes" that entered the cytoplasm and controlled the rates and kinds of protein enzymes synthesized.[24] Subsequently, he turned to bacteria and its viruses to investigate the relations among DNA, RNA, and proteins. In 1961, he and Benjamin Hall developed a DNA–RNA hybridization technique using *E. coli* and bacteriophage T_2 to demonstrate "the existence of complementary RNA and its possible role as a carrier of information from the genetic material to the site of protein synthesis."[25] It was a messenger like that conceived by Brenner, Jacob, and Meselson.

In Paris, Spiegelman and Woese had a discussion about the evolution of the code and Woese's aim to find specificity in the nucleic acid–amino acid interaction. The following fall, Spiegelman invited Woese to the University of Illinois for a visit. He was offered a position with immediate tenure to begin in 1964.[26] That position gave him the freedom to pursue high-risk problems far off the main paths of biology. Why did UUU or UUC encode phenylalanine, and CCC proline? "While it is important to know what the genetic code codon assignments are," Woese wrote in 1965, "it is more important to know *why* they are, i.e. to know the mechanisms giving rise to the particular assignments observed. Only when the latter question is answered can we truly claim to begin to understand the genetic code."[27] Though one could not see physical evidence for why nature had adopted this particular correspondence rather than another, he reasoned that perhaps primitive organizations had some constraints of structure that molecular biologists knew nothing about. Nucleic acid—amino acid relationships may have existed that forced the genetic code to evolve as it did.[28]

Frozen Accident Theory

Fundamental principles of biology were to be discovered from an understanding of the evolutionary relationship between codon and amino acid, in Woese's view. "There seems little awareness," he wrote in 1970, "that the concepts that will eventually emerge here may be as novel to the present biological thinking as was the concept of the gene to the biology of its day."[29] There were essentially two reasons for this lack of interest, as he saw it. One was technological: no experimental system had been developed for examining *why* codon assignments exist. The second was conceptual: the common view was that the relationship between a codon sequence and amino acid had emerged by pure chance; codon assignments were meaningless—the codons assigned to amino acids were due to a series of historical accidents. There was no causal relationship between an amino acid and its codons—in other words, were the genetic code to evolve again (under the same conditions), the codon assignments would be different.[30]

This was called the "frozen accident theory." It had become "one of the central dogmas of the coding field," as Woese saw it, and it predicted the answer to the "why" question "to be entirely uninteresting and trivial."[31]

Crick was its chief advocate. A genetic code was conceived of as a dictionary of words, and as he explained in 1968, even if it were due to chance, once the relations between sign and meaning were established, they would be difficult to change:

> The frozen accident theory states that the code is universal because at the present time any change would be lethal, or at least very strongly selected against. This is because in all organisms (with the possible exception of certain viruses) the code determines (by reading the mRNA) the amino acid sequences of so many highly evolved protein molecules that any change to these would be highly disadvantageous....

This accounts for the fact that the code does not change. To account for it being the same in all organisms one must assume that all life evolved from a single organism (more strictly, from a single closely interbreeding population). In its extreme form, the theory implies that the allocation of codons to amino acids at this point was entirely a matter of "chance."[32]

Before turning to microbial phylogeny as a framework for understanding the evolution of the code, Woese confronted a related conception of Crick's: how the translation from nucleic acid message into protein structure works. Called "the adaptor hypothesis," it remains a pillar of molecular biology today. It was born of the frozen accident theory and was embodied in the accepted function of tRNA, the adaptor that carries amino acids to the ribosomes. It is based on the assumption that nucleic acids could not in any way recognize amino acids. Each amino acid would have to be first recognized by a specific protein enzyme before it could be attached to a specific tRNA molecule, which then carries the amino acid to the ribosomes.

Crick proposed his "adaptor hypothesis" informally in 1955 in an open letter to Gamow's "The RNA Tie Club," and then published it in 1958.[33] The mechanism he proposed was based on templating, the concept that explained how the double helix of DNA replicated and was transcribed to RNA. Cleave or unzip the hydrogen bonds between base pairs (between G and C, and between A and T) that hold the strains together, and each strand would act as a template for the formation of a complementary strand. There were two aspects to templating: recognition, as manifested in the complementarity (like lock and key) of nucleic acid basis, and alignment, by which the units were juxtaposed in a way that promote their being joined together.

It was unlikely that nucleic acids in themselves could recognize the appropriate amino acid and template them in protein synthesis, Crick reasoned. Nucleic acids could recognize only nucleic acids, not amino acids. It was a "naive idea" to suggest that RNA would "take up a configuration capable of forming twenty different 'cavities' one for the side chain of each of the twenty amino acids."[34] To resolve that problem, he conjectured that each amino acid would be outfitted with its own small "adaptor" molecule before it could be keyed into its corresponding codon along the mRNA.[35]

Each adaptor would contain two or three nucleotides so as to "enable them to join to the RNA template by the same 'pairing' of bases as if found in DNA."[36] There would be 20 adaptors, one for each amino acid. "If the adaptors were small molecules," Crick said,

> one would imagine that a separate enzyme would be required to join each adaptor to its own amino acid and that the specificity required to distinguish between, say, leucine, isoleucine and valine would be provided by these enzyme molecules instead of cavities in the RNA. Enzymes, being made of protein, can probably make such distinctions more easily than can nucleic acid.[37]

One of the key points of the adaptor hypothesis, Crick said, was that "it meant that the genetic code could have almost *any* structure, since its details would depend on which amino acid went with which adaptor. This had probably been decided very early in evolution and possibly by chance."[38]

In textbooks and classrooms since the 1960s, it is said that Crick's adaptor hypothesis was confirmed when "transfer RNA" (formerly known as soluble RNA) was discovered: tRNA served as the adaptor that Crick had so presciently foreseen.[39] The discovery of tRNA, and the activating enzymes that catalyze the linkage of that RNA molecule to its corresponding amino acid during protein synthesis, was made by biochemists Mahlon Hoagland and Paul Zamecnik in 1955.[40] Although it has been said that tRNA represented the confirmation of the adaptor that Crick had envisaged, that claim is flawed.

Actually the size and molecular complexity of tRNA were nothing like what Crick had proposed or would have required for his model. Whereas he had pictured adaptors to be the size of about three nucleotides in length, just enough to recognize a codon, tRNA molecules were large enough to contain between 25 and 200 nucleotides.[41] In fact, Crick was initially skeptical that those molecules *could* be the adaptors that he had hypothesized. As he commented in 1958, they were "too short to code for a complete polypeptide chain, and yet too long to join onto template RNA...by base pairing."[42]

The next year, Hoagland noted "several discrepancies and peculiarities" that would not have been anticipated by the adaptor hypothesis.[43] The large size, "some 25 to 200 nucleotides," he said, "seems a rather large molecule if its sole function is to code for a single amino acid. One is tempted to guess that the large molecule serves a purpose unknown."[44] He noted other peculiarities in the base composition of tRNA when compared to other RNAs. "These intriguing mysteries notwithstanding," he said, "it may be concluded that the adaptor hypothesis has thus far stood up satisfactory in the face of several experimental tests of its validity. It is safe to say that it remains an adequate and useful framework in which to design new experiments to further the understanding of protein synthesis."[45] Thus, the adaptor hypothesis was coupled with tRNA; any difficulties with their fit were effectively dropped.[46]

The adaptor–template hypothesis was antithetical to Woese's search for an evolutionary bridge between the gene and protein, and the reason for codon assignments.[47] As he saw it, the adaptor model not only was based on an erroneous assumption about the way codon assignments originated, as "frozen accidents," but was belied by the anomalous size and complexity of tRNAs. Woese doubted that tRNA functioned passively merely as a static adaptor, and he developed an alternative view based on the idea that tRNA was active and, like protein enzymes, worked according to different functional states.

Molecular Metaphysics

Molecular biology was an engineering discipline ensconced in misguided reductionism and devoid of evolutionary explanation, as Woese came to see it in the early 1970s.[48] It tended to treat the translation apparatus as a given, a *machina ex deus*.[49] If the principles of molecular biology were correct, he commented in 1972, "evolution would indeed be a mixed bag of peculiarities, a basically unrelated collection of 'historical accidents,' an unordered wandering though an immense evolutionary phase space."[50] In his view, to conceptualize the evolution of translation as an integrated whole would require "nothing short of a metaphysical purge, a rejection of the reductionist materialism that pervades biology today."[51]

Woese pointed to the holistic philosophy of Alfred North Whitehead, who sought "a system of thought basing nature on organisms not upon the concept of matter."[52] Between "the *material* on the one hand, and on the other hand *mind*," Whitehead wrote in 1925, "there lie the concepts of life, organism, function, instantaneous reality, interaction, order of nature, which collectively form the Achilles' heel of the whole system."[53] For Woese, Whitehead's philosophy of organism stressed "*process* as fundamental, a view in which existence and evolution begin to fuse," he commented in 1972. "In terms of a Whiteheadian metaphysics, evolution seems to define itself as the problem of how order at one 'level' of the universe relates to order at the adjacent level."[54]

During his first years at Illinois, Woese wrote *The Genetic Code: The Molecular Basis for Genetic Expression* (1967). Therein, he envisaged translation in terms of cellular tape reading. Information molecules were the tapes, and the molecules that bring about the transfer of information were tape readers.[55] According to the adaptor hypothesis, the ribosomes did the reading. Shuffling adaptors in and out, the ribosomal particles were visualized as moving from end to end of the mRNA like the reading head of a tape recorder passing over the tape. The protein chain was thus synthesized in a stepwise fashion from one end to the other. In Woese's model, the tRNA played the active role in tape reading, and "it could well have a say in specifying the amino acid it is to carry."[56]

To appreciate those possibilities Woese speculated on how the code might have evolved. The modern translation apparatus was complex, requiring participation of a set of activating enzymes (later called aminoacyl tRNA synthetases), the set of tRNAs, the ribosome, and the mRNAs.[57] But a primitive system would have lacked activating enzymes and mRNA. He suspected the proto-tRNA probably functioned like an enzyme in a close relationship with amino acids. tRNA was an odd and elaborate molecule, "a kind of molecular misfit," as he put it. It possessed a complicated (cloverleaf) molecular geometry, almost as complicated as a protein. Indeed, he said,

> tRNA resembles a protein more than an RNA.... On the grounds that it is easier to adapt an existing structure than to evolve one *de novo*, it could be argued that tRNA owes its present role more to the fact that some tRNA ancestor was present in translation very early than to the fact that tRNA is peculiarly suited to the role that it now has.[58]

The proto-tRNA might have possessed one site for selecting amino acids and one for holding them. At first, a peptide chain might have grown simply by adding and bonding one amino acid to another of the same or related kind (without any message reading), resulting in a homopeptide. In a subsequent step in evolution, amino acids would be transferred from one proto-RNA to another, resulting in various mixed polypeptides.[59] This would not represent translation per se. That would occur when amino acids were "aligned on a nucleic acid template by codon-amino acid pairing."[60] That "aboriginal RNA" would have been "some evolutionary precursor" of ribosomal RNA. "In fact," he wrote in 1967, "proto-RNA could have been the original genome." tRNA would have evolved later.[61]

In 1970, Woese proposed an alternative to the adaptor hypothesis, which he called the "reciprocating ratchet mechanism," according to which tRNA would undergo conformational changes that pulled mRNA through the ribosome during translation.[62] According to the adaptor model, tRNA was passive and appeared in only one conformational form, and the ribosome was the actual translation mechanism that took the process down the assembly line. The ratchet model reversed that: the essential mechanism of translation was defined allosteric transitions in the arms of tRNA; the ribosome served only to enable and refine the mechanism inherent in the tRNAs. This was how translation had evolved and works today, Woese wrote in 1972. "The 'problem of the ribosome' is precisely the 'problem of the *evolution* of the ribosome.'...The movement that pulls the mRNA tape in translation was and still *is* an inherent property of a simple tRNA machine."[63]

Thus, Woese theorized on what he called the "era of nucleic acid life," wherein programmed protein synthesis gradually emerged.[64] It was an ancient evolutionary era when biopolymers, including simple polypeptides flourished "but translationally produced proteins had yet to arise, it was an era dominated by nucleic acids."[65] The translation apparatus would have evolved from interactions between RNA and amino acids before the origin of ribosomes. In such a primitive translation apparatus, codon recognition and the ordering of amino acids in protein synthesis would have been highly imprecise, and accomplished using only a few primitive tRNAs.[66] The conception of RNA functioning like an enzyme was fundamentally at odds with the canons of classical molecular biology until the mid-1980s, when studies demonstrated that some RNAs have catalytic properties.[67] Molecular biologists then began to speak of an ancient "RNA world."[68]

Woese's book *The Genetic Code* appeared two years after Watson's *Molecular Biology of the Gene* (1965). Watson's became standard reading in courses, as did its sequel, *Molecular Biology of the Cell* (1983). It helped to set the conceptual framework for molecular biology and mapped the perimeters of the field. It made virtually no reference to evolution in its 494 pages, except to say on the first two pages that humans and apes have a common ancestor, that bacterial-like life evolved billions of years ago, and that "evolutionary theory further affects our thinking by suggesting that the basic principles of the living state are the same in all living forms."[69] Evolution was not considered to be important for molecular biological understanding. Theodosius Dobzhansky's famous dictum of 1973, "Nothing in biology makes sense except in the light of evolution,"[70] did not seem to apply. And leading molecular biologists had come to understand that molecular biology was virtually complete, finalized; there would be no new principles forthcoming, only details to work out and new technologies to apply.[71] There were but a handful of papers discussing the evolution of the genetic code.[72]

The quest to find life's origins and the evolution of the genetic code seemed to be quixotic. "But then how did it all begin? And with what?" asked François Jacob in *The Logic of Living Systems* in 1970:

> The genetic message can be translated only by the products of its own proper translation. Without nucleic acids, proteins have no future. Without proteins, nucleic acids remain inert. Which is the hen, which the egg? And where can traces be found of this precursor, or some precursor of the precursor? In some still unexplored corner of the globe? On a meteorite? On another planet of the solar system? Without any doubt the discovery somewhere or other, if not of a new form of life, at least of somewhat complex vestiges, would be priceless. It would transform our way of envisaging the origin of genetic programmes. But as time passes, the hope of this diminishes.[73]

The evolution of the genetic code lay beyond the realm of science. Jacob put it in a nutshell:

> If the genetic code is universal, it is probably because every organism that has succeeded in living till now is descended from one single ancestor. But it is impossible to measure the probability of an event that occurred only once. It is to be feared that the subject may become bogged down in a slough of theories that can never be verified. The origin of life might well become a new center of abstract quarrels, with schools and theories concerned, not with scientific predictions, but with metaphysics.[74]

Need for a Phylogenetic Framework

By the time Woese completed *The Genetic Code*, he had come to understand that one could not study these problems of deep evolution without a phylogenetic framework. With methods for a universal phylogenetic system, one might be able to follow the translation machinery's evolution back to a stage before cells reached their present sophisticated complexity. A universal tree would therefore hold the secret to its own existence, as well. Woese explained his intentions in a letter to Crick in the summer of 1969:

> If we are ever to unravel the course of events leading to the evolution of the prokaryotic (i.e. simplest) cells, I feel it will be necessary to extend our knowledge of evolution backward in time by a billion years or so— i.e. backward into the period of actual "Cellular Evolution." There is a possibility, though not a certainty, that this can be done by using the cell's "internal fossil record"—i.e., the primary structures of various genes. Therefore, what I want to do is to determine primary structures for a number of genes in a very diverse group of organisms, on the hope that by deducing rather ancient ancestor sequences for these genes, one will eventually be in the position of being able to see features of the cell's evolution—i.e., by knowing what features of the primary structures are "locked-in," what regularities (repeats, etc.) existed, and how one ancient primary relates to another ancient primary structure(s) [*sic*] (which gave rise to some different cellular function).[75]

Woese set out with the hope of tracing cell life back to some kind of universal ancestor that might not possess the modern translation machinery.[76] The study of evolution and phylogeny necessarily required a comparative method. Cut and dried in a textbook table in which 20 encoded amino acids are matched with symbols for the 64 possible triplet codons, there were few if any differences in the code of all organisms. There was virtually nothing to compare for phylogenetic purposes in that catalog of codon assignments. There was nothing to discern about its evolution except that perhaps the common ancestor of all organisms possessed the same code. But the black-box conception of the code was deceptive. Understood as process, the genetic code might not be universal at all or an "event" that occurred only once. One might be able to discern differences in the mechanisms of translation among such widely divergent organisms as prokaryotes and eukaryotes. Based on such ancient differences, one could make inferences about the translation machinery of their common ancestor.

There was, of course, an alternative premise—that such differences in the translation apparatus might not be evidence of an ancestral state but rather evidence that prokaryotes and eukaryotes had altered their coding systems after they diverged from a common ancestor, which itself possessed a fully evolved genetic code. That possibility was less plausible, Woese argued, because once formed, such a system of extreme intricacy would not easily be subjected to appreciable change.[77] Once evolutionarily optimized, it would be highly conserved and not easily modified.

In order to bring the evolution of the code into connection with the evolution of the cell, the prokaryote and the eukaryote concepts had to be revised. That split had been made on grounds of cytological visible cell structure: the eukaryote possessed a membrane-bound nucleus and cytoplasmic organelles, the mitochondria (the chloroplasts in plants); the prokaryote lacked these structures. Woese would begin what would be a lifelong task of conceptualizing them in terms of fundamental differences in their translation machinery.

In 1970 Woese pointed to differences in the gross composition of the ribosomal RNAs in eukaryotes and prokaryotes, as well as differences in antibiotic sensitivity and a striking difference in a particular "loop" in the tRNA molecule.[78] From this molecular evolutionary perspective, he suspected that (1) the split between eukaryotes and prokaryotes was much more ancient than biologists typically assumed, and (2) prokaryotes did not give rise to eukaryotes in the accepted sense, but rather the two main lineages had diverged much earlier from a nonprokaryotic lineage. The fundamental differences in translation apparatus between the two life forms, he wrote in 1970, suggest "that the final stages in the evolution of the genetic code may have occurred independently in the two lines."[79]

Back to Evolution's Core

Much of the conceptual ground work for the program had been laid out by 1967. It was then a matter of putting it into experimental effect. The key to a deep microbial phylogeny was in choosing the right molecule for the job. Woese looked to those RNAs that, together with proteins, comprised ribosomes. Ribosomal RNA had universal attributes. The cells of all organisms from bacteria to elephants needed them to construct proteins. Ribosomes were also abundant in cells; there were thousands of them in each cell, and their RNA was easy to extract. The ribosome was the ideal cell structure for following the course of evolution: if it could be followed at the level of molecular sequence differences.

The RNA of ribosomes was also far removed from the usual vicissitudes of phenotypic characters. They would be among the most "conserved" elements in all organisms (evolving far more slowly than most proteins) and therefore would make the best recorders of life's long evolutionary descent. Recall that DNA–RNA hybridists of the 1960s maintained that the ribosomal RNAs and tRNAs, as fundamental to the translation machinery, were at the "core" not subject to the vagaries of adaptive evolution. But Woese's approach differed: it would be based on actual nucleotide sequences, not on overall similarity of sequences.

Two events occurred in the 1960s that were vital to Woese's research agenda. First, in 1965 Frederick Sanger and collaborators announced methods for sequencing and cataloging oligonucleotides, short fragments of RNA (Greek *oligo*, few or small). They applied them to ribosomal RNA and to tRNA of *E. coli* and yeast. They noted, for example, that some common sequences, such as GCUCAG, were present in more than half of the different tRNAs in *E. coli*. It is not common, however, in the tRNAs from yeast, and that was "one of most obvious distinctions between them."[80] Their method could thus be used as a "fingerprinting technique to characterize RNAs and perhaps to detect small differences between species and mutants."[81]

Second, in 1968 Spiegelman invited one of Sanger's graduate students, David Bishop, to work as a postdoctoral fellow to set up Sanger's system for sequencing viral RNA in Spiegelman's laboratory in Urbana. There was vigorous research on RNA in the department of microbiology at that time—recall that Spiegelman's team had created the hybridization methods for studying the complementarity between DNA and ribosomal RNA in 1961. Over the next few years, he and his team improved and simplified the DNA–RNA hybridization technique, making it accessible to a wider circle of researchers.

The sequencing of ribosomal RNA was the logical next step. A year after Bishop's arrival, Spiegelman accepted a position as director at the Institute of Cancer Research at the Columbia University College of Physicians and Surgeons, where he would explore the possible role of RNA tumor viruses in certain human cancers. When Spiegelman left for New York in 1969, Woese asked his student, Mitchell Sogin, who was just beginning his doctoral research, to learn the RNA sequencing techniques from Bishop before he left. Woese then inherited Spiegelman's setup, and established a vibrant team of students and technicians.

The sequencing technology was slow and arduous, but Woese's team had the field virtually to themselves for a decade. Using the Sanger technique, many small pieces of an RNA molecule could be electrophoretically separated and thus cataloged. The long RNA molecule was broken into small fragments several nucleotides long by cutting at every G residue with the enzyme T_1 ribonuclease. Those fragments were then purified by electrophoresis (a technique for separating molecules on the basis of electrical charge). Each of the fragments of significant size was, if needed, then broken into subfragments with enzymes that cleaved at some other nucleotide(s). This method allowed them to reconstruct the nucleotide sequence of each original ribosomal RNA fragment.

They made catalogs of oligonucleotide sequences for various taxa and compared them to one another. It was like cutting a book into isolated words and ending up with word lists. From those lists, one could discern what words are characteristic of, unique to, the ribosomal RNA for that particular group of microbes. Sogin was crucial in getting the laboratory work up and running. He prized technology and pushed Woese to increase the capacity of the laboratory (from two to six Sanger tanks). He designed and built a translucent Plexiglas wall transilluminated by florescent lights behind it for reading the sequences. The RNA fragments were visualized as fuzzy spots on film, which Woese learned to "read."[82]

The work required considerable expertise, and it was expensive. To obtain the radioactive RNAs, the bacteria had to be grown in media from which the phosphate was removed as much as possible and to which high levels of ^{32}P-phosphate were added. The radioactive rRNA was then separated from the other RNA components by polyacrylamide gel electrophoresis. Once isolated, the rRNA could then be digested with a particular enzyme into smaller pieces and then catalogs—lists of oligonucleotides from diverse organisms—were compared. The more similar the catalogs were, the more closely related were the taxa. During the first few years, Woese and his technician Linda Bonin worked to improve the technique. It took several years of work to get to the point where it was really useful.[83]

Ribosomes are composed of two subunits: a smaller one slightly cupped inside a larger one; both contain RNA and protein. The smaller subunit of ribosomes contains 16S rRNA (about 1,540 nucleotides long), which forms the scaffolding onto which ribosomal proteins are attached and positioned; the larger subunit is composed of 23S rRNA (about 2,900 nucleotides long) and the much smaller 5S rRNA (about 120 nucleotides), to which some 35 ribosomal proteins are bound altogether. (S, for Sverberg units, refers to the relative rate at which a molecule of given shape and molecular weight sediments in an ultracentrifuge.)

They began with 5S rRNA of the large subunit, but the RNA of the small subunit (referred to as SSU rRNA, or as 16S rRNA) proved optimal for the available technology.[84] The oligonucleotide fragments were usually quite short; they ranged in size from 5 to 20 nucleotides and up, but they were long enough that almost all oligonucleotides of six or more nucleotides occurred only once in a typical rRNA molecule. So Woese could search for matching oligonucleotides (homologs) in different bacteria to determine how closely two 16S rRNAs were related. The "comparative cataloging" approach provided information from only about 35% of the 16S rRNA. Sequencing of the entire 16S rRNA gene was not feasible until the late 1970s.[85]

Woese got bacteria from anywhere he could. He sought microbiologists, one by one, to team up with him, grow bacteria radioactively, and send them as frozen cell pellets. Some bacteria were exotic enough that only a handful of microbiologists could easily handle them. Microbiologists cooperated, though many were reluctant to use the extremely high levels of radioactive materials that were required.

The collaboration between Woese and George Fox (b. 1945) was especially vibrant for several years. Fox had completed his doctoral work in chemical engineering in Woese's home town, at Syracuse University.[86] In the winter of 1973, when he was finishing up his doctoral work, he decided to abandon chemical engineering. He wrote to James Watson expressing his interest in a career in biological research, especially theoretical biology. Watson suggested that he apply to one of the summer courses at the Cold Spring Harbor Laboratory where Watson was director.[87] Fox followed his advice, but instead enrolled in the microbial ecology summer course at Woods Hole Marine Biological Laboratory. The lead instructor was renowned microbiologist Holger Jannasch, best known today for his work on deep sea microbiology.[88] Fox wrote to Woese after reading one of his review papers on the ratchet and the evolution of translation.[89]

When Fox arrived in Woese's laboratory, he carried out a formal analysis of the data, and produced the first phylogenetic trees, or dendrograms: branching diagrams showing clustered hierarchies of taxa. He developed a similarity coefficient, S_{AB}, to characterize sequence similarity. The nucleotides in the sequences that were common between catalogs from organisms A and B were summed, and that sum was multiplied by 2. Then *all* the nucleotides in all the oligonucleotides in the same two catalogs were added up, and that sum was divided into the previous sum to arrive at the similarity coefficient: S_{AB} = (total nucleotides in shared sequences × 2) ÷ (sum of all nucleotides). If two catalogs were identical, the coefficient would be 1.0. As two organisms become more distant phylogenetically, S_{AB} decreases—down to a small number. Woese also classified organisms by grouping them by what he called "signatures." "Good signatures" were obvious. If he found an oligonucleotide sequence that was present in, say, about 95% of some cluster of bacteria and was found almost nowhere else, then that oligonucleotide would be a characteristic of the group; all oligonucleotides of this nature would then compose a signature for that cluster.

By 1976, Woese's lab had characterized and cataloged the 16S rRNA of a broad range of about 30 different kinds of bacteria, including enterobacteria, spiral bacteria, cyanobacteria, and the purple bacteria.[90] Each group had a distinctive signature. There were also highly conserved regions of 16S rRNA, some of which were shared by more than 90% of the bacteria taxa examined, and other signatures that were shared by, say, only 65% of them. Other sections of 16S rRNA were highly variable.[91] These two types of regions were, as Woese put it, comparable to an hour hand and a minute hand in a phylogenetic clock. The highly conserved regions would be used to measure distant phylogenetic relationships, and the variable regions to determine close relations.

Then, in 1977, Woese and Fox made the electrifying claim that prokaryotes were not a genealogically coherent group. They had discovered bacteria that did not have the telling prokaryotic signature. They named them "archaebacteria" and said that they were no more related to other bacteria than to eukaryotes.[92]

TWELVE | # A Third Form of Life

But with his departure an era had definitively come to an end; new,
epoch-making finds of the kind that Beijerinck had made, could scarcely be
expected any more. A few years later Kluyver himself was to formulate this
situation by saying that the discovery of a truly novel type of bacteria would
cause no less a sensation than that which the "Loch-Ness monster" threatened
to do at that time.

—A.F. Kamp, J.W.M. La Rivière, and W. Verhoeven,
Albert Jan Kluyver: His Life and Work (1959)

Dividing the living world into Prokaryotae *and* Eukaryotae *has served, if*
anything, to obscure the problem of what extant groupings represent the var-
ious primeval branches from the common line of descent. The reason is that
eukaryote/prokaryote is not primarily a phylogenetic distinction, although it
is generally treated so.

—Carl Woese and George Fox, "Phylogenetic Structure of the
Prokaryotic Domain: The Primary Kingdoms" (1977)

THE 16S RRNA RESULTS were received quietly until the announcement in 1977
of another primary kingdom of organisms, the Archaebacteria. It was a diverse
group of odd organisms dispersed throughout bacterial taxonomy: methanogens, a morphologically varied group that live in anaerobic environments where
even trace amounts of oxygen were lethal to them; extreme halophiles, which

live in brines five times as salty as the oceans; and extreme thermophiles, which live in geothermal environments that would cook other organisms. The conceptualization of this urkingdom figures as one of the key developments shaping the history of modern microbial evolutionary biology.

Cultivating the Unusual

Woese requested advice from many microbiologists about what organisms to analyze; and some agreed to send radiolabeled cell masses to classify with the new technology. As his group became sure of the methodology, they looked far and wide to examine diversity within the bacteria world. Extreme halophiles had been on Woese's to-do list since 1973; they were characterized in August 1977. Mycoplasmas, organisms that lack cell walls, were of great interest. In 1975, Woese began collaboration with Jack Maniloff at the University of Rochester to grow them in radioactive batches. They made a list of six mycoplasmas for testing; among them were the thermophilic *Thermoplasma*, characterized in December 1977. The first organisms to be classified as archaebacteria were the methane-producing bacteria. Their phylogenetic analysis began in June of 1976.

Methanogens were found in a range of anaerobic niches: from the gastrointestinal tract of humans and animals, to marshes and sewage treatment plants, to oceans. They were by and large chemo-autotrophs that derived their energy by reducing carbon dioxide to methane using electrons from hydrogen. Woese's colleague Ralph Wolfe had been studying the biochemistry of methanogens for a number of years; he recommended them to Woese in 1974. Wolfe had a classical microbial-biochemical education, completing his Ph.D. in 1953 on the enzymes and metabolic steps involved in the oxidation of pyruvate and the production of ATP in *Clostridium butyricum*, a species common in soured milk, fermented plant substances, and soil.[1] His turn to study the biochemistry of unfamiliar organisms resulted directly from his appointment that year in the Department of Bacteriology at the University of Illinois. Spiegelman, Luria, and I.C. Gunsalus had arrived three years earlier, and with H. Orin Halvorson, and Elliot Juni, the department was one of the best in the country.[2]

From the outset, Wolfe was encouraged to fit into a specific niche within the department. He recalled a critical moment when Halvorson, the department head, conveyed to him one of the great truths of academic life: "I just want to tell you one thing," he said. "You are paid to teach; you get promotions for doing research."[3] The department wanted someone who had a real interest in bacterial diversity and who would want to teach the kind of microbiology course that van Niel taught at Stanford University. "The message was clear," Wolfe recalled: "I had better begin visibly studying unusual organisms or I did not have a future at Illinois."[4] Van Niel accepted him as an observer in his famed summer course at Pacific Grove in 1954.

Wolfe turned to the methanogens six years later. Though they were known for decades, detailed knowledge of their biochemistry had lagged because, as strict anaerobes living on H_2 and CO_2, they were difficult to cultivate. Wolfe focused on *Methanobacillus omelianskii*. In 1963 his lab showed that pyruvate provided electrons, carbon dioxide, and ATP for the production of methane.[5] But there was a problem: they were not certain that they were dealing with a pure culture. Slightly differing cell shapes were sometimes seen in cultures, and it was possible that they were dealing with two species. Horace Barker at the University of California–Berkeley pioneered the research on the basic chemical pathway of methanogenesis using *M. omelianskii*.[6] He obtained it from mud in the canals outside Kluyver's laboratory in Delft. In 1940, when Barker reported that methane production resulted from the reduction of carbon dioxide, he said that he used pure cultures.[7] In 1965, Wolfe teamed up with Marvin Bryant from the Department of Dairy Science at the University of Illinois to resolve the issue. Bryant had specialized in the microbiology of ruminants. He was responsible for reclassifying the methanogens in the eighth edition of *Bergey's Manual* of 1974, bringing them together as a group based on their biochemistry regardless of their morphological diversity.[8]

As it turned out, *M. omelianskii* was not one kind of organism; it was a mixed culture, a symbiosis or consortium of two unrelated organisms: one produced hydrogen, which was transferred to the other, a methanogen.[9] That symbiosis was a mixed blessing, as Wolfe came to see it. On the one hand, with knowledge of interspecies hydrogen transfer, they had discovered "one of the first principles of anaerobic microbial ecology." On the other hand, the roof of his research program "more or less collapsed"[10]—five years of research needed to be reinterpreted. Enzymes had been isolated, and characterized, but it was uncertain to which organism they belonged. The bulk of Wolfe's students' results could not be published until they developed a technology for growing pure cultures of methanogens on H_2 and CO_2.

Then, in 1971 Wolfe and his student Barry McBride discovered a unique coenzyme involved in the formation of methane; they called it "coenzyme M."[11] They determined its structure and learned that it was a "vitamin" essential for *Methanobacterium ruminantium* that lived in the rumens of cows.[12] The next step was to determine how widely distributed the coenzyme was in nature. To establish what other organisms possessed the coenzyme, they would have to grow different organisms in sealed tubes and then add specific amounts of the coenzyme for assay. The sealed tubes would have to be opened and replenished with hydrogen and carbon dioxide twice a day for four or five days. Because of the negative pressure developed inside the tubes, it was difficult to prevent contamination by oxygen and airborne bacteria when the stopper was removed.

To overcome the problems associated with negative pressure, Wolfe suggested to a new doctoral student, William Balch, that he should try to develop a system in which cells could be grown in a pressurized atmosphere. Balch invented a whole new technique in 1976: he pressurized the hydrogen and carbon dioxide to two atmospheres and inoculated the growth medium with *M. ruminantium* by use of a syringe inserted through a rubber stopper, capped with an aluminum seal like a serum bottle. As the organism used H_2 and CO_2, the atmosphere could be repressurized aseptically.[13] In Balch's hands, the coenzyme M growth-dependent assay became routine. Over the next two years, Wolfe's laboratory tested a wide range of organisms.

To their dismay, coenzyme M was not widely dispersed in nature; it was present only in methanogens. Wolfe had been imbued with the concept of the biochemical unity of life—that a core of essential biochemical processes, common to all organisms, appeared early in the evolution of life.[14] Balch's results were disheartening. It all seemed to have been a waste of time, a minor curiosity; coenzyme M was of little general biological significance. "I was disappointed," Wolfe recalled; "not only had the unity of biochemistry thesis let me down, but it appeared as if we had spent two years on a fruitless endeavor."[15]

The Strange Signatures of Delta H

Wolfe had been educated in a world view framed not only by the unity of biochemistry thesis but also by the futility of phylogenetic classification. But he had become intrigued by the new molecular methods that Woese's lab had developed. "I was fascinated by this new approach," he recalled,

> for as a graduate student I had been prejudiced against "bug sorting" and its tenuous results by my professor, and I resolved never to get involved in taxonomy with its constant reshuffling and renaming of species. But here was something I could believe in, for it had a sense of permanence. I was especially impressed by results of Woese's initial study of the genus, *Bacillus.* . . . I became a believer.[16]

As Woese recalled, "I was most taken by Wolfe's description of the methanogens. He sat me down in his office and in detail explained to me that there was an interesting problem here because the methanogens showed this unique uniformity in metabolism but incredible variety in morphology, saying they'd be good candidates for my method."[17] Woese and George Fox discussed the idea of testing the methanogens, but in 1974 there had been a severe experimental limitation in labeling the nucleic acids of methanogens with the high levels of radioactive phosphorus-32 (^{32}P). The block was in culture manipulation—the fear of ^{32}P spills. But that problem was overcome when Balch developed a reliable sealed growth procedure for growing methanogens under pressure.

Fox went to see Balch. Although he was more of a theoretically oriented biologist than an experimental one, after three years as a postdoctoral fellow in Woese's lab, he was nearing the time when he would need to obtain a university professorship. In Woese's lab he had acquired some of the skills required for experimental work, for example, growing bacteria in labeled medium, harvesting them, and extracting and purifying the ribosomal RNA. Fox and Balch had first met while taking a microbial ecology course at the Marine Biological Laboratory in Woods Hole, Cape Cod, in the summer of 1973.[18] Three years later, in May, they began a close working relationship in collaboration with Woese's technician Linda Magrum.

The formal name of the organism they tested was *Methanobacterium thermoautotrophicum*; it was always shortened to "delta H," its strain designation. It was thermophilic and grew optimally around 65°C. Their job was to "work up the fingerprints," as they called it. Balch grew the organisms in the presence of massive amounts of ^{32}P. Fox and Kenneth Leursen separated and identified the RNA by polyacrylamide gel electrophoresis, and Magrum created the 16S rRNA fingerprint.[19] The films were then given to Woese, who analyzed the patterns, designed the secondary cuts, and inferred the sequences of the oligonucleotides.

The 16S rRNA patterns were ready for Woese in June. By that time, he had a good grasp of the key bacterial "signatures":

> There were two spots on the primary films of all procaryotes that easily caught one's eye, for they contained modified bases and, so, were located at places in the Sanger pattern where there should be no oligonucleotide. These "odd" oligos allowed one to recognize bacterial 16S rRNA at first glance.[20]

When Woese analyzed the oligonucleotide sequences from the methanogens, they did not register as bacteria. They were missing almost all the signature sequences. Accounts of what transpired next vary. According to Wolfe, Woese was cautious. At first he doubted the results: perhaps Fox and Balch had extracted the wrong nucleic acids or some other mix-up had occurred. He asked them to repeat the experiment. "When I asked Woese about the results of the first attempt to label the 16S rRNA of a methanogen," Wolfe recalled, "he replied that something had gone wrong with the extraction—perhaps they had isolated the wrong RNA. The experiment was repeated with special care, and this time Carl's voice was full of disbelief when he said, 'Wolfe, these things aren't even bacteria.' "[21]

Fox remembered things differently on three counts. First, Wolfe was on sabbatical at the Pasteur Institute in Paris at the time, yet the story persists of Woese bursting into the room and telling Wolfe about the discovery:

> This is true in that he burst into my room in the adjoining lab where I was with just such a statement and probably actually did the same with

everyone he saw that day. If Wolfe was still away then why would Wolfe remember this? I assure you that Carl saved it for him and went to see him and tell him the news as soon as he heard Wolfe was back—enthusiasm not lessened even a bit!! So from Ralph's perspective it almost certainly sounded like something that Carl had done five minutes ago! But it wasn't.[22]

Second, Woese was not dumbfounded when he saw the results, and he recognized them immediately when he had just a glimpse of the data in hand:

Carl knew that it was unique the moment he saw the first good picture. It lacked the key modified small oligos that stuck out like a sore thumb on *every* SSU fingerprint....The discreteness of the fingerprints made the "third form" jump off the page. Also I suspect, but do not know for sure, that Carl was hoping/expecting to find something totally different from the very beginning but he didn't really fully understand what different meant until later.[23]

Third, there was no experimental replication as such:

When Carl saw the first data and calmed down a bit, it was decided to continue with a second prep from a different methanogen rather than to do a redo....

Carl might very well have played it down with a colleague until he had all the data, but definitely we internally would not have just assumed it was some kind of screw up! The Woese lab thrived on ideas and theory.[24]

Woese remembered the discovery similarly:

This methanogen rRNA was not feeling procaryotic. The more oligos I sequenced, the less procaryotic it felt, as signature oligo after procaryotic signature oligo failed to turn up....Then it dawned on me. Was there something out there other than procaryotes and eucaryotes—perhaps a distant relative of theirs that no one had realized was there? Why not?[25]

By December 1976, they had analyzed the 16S rRNA of *Methanobacterium ruminantium* and several other methanogens. All possessed the same non-prokaryotic signature and were closely related, possibly of the same genus or family, they thought. Fox was quick to catch onto the idea of the phylogenetic distinctiveness of the group; Wolfe and Balch did not lag behind. For Wolfe, the ribosomal RNA analysis made some sense of why coenzyme M was unique to methanogens.[26]

A year would pass before the work on methanogens was published and the announcement of "a third form of life" was made. As Fox recalled, there were three aspects to the discussions between himself and Woese that immediately followed the first two methanogen catalogs.[27] First, what did it mean to "actually not be a bacterium"? It was clear to them that if the methanogens represented a "third" form, it should be more-or-less equally different from prokaryotes and eukaryotes, or else it would be related to one or the other. To that end, they immediately concluded that it was necessary to compare the ribosomal RNA catalogs to those of some eukaryotic organisms.

Obtaining eukaryotic catalogs was an extremely difficult undertaking because not only was the eukaryotic SSU ribosomal RNA bigger (18S vs. 16S), it also contained very large numbers of post-transcriptionally modified bases. Convinced of the importance of the three-way comparison, Woese arranged to get labeled RNA from three eukaryotes, yeast cells, primate cells, and plant cells, and did crude catalogs. As Fox commented decades later,

> This was a horrendous task and truly heroic that he pulled it off. In the end, although large numbers of oligos could not actually be sequenced, they could at least be determined to the extent that one could be sure they were different from things seen in either the Archaea or the Bacteria. Thus, the first test was passed.[28]

The second matter of immediate concern was that if there were a third form of life, it should be supported by other data, not just 16S rRNA sequence information. Molecular characteristics such as modified nucleotide patterns and transfer RNA similarities were sought and found, as detailed in later chapters. The big thing, however, was diversity. If there was a third form of life, there surely would be other kinds of organisms. In the midst of this discussion came the now famous visit of Otto Kandler to Urbana in January 1977 (see chapter 15). Woese and Fox learned from him that methanogens lacked peptidoglycan, as did extreme halophiles. This struck a special chord for Fox, who remembered from his microbiology class that the occurrence of peptidoglycan in bacterial cell walls was the only unifying positive trait of prokaryotes. (Stanier and van Niel had referred to prokaryotes possessing a "specific mucopeptide" [chapter 7].) He went to the library and searched for other organisms that lacked peptidoglycan and discovered *Thermoplasma* and *Sulfolobus*. Thus, these organisms made it to the "hit list" for cataloging; it was fortuitous that a mycoplasma collaboration was already in place with Maniloff at Rochester. Woese and Fox also learned, from the papers that they unearthed while searching for data, that all of these organisms had unusual membrane lipids (see chapter 14).

The third problem was writing the paper introducing what they called a third "urkingdom." They went through numerous drafts and discussions necessarily confronting the prokaryotic–eukaryotic dichotomy to fully understand what a third urkingdom meant. In the end, Woese and Fox wrote a paper regarding pre-cellular entities that would have been in the processes of developing the mechanisms of DNA replication, transcription, and translation. Once that was written, they reached a point where they were able to articulate the concepts of three primary kingdoms at the base of the tree of life. All of these discussions occurred before the interactions with the outside world began; all became, as Fox put it, "core ideas of the Woese research program immediately after the eureka moment."[29]

An Ancient Divergence

In January 1977 Woese sent a paper titled "An Ancient Divergence among the Bacteria," coauthored with Balch, Magrum, Fox, and Wolfe, to the *Journal of Molecular Evolution* (*JME*).[30] Methanogens received increasing attention in the 1970s because of their possible use as an alternative fuel source converting energy from organic waste. But the uniqueness of the methanogen ribosomal RNA together with their anaerobic biochemistry supported the idea that these organisms may have played an important part in the early evolution of the planet. As Woese and colleagues saw it, the methanogens represented the earliest phylogenetic divergence so far detected, preceding the divergence of the blue-green algae from the main bacterial line, which according to the fossil record was at least 2.5 billion years old. "The evolutionary divergence between the methanogenic bacteria and the procaryotes" they said, "appears to be extremely ancient—antedating any others so far detected."[31]

Woese sent three other papers to *JME* that year. The first was "A Comment on Methanogenic Bacteria and the Primitive Ecology," in which he argued that the phenotype of the methanogens was well suited to the environment thought to have existed on primitive Earth.[32] Earth's original atmosphere would have been formed by out-gassing of its interior, and based on volcanic gasses today, it would have for the most part contained carbon dioxide, water, nitrogen, and hydrogen. Given that carbon dioxide and hydrogen would slowly convert to methane, Woese argued, methanogens could have easily lived by catalyzing this reaction. He suspected that this idea was not novel and was "too simple not to have occurred to those familiar with the facts."

Certainly there had been proposals, early in the century, that bacteria that oxidize methane and convert it to carbon dioxide and hydrogen were among the primitive organisms, along with organisms that have the ability to convert CO_2 to organic compounds. The general paradigm of the times was that chemo-autotrophs that lived on inorganic compounds would have been the progenitors of life on Earth (chapter 5). But the origin of life paradigm subsequently shifted from autotrophs-first to heterotrophs-first, emerging from and feeding on an energy-rich ocean containing the required amino acids to sustain life (see chapter 7). In 1940, when Barker and colleagues showed the reduction of carbon dioxide by methane-producing bacteria, they confirmed van Niel's hypothesis that methane production was "essentially an anaerobic oxidation process in which carbon dioxide acts as the ultimate hydrogen acceptor (oxidizing agent) and is reduced to methane."[33] There was no suggestion that they might be an ancient form.

Stemming from the Progenotes

The second paper submitted to *JME* that year was critical to Woese's original purpose:[34] understanding how the translation apparatus might have evolved. In it, he and Fox elaborated on the nature of the hypothetical common ancestor of prokaryotes and eukaryotes that lived in a premodern epoch when the translation machinery was not fully evolved. They introduced the term "progenotes" for those hypothetical organisms "in the throes of evolving the genotype-phenotype relationship." In the age of progenotes, they hypothesized, the degree of translation inaccuracy ("a noisy genetic transmission channel") would be extremely high, and the "primary constraint was on the size and properties of the proteins that could be evolved."[35]

Woese and Fox also emphasized that prokaryotes may not have led to eukaryotes in the manner generally assumed. Rather, those two primary lineages could have diverged at the progenote stage when primitive organisms were in the process of evolving the translation apparatus. If so, they reasoned, then features relating to storage and processing of genetic information that had not yet evolved, or incompletely so in the progenotes, would have evolved independently in the two lines of descent:

> Thus genome organization, control hierarchies, (some) repair mechanisms, certain enzymes involved in DNA replication, should appear quite dissimilar in the two cases. Likewise, those aspects of translation, having to do with the final "fine tuning" of that mechanism would appear idiosyncratic. The more (functionally) subtle of the ribosomal proteins, patterns of base modification in the RNAs, the detailed aspects of initiation and termination, and so on, would seem dissimilar.[36]

In keeping with this view, Woese and Fox noted that bacterial ribosomes were smaller than eukaryotic ones. The patterns of base modification in the two types of ribosomal RNAs had little in common. Bacterial ribosomal RNA all exhibited considerable sequence homology among various species, as did eukaryotic ones; sequence homology between the two groups was far less. Bacterial 5S rRNA exhibited a constant structural feature absent in eukaryotic 5S rRNA. Eukaryotic 5.8S rRNA appeared to have no counterpart in bacteria. Each group exhibited characteristic antibiotic sensitivities.

According to their conception, the great differences in genomic organization of prokaryotes and eukaryotes did not result from an evolutionary leap after the emergence of prokaryotes; it represented differences that evolved contemporaneously as the two lines of descent diverged from the progenotes.[37] The lineage that gave rise to the eukaryotic nucleocytoplasm, "the urkaryote," never was prokaryotic. The urkaryote diverged at a prebacterial stage before the completion of the final stages in the evolution of the genetic machinery.[38] They suspected that mitochondria and chloroplasts evolved from acquired symbionts. Far from being rare, they suspected that "endosymbiotic events and the like" would be "normal, common occurrences." In light of the long parallel evolution of prokaryotic and eukaryotic lines of descent, they suggested that symbiosis and similar phenomena would have a long history: "endosymbiosis should be considered an aboriginal, not an acquired trait."[39]

There was plenty of circumstantial evidence supporting the idea that chloroplasts and mitochondria were symbionts (see chapter 9), but definitive molecular phylogenetic evidence was not provided until the late 1970s (see chapter 17). In 1975, Woese and two of his students reported that the 16S rRNA of the chloroplast of the unicellular alga *Euglena gracilis* was unrelated to its counterpart encoded in nuclear genes of eukaryotes and was quite likely related to blue-green algae.[40] Symbiotic theories generally assumed that mitochondrial symbiosis had originated as adaptations about 1.5 billion years ago for functions of ATP generation and respiration that they exhibit today. Woese offered an alternative suggestion in the third paper he sent to *JME*—that mitochondria arose by endosymbiosis in a much earlier, anaerobic period, and that it was initially a photosynthetic organelle, analogous to the modern chloroplast, and only later evolved into a respiratory organelle.[41]

Untangling the Prokaryote

Of all three papers, only that on cellular evolution from "the progenotes" received harsh criticisms from peer reviewers. Certainly, there had been recognition of precellular organisms by others, but microbiologists had lumped them into the prokaryote concept. Ernst Haeckel had offered Monera as a concept for precellular entities, but that word was carelessly modified to mean simply prokaryotes. Verne Grant even suggested a kingdom for hypothetical precellular entities in 1963 (see chapter 8). But the question of origin and evolution of the genetic code was generally ignored by microbiologists and molecular biologists. The progenotes concept, at first, seemed to fair no better in the hands of reviewers.

The first reviewer did not find fault with the logic of the argument for progenotes in the throes of evolving the translation machinery, but found no compelling evidence supporting it:

> The argument turns on a subjective assessment of the degree of "completion" of the "channel for transmission of information between the genotype and the phenotype. It could just as well be argued that all essential

details of the "channel" had evolved completely. One could emphasize the universality of the genetic code, the use of GTP in protein synthesis, similarities in sensitivities to some antibiotics, etc. etc. to make an equally impressive case along these lines.[42]

Emile Zuckerkandl, the editor of *JME*, made himself the second reviewer.[43] He would be instrumental in supporting Woese's research and facilitating its publication. At first, he, too, thought that, one way or the other, the lineage to the eukaryote would have been prokaryotic, lacking chromatin surrounded by a nuclear membrane. All one could really say was that "available evidence strongly suggests that the split between prokaryotes and eukaryotes is extremely ancient, more ancient than many are ready to believe, and took place at a time when the apparatus for regulation of gene transcription and of translation of the transcript were not definitively fixed." As he wrote to Woese,

> What an extraordinarily interesting series of little papers—not at all little by their content!...
>
> Does the following view not remain plausible: the most basic eukaryote character is the eukaryote nucleus, with its various properties. The eukaryote nucleus has evolved in some prokaryote. Whether some or all of the endosymbionts joined the prokaryote cell before or after it evolved a nucleus probably cannot now be stated. Yet whether the one or the other obtains, it does not change the "fact" (says this theory...) that the eukaryote cell evolved out of a prokaryote.
>
> This view and yours seem however, finally, to be separated only by words. <u>Within</u> the "classical" view I resort to here, I am indeed quite ready to recognize that the prokaryote that gave rise to something like chromatin surrounded by a nuclear membrane did so in extremely ancient times. This assumption is necessary to account for the differences in the translation apparatus that you point out. Here you will say: sorry, this very old cell wasn't a prokaryote, but a progenote, there was, at one time, a more prokaryote like one and a more eukaryote like one. You don't want to call the prokaryote like progenotes a prokaryote, because some traits of the translation machinery and important traits of the apparatus for regulation of transcription were still being fashioned. But is it not equally possible that the transformations, in this respect, occurred mainly in the eukaryotic line? If so, the common ancestor of prokaryotes and eukaryotes would still have been some sort of prokaryote, even if a progenote.[44]

The merit of the progenote concept, Woese replied, was that it opened up to inquiry unquestioned (erroneous) phylogenetic assumptions embedded in the prokaryote concept. The "classical" view concealed problems of the prokaryote's own origin. The progenote concept as the universal ancestor allowed for the possibility that there was as much phylogenetic distance within prokaryotes as between prokaryotes and eukaryotes:

> What eucaryote, procaryote, and progenotes are basically, are three different <u>levels</u> of complexity of living systems.... The three terms should be used with <u>no phylogenetic connotation</u> whatever.
>
> The problem with the current conventional view, outside of its being muddled, is that the prokaryote *is* given phylogenetic connotation. See Margulis' work for a prime example of this. In the proper view, one could have in principle some procaryotes more closely related to eucaryotes than they are to other procaryotes. With such a view the evolutionary questions become more clear. Was the (most recent) "common ancestor" a prokaryote; i.e. did all life arise from a common <u>procaryotic</u> ancestor? If not, how many times did the prokaryotic state evolve (from the progenote level)? How many times, by analogy then, did the eucaryotic level arise (from the prokaryotic level)? These questions are not generally asked explicitly.
>
> You see, Emile, it is a very different thing if you feel obliged (and don't know it) to derive your eucaryote from species all of which are <u>phylogenetically related</u> procaryotes, than if you don't feel so obliged.

Woese explained that symbiosis did not have to arrive late on the scene after the emergence of bacteria and eukaryotes: "Endosymbiosis seems ad hoc; it needs to be excused today," he said. But when progenote versus procaryote is recognized, as an organizational distinction, one no longer "takes this narrow and misleading view of endosymbiosis":

> Endosymbiosis becomes an <u>aboriginal</u> property of progenotes, not an <u>acquired</u> property of procaryotes. The "ancestor" has no cell wall; this evolved separately in typical bacteria (peptidoglycan walls) and eucaryotes (plant and fungal walls). Methanogens also have their own versions of the cell wall. Endosymbiosis then suddenly becomes an interaction that is widespread and diverse. What we now take as endosymbiosis is only the tip of the iceberg.... Endosymbioses have been a major force in evolution for over three billion years....
>
> Now when you take on the problem of the vast differences in eucaryotic genome organization vs. typical bacterial genome organization, you again see it in a new and clearer light. You are no longer obliged to derive the eucaryotic nuclear organization from bacterial organization. It is no longer a question of an evolutionary jumping of this gap. The gap was not spanned, it was created by the systems separately evolving from an ancestral progenote whose genome organization was far simpler (because it had far fewer genes than procaryotes even).
>
> ...It always astounds me how little attention is paid by the usual writers in this area to the evolution of translation. They totally ignore

it, including the facts that are known. Take Margulis for example. She will talk at great length about prebiotic this and that, biochemistries, structures, and so on. She will even talk about the evolution of "self-replicating" nucleic acid. Then she will simply state that the "genetic code evolved" and pass on to more important things. In this she merely epitomizes the general attitude.

...When you have a muddled paradigm, nothing has meaning.[45]

Zuckerkandl was convinced. He took it further and developed the argument about symbiosis and the like as an aboriginal property of life, not merely as something that was added onto the mechanisms of evolution after prokaryotes and eukaryotes were formed:

Three <u>levels</u> of organization; <u>a priori</u> no phylogenetic connotation: this cleared it all up for me. Long live the progenote,—and I think long live it will!...

First define the progenote, then look at phylogeny: as you say, all of a sudden the phylogeny of unicellular organisms makes sense. An important point is that the progenote is a much more <u>fluid</u> cellular entity than either the prokaryote or the unicellular eukaryote, in the sense that exchanges of components between different evolutionary lineages of cells was much more generalized than ever after. In fact, I wonder whether the identity of a progenote "species" was anything definable and real. In case of very frequent inter-cellular exchanges of genetic materials, there may not have been progenote species, but progenote societies, each society having a composition defined by an ecological niche. Contemporary exchanges of genetic material between organisms would appear as a remnant of a pervading phenomenon that characterizes your progenote era. If so, the concept of species became applicable relatively late (even though a long time ago) and perhaps the beginnings of its strict applicability coincide with the end of the progenote era.

I very much like your revised version. It is a fundamental paper. Thank you for JME, Carl, and congratulations.[46]

Introducing the Archaebacteria

In the summer of 1977, Woese and collaborators sent two papers on the methanogens to the *Proceedings of the National Academy of Sciences*, a journal that publishes articles of broad interest to the scientific community. Members of the academy could sponsor papers of non-members. The first paper dealt mainly with the molecular methods and data setting the methanogens phylogenetically far apart from other bacteria; it was sponsored by Barker, the pioneer of methanogen research.[47] By that time, the ribosomal RNA of 10 species of methanogens had been characterized. They claimed a phylogenetic coherence of the group based "on cellular information processing systems" independent of particular biochemistry or any other overt phenotypic property."[48]

The ancestral molecular traits, they asserted, were not adaptations regarding the production of methane, nor to the requirement of a strictly anaerobic niche.[49] The transfer RNAs of methanogens possessed a unique sequence structure. Transfer RNAs in nearly all other organisms previously investigated shared the common base sequence: thymidine–pseudouridine–cytidine–guanosine (TΨCG) in one arm. The transfer RNA of methanogens lacked this sequence.[50] Kandler in Germany showed that the cell walls of methanogens lacked peptidoglycan, a defining characteristic of the prokaryote (see chapter 15).[51]

In the second paper, titled "Phylogenetic Structure of the Prokaryotic Domain: The Primary Kingdoms," Woese and Fox wrote more boldly about a "third kingdom of life," which they called the Archaebacteria.[52] That paper was sponsored by the eminent geneticist Tracy Sonneborn at Indiana University. He was among the few who had addressed the evolution of the genetic code, and he was appreciative of Woese's theoretical writings on the evolution of macromolecular complexity.[53] Woese and Fox's paper expressed five interrelated concepts that defined the archaebacterial research program.

First, there was a fundamental trifurcation of life on Earth, "primary kingdoms" or "urkingdoms" (Greek *ur*, original, primitive, or earliest). Each urkingdom represented a separate line of descent. The typical bacteria they called the eubacteria (Greek *eu*, true; this term was formerly used for one of five bacterial orders, comprising "the least differentiated and least specialized" of bacteria—see chapter 5). The "major ancestors" of the eukaryotes, the urkaryotes, represented the hypothetical hosts that would have lacked symbiotically acquired organelles: the line of descent manifested in eukaryotic cytoplasm. The third urkingdom, the Archaebacteria, was represented thus far solely by the methanogenic bacteria.[54] "The apparent antiquity of the methanogenic phenotype," they said, "plus the fact that it seems well suited to the type of environment presumed to exist on Earth 3–4 billion years ago led us tentatively to name this urkingdom the Archaebacteria. Whether or not other biochemically distinct phenotypes exist in this kingdom is clearly an important question."[55]

Second, eukaryotes did not evolve from prokaryotes, as had been assumed. Each urkingdom evolved from the hypothetical "progenotes," precellular entities that predated the modern translation mechanism. Because the distances measured are actually proportional to the number of mutations and not necessarily to time, they could not say for certain whether the three lines of descent branched from the progenotes at about the same time.

Third, the translation complex was at the core of the organism. The differences in ribosomal RNA they used to distinguish the methanogens as a distinct urkingdom were not related to the niche that those microbes occupy.[56] Instead, "features such as RNA base modification represented the final stage in the evolution of translation." If those features evolved separately in different lines of descent, then "their common ancestor, lacking them, had a more rudimentary version of the translation mechanisms and consequently, could not have been as complex as a prokaryote."[57]

Fourth, the prokaryote and eukaryote dualism was phylogenetically erroneous: phylogenetically, life was not "structured in a bipartite way along the lines of the organizationally dissimilar prokaryote and eukaryote. Rather, it is (at least) tripartite."[58] And fifth, eukaryotes were organized in a different and more complex manner than were eubacteria and archaebacteria. The evolution of that complexity did not result from the usual Darwinian mechanisms. It reflected its composite origin as a symbiotic collection.[59] "The organizational differences between prokaryote and eukaryote and the composite nature of the latter, indicate an important property of the evolutionary process," Woese and Fox wrote. "Evolution seems to progress in a 'quantized' fashion. One level or domain of organization gives rise ultimately to a higher (more complex) one."[60]

16S rRNA analysis left "no doubt that the chloroplast is of specific eubacterial origin." The case for mitochondria was not as certain, and "it was even conceivable that some endosymbiotically formed structures represent still other major phylogenetic groups; some could even be the only extant representation thereof."[61] The nature of the engulfing species, "the pure urcaryote" (represented by the cytoplasmic component 18S rRNA), was unknown, but Woese and Fox pointed to anaerobic amoebae such as *Pelomyxa palustris*, which seemed to lack mitochondria and did not divide by mitosis.[62] These concepts would shape archaebacterial research over the next three decades. None would be accepted outright, least of all the claim for the third urkingdom of life.

| # A Kingdom on a Molecule

Though scientists have established that the earth was born some 4.6 billion years ago, formed from debris orbiting the sun, they are less certain about when—and under what conditions—life began on the planet.... Now biologists working under grants from NASA and the National Science Foundation have identified living creatures that may be little changed from organisms that lived during the first billion years of the earth's existence.

The new candidates for the oldest-form-of life title are organisms that scientists have dubbed "archaebacteria."

—"The Dawn of Life," *Time* (1977)

IN THE FALL of 1977, Carl Woese took the archaebacteria to the world outside of science. Publicity for the new form of life began when the National Science Foundation (NSF) and the National Aeronautics and Space Administration (NASA) released a joint public statement on November 3, 1977, a few days before the papers in the *Proceedings of the National Academy of Sciences* (*PNAS*) appeared. Woese's research had been funded by both. He had met Richard Young, "a very enlightened administrator" from NASA, four years earlier at a roundtable conference on the "Origin of Life" held in Paris.[1] After hearing Woese's presentation, Young encouraged him to apply to NASA's exobiology program.

Founded in 1958, a year after the Soviet Union launched Sputnik, NASA broadened its mission to search for extraterrestrial life, developing a new branch in "exobiology," the term Lederberg had coined in 1961.[2] To search for life elsewhere, one needed to know what kinds of things to search for. The search for life beyond Earth was therefore intricately bound to studies of the origins of life on Earth, which had become a field in its own right with journals, societies, conferences, and funding programs devoted to it.[3] There was especially interest in the possibility of life on Mars with the NASA's Viking program; a pair of space probes had been launched in the fall of 1975. The announcement of a fundamental, phylogenetically distinct form of life whose biochemistry was well suited to the conditions thought to exist on the ancient Earth was of considerable interest. Woese explained to Young that some publicity would benefit both of them.

About Going Public

The discovery of archaebacteria as an ancient "third form of life" was announced in prominent newspapers and weekly magazines in the United States, and summary reports were written for *Science* and *Nature*.[4] The discovery of the archaebacteria made front-page news in the *New York Times*. Methanogens were a "third kingdom" of the most primitive forms of life, Woese explained; they may have arisen some 3.5–4 billion years ago, and there may be other forms of life that emerged before bacteria.[5] Cyril Ponnamperuma, director of the Laboratory of Chemical Evolution at the University of Maryland, was quoted as saying that the results were "very exciting, even fantastic." They fit "into the general idea of evolution under nonoxygen conditions," and Spiegelman commented further that "the research results look O.K. Dr. Woese is a substantial scientist of international reputation who has contributed a number of ingenious ideas to science."[6]

The media reports of a "third form of life" caught the attention of biologists throughout the world, just as Woese had hoped it would. Ralph Wolfe, on the other hand, viewed that publicity with trepidation. Reflecting back, decades later, he recalled that Woese's going public with the "third form of life" was met with "disbelief and much hostility, especially among microbiologists." "Scientists" he said, "were suspicious of scientific publication in newspapers."[7] He had received many phone calls on the morning of November 3 when the *New York Times* article appeared. Among them, the one from Salvador Luria was "the most civil and free of four letter words."[8] Luria shared the Nobel Prize in Medicine and Physiology with Max Delbrück and Alfred Hershey in 1969 for their discoveries concerning viral genetics. Wolfe remembered the conversation: Luria said, "Ralph, you must dissociate yourself from this nonsense, or you're going to ruin your career!"[9]

In Wolfe's view, Woese's going public only served to interfere with the acceptance of the discovery:

> In Philadelphia, I explained in dismay to my father-in-law what my colleague had done, and his response was: "You know, in my long life I have

observed something, if you don't overstate your case, no one will listen." I felt better. However, in hindsight, the press release polarized the scientific community, and the majority refused to read the literature, delaying acceptance of the Archaebacteria for perhaps a decade.[10]

Wolfe's recollection of the antagonism to the concept of a "third form of life" is telling. But his claim that the press release was responsible for delaying the acceptance of the Archaebacteria for a decade is questionable. First of all, it implies that such press releases were uncommon. Yet, there had been press releases in the *New York Times* when the genetic code was being cracked, and there had been press releases in the media when ancient microbes were discovered in the fossil record. In fact, a month before the announcement of the Archaebacteria, the *New York Times* had reported the discovery of microfossils in South Africa determined to be photosynthetic aerobes, resembling blue-green algae, that existed about 3.4 billion years ago, only a billion years after Earth was formed. That such sophisticated organisms had evolved so early, the reporter noted, suggested that life's origin "was not the consequence of a highly improbable chemical 'accident' as many people once imagined."[11]

Scientists and the public were also long accustomed to headlines of medical breakthroughs in the press before they were published in scientific journals. In fact the same month that the third form of life was announced to the press, Philip Handler, president of the National Academy of Sciences, released a press statement about the cloning of the human growth hormone somatotropin. That report was on the back pages of the *New York Times* the day that the discovery of "archaebacteria" appeared on the front page.

Second, Wolfe's statement implies that the conclusions drawn from the SSU rRNA data would have been obvious to anyone who cared to look at them— that the facts would "speak for themselves." That perspective belies the fact that the molecular phylogenetic techniques and the data were arcane; they were as novel as their interpretations. The disbelief that 16S rRNA could be used to construct phylogenies of bacteria, and that the archaebacteria represented an urkingdom distinct from other bacteria, was indeed deep.

Third, contrary to Wolfe's statement that "the majority" refused to read the publications, many biologists wrote letters to Woese asking about the data behind the media claims and requesting reprints of the impending scientific publications. Christian Sybesma at the University of Brussels had not suspected the kind of evidence involved when he wrote to Woese about the news in the French paper *Le Soir* and the Flemish *De Standaard*. He was certainly irritated by the media representations but wanted more information "Both newspapers made sensation out of it in a way I don't understand or appreciate," he said.

The French headline talks about a "new form of life" and continues this way, pointing out that "the newly discovered bacteria constitute a third form of life." … The Flemish newspaper has essentially the same but its headline adds "discovery upsets theory of evolution."...Radio and TV also had the story and someone I know at the University of Ghent (Sjef Schell) was consulted....We both found the sensation made of it somewhat annoying, especially because there was a conspicuous absence of any data or evidence in the press release.

As you can imagine, the thing intrigues me very much. I would appreciate it very much if you could tell me more about it, perhaps by sending me the communication proper of NSF and NASA or some preprint if you have any. Do you, indeed, see in these bugs the "ancestral prokaryote" and, if so, on what grounds? I would be grateful if you could send me that information.[12]

Woese explained to Sybesma the science behind the popular reports: his lab had

developed a technique, based upon partial rRNA sequencing that allows the determination of phylogenetic relationships among organisms. The approach is sufficiently powerful that one can detect the relationship between any two organisms, regardless of how distant they are. [In other words, one can "see" the so-called universal common ancestor.] We began this work almost ten years ago, and got into high gear about five years ago. To date, we have characterized nearly ninety organisms by the rRNA cataloguing method.[13]

Few microbiologists understood the nucleic acid sequencing approach, and those who did had grave doubts that the results in Woese's lab would be sufficient for a establishing the existence of an ancient kingdom. Lederberg raised a critical question while he waited for the *PNAS* papers to appear. Perhaps the unique rRNA sequences of methanogens did not reflect their age and early divergence from typical bacteria, but rather the extreme environment in which they lived. "I am sure you saw the stuff in the papers about the methanogens," he wrote to his colleague Dennis Smith on November 9,

but as far as I can tell (the PNAS article is on the way) he has simply shown an evolutionary divergence of the ribosomal-RNA sequences of these bugs; and they may well be neither primitive, nor as unrelated as he supposes....He leaves out of account the fact that they live in unusual environments that select for idiosyncrasies; and once the RNA starts to evolve, there is no reason for it to remain quasi-stable in information content—so his clock may be non-linear.[14]

Were the unique signatures of 16S rRNA of methanogens a function of their unusual environments, and not of their ancient nature? That the 16S rRNA was insulated from the vicissitudes of life, that its evolution was not prone to selection, was at the core of Woese's research program. Even those who admired

Woese's innovative thinking doubted that a universal phylogeny could be balanced on the head of one molecule (and only part of one at that). Geneticist Tracy Sonneborn, who sponsored the paper's submission to *PNAS*, appreciated Woese's decision to use "protein synthesis as the basis for conclusions about early evolution," and he was pleased to see that Woese was giving attention to transfer RNA as well as rRNA, but as he wrote to him,

> conclusions based on any one of these molecules, no matter how evolutionary stable it may be, will have to remain tentative until comparable data are obtained on as many as possible of the basic molecules involved in protein synthesis. I realize that this is a large order, but the great questions you are attacking are gigantic questions. More to the point, you have clearly made far more than a small start in the attack.[15]

Confronting Classical Microbiology

The reaction to the archaebacterial concept was complex as theory, technique, and data, intermingled indissolubly. While some doubted that the 16S rRNA comparisons were in any way phylogenetic, others insisted that, in any case, phylogeny should not be the basis of sound taxonomy or systematics. Both objections were raised in the peer reviewer reports of the two manuscripts submitted to *PNAS* in the summer of 1977. Those papers received mixed and conflicting assessments. Every one of their major claims was rejected by one peer reviewer or another.[16]

One peer reviewer recommended that Woese and George Fox's paper introducing the archaebacteria not be published because most of the conclusions "appear to be based on rather dogmatic statements on issues which are far from decided." Woese and Fox had claimed that eukaryotic organelles are symbiotic in origin. "They may well be," the reviewer commented, "but the origin of eukaryotes and their organelles is still a matter of some dispute." Then there was the kind of data they used:

> The data offered in support of the three urkingdoms consists of rRNA oligonucleotide differences. I think it's a little risky to erect a kingdom on one character. Again the conclusion may be correct, but the criteria are inadequate and possibly as misleading as would be the erection of an animal phylum containing insects, birds and bats on the basis of their possession of wings.[17]

Recall that this kind of criticism had been made early in the century against Walter Migula's attempt at a morphological classification of bacteria based on the presence and absence of flagella (see chapter 5). The reason that classification could not be based on wings is, of course, because wings had originated as adaptations several times in the course of evolution. That 16S rRNA was just another arbitrarily chosen trait no different than any other (phenetic) character was assumed by another peer reviewer, who commented on the second paper, which dealt mainly with the molecular methods and data: "The dendrograms show phenetic relationships but not a phylogenetic analysis.... Whereas the great differences between the methanogens and other bacteria are well documented, phylogenetic relations are not."[18]

For Woese and Fox, ribosomal RNA was a highly conserved nonadaptive structure at the core of all organisms. Still, even Darwin, who had argued for the importance of essential characteristics in classification, insisted that no single character could ever serve as the basis of a natural classification. "Hence, also, it has been found," he wrote in the *Origin*, "that a classification founded on any single character, however important that may be, has always failed; for no part of the organisation is universally constant."[19]

Molecular phylogenetic methods and concepts constituted a paradigm apart from classical evolutionary biology. Molecular phylogenetics did not grow out of Darwinism and its interdisciplinary articulations of the 1930s and 1940s. It represented a radically new departure in evolutionary biology. It changed the whole meaning of what a "character" is. Molecular phylogenetic reconstructions were based on non-Darwinian, non-adaptive changes in a molecule. Classical taxonomists knew no more than did Darwin: different molecular sequences can correspond to the same function, and one can measure variability within a molecule without phenotypic change. The substitutions in nucleotides of RNA or any other "semantide" did not affect the function of the molecule or the behavior of the organism. The nucleotide components of a nucleic acid sequence (or the amino acid components of a protein) were characters in an entirely new sense.

That one molecule (or many) could be used to discern phylogenetic relationships was more than most microbial taxonomists could accept. The deep skepticism toward microbial phylogeny was well inscribed in the minds of microbiologists and spelled out in their leading texts. In the third edition of *The Microbial World* (1970), Stanier, Doudoroff, and Adelberg continued to argue against the possibility of a phylogenetic classification of bacteria just as they had in the second edition seven years earlier. Actually, they went further and asserted that the lack of a detailed fossil record in Precambrian rocks more than 400 million years old belied any possibility of a general phylogeny of microbes:

By the beginning of the Cambrian period, most of the major biological groups that existed had already made their appearance; vertebrates and plants are the principal evolutionary newcomers in Postcambrian time. For these two groups, the fossil record is, accordingly, reasonably complete, and the main lines of plant and vertebrate evolution can be retraced with some assurance. For all other major biological groups, the general course of evolution will probably never be known, and there is simply not enough objective evidence to base their classification on phylogenetic grounds.[20]

Woese kept Stanier informed of the methanogen data, and his phylogenetic classification of eubacteria based on SSU RNA. On February 9, 1977, Stanier wrote to Woese that in his recent visit to Munich, Otto Kandler informed him that the methane bacteria "stick out as an isolate group in terms of the 16S rRNA criterion and cell-wall chemistry." Stanier wrote that Woese's results regarding the classification of the typical bacteria seemed "to fit reasonably well with the picture that could be constructed on the basis of other but very fragile evidence concerning evolutionary filiations":

> You may know that R.G.E. Murray and N.E. Gibbons are in the process of proposing major taxa for prokaryotes, to be used in the next edition of Bergey's Manual.... Since I shall probably be engaged in a good deal of discussion with Gibbons and Murray about the problem of major taxa, I'd much appreciate receiving a dendrogram for 16S rRNAs, if you've got one prepared.[21]

Woese replied, explaining the concept of the three primary lineages:

> These organisms are sufficiently unlike other prokaryotes that I do not want to call them prokaryotes in any but an organizational sense.... It seems to me that methanogens, typical bacteria, and the pure line of descent that ultimately gave the cytoplasm, constitute three separate lines, each independently achieving prokaryotic status (again in an organizational, not phylogenetic sense), from a simpler level of organization. How do you react to this interpretation?[22]

Stanier did not reply. But when Woese learned that the ninth edition of Bergey's Manual was in the planning stage, he wanted his new classifications to be included. Recall from chapter 8 that R.G.E. Murray, one of the editors of Bergey's Manual, had been one of the first to celebrate the taxonomic significance of the eukaryote–prokaryote dualism, insisting that prokaryotes be given the rank of superkingdom, Procaryota—as they were in the eighth edition of Bergey's Manual of 1974. Woese recognized prokaryotes in an organizational sense—although he would change his views in that regard, too, in the 1990s (see chapters 21–23). He explained to Murray in July 1977 that the prokaryote and eukaryote were organizational distinctions only, not phylogenetic domains, although they were confused as such. To recognize that confusion was to bring into the open unresolved fundamental problems of deep evolutionary biology that lay concealed. "A priori," he wrote to Murray, "one has no idea how many kingdoms (urkingdoms) exist in the procaryotic domain." No one really knew how many times the eukaryotic cell had arisen, or the prokaryote for that matter, "for the procaryote is sufficiently complex that it in turn must have arisen from yet a simpler class of entities (apparently no longer extant)":

> There exists however, a third kingdom as distinct from the other two as they are from each other. This kingdom I tentatively call *archaebacteria*, because of the seemingly primitive nature of the representative phenotypes. At present the archaebacteria contains only methanogens (eleven species, done in collaboration with R. Wolfe's lab). How many more (phenotypically distinct) genera will ultimately be in this kingdom, if any, I have no idea. Whether more kingdoms will be found in the prokaryotic domain, I also have no idea.[23]

Murray replied with two letters that fall. In the first, he appeared to appreciate Woese's approach; he recognized it to be phylogenetic, and that the Prokaryotae was not a phylogenetic distinction, but one based on cellular organization only.[24] But when Woese sent him further data in support of the archaebacteria, Murray echoed the statements in *The Microbial World*, pointing to the absence of a bacterial fossil record against their phylogenetic classification. He suspected that in the new edition they would "have to comment in some place upon the growing molecular evidence for patterns of relationship that cut across the old assumptions," but explained to Woese "that classifications are human estimations and seldom have a phylogenetic basis unless there is a remarkable fossil record."[25]

The ribosomal RNA approach did not just confront the skepticism regarding a phylogenetic ordering of bacteria. It was not just a matter that bacteriologists did not *believe in* the approach—those new molecular methods clashed with microbiological tradition: what it was to be a microbiologist. Adopting the molecular techniques developed in Woese's laboratory would be an onerous task. It required new equipment, new laboratory design, and skill in deducing oligonucleotide sequences from the images obtained. Employing those methods elsewhere required first-hand experience in Woese's laboratory (see chapters 15 and 18).

In 1977, Woese and collaborators emphasized that theirs was not a monopoly of technique:

> Comparative cataloging is not, however, a panacea. Compared to traditional techniques it is relatively expensive and time-consuming, and it requires considerable specialized expertise. Thus, it is appropriate to view comparative cataloging of 16S rRNA as an adjunct to and not a replacement for the more usual approaches.[26]

Monopoly of technique or no, one also had to weigh the value of renaming many bacteria to conform to the rRNA phylogenies against the practical importance of a stable taxonomy. Some microbiologists were dead against it; even if one could know the phylogenetic relations among microbes based on 16S rRNA, they insisted that classification based on such relations would not necessarily be appropriate.

Phenotype or Phylogeny?

Should one classify in terms of overall similarity of organisms, based on those organismal characteristics that are expressed and understood in terms of diversity—chemistry and morphology? Or should one classify based on phylogeny, regardless of the amount of evolutionary change that has occurred? This issue was raised by one of the peer reviewers of the *PNAS* papers of 1977, who congratulated Woese and collaborators for making "a fascinating discovery." Should taxonomic rank be based primarily on "organismal resemblance" or on nucleic acid or protein sequence resemblance?

> Sequence evolution in proteins and nucleic acids appears to go on rather steadily with time, whereas a phenotype may be highly conserved, as in the methane bacteria, or very unstable, as in the lineage leading to humans. So, it is doubtful whether taxonomic rank should be based on sequence resemblance. The decision whether the methane bacteria should constitute a separate kingdom, must be based primarily on phenotypic criteria.[27]

When push came to shove, many systematists agreed with the view expounded by Ernst Mayr, which privileged general phenotypic characters, that is, the amount of evolutionary change that had occurred, over phylogenetic divergence from a common ancestor. That is why humans were classified as hominids, not as apes, and alligators as lizards, not as birds. Although humans were known to be more related in time and at the macromolecular sequence level to chimpanzees than chimpanzees are to orangutans, humans are classified in the Hominidae family, separate from the ape family, Pongidae, which includes chimpanzees and orangutans. This was because humans are so distinctive in phenotype and way of life. Similarly, alligators were known to be closer in time and at the molecular level to birds than to other reptiles, yet taxonomists and ordinary people unhesitatingly classify alligators with reptiles, on phenotypic grounds. Woese was stepping into hot taxonomic water. "Be rather careful about the use of formal taxonomic terms such as kingdom," the peer reviewer warned Woese.[28]

A Revolution's Afoot

Fox had moved to the Department of Biophysical Sciences at the University of Houston by the time that the final draft introducing the archaebacteria was sent to *PNAS*. In November, Woese informed him that he had received many requests for their papers of 1977, especially for their *JME* paper on the hypothetical progenotes.[29] François Jacob wrote to say that he had found their

> arguments concerning a very early divergence between procaryote and eucaryote fairly convincing.... I completely agree with you. In particular

in the point that it is mainly the genotype phenotype relationship which is the most important matter to understand. It is a long time we did not see you. Is there a chance for you to come to Paris?[30]

There was also great interest among molecular biologists and chemists involved in studies of the origin of life in the proposal that methanogens represented a new primary kingdom, just as there was in the popular media. Cyril Ponnamperuma, editor of the journal *Origins of Life*, asked Woese and Fox to write a review article on the methanogens,[31] as did the associate editor of *The Sciences*.[32] Fox sent preprints of the 1977 papers to James Watson, who replied that he "most enjoyed the evolutionary insights, particularly as to the methane-producing bacteria" and he invited Fox to a meeting on transfer RNA that he was organizing at Cold Spring Harbor.[33]

Short reports titled "Are Methanogens a Third Class of Life?" and "Methanogenic Bacteria: A New Primary Kingdom?" appeared, respectively, in *Science* in November and in *Nature* in February.[34] Journalist Thomas Maugh interviewed Walter Fitch, one of the pioneers of molecular phylogenetics, for the report in *Science*, who commented "that if all of the Illinois group's evidence is correct, then their interpretation is probably also valid." Maugh concluded, "Ultimately, it would appear that acceptance of the Illinois group's arguments will depend on the strength in one's faith in phylogenetics."[35] J.F. Wilkinson, microbiologist at the University of Edinburgh, wrote the overview in *Nature*, dealing more carefully with the evidence and issues in evolutionary biology. "It is clear," he concluded, "that the recognition of these three lines of descent represents an exciting phase in the history of biology and there remains much to be done in clarifying the position."[36] Other microbiologists felt the same; they sensed an emergent revolution in microbial evolutionary biology.

| Against Adaptationism

It turns out that a number of people are getting inklings that something is strange with certain organisms, and, if we didn't tell them what the situation is, they would publish a lot of muddled speculations about them— further obfuscating an already conceptually diffuse area.

—Carl Woese to Emile Zuckerkandl, February 2, 1977

WERE THE SIGNS of urkingdoms inscribed as signatures in one molecule? The archaebacteria program in Woese's lab was two-pronged: (1) to explore the expanse of the newly proposed kingdom by analyzing the 16S rRNA of other unusual kinds of microbes, and (2) to search for phenotypic traits to corroborate the rRNA evidence. Extreme halophiles that thrived in conditions salty enough to kill typical bacteria and the thermoacidophile *Thermoplasma* were on Woese's research agenda. These organisms were considered to be completely unrelated, their unusual chemical features thought to have arisen as independent adaptations to life in "extreme" environments. Their 16S rRNA analysis led to a radically different understanding. Together with the methanogens, they were a phylogenetically coherent group. To thus expand the urkingdom was to confront pan-adaptationist thinking that permeated biology.

Biologists of the 1970s tended to treat every evolved trait as if it were an adaptive trait that had originated for its present-day function. This pan-adaptationism was the main target of Stephen Gould and Richard Lewontin's influential paper of 1978, "The Spandrels of San Marco and the Panglossian Paradigm: A Critique of the Adaptationist Program."[1] To caution evolutionists from assuming that every phenotypic trait is an adaptive trait designed by natural selection, they drew an analogy with the architecture of the basilica of San

Marco in Venice. The tapered spaces, the spandrels (or pendentives), between the archways supporting the domed roof were beautifully decorated in a way that made splendid use of them, almost as if they had been made for that very artistic purpose. However, they were not designed for that purpose at all; they were really just an architectural by-product of employing arches to support a domed room.

Even if one assumed that a character arose by natural selection in the remote past, one could not assume that it arose for its present purpose. Appreciating this point was important for understanding the early evolution of complex organs, such as the wings of a bird or insect before they were used for flying; proto-wings may have had other purposes.[2] Evolutionists early in the century invoked Voltaire's Dr. Pangloss to reveal the pan-adaptationist thinking underlying much of Darwinian evolutionary biology:

> Things cannot be other than they are, for since everything was made for a purpose, it follows that everything is made for the best purpose. Our noses were made to carry spectacles, so we have spectacles. Legs were clearly intended for breeches, and we wear them.[3]

Adaptationist thinking was a central issue in taxonomy. Darwin had explained in the *Origin* that for phylogenetic purposes those characteristics most closely related to "habits of life" and "the general place of each being in the economy of nature" were the least useful (chapter 3). Such external resemblances of a mouse to a shrew or whale to a fish were, as he put it, "merely adaptive or analogical characters" of no use in a natural genealogical classification.[4] He advocated "as a general rule that the less any part of the organization is concerned with special habits, the more important it becomes for classification."[5] He emphasized the importance of embryonic characters in animals, which were considered to be highly conserved.

Similarly, the ribosomal RNAs (rRNAs) and other features of the translation apparatus were held by molecular phylogeneticists to be non-adaptive, but critics of Woese's rRNA approach considered it to be an arbitrary adaptive trait like any other (chapter 13). Adaptationism also underlay the assumption that the odd features of halophiles, methanogens, and thermophiles arose independently in response to the unusual environments in which those organisms live today. It was at the basis of the notion that those organisms were unrelated.

Expanding the Urkingdoms

HALOPHILES

The most extreme forms of halophiles were discovered in the 1940s in the Dead Sea, which contains about 25% salt. Similar types were well known to cause spoilage of fish and raw hides salted for preservation. These halophiles were known to have another distinct feature: they were sensitive to hypotonic solutions. Most bacteria were relatively insensitive to whether or not the environment has a higher osmotic pressure (i.e., a higher concentration of solutes than does the interior of the cell). Extreme halophiles were different; they would undergo rapid osmotic lysis (burst) if suspended in distilled water.[6] The chemistry of their cell walls was strikingly different from that of typical bacteria. Surrounding the cell membrane, the wall maintains the bacterium's characteristic shape. It counters the effects of osmotic pressure—the wall keeps the cell from bursting when the pressure inside is greater than the pressure outside the cell. The strength and rigidity of the bacterial cell wall were known to be due to a huge macromolecule, a heteropolymer composed of two constituents: glycan strands cross-linked by short amino acid chains. Since the mid-1960s, it was called either "murein" (Latin *mur*, wall) or "peptidoglycan" to indicate its chemical nature. Typical bacteria had peptidoglycan in the cell walls; halophiles did not.[7]

The lipids in the cell membranes of halophiles were also unlike those found in any of the typical bacteria. Some well-known species of extreme halophiles also possessed a form of photosynthesis; they contained the photosynthetic pigment bacteriorhodopsin, which acts as a proton pump; it captures light energy and uses it to move protons across the membrane out of the cell. The resulting proton gradient is subsequently converted into chemical energy. *Bergey's Manual* classified the halophilic bacteria as a genus, *Halobacterium*, comprising five species.[8]

Woese and Fox mentioned the halophiles when they introduced the methanogens as archaebacteria in November 1977. "Because of the great phenotypic diversity of the eubacteria," they wrote, "it would be "unlikely that many, if any, of the yet uncharacterized prokaryotic groups will be shown to have coequal status with the present three. Conceivably, the halophiles, whose cell walls contain no peptidoglycan, are candidates for this distinction."[9] Were the halophiles a group of their own, or member of the Archaebacteria?

In fact, Woese had startling preliminary 16S rRNA data on halophiles the week before he submitted that first archaebacteria paper. "*Halobacterium halobium* is a specific relative of the methanogens," he wrote to Otto Kandler in Munich on August 11, 1977.

> In addition to having significant number of common sequences exclusive to the two groups, the two have the same distribution of modified bases in 16S rRNA. Relatedness then predicts that the halophile will manifest the same peculiarity in the common arm of tRNA found in methanogens, and one (badly done) experiment (nevertheless) shows this to be the case.[10]

The very idea that halophiles and methanogens with their radically different phenotypes could be members of the same group was striking and unexpected.

"The only strong feeling I have so far," wrote Woese, " is that both phenotypes feel primitive—the methanogens because of their use of CO_2, the halophiles because of their simple photosynthetic "proton battery.""[11]

The idea that halophiles were ancient organisms phylogenetically related to methanogens, and distinct from typical bacteria, had been far from the minds of bacteriologists. But Woese and Fox's paper introducing the archaebacteria turned the light on for a few biochemists who studied halophiles. Stanley Bayley at McMaster University in Canada reported in 1966 that the ribosomal proteins of *Halobacterium cutirubrum* were acidic rather than basic, as had been presumed to be the case for all bacteria.[12] He wrote to Woese on January 6, 1978, explaining how he agreed that the extreme halophiles may be "distinctly different from eubacteria."[13]

Alistar Matheson at the National Research Council in Ottawa also honed in on the phylogenetic uniqueness of the halophiles. Matheson, Ross Nazar, and Guy Bellemare were engaged in comparative studies of the 5S rRNA–protein complex of extreme halophiles, mesophiles, thermophiles, and typical bacteria. Their interest was in how the proteins and the RNA of ribosomes interact; they used "odd bacteria" to detect functionally conserved features of all bacteria.[14] They discovered several unique ribosomal characteristics of extreme halophiles, all of which they considered to be adaptations to extremely salty environments.[15]

That adaptationist interpretation was perfectly reasonable if one assumed that halophiles arose from typical bacteria. Indeed, it was the only possible explanation—if bacteria arose from a common stock as virtually every biologist had assumed. "Actually we somewhat misinterpreted the halophile 5S RNA structure," Nazar recalled decades later. "It turns out that essentially all the structure was widely conserved."[16] When the archaebacteria paper appeared, Nazar remembered being neutral to the new interpretation, whereas Matheson was rather keen on it.[17] Indeed, he saw the implications of their 5S rRNA comparisons in a radically new way.

Ironically, in January 1978, a month before Matheson and colleagues submitted their paper interpreting the unique features of *Halobacterium* ribosomes as adaptations, Matheson wrote to Woese suggesting that they submit a joint publication making the announcement that halophiles and methanogens were phylogenetically related as a group apart from typical bacteria. Woese explained that such a joint statement would be inappropriate for several reasons. His student Kenneth Luehrsen had been analyzing and comparing the primary and secondary structural properties of 5S rRNA of methanogens and halophiles, and over the previous eight months, his laboratory had completed 16S rRNA catalogs of seven halobacteria. "The relationship between methanogens and halophiles is blatantly clear from these data" he wrote Matheson. "In addition, halophiles have patterns of base modification in both SSU and tRNAs that are characteristic of the methanogens."[18] It was clear to Matheson that the discovery of the methanogen–halophile relationship had been made by Woese's laboratory. There was no contestation.[19]

Still, the cat had been let out of the bag, and Woese was worried about being scooped. That month, he sent a paper on the halophiles, coauthored with Woese's technician Linda Magrum and Luehrsen, to the *Journal of Molecular Evolution* (*JME*) asking Emile Zuckerkandl, *JME*'s editor, for quick publication: "I've let the word out, and there is considerable data on halophiles now that suggest them to be, at least, very different organisms—all that was lacking was conceptual framework."[20] Zuckerkandl, perhaps more than anyone else outside of Woese's laboratory, understood the conceptual edifice that was emerging. "This is again such an interesting paper and one more piece of the thrilling puzzle whose outlines you have recently revealed," he replied. "I made myself into a referee and found the paper so clear and straightforward that I sent it immediately to the printers."[21]

Woese, Magrum, and Luehrsen's announcement that *Halobacterium halobium* was a member of the Archaebacteria urkingdom appeared in the spring 1978 issue of *JME* with the provocative title "Are Extreme Halophiles Actually 'Bacteria'?" They aimed to overturn the adaptationist thinking and the concept of the monolithic prokaryote that inspired it:

> Comparative cataloging of the SSU rRNA of *Halobacterium halobium* indicates that the organism did not arise as a halophilic adaptation from some typical bacterium. Rather, *H. halobium* is a member of the Archaebacteria, an ancient group of organisms that are no more related to typical bacteria than they are to eucaryotes.[22]

THERMOPHILES

By the time Woese sent the paper on halophiles to *JME*, his group had tested the thermoacidophiles *Thermoplasma* and *Sulfolobus*. As he wrote to Zuckerkandl, "Archaebacteria are beginning to look genuinely primitive."[23] Conceiving the thermoacidophiles as ancient members of the Archaebacteria also confronted commonsense adaptationist assumptions regarding their unusual characteristics. They live in geothermal environments that are acidic and so hot they would cook other organisms.

Ever since Pasteur, it had been generally understood that the temperature range for life is very narrow: from about –5°C to 80°C.[24] The lower limit is set by the temperature at which water freezes, but it is slightly below 0°C in a cell, owing to dissolved organic and inorganic compounds. Of course, exposure to lower temperatures does not necessarily result in death; many microbes can survive in a frozen state for long periods of time even though all metabolic activities are arrested. The upper temperature limit is set by proteins and nucleic acids, both of which are typically destroyed at temperatures ranging between

50°C and 90°C.[25] Bacterial spores were known to survive in boiling water, but not bacteria themselves.[26] Bacteria such as *E. coli*, that have temperature optima between 20°C and 45°C were known as mesophiles; those whose temperature optima lie below 20°C, as psychrophiles; and those that thrive above 45°C, as thermophiles.

Since the 1830s, there were reports of "algae" growing in hot springs and geysers with temperatures ranging from 70°C to 83°C.[27] In 1875, Felix Hoppe-Seyler at Strasbourg noted two problems related to those reports: the evidence that they were actually alive and the precise temperatures in which they lived. His own studies indicated that the upper limit of "algal" growth was actually about 60°C. He speculated that as Earth cooled, chlorophyll-containing organisms could have lived when the temperature was about 60°C.[28]

That life might have originated in hot water was advocated by microbial ecologist Thomas Brock of Indiana University when in the 1960s he and his coworkers set out to culture and characterize microbes from hot springs derived from volcanic activity, ancient habitats at least as old as life on Earth.[29] Far from the cozy temperate conditions of Darwin's "warm little pond," he commented in 1967, "it is not inconceivable that thermophilic microorganisms are related to primordial forms which gave rise, through many mutations followed by selection, to mesophilic and psychrophilic forms."[30]

Brock's team isolated the organism they named *Thermoplasma acidophilum* in a coal refuse pile at the Friar Tuck mine in southwest Indiana in 1970.[31] Its temperature range was 45°C to 62°C, and it lacked a cell wall. Brock had little doubt that *Thermoplasma* was related to the mycoplasmas, the general name of a group of bacteria that were nonmotile and lacked a cell wall.[32] They were known to cause a variety of diseases in animals—from mice to humans.[33] The discovery of *Thermoplasma* was thought to broaden considerably the range of habitats in which "mycoplasma-like" organisms lived, and mycoplasmas themselves were considered to be primordial forms. "The structural simplicity of these free living mycoplasmas," Brock and colleagues wrote, "suggests that they might be the homologs of a primordial organism, and hence the study of these organisms may provide some insight into aspects of the origin of life."[34] Others followed them in arguing that, because of their small genomes and general simplicity, mycoplasmas might be descendants of primitive organisms that preceded the typical bacteria in evolution.[35]

Brock's team isolated and characterized another genus of thermoacidophilic bacteria living in highly acidic hot springs of Yellowstone Park, which in 1972 they named *Sulfolobus acidocaldarius*.[36] It oxidized sulfur and grew at a temperature optimum of 70–75°C.[37] And, as they saw it, it was completely unrelated to the wall-less *Thermoplasma*.[38] "Sulfolobus apparently has no close relationship with any previously described bacteria, either heterotrophic or autotrophic," they said.[39] "The two genera are quite unrelated, and can be easily distinguished both in nature and in culture."[40] *Thermoplasma* formed typical "fried-egg" mycoplasma-like colonies, whereas *Sulfolobus* formed smooth, glistening colonies.[41] *Thermoplasma* appeared to reproduce by budding or by narrow hyphae, whereas *Sulfolobus* apparently reproduced by septation of lobes. "Probably the best characteristic for distinguishing the two organisms is sphericity, *Thermoplasma* always being evenly spherical and *Sulfolobus* being lobed spheres."[42]

The 16S rRNA comparisons of extreme thermophiles pointed to a fundamental different order of things: (1) The mycoplasma was not a group of ancient primordial organisms, at all; they were actually related to typical bacteria but had lost their walls in the course of evolution. (2) *Thermoplasma acidophilum* was not phylogenetically related to the mycoplasma.[43] Its lack of a cell wall also resulted from a loss; it was a relatively recent adaptation. Wall-lessness was not a phylogenetic trait. Thus, the commonality of the *Thermoplasma* and *Mycoplasma* was an illusion, reflecting convergent evolution; it held no phylogenetic meaning.[44] (3) *Thermoplasma* and *Sulfolobus* were phylogenetically related as members of the archaebacterial group. (4) Extreme thermoacidophiles did not give rise to mesophilic and psychrophilic bacteria as Brock had supposed; they were, phylogenetically speaking, worlds apart.[45]

In the spring of 1978, Woese sent a paper to *Nature* with the news about *Thermoplasma*. To his bewilderment, and despite positive peer review reports, it was rejected as being "of insufficient interest to biology."[46] In reaction to *Nature*'s rejection, he did not publish the rRNA data on *Thermoplasma* and *Sulfolobus* until 1980, by which time he constructed the outlines of a universal tree of life that also included extensive work classifying the typical bacteria on the basis of 16S rRNA comparisons (see chapter 17).

Searching for the Phenotype

In the meantime, the search for other features to corroborate the archaebacterial grouping continued. The unique coenzyme M found in methanogens was not found in the extreme halophiles or thermoacidophiles. Woese and his student Ramesh Gupta showed that unusual structural features of the tRNAs of methanogens were also features of halophiles and thermophiles.[47] By the end of the 1970s, methanogens, extreme halophiles, and thermoacidophiles were shown to share a unique phenotypic property in the structure of the lipids in their membranes. Demonstrating that this characteristic was a conserved trait again entailed confronting adaptationist thinking in microbiology.

UNIQUE LIPIDS

The membrane is a regulatory structure; it is selectively permeable and controls what enters and exits the cell. Research on the composition of bacterial membranes was important for understanding many practical problems, including how viruses and antibiotics could penetrate them. Biochemists had shown that the membranes of extreme halophiles and those of thermoacidophiles *Thermoplasma* and *Sulfolobus* had a unique branch-chain, ether-linked lipid structure not found in any of the typical bacteria (or in eukaryotic cells) whose membranes contain ester-linked lipids: glycolipids and phospholipids. The lipids of these "extremophiles" were based on isopranoid-branched chains (five-carbon subunits). Their unusual membranes were understood to be the means by which these organisms adapted to the unusual environments. Their common membrane type was thought to be a case of convergent evolution; it was held to be of no phylogenetic significance.

The unusual lipids in the membranes of *Halobacterium cutirubrum* were first reported in 1962 by Morris Kates at the National Research Council of Canada in Ottawa, who examined them on the premise that they would have evolved differently to cope with saturated salt environments.[48] The same adaptationist explanation was applied to the unusual membrane lipids of *Thermoplasma* and *Sulfolobus* when their unique chemistries were first reported in 1972 and 1974.[49] Their lipid structures were assumed to be the result of independent adaptations to life at high temperatures and to low pH.[50] The membranes were not considered to be a common phylogenetic trait, and *Thermoplasma* and *Sulfolobus* were not considered to be related to one another any more than they were related to halophiles. Brock wrote the book on *Thermophilic Microorganisms at High Temperatures* in 1978, in which he made the claim plain:

> The similarities between the lipids of *Thermoplasma* and *Sulfolobus* provide strong evidence for the importance of these structures in the stabilization of the membranes of these organisms under hot, acid conditions. Because these two organisms are clearly unrelated we have here an excellent example of convergent evolution.[51]

Again, he commented,

> The fact that *Sulfolobus* and *Thermoplasma* have similar lipids is of interest, but almost certainly this can be explained by convergent evolution. This hypothesis is strengthened by the fact that *Halobacterium*, another quite different organism, also has lipids similar to those of the two acidophilic thermophiles.[52]

The 16S rRNA approach functioned like a theory, a cognitive map, a phylogenetic forecaster to determine which organismal traits were adaptations and which were conserved.[53] In this case, it predicted that methanogens would also have the same unique lipids as *Thermoplasma*, *Sulfolobus*, and halophiles. If so, membrane lipids would represent the first organismal trait shown to be common to the newly proposed urkingdom.

Thomas Tornabene at Colorado State University and Thomas Langworthy at the University of South Dakota tested the membrane lipids of methanogens.

Marvin Bryant, a methanogen expert at the University of Illinois who had worked with Ralph Wolfe on culturing methanogens, approached Tornabene at a symposium of the United Nations Institute for Training and Research held in Göttingen in October 1976. The theme of the conference was alternative fuels. Tornabene presented a paper on microbial formation of hydrocarbons. Bryant explained to him that "they had methanogens that appeared to have no classical lipids (glycerol acyl fatty acid derivatives)."[54]

Tornabene had been immersed in the principles of evolution and the origin of life as a graduate student.[55] He had completed his doctoral research in 1967 under the direction of John Oró at the University of Houston, well known for his research on the origin of life. Following Stanley Miller, Oró synthesized amino acids from hydrogen, cyanide, water, and ammonia. Subsequently, he synthesized the nucleotide base adenine from those chemicals, which had been identified to have existed on primordial Earth.[56] Those experiments led to the laboratory synthesis of the rest of the components of nucleic acids. Oró received some of the first lunar samples that were released by NASA for analysis, and he helped to design experiments and built equipment used during the Viking mission to investigate the existence of life on Mars. After arriving in Houston, Fox informed Oró about their theory that the methanogens would have ether lipids. Oró was excited about the archaebacterial concept; they worked to get methanogen cells to Tornabene from Wolfe.

Tornabene's research program on lipids had begun after he completed his Ph.D., when he worked for one year as a postdoctoral fellow with Kates in Ottawa.[57] In his view, the evolution of fundamental cell processes was not a random affair driven by natural selection; it was an orderly process governed by physical and chemical laws. One had to consider colloid properties, energy states, and principles of self-assembly.[58] Tornabene reasoned that for an organism to have the unusual isopranoid lipids required, it had to possess a gene cluster that had followed a long evolutionary path. All organisms that had those lipids evolved along the same related evolutionary tracks. The lipid contents were therefore "key taxonomic markers, more significant than the often accepted classification markers of nutrition, habitat, and morphology."[59] Nonetheless, Tornabene was initially skeptical of Woese's rRNA approach and had little understanding of the concepts underlying nucleic acid sequencing. "I was shocked at the early successes of SSU RNA as a taxonomic tool," he commented decades later.[60]

When Tornabene began his studies of the lipids of methanogens, those organisms seemed to be worlds apart from *Halobacterium*. Methanogens were anaerobic, and the extreme halophiles were reported to be strictly aerobic. As a first step to establishing a link between them, Tornabene showed in the fall of 1977 that the extreme halophile *Halobacterium cutirubrum* was in fact not strictly aerobic: it could also be cultured in anaerobic conditions. He wrote up the results in January 1978 and sent them to Oró, who asked Fox to comment.

The fact that this extremely halophilic bacterium was not obligately aerobic, Tornabene argued, accentuated its already recognized uniqueness:

> The data support the proposal that the halophiles have followed a long and distinct evolutionary path. However, the data are still insufficient for recommending that the organism be separated from the prokaryotes and put into a third kingdom as recently suggested for methanogens or that it is a primitive form of life.[61]

Wolfe sent Tornabene pellets of the thermophilic methanogen *Methanobacterium thermoautotrophicum* (delta H) for lipid membrane analysis, which he completed after six months of concerted effort.[62] It possessed the same unusual diether-linked lipids as found in *Halobacterium cutirubrum*, and those lipids were akin to those of the *Thermoplasma* and *Sulfolobus*.[63] When, in early March 1978, Woese learned the news from Fox in Houston, he wrote to Oró:

> That is great; it adds strong confirmation to our discovery of their relatedness, in terms of rRNA and tRNAs. Kandler's group now finds similarities in wall structures also. So, the archaebacterial kingdom is shaping up nicely. As I believe I told you over the phone (and George has told you in any case) we have shown Sulfolobus and Thermoplasma to be in the group as well.[64]

That finding did indeed support the concept that the methanogens, halophiles, and thermoacidophiles shared a deep phylogenetic lineage. But, by itself, it was not decisive.

Not all methanogens lived in extreme environments. Some lived in moderate temperature ranges: the mesophilic methanogens. Did they also have the unusual lipid membrane? If the odd lipid structure of the thermophile *Methanobacterium thermoautotrophicum* were an adaptation to high temperature, one might not expect to find that same unique lipid structure in mesophilic methanogens. Woese contacted Langworthy, who in the early 1970s had shown that the membranes of *Thermoplasma* and *Sulfolobus* contained the same unusual lipid membranes as did halophiles.[65]

Langworthy had completed his doctoral research at the University of Kansas in 1971 on the effect of T_4 phage on the bacterial surface receptor lipopolysaccharide composition. Once infected, a bacterium became resistant to further infection. Why? The field became crowded, and after completing his thesis he went to work as a postdoctoral fellow at the University of South Dakota with Paul Smith, a specialist in *Mycoplasma*.[66] Over the next year, Langworthy had searched for peptidoglycan precursors in the wall-less mycoplasmas to resolve the question of whether they were really degenerate bacteria or ancient ancestors to the bacteria.[67]

Langworthy and his coworkers reported in 1972 that the abnormally long tetraether lipids of *Thermoplasma* were the result of adaptation to their extreme environment, just as were the diether lipids of halophiles.[68] The next year he obtained a culture of *Sulfolobus* from Brock. When he and coworkers reported their results in 1974, they again saw them to be adaptations to life in the extremes:

> The occurrence of long-chain glycerol diethers in *Sulfolobus* lends support to the suggestion that thermophily can be related to the long isopranol chains of the lipids and acidophily can be related to the sole presence of ether lipids in these obligatory thermophilic, acidophilic organisms.[69]

By the mid 1970s, things seemed pretty clear regarding membrane structure and function of microbial life in extreme environments. The common isopranoid-ether lipids in the membranes of *Thermoplasma*, *Sulfolobus*, and halophiles were independent adaptations to life in extreme environments. Langworthy put it in a nutshell: "This all seemed to be a case of convergent evolution."[70] If the odd lipid structure of the thermophiles was an adaptation to high temperature, one might not expect to find that same unique lipid structure in mesophilic methanogens. Indeed, Langworthy and Tornabene suspected that the membranes of mesophylic methanogens would *not* possess the abnormal lipids.[71] Wolfe sent them eight different strains. All of them possessed the unique lipids composed of either diphytanyl glycerol diethers or of tetraethers.[72] The unique isopranol glycerol ethers were common to all methanogens, as well as halophiles, *Sulfolobus*, and *Thermoplasma*.

Tornabene and Langworthy changed their adaptationist views. As they now saw it, the similarities in those lipid structures were due to common ancestry, not convergence. They made the issue plain when they published their results in *Science* in January 1979:

> The data reported here establish that the mesophylic methanogens also contain the same ether lipids, which must represent a long evolutionary relationship between methanogens, *Halobacterium*, *Thermoplasma*, and *Sulfolobus*; this argues against the possibility that these ether lipids reflect environmental adaptation.[73]

The biochemical work on the lipids of mesophilic methanogens was important, but it was certainly not decisive; it could still be argued that thermophilic methanogens preceded mesophilic forms and passed along the unusual trait. Theories about the evolution of biochemical pathways and genetic evolution aside, all these characteristics could still have been adaptive, unrelated adjustments to life in extreme environments—remarkable examples of convergent evolution, as Brock and many others had supposed. Woese wrote to John Bu'Lock at the University of Manchester in March 1978, when explaining the latest results: "We feel archaebacteria are of very ancient origin, perhaps predating the common ancestor of typical bacteria, and the early Archaen Era was an 'Age of Archaebacteria.'"[74]

Still, there was great skepticism regarding the rRNA phylogenetic approach, and especially the claim that methanogens represented a deep divergence, an organismal lineage apart from other bacteria. Lynn Margulis was among many microbiologists who were completely unconvinced of both. In 1978, she teamed up with ecologist Robert Whittaker in advocating five kingdoms: Monera, Protista, Fungi, Planta, and Animalia.[75] When Woese sent reprints of the papers on methanogens to her and informed her of results indicating that the halophiles and thermophiles were also members of the Archaebacteria, she replied in January 1978: "The 'Archaebacteria' question needs a discussion."[76] Molecular biologists in the United States also (initially) considered it likely that the unusual 16S rRNA results, like the odd lipids, indicated adaptation to high temperature, low pH, high salinity, or some combination thereof.[77] In other words, the shared traits of the methanogens, *Sulfolobus*, *Thermoplasma*, and *Halobacterium* would be the result of convergent evolution.

Outside the United States, the reception to the archaebacterial concept seemed to be more favorable. Nowhere was the support greater than in Germany. There, a small band of biologists warmly embraced the new method as providing a phylogenetic tree on which they could order corresponding fundamentally conserved traits. For them, the 16S rRNA approach and the archaebacterial concept rejuvenated microbial taxonomy, transforming it from descriptive cataloging to a dynamic experimental question-driven endeavor of the greatest evolutionary significance.

FIFTEEN

In the Capital of the New Kingdom

Weissbier can not be found in Urbana. Too bad.... However, Munich will
soon be even more famous as the world capital for archaebacterial research
than for its beer.

—Carl Woese to Wolfram Zillig, August 6, 1979

ARCHAEBACTERIAL RESEARCH PROGRAMS got started in Germany right off the
bat. They emerged from the study of bacterial walls led by Otto Kandler (b. 1920).
Director of the Botanical Institute at the University of Munich, he was instrumental in the development of archaebacterial research and of 16S rRNA phylogenetics more generally. He organized the first meeting on the archaebacteria in
Munich in 1979; two years later he co-organized the first international workshop
on the archaebacteria, and he edited the first book on the archaebacteria, based
on that meeting.[1] Kandler was also editor-in-chief of the journal *Systematic and*
Applied Microbiology, which provided a forum for papers on microbial phylogeny.

Kandler developed the archaebacterial concept from the perspective of cell
wall structures, and he encouraged others in Germany to begin archaebacterial
programs from other viewpoints. His former student Karl Stetter (b. 1941),
together with Wolfram Zillig (1925–2005), launched an extensive archaebacterial research program based on molecular biology and natural history.

Walls for the Kingdom

Kandler was informed of the 16S rRNA evidence that methanogens were
unique five months after that data had been analyzed. On November 11, 1976,
he received a letter from Ralph Wolfe. They had met briefly in Woods Hole

a few years earlier when Wolfe was teaching in the microbial diversity course there, and Kandler was visiting his friend Holger Jannasch. In his letter, Wolfe explained that he had been collaborating with Woese's group, who established "that methanogenic bacteria are clearly set apart from other prokaryotes" and that methanogens possessed "two unique coenzymes, coenzyme M and coenzyme F_{420}." He explained that they had the expertise to mass culture the methanogens; they had almost all of them under cultivation and could supply Kandler with "lyophilized cells in reasonable amounts" to examine the chemical structure of their walls.[2]

Kandler needed no encouragement. The previous year his group had analyzed the walls of the extreme halophile *Halococcus morrhuae*.[3] And just two weeks before he received Wolfe's letter, he had analyzed the walls of two strains of *Methanosarcina*.[4] His colleague Hans Hippe in Göttingen had learned to grow them a few years earlier. Kandler demonstrated that the walls of those methanogen strains lacked peptidoglycan, and their chemistry seemed to resemble the walls of *Halococcus*. Coincidentally, at the time that Wolfe's letter arrived, Wolfe's colleague Marvin Bryant was visiting Kandler in Munich and was to give a seminar on methanogens; his own specialty was anaerobic degradation and methanogenesis of sewage.[5] He and Kandler had already been discussing the importance of testing other methanogens. Recall that Bryant had suggested that Thomas Tornabene examine the lipid membranes of methanogens the month before, at a conference in Göttingen centered on biofuels (see chapter 14). Kandler replied to Wolfe on November 17, 1976, suggesting that both methanogens and halophiles might be "ancient relics" that may

> have branched off from the bulk of the procaryotes before the peptidoglycan had been "invented." Halobacteria as well as Methanobacteria would be very appropriate candidates for such ancient relics, as also indicated by the finding of Carl Woese, mentioned in your letter, that the Methanobacteria are also set apart from other procaryotes by the oligonucleotide pattern of S 16 RNA. The situation may be compared with that in the algae, where the different classes developed quite different cell wall polymers, too.
>
> Therefore it would be very important to investigate all the other strains of Methanobacteria available at present to demonstrate that the lack of peptidoglycan is a general feature of this group. Dr. Bryant promised to send some material of M. ruminantium which he can grow in Göttingen, where he spends some time as a guest.
>
> I would certainly be interested to get also material from your collection.[6]

Woese knew nothing of Kandler or of his new collaborations with Wolfe and Bryant to survey the methanogens' walls.[7] Woese and Kandler first met when Wolfe introduced them in Urbana on a cold winter's day in January 1977. Kandler was attending a Gordon Conference on bacterial cell walls in Santa Barbara and made arrangements to visit Wolfe on the way back to Europe.[8] It was as if Kandler had come out of the blue. Woese and Fox explained to him the story that was emerging. He required no convincing.[9]

Kandler had come to microbiology with great difficulty in the troubled circumstances of war-torn Germany. He had been a radio operator during the Second World War, listening to the British and recording information from decoders, mainly on the Eastern Front. He was in Austria at the war's end, and to avoid capture by the Russians, he bicycled to the Western Front to be captured by the Americans. After spending a few months in prison camp, he went home to help his father in the rural Bavarian town of Deggendorf. His father was a gardener who grew and sold vegetables at the local market, and he had the intention of taking over the business. But his father was not enamored with his livelihood and recommended that his son look elsewhere.[10]

In 1946, Kandler enrolled in botany at the University of Munich, earning money by growing and selling cabbage and flowers in Deggendorf. To be admitted, he and other students had to remove rubble from the ruins from the university by wheelbarrow. Much of the university had been bombed. There were few professors, no diploma programs, and no one to supervise his research. But he met a medical doctor who had been in China during the war, as the personal physician to Chaing Kai-shek. He was able to obtain recent scientific journals, and as a member of the botany society in Munich, he went on small botanical expeditions, and took Kandler with him. He suggested that Kandler work on plant tissue culture, then a rather new field. Kandler became involved in research on photosynthesis and biochemistry.[11]

After receiving his doctorate in 1949, Kandler remained at the University of Munich until 1957, teaching and studying various aspects of plant physiology, as well as the wall-less mycoplasmas and the cell walls of other bacteria.[12] The research facilities were extremely poor at the university; even obtaining such basic equipment as a pipette required a special request. As a result, in 1957 he accepted a position as director of the Bacteriological Institute of the South German Dairy Institute. There he found better research conditions. He developed and introduced new starter cultures for the production of fermented milk products, and also for vegetable (sauerkraut) and meat (dry sausage) fermentations.[13] He also improved and optimized biofuel production by utilizing thermophilic methanogenic bacteria, and he studied the efficiency and stability of methane fermentation of wastes. He also studied the physiology and taxonomy of several other groups of bacteria, particularly lactobacilli, leuconostocs, and bifidobacteria. Three years later, Kandler was appointed chair professor of applied botany at the Botanical Institute at the famed Munich Polytechnic Institute, renamed in 1970 Technische Universität München (Technical University of Munich).

In 1973, Kandler accepted the chair of botany at the Botanical Institute of the University of Munich.[14] At the time he met Woese, he was already a distinguished botanist, microbiologist, and comparative biochemist.[15] He was professor in the old German sense—when biologists were much less specialized than today. He called himself a "biologist" and taught a diversity of courses, including plant and bacterial taxonomy.[16] His evolutionary disposition combined with his study of cell walls made him most receptive to the archaebacterial concept in the winter of 1977. As he wrote to Woese the following year,

> It was a very fortunate coincidence…that I met you just at the time, when you got definit [sic] results on the methanogens and I had the first information about the lack of peptidoglycan and the presence of different cell wall polymers in these organisms. It was not too difficult to recognize the phylogenetic implications.[17]

Indeed, Kandler's visit pointing out the lack of peptidoglycan in the walls of methanogens and halophiles was important to the research program in Woese's laboratory in searching for other organisms that might be members of the urkingdoms based on the 16S rRNA probe (see chapter 12).

Studies of cell walls had a long history in biology and taxonomy. The discovery of the walls in plant tissue had given rise to the term "cell" itself, which, in the seventeenth century, Robert Hooke called those hollow structures he believed to be responsible for the buoyancy of cork. In the early nineteenth century, when cells were conceived of as structural entities only, the wall was considered to be the most important part. During the second half of that century, the word acquired a different and contradictory meaning when biologists recognized that, far from being hollow chambers, the so-called "cells" were alive, when the study of their internal organization took on special importance. Not all cells had walls. The recognition that bacteria possessed walls had supported their inclusion in the plant kingdom.

Peptidoglycan was known to be responsible for the rigidity of the bacterial wall (see chapter 14). It is a large macromolecule composed of two constituent glycan strands (polysaccharides) cross-linked by short amino acid chains. In 1962 Stanier and van Niel pointed to the presence of peptidoglycan in the cell wall as the only positive character defining the "prokaryote" when they referred to a "specific mucopeptide" (see chapter 7).[18] The demonstration that the walls of blue-green algae possessed peptidoglycan corroborated their inclusion among the bacteria as "cyanobacteria."[19]

Kandler and his student Karl-Heinz Schleifer had been studying bacterial classification based on differences in the amino acid composition of peptidoglycan.[20] Gram-positive bacteria retained the Gram stain (a dye-iodine complex) after treatment with alcohol, whereas the Gram-negative bacteria were decolorized. Gram-negative bacteria had diverse and complex cell walls of several layers consisting of only 10% peptidoglycan. Gram-positive bacteria had one thick layer in their cell wall consisting of 30–70% peptidoglycan. The amino acid composition of peptidoglycan did not vary much among Gram-negative bacteria but varied greatly among Gram-positive bacteria.

There were relatively few comparative studies of the amino acid sequences of peptidoglycan to classify the Gram-positive bacteria.[21] About 50 peptidoglycan types had been identified by 1970. But the methods then were too laborious for routine application.[22] Schleifer and Kandler aimed to improve them. Peptidoglycan had great value as a "taxonomic marker": it was widespread among bacteria, and it was determined by at least 20 genes. Schleifer and Kandler argued that the polygenic character of the different peptidoglycan types made it rather unlikely that they would have arisen by convergent evolution. The very diversity of the peptidoglycan types conserved over the course of evolution suggested that they were not especially adaptive; they were not connected with any remarkable "advantage or disadvantage of selection."[23]

Complementarity

The structure of the walls was good for clustering organisms into related groups, but it was of no value in determining relationships between groups. One could not recognize derived sequences so as to determine if a particular structure is primitive or highly developed. In 1972, Schleifer and Kandler offered a speculative phylogenetic sketch of bacterial evolution based on the assumption of an evolutionary trend toward chemical simplification and loss of variability in cell walls.[24] The Gram-positive non-spore-forming bacteria were at the base of their tree. There had been reports since the 1960s that extreme halophiles such as *Halobacterium halobium* lacked peptidoglycan in their walls.[25] Kandler's group had shown that this was also true of another extreme halophile that he and Schleifer assigned to a new genus *Halococcus*.[26]

That such abnormalities in bacterial walls were of any taxonomic importance was certainly not accepted by all. In 1972, Thomas Brock and his colleagues reported that *Sulfolobus* also lacked peptidoglycan in its walls, just as did extreme halophiles.[27] But, as they saw it, those abnormalities were adaptations to life in extreme environments. They were of no more taxonomic value than were the similarity of the unusual lipids of *Sulfolobus* and *Thermoplasma*—which they suggested was due to convergent evolution. "The wall, and perhaps the membrane," they said, "are the only cellular structures which must be exposed to high hydrogen ion concentration of the environment and conceivably the unusual wall of *Sulfolobus* may be a factor permitting adaptation to this extreme condition."[28]

A few weeks after meeting with Woese and Fox, in January 1977, Kandler drafted a paper emphasizing that the structural components of the cell wall of *Methanosarcina barkeri* resembled that of *Halococcus morrhuae*:

> In addition to the unique metabolism of the methanobacteria and their very aberrant ribosomal RNA, the oligonucleotide pattern of which resembles that of the typical procaryotes no more than it does that of eucaryotes (Woese and Wolfe, pers. comm.), the lack of peptidoglycan supports the opinion, that the methanogenic bacteria form a divergent systematic group clearly separated from the other bacteria.[29]

He sent a draft to Woese.[30]

When Woese's lab subsequently analyzed the 16S rRNA signatures of halophiles, and included them in the archaebacteria, he wrote to Kandler six months later, "Your paper on methanogen walls suggests that this possible relationship has been in your mind."[31] The next year, Kandler and Helmut König analyzed the walls of seven strains of methanogens. None of them possessed a cell wall composition of the typical bacterial peptidoglycan. In the evolutionary scenario they envisaged, peptidoglycan would have evolved in typical bacteria after a wall-less ancestor had branched. Beginning with a wall-less ancestor, there would be two primary branchings of the bacteria:[32] the typical bacteria, as characterized by peptidoglycan wall structure, and the lineage represented by the methanogens and the extreme halophiles that would have branched early before the "invention" of peptidoglycan.[33]

Woese and Kandler worked in collaboration, sharing cultures, data, and ideas to complement each other's work as the archaebacterial urkingdom expanded and as the phylogenetic relationships within the two primary prokaryotic kingdoms were being sorted out. As Woese's laboratory identified members of the archaebacteria and their phylogenetic relationships by 16S rRNA, Kandler surveyed their cell wall composition and identified characteristic components. Chemical analysis of the walls of methanogens revealed an amazing diversity of structure and chemistry.[34] Kandler was able to match that diversity in structure and chemistry with the proposed phylogenetic relationships based on comparative cataloging of the 16S rRNA. In 1978, he mapped the distribution of cell wall components onto the first dendrogram of methanogens.[35] No common cell wall component, such as peptidoglycan, was found within the archaebacteria. But Kandler's work showed that the walls of all those organisms possessed a remarkable structural and chemical diversity.

As the 16S rRNA catalog work expanded to construct a large-scale phylogenetic ordering of what Woese and Fox referred to as the eubacteria, there, too, many of the previously proposed taxa based on morphology had to be changed and new taxa created. When Woese's laboratory identified a new taxon, Kandler would try to find a corresponding phenotypic character for it. It was a relationship of complementarity. On the one hand, the characterization of the cell walls put "flesh on the bones of a phylogenetic skeleton derived from sequence data."[36] On the other hand, the 16S rRNA sequence data offered a deeper insight into cell wall evolution.

Transforming the Eubacteria

The 16S rRNA technology was transferred to Germany by Erko Stackebrandt, who spent an extraordinary year in Woese's laboratory in 1978. He had completed his thesis four years earlier with Kandler on "Biochemical and Taxonomic Investigations of the Genus *Cellulomonas.*" Then he worked at the German Collection of Microorganisms and Cell Cultures that was partly housed in Kandler's institute. Stackebrandt classified bacteria based on peptidoglycan types and on DNA homology: DNA–DNA similarities were useful for classifying closely related taxa. But he had become disenchanted with his job and accepted a position in Pretoria, South Africa, to work on virus molecular biology.

But when he came across a paper from Woese's laboratory in 1976 describing 16S rRNA homologies, Stackebrandt changed his mind and wrote to Woese asking to work with him to learn his techniques: "Since DNA homology studies are of taxonomic value only for closely related taxa, i.e. for the differentiation of species and subspecies, it would be very important for my further scientific career to get acquainted with the method of comparing the structural homology of the 16S rRNAs, which allows the recognition of the phylogenetic relationship between more distantly related species, genera or families."[37] Kandler recommended Stackebrandt as "a gifted researcher."[38]

Woese and Stackebrandt mapped out an ambitious research plan in the summer of 1977. Their aim was to examine those taxa of eubacteria that might be polyphyletic and then separate out the ones that did not belong.[39] Then, in November, the news arrived in the media about the third form of life. Stackebrandt wrote to Woese: "Each newspaper and magazine write about the successful 'discovery' of a new group of organisms by the team of Woese, Wolfe and Fox. Of course the daily papers exaggerate and speculate a little bit but there is much interest in this subject in Germany at the moment."[40] He arrived in Urbana in the next month.

Stackebrandt had little idea that he would be part of a broad revolution in bacterial taxonomy. Species of eubacteria defined by morphological characters often turned out to be unrelated; the most notable cases involved the class Actinomycetes. The majority of the previously proposed taxa (above the genus) had to be changed. "We have completed our 'phylogenetic dissection' of the Micrococcaceae," Woese wrote Kandler at the end of the year.

> Perhaps Erko has told you already. The Family is now defunct as a phylogenetic unit: the three recognized genera are unrelated to one another. In fact, only Staphylococcus has any chance of remaining intact as a genus! Any reasonable group of *Micrococcus* species also includes some *Arthrobacter* species, which makes the micrococci no more than degenerate forms of the arthrobacteria.[41]

Stackebrandt's work far exceeded expectations, as Woese wrote to Kandler on December 22, 1978: "His year here has been extremely productive for the laboratory. I am most grateful to him. He is a superb young scientist, and I look for him to do well. I have never encountered a harder working man. He is also blessed with a wonderful disposition."[42] Stackebrandt and Woese wanted to continue their collaboration and to have the 16S rRNA technology transferred to Germany. Stackebrandt was to work as a research associate in a new institute of microbiology at the Technical University of Munich headed by Schleifer. Kandler had left that university in 1973 to accept the chair in botany at the University of Munich. But before leaving, he convinced the administration to establish a new chair of microbiology there. Schleifer was appointed to that new chair the following year. His laboratory was deeply involved with the DNA–RNA hybridization when Stackebrandt arrived.

Transferring the 16S rRNA technology to Germany required funds and new equipment; it also entailed transforming the procedures—the German government's policy restricted the use of radioactivity at the level that Woese's laboratory used for labeling RNA.[43] Stackebrandt and coworkers developed an alternative method that used lesser amounts of radioactivity. Instead of growing radioactive cultures, they established a method to label the oligonucleotides after they were generated. The way was open to produce ^{32}P-labeled preparations in Germany.[44] "We are busy now in our sequencing work," Stackebrandt wrote Woese on May 26, 1979.[45] He worked as a research associate in Schleifer's laboratory for five years before he was appointed head of the Department of Microbiology at the University of Kiel.[46] During that time, he and Woese collaborated on about 20 papers, which including reordering the purple bacteria (later called proteobacteria) in terms of alpha, beta, and gamma groups.[47]

Kandler's Engine

When Stackebrandt returned to Munich at the start of 1979, he was surprised by the great interest in the archaebacteria there. "A lot of people are now working with—even isolating—archaebacteria," he reported to Woese."[48] Kandler and Woese had corresponded regularly since the time they met in Urbana in mid-January 1977, sharing ideas and strains and informing each other of research progress in their laboratories. "I am pleased that there is so much active interest in archaebacteria in Munich" Woese replied. "I hope many new ones are isolated. I am both pleased with and happy for Dr. Kandler—his intense interest in archaebacteria will help get the concept accepted and work done."[49] The archaebacterial kingdom was indeed gaining in strength with the evidence that those organisms that Woese's laboratory identified as members possessed unique characteristics, including the structure of their transfer RNA and the lipids in their membranes. All of them lacked peptidoglycan in their walls, and Kandler had shown they possessed such a remarkable diversity in their wall structure.

Yet another remarkable feature of the archaebacteria was discovered in the late 1970s, at the heart of the genetic machinery of the cell. The composition of the enzyme responsible for the transcription of DNA to RNA, DNA-dependent RNA polymerase, was strikingly different in the archaebacteria compared with typical bacteria. That important discovery, which convinced many of the uniqueness of the Archaebacteria, was made by Karl Stetter and Wolfram Zillig. They developed a vibrant research program of their own combining microbiology, molecular biology, and natural history and expanded the kingdom by culturing new forms that they collected in field trips to "extreme" environments.

Stetter started off as a machine engineering student at the famed Munich Polytechnic Institute in 1964. After two and a half years, he switched to work toward a diploma in biology and went to work under Kandler's direction. Those were turbulent years as a wave of student protest movements rolled across Europe, Japan, and the United States. Strikes and marches in the late 1960s were aimed at the "establishment," at the war in Vietnam, at the state, and at the universities, which were perceived to be complicit "knowledge factories" that did not encourage independent critical thinking. Students in Europe demanded control over course content and the selection of new faculty members, and they protested against the elitist power of the professors.

In Germany, as elsewhere in Europe, authority in the universities was steeply hierarchical. German institutes were essentially the domain of one chaired professor who possessed the bureaucratic power to allocate facilities and funds to junior staff, thus determining the direction and nature of research in his institute. Sociologists of the 1970s emphasized that this university system tended to be conservative and that the creation of new specialties was often difficult.[50] The Ministries of Culture in the German states were reluctant to establish a new chair because of financial and cultural considerations, and chair professors often preferred to accommodate new specialties within existing institutes by offering temporary teaching contracts.

Kandler certainly wielded great influence in microbiology in Germany, and he did his best to ensure that his field prospered and that his talented students were well situated. But his attitude was as far from conservative as his effect was to stifle innovation. He was the connector, the hub of the microbiology network in southern half of Germany, organizing meetings and suggesting research directions. Stetter did not participate in the student protests, and the changes ensuing from those movements were not for the better, as he later saw it.[51] They did not entail more student involvement; they only resulted in authority being taken from the professoriate and given to the bureaucratic administration of the university.[52] As a young student, he did not see the abuse of professorial power. Kandler mentored and encouraged him and taught him biochemistry and microscopy and how to culture organisms in isolation. And after completing his doctoral research, Kandler suggested that he work on the archaebacteria. Indeed, the thrust toward archaebacteria in Germany, Stetter said, "was really the engine of Kandler; he was very influential and critical; people trusted him."[53]

Into the Cave of the Lion

Stetter's collaboration with Zillig stemmed from his doctoral research on gene regulation in *Lactobacillus*, bacteria that form lactic acid from sugar. Lactic acid fermentation had been used since Roman times to preserve food, as it is today. It is responsible for the sour taste of dairy products such as cheese, yogurt, and kefir. It also gives the sour taste to fermented vegetables such as traditionally cultured pickles and sauerkraut. Stetter's research was on the development of the lactic acid isomers (compounds that have the same number and type of each atom but a different arrangement of atoms). There are two kinds of lactic acid isomers, D and L. They are the same in every way except being non-superimposable mirror images: D for Latin *dextro*, right, and L for Latin *levo*, left (see figure 15.1).

Some strains of *Lactobacillus* produce the L-form, some the D-form, and others, a D,L-form—a mixture of equal amounts of D- and L-forms. Stetter's experiments on the regulation of D- and L-lactic acid revealed what is called "the manganese effect."[54] During its exponential growth phase, *Lactobacillus curvatus* produces L-lactic acid, but in the stationary phase it converts all the lactic acid to the D,L-form. He demonstrated that the buildup of the L-form turned on a gene responsible for converting L-lactic acid to the D,L-form. Manganese also seemed to be involved: the D,L-form was produced only in the presence of manganese in the medium. Stetter concluded that manganese was a co-inducer with L-lactic acid of the transcription of the gene that encoded the enzyme involved in D,L-lactic acid formation. But that interpretation contradicted the accepted view, based on *Escherichia coli*, that magnesium stimulated transcription. Stetter went to see Zillig, one of the world's experts on transcription

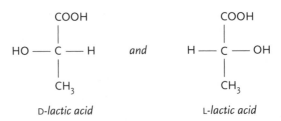

Figure 15.1 Lactic acid isomers D and L.

enzymes of bacteria and a director at the Max Planck Institute for Biochemistry, then located in downtown Munich.

Zillig was one of the founders of molecular biology in Germany. He was a student of the famed biochemist Adolf Butenandt, director at Kaiser-Wilhelm Institute for Biochemistry in Tübingen (renamed the Max Planck Institutes in 1948).[55] Butenandt shared the Nobel Prize in Chemistry in 1939 for his research on sex hormones. His research led to a number of patents and to close relations with the German chemical industry. During the war years, he used those funds to support viral research as a model system for studying the gene.[56] Tübingen had been largely unaffected by allied bombings. Tübingen University reopened in the fall of 1945; Butenandt held a professorship there, as well.[57] In 1950, a new building was completed for an Institute of Virology Research headed by his protégé, Gerhard Schramm. Zillig completed his doctoral research in 1949 on isolating compounds and amino acids in silkworm pupae and in 1952 went to work as an assistant with Schramm on the nature of the tobacco mosaic virus.[58]

Was the hereditary molecule a protein or nucleic acid? Schramm had been convinced that nucleoprotein, not nucleic acid, was the genetic material and that the RNA of the virus was on the outside. The correct interpretation came in 1955, after a visit from James Watson to Tübingen, and after a series of experiments by Zillig and Schramm, who showed that the RNA of the virus was the infectious agent, not the protein coat.[59] Zillig subsequently turned to study *E. coli* and its viruses.[60] He and his collaborators focused on DNA-dependent RNA polymerase, the key enzyme involved in the transcription of DNA to messenger RNA.[61] Zillig was appointed director of a new department (Abteilung) for molecular biology at the Max Planck Institute for Biochemistry in 1972.

The following year, Stetter went into "the cave of the lion," as he called it, to see Zillig.[62] "Later he told me that he did not believe a word of what I said," Stetter recalled. "Something was probably wrong with my experiments."[63] Zillig invited Stetter to work with him to establish if it was manganese or actually magnesium, as usual, that was a co-factor in stimulating transcription. At that time, the Max Planck Institute for Biochemistry was in the process of moving to new facilities. New buildings for it were almost completed in Martinsried about 20 minutes outside Munich. There would be space for Stetter there.

Their aim was to purify the RNA polymerase of lactobacilli. Zillig was one of a about two or three people in the world who could do it. He had established an *in vitro* protein-synthesis system out of *E. coli*, from which he was able to characterize many of the components involved. Zillig was a master craftsman in the laboratory constantly improving techniques. Work that normally took fifteen hours, Zillig could do in three, Stetter said. "He was so damned fast no one could beat him."[64] Four weeks after they began the work, they purified the RNA polymerase of *Lactobacillus curvatus*.[65] Stetter was right—its RNA polymerase did require manganese, not magnesium.[66] He and Zillig then turned to extend those studies of transcription to the methanogens, the extreme halophiles, and the thermoacidophiles *Sulfolobus* and *Thermoplasma*.

Unique Transcription Enzymes

In the winter of 1977, Stetter was teaching as an assistant in Kandler's Institute at the University of Munich. He ran a lab there as well as at the Max Planck Institute in Martinsried, traveling back and forth every day. In January, Kandler informed him about his inspiring visit with Woese and Fox in Urbana. He suggested that Stetter and Zillig extend their analysis of transcription by examining the RNA polymerase of the proposed "third kingdom."

The group in Martinsried held their weekly Friday seminar to discuss research and ideas. Stetter informed them about Woese's "third kingdom." Zillig had known of Woese from his work on the genetic code, but he knew nothing about him since those early days. He had not heard of the archaebacteria, and he certainly did not at first take the idea of a third kingdom seriously. "A Third Reich?" he quipped, "We had enough of the Third Reich!"[67] "No one believed in it. It required convincing," Stetter later said. "Carl Woese was reaching for the stars with little evidence. It was based on a few base pairs. And he claims a third Reich. They were behaving like scientists, not like people in a church. He had incredible luck, he was ingenious, of course, but he was also lucky."[68]

Wolfe had sent Kandler a whole treasure of *Methanobacterium*, about 50–100 grams of freeze-dried cells. Stetter purified the RNA polymerase of the methanogen in his lab in Munich. But again, he found something unusual. To determine that one had purified RNA polymerase, the protocol was to use the antibiotic rifampicin as a control. It binds to the bacterial RNA polymerase beta subunit and blocks the enzyme, making it inactive. But when Stetter added the rifampicin, the RNA polymerase activity remained active. He went to Zillig with his results. Zillig responded that if that were the case, it simply was not an RNA polymerase.

The turning point occurred in the summer of 1977 when Stetter learned from Kandler that Woese's laboratory had determined that halobacteria were also archaebacteria. When Stetter relayed this news, it sparked Zillig's curiosity. Six years earlier, two Canadian biochemists, Gregory Louis and Peter Fitt, had reported that *Halobacterium* possessed a completely different RNA polymerase.[69] The RNA polymerase of typical bacteria was a large macromolecule with five subunits: beta, beta-prime, sigma, and two alpha subunits. This structure was thought to be highly conserved throughout the bacterial world. But Louis and Fitt had reported that the polymerase of *Halobacterium* was made up of only two subunits, and that it was about 1/20th the size of the *E. coli* enzyme.

Zillig doubted their claims. But now in light of "a third urkingdom," he asked Stetter to repeat the experiments of Louis and Fitt. Stetter followed their methods, but he could not purify the enzyme. He suspected their results were artifacts.[70] Then he and Zillig went after "the real RNA polymerase" of *Halobacterium halobium*. Halophiles were difficult to handle because they have a salty interior. Purifying a protein typically involved ammonium sulfate precipitation, which was done by adding salt. As salt is added slowly, one protein after another precipitates, depending on the protein's isoelectric point (i.e., the pH at which the protein has an equal number of positive and negative charges)—each protein has a different isoelectric point. But because halophiles have internally saturated salts, one had to do it the opposite way: dilute out the protein. Zillig had great technical experience. He was the only one who could have done it, in Stetter's view.

When the enzyme was purified, Zillig and Stetter then separated out its subunits using polyacrylamide gel electrophoresis, which took a whole day. This was in the late fall of 1978, so Zillig and Stetter went cross-country skiing to pass the time. When they returned to find the completed gels, as Stetter recalls, "the pipe almost fell out of his mouth. 'Ah look at this! It is not bacterial!' The beta and beta-prime subunits were different. Then he caught fire in the archaebacteria, this was a very big fire."[71] Far from being smaller, the RNA polymerase of *Halobacterium halobium* was much larger and much more complex than that of typical bacteria.[72]

Then came word that winter from Woese's laboratory via Kandler that *Sulfolobus* was a member of the archaebacteria. Zillig asked Stetter to grow them. *Sulfolobus* is an extreme thermopile: it grows at 80°C—the pasteurization temperature. It is autotrophic and grows on sulfur, a completely different medium than Stetter had been used to, but he got the strain to grow on peptone and yeast extract. *Sulfolobus* is a stinky bug; it produces an ugly "cheese-socks" odor, and the whole institute was stinking. Halophiles do not smell nice, with their complex medium, but *Sulfolobus* smells worse. Botanists at the institute would walk by and sneer at Stetter, "You and your dirty stuff."

Then there was the problem of growing *Sulfolobus* in large quantities. Zillig always wanted mass quantities, to have a lump of cells in the deep freezer so that he would have the same material for all experiments. When Zillig and Stetter purified the RNA polymerase of *Sulfolobus acidocaldarius*, they found that it, too, had the unusual structure. It turned out to be even more complex than that of halophiles and methanogens, composed of 10 subunits; and it resembled the RNA polymerase of eukaryotes.[73]

Life in the Extremes

In the winter of 1978, Zillig and Stetter examined the RNA polymerase of another extreme thermoacidophile. Mario De Rosa and his colleagues at the University of Naples had called it the genus *Caldariella*.[74] Zillig and Stetter reinterpreted *Caldariella* as a species of *Sulfolobus*. That taxonomic reorganization also entailed experimental twists and turns. Like their work on the RNA polymerase of *Halobacterium halobium*, it involved repeating the experiments of others and refuting previous claims.

Caldariella and *Sulfolobus* were understood to be unrelated. Brock and his collaborators reported that the GC content (see chapter 10) for *Sulfolobus* was 60%, but the GC content for *Caldariella* was reportedly 31%.[75] Zillig wrote to De Rosa to obtain the strain in order to do the analysis of its RNA polymerase. De Rosa replied that the strain could not be sent because it would die during transport. It grew in volcanic hot springs optimally at 87°C and low pH (3.5). Zillig and Stetter knew from their work with *Sulfolobus* that all De Rosa needed to do was adjust the pH to neutral. They suspected that he simply did not want to send the strain. When Zillig explained that he wanted to purify the RNA polymerase, De Rosa said that his group had already done that, and sent a paper in which he reported that the RNA polymerase of *Caldariella acidophila* was normal.[76]

Zillig and Stetter suspected that what the Naples group reported to be transcription enzymes were actually band impurities; they had simply ignored any subunits that were different. They drove to the acidic hot spring near the town of Pisciarelli, which lies in the center of Campi Flegrei ("burning fields"), a volcanic caldera north of Naples, where De Rosa said he had isolated the organism. They found the stream with the steam coming off, and took samples. They let the samples cool down, adjusted the acid environment to neutral pH, took them back to their lab, and cloned the microbe. *Caldariella* had the same culturing properties that De Rosa and colleagues had described; Zillig and Stetter assessed its GC content as 31%, just as De Rosa et al. reported it to be. Otherwise, it was just like *Sulfolobus*, except that its growth temperature was a little higher. Zillig and Stetter then examined the GC content of *Sulfolobus acidocaldarius* and found that it was actually 31%, not the 60% that Brock and his collaborators had reported. And *Caldariella* possessed the unique complex RNA polymerase.[77] Zillig and Stetter renamed it the new archaebacterial species *Sulfolobus solfataricus*.

Zillig and Stetter subsequently showed that the transcription enzymes of methanogens and halophiles also had a structure that was much more complex than typical bacteria, and that those enzymes were also resistant to the antibiotics rifampicin and streptolydigin, known inhibitors of the RNA polymerases of typical bacteria. Here, then, was another "common feature of *Archaebacteria*."[78]

Blares of Trumpets

Archaebacterial research thus expanded quickly in Germany. Kandler had already organized an archaebacterial conference in Munich in the spring of 1978. Twelve professors from across Germany participated. Such support for the archaebacterial concept stood in stark contrast to the position of archaebacterial research in the United States. When, in February 1978, Kandler invited Woese to attend the 12th International Congress of Microbiology in Munich, Woese declined while explaining the poor funding for his research:

> Right now the state of my research support can only be described as "pitiful." (NASA and NSF are quick to enjoy publicity of my work, but not to fund it adequately. USPH [the U.S. Public Health Division of the National Institutes of Health] doesn't seem to want to fund it at all.) I am not telling you this as an attempt to obtain special treatment for myself. In fact, all things being equal, I most likely would have decided not to attend anyway. I am reluctant to do a lot of traveling from Urbana. Any more than a little traveling has a detrimental effect on the research—and I have already accepted a number of seminar invitations for this spring.[79]

When Kandler again invited Woese to Munich 10 months later, Woese agreed.[80] Kandler was in charge of organizing a joint meeting of the Deutche Gesellshaft für Hygiene und Mikrobiologie and the German branch of the American Society of Microbiology in the spring of 1979.[81] He arranged funds for Woese's travel and planned the program: Woese's lecture would take place on the evening of the meeting, as an opening event, on March 29, 1979.[82] Zillig also invited him to give a seminar at the Max Plank Institute: "Reading your interesting papers on Archaebacteria gave our work a new direction. We would be happy if you could give us a seminar at our institute. . . . We would be pleased if it could be 'Archaebacteria.' "[83]

When Woese arrived in Munich, he was met with fanfare and celebrated at a dinner in the main hall of the Botanical Institute, with blares of trumpets that Kandler had arranged from a church's brass choir. He and Kandler were in a "complementary relationship," as he wrote to him on his return in early April 1979:

> Without doubt that was the most enjoyable and memorable scientific trip I have ever taken. And it is obvious to whom I owe the occasion. . . . What I perhaps value most is the interchanges between ourselves. We exist in some sort of a complementary relationship to the archaebacterial problem, and it is therefore important that a genuine scientific rapport exist between us. . . . Without doubt München is the scientific capital of the "third kingdom" (which Zillig tells me does not translate easily into German).
> . . . A bit of my heart still remains in München.[84]

The fundamental camaraderie established at that meeting in Munich reinforced the international collaboration that followed for many years. "We are all very much impressed by your work" Kandler replied, "and your visit was an important impulse for further research on Archaebacteria in München."[85]

Woese was equally impressed by the work of Stetter and Zillig, with whom he began a close correspondence. The unique transcription enzymes provided strong evidence of the uniqueness of the archaebacteria. Their analysis of the transcription enzymes seemed to indicate that the archaebacterial and eubacterial lineages had split before those subunit structures were established. "Your polymerase work is an especially important study," Woese wrote to Zillig upon returning to Urbana that spring.[86] And again that summer: "It was a pleasure to receive your letter and preprints. You have produced most convincing evidence of the uniqueness of archaebacteria," Woese wrote Zillig on August 6, 1979. "Not only that, but you seem on the verge of refining or extending the concept."[87]

Indeed, those RNA polymerases had diverse and characteristic component patterns that Zillig and Stetter compared to make groupings within the archaebacteria: the polymerase of *Thermoplasma* resembled that of *Sulfolobus*; the polymerase of *Methanobacterium* resembled that of the *Halobacterium*.[88] The structure of the RNA polymerases also hinted at the possible evolutionary relations among the archaebacteria, eubacteria, and eukaryotes. Zillig and Stetter would subsequently show by the early 1980s that complexity of archaebacteria RNA polymerases was strikingly similar to those of eukaryotes, which were also resistant to the antibiotics rifampicin, streptolydigin, and α-aminitin.[89] Wolfe would comment decades later that the results concerning the unique transcription enzymes were "so astounding that many viewed them as conclusive evidence, alone, that Archaebacteria indeed represented a third domain of life."[90]

Zillig and Stetter formed a dynamic partnership and turned the archaebacterial concept into a vigorous research program of their own, searching for new organisms in hot acid environments near volcanoes and hot springs, characterizing the organisms in terms of growth properties and form and by their diverse RNA polymerase component patterns.[91] In their trip to the hot springs of Iceland in 1981, they found a novel type of extremely thermoacidophilic anaerobic archaebacteria that they named *Thermoproteales*, as well as an extremely thermophilic methanogen.[92] The next year, Stetter extended the known range of life to beyond the boiling point at 105°C.[93]

Archaebacteria research and the study of thermophiles more generally had an important applied aspect. The enzymes of thermophilic bacteria were prized for their high stability and important use in biotechnology and other industries. The heat resistance of the enzyme Taq polymerase obtained from *Thermus aquaticus*, for example, proved to be vital for one of the most revolutionary biotechnologies of the twentieth century: the polymerase chain reaction (PCR), developed in 1983. PCR allowed biotechnologists to make many copies of short pieces of DNA. It became a common technique in genetic screening in diagnosis of some human diseases, DNA fingerprinting in criminal situations, and the creation of transgenic organisms. The enzymes of thermophiles held promise for a new generation of biocatalysts and organic compounds, and of more efficient industrial processes. Polymer degrading enzymes could be important for food, chemical, pharmaceutical, paper, pulp, and waste treatment industries. Even the membranes of extremophiles could contain unique stable compounds that might prove useful in pharmaceutical formulations.

Stetter was appointed professor at the University of Regensburg, about 200 km north of Munich in 1980. There he constructed big fermenters to grow microbes on a massive scale. Deploying his early education as an engineering student, in collaboration with a Swiss engineering company, he invented a high-temperature fermenter that is commercially available today to grow microbes, especially the highly corrosive ones, such as the new genus he discovered called *Pyrodictium*.[94] Stetter would later co-found a California-based biotech company of his own, the Diversa Corporation.

During the 1980s, Zillig was in search of species and strains of archaebacteria that carried viruses or plasmids. Molecular biology had been based on the study of *E. coli* and its viruses. Zillig aimed to create a parallel genetic system for the archaebacteria and their viruses to study transcription and the nature of their genes. He discovered *Sulfolobus* phage obtained from Iceland in 1981, and by the end of the decade, he and his coworkers had characterized four new families of archaebacterial virus.[95]

By the end of the 1970s, several lines of evidence supported the fundamental uniqueness of the archaebacteria: their characteristic ribosomal RNA signatures, unique transfer RNAs, diverse and unusual walls lacking peptidoglycan, unique membrane lipids, and distinctive RNA polymerase structures. While the new urkingdom concept was strengthened, two major questions remained unanswered: What was the relationship of the archaebacteria to the typical bacteria and the eukaryotes? Were archaebacteria as old as their name implied?

| # Out of Eden

The supposed great antiquity of the archaebacteria remains an unproven prejudice, but it is a plausible one.

—Carl Woese, "Archaebacteria" (1981)

THERE WAS NO DIRECT PHYLOGENETIC evidence to support the view that the archaebacteria were actually older than other bacteria, the eubacteria, but the conjecture that the archaebacteria were ancient persisted as that grouping grew to include the extreme halophiles and thermoacidophiles. Their unique lipid membranes, odd walls lacking peptidoglycan, and the unique transcription enzymes were all compatible with the concept that the archaebacterial lineage diverged deeply in the tree of life before peptidoglycan of the eubacteria had evolved. John Bu'Lock and Thomas Tornabene informed Woese that the archaebacterial lipids were more similar to the hydrocarbon in ancient sediments than were those of any other organisms. "Perhaps they really are 'arche,'" Woese wrote to Kandler in August 1979.[1]

On the other hand, the archaebacteria had molecular characteristics in common with eukaryotes, and it was possible that the eukaryote lineage evolved from them, that is, that the archaebacteria were "proto-eukaryotes." Woese favored his own conception that all three lineages stemmed from progenotes in the throes of developing the genetic translation system. "It seems as though many biologists are ready to jump on the band wagon that claims archaebacteria, even certain of the archaebacteria, are 'proto-eucaryotes,'" Woese wrote to John Wilkinson in Edinburgh in March 1978:

The waters may consequently be muddied for a while. Archaebacteria do have some things in common with eucaryotes, but so do eubacteria (tRNA common arm sequence, ester linked lipids, FMN-Fad, etc.) When it all settles, Archaebacteria should remain as a distinct third primary kingdom. But the debate will be for the good. Biologists will focus on the evolutionary issues, define the evolutionary stages (eucaryote, prokaryote, and the one I named "progenote"), and have a clearer concept of the "common ancestor state."[2]

By the end of the year, Stetter and Zillig had data indicating that the DNA-dependent RNA polymerase of *Sulfolobus* was similar to that of yeast. But Woese remained unconvinced that the eukaryotic lineage had branched from the archaebacterial lineage. Instead, he considered an alternative conception: perhaps the eukaryote was wholly chimeric. The similarities between eukaryotes and archaebacteria could be explained in terms of symbiosis deep in the roots of biological evolution. "The polymerase similarity between Sulfolobus and yeast is intriguing, but does not necessarily show a specific relationship," Woese wrote to Stackebrandt on January 9, 1979:

My attitude is that the eucaryotes are endosymbiotic chimeras, and this is not confined to large units, like mitochondria; it goes all the way down to single genes. Thus it is not unreasonable that a polymerase gene from the Sulfolobus line finds its way into the chimera. There's a lot to be sorted out, and it will be most interesting.[3]

On the Metabacteria

Woese's collaborations with Fox and Stackebrandt and his relations with Kandler, Zillig, and Stetter represented an almost idyllic image of science working in a complementary fashion for a common good: sharing unpublished data, opinions and advice, updates of work in progress, and laboratory aims. While the relationship between Germany and Woese's laboratory exemplified international cooperation at its best, in Japan, the situation was the contrary. In 1979, Woese came to learn that Horishi Hori and Syozo Osawa were attempting to undermine his priority in the discovery of the archaebacteria. They used the term itself as a pivotal rhetorical device.

Hori and Osawa argued that the archaebacteria were not ancient as the name implied. They constructed a phylogenetic tree of 5S rRNA sequences that were available in the published literature and concluded that *Halobacterium* was phylogenetically closer to eukaryotes than to eubacteria. They suggested that archaebacteria arose *after* the eubacteria, and were probably the ancestors of eukaryotes; they proposed the name "metabacteria" as a replacement for "archaebacteria."[4]

Osawa is perhaps best known today for his work with Thomas Jukes in 1989 showing that the genetic code was not entirely universal. Variations, results from reassignment of codons, especially stop codons, were found in mitochondria, *Mycoplasma*, and some ciliated protists.[5] Osawa worked at the Institute of Nuclear Medicine and Biology at Hiroshima University, and he had investigated the molecular biology of the translational apparatus since the early 1960s.[6] Hori was Osawa's former student and an assistant professor of biochemistry there.[7]

The problem occurred in the fall of 1979 at a symposium in Tomakomai on "Genetics and the Evolution of the Transcriptional and Translational Apparatus." "Their tree is in good agreement with yours," Kandler wrote to Woese.

> However, they call the archaebacteria metabacteria and do not refer to your recent papers, which is pretty unfair. The documentation of the position of archaebacteria is extremely poor, anyhow. The slight tendency toward the Eukaryotes may be real as you have discussed earlier, but it does not change the basic concept of 3 more or less independent branches of organisms.[8]

Woese had no idea of what Hori and Osawa were putting forth. But he had an experience with Hori a few years earlier, in 1975, when he peer reviewed a paper that Hori submitted for publication in *Journal of Molecular Evolution*. It was on sequence homology analysis based on 5S rRNA of 17 organisms ranging from humans to bacteria.[9] Woese mentioned that Hori should make use of the RNA secondary structure of 5S RNA, about which he and Fox had just published a paper in *Nature*.[10] When Hori subsequently revised that paper, he included a discussion of the secondary structure of 5S rRNA but made no mention of Fox and Woese. The next year he published his own account of the secondary structure of 5S rRNA that was similar to that of Fox and Woese, but he claimed that his own work was independent of theirs.[11] "What you report of Hori and Osawa is somewhat depressing for two reasons," Woese replied to Kandler, "one specific, one general":

> This is the second time Hori has tried something like this. Back in 1975, I reviewed his first 5S comparison work. As I recall, the mss aligned sequences without any apparent awareness of 5S RNA secondary structure, a model for which George Fox and I had just published. I pointed out to him in review that he should utilize the secondary structure. He then turned around and published a secondary structure for 5S RNA without reference to us. [His claim of 70 5S sequences is ludicrous; 5–10 are from *E. Coli*.][12]

Hori and Osawa were late in recognizing the uniqueness of the halophiles, even though Woese and Fox had mentioned them in their paper in November 1977 announcing the new urkingdom. In October 1978, Hori and Osawa submitted a paper on 5S rRNA phylogenies in which they had a halophile 5S sequence that was their deepest branch, but they made no note of it.[13] The 16S rRNA analysis of several extreme halophiles had been done in Woese's laboratory by August 11, 1977, but the paper on the halophiles as archaebacteria appeared in the spring of 1978.[14] As Fox commented decades later, Hori and Osawa "did later claim to have discovered 'metabacteria' after we explained it to them."[15]

Hori and Osawa continued to use the term "metabacteria" instead of archaebacteria, maintaining throughout the 1980s that the metabacteria were closely related to eukaryotes and evolved much later than the eubacteria.[16] To be sure, Woese and Zillig also had doubts about whether the archaebacteria were actually as primitive as their name implied. Zillig's and Stetter's work on the structure of transcription enzymes suggested that the archaebacteria might be closely related to eukaryotes.

But Zillig deplored the style of Hori and Osawa in the fall of 1979 when he attended the symposium in Tomakomai. He thought the evidence for their tree was extremely weak and perhaps based on false comparisons. "I was somewhat shocked, however, by the style of the paper of Hori and Osawa...which is clearly an attempt to bypass you in a manner which I cannot consider fair," he wrote to Woese.

> They have investigated only one lousy species, Halobacterium. What worries me more is that they compare 5S RNA of prokaryotes with 5S and not with 5.8S RNA of eukaryotes. I always thought that much evidence speaks for a homology of prokaryotic 5S with eukaryotic 5.8S RNA. This might, however, be wrong.
>
> In any case I regret that the authors neglect your central role in recognizing this group as a unique one and, instead, overemphasize the name Archaebacteria, which they consider wrong (though it has priority). I must admit, that I am not too happy with that name myself and I remember that in our discussion on this point in Munich you yourself were not sure if Archaebacteria are really "older" than Eubacteria. I wonder if you could propose an even better new name which does not at all contain the designation bacteria? But I hate the way in which Hori and Osawa deal with the problem trying to conceal your discovery by setting wrong accents.
>
> ...One of the highlights of the meeting, though missing the theme, was a paper by Joan Steitz...She really has become a believer....
>
> From a copy of a letter you wrote to Kandler I see that he already told you of the Hori and Osawa affair. I assure you that other people in Tomakomai, especially Matheson, felt the same way as we do.[17]

Woese replied,

> As you well know, you are not alone in your uneasiness with the word *archaebacteria....* I have an intuitive feeling that the name will ultimately

be justified by the organisms that bear it. We have a long trail ahead of us regarding the archaebacteria-eucaryote connection... I do not understand the evolutionary makeup of the eucaryote. It may not have arisen as a simple endosymbiotic ensemble—i.e. an engulfing species that takes in a small number of endosymbionts. The eucaryotic genome could be a "radical ensemble" reflecting hundreds of gene capture events, some from eubacteria, some from archaebacteria, some perhaps from elsewhere. There may never have been an "engulfing species" that represents the eucaryotic "cytoplasm" today. The "cytoplasm" itself could have evolved outward from a "nucleus", which represents the heart of the symbiotic evolution. We need many gene families to be characterized and traced back to their prokaryotic (or simpler) roots before eucaryotic origins will become clear.[18]

Woese wrote similarly to Kandler the following week. "The questions really concern the relationships of eucaryotes to eubacteria and eucaryotes to archaebacteria; not the status of archaebacteria. (This is a subtle, but essential shift of emphasis.)"[19]

An Aquatic Eden

Woese's adherence to the ancient status of the archaebacteria was also rooted in his view that autotrophs were the first organisms. In this, he was in agreement with Thomas Brock and a few others of the 1960s and 1970s.[20] In 1979 he directly confronted and debunked the conception of Oparin that primordial life forms were heterotrophic anaerobes that fed in a rich hot primordial soup of organic molecules that had been synthesized abiotically. That autotrophs were primordial, of course, had been proposed in taxonomic schemes soon after it was discovered that some microorganisms could derive their energy and nutrition directly from inorganic chemical compounds through the oxidation of ammonia to nitrous and nitric acid, of sulfur to sulfurate, of iron to iron oxide, and of methane to carbon dioxide and hydrogen.

In Oparin's view, all these models were flawed in their assumption that CO_2 was the sole source of carbon when life emerged.[21] His own assumption that the first organisms were anaerobes living in the absence of free oxygen was crucial, because if there were free oxygen, all organic compounds would be rapidly oxidized to CO_2 and H_2O. But if the primordial environment lacked oxygen, then simple organic compounds could accumulate.

In Oparin's model, hydrocarbons, formed in the superheated aqueous vapor of a primitive hot Earth, would have given rise to such derivatives as alcohol, aldehydes, ketones, and organic acids, which would react with ammonia to produce amides, amines, and other nitrogenous derivatives.[22] Subsequently, when the planet cooled off, the aqueous vapor condensed and hot oceans formed, which would contain organic substances consisting of carbon, hydrogen, oxygen, and nitrogen, from which proteins originated. "Special formulations" resulting from their mixing formed bits of semi-liquid colloidal gel droplets.[23]

Those droplets represented the first momentous step in the evolution of life. Gradually, the droplets would acquire some structure, and divide into daughter droplets (like bubbles).[24] There would then be a competition to increase the efficiency of their chemical reactions by developing enzyme catalysts that would speed up reactions hundreds of thousands of times faster than if left only to the physical properties of the organic substances themselves.[25] Thus life arose "as a venture in colloidal systems separating from the 'hot thin soup.'"

Haldane's proposal was similar. He suggested that energy from ultraviolet light, from volcanic heat, or from lightning could help in synthesizing organic molecules (e.g., sugars and amino acids) from carbon, hydrogen, oxygen, and nitrogen contained in water vapor.[26] These organic compounds would accumulate in Earth's oceans until they "reached the consistency of hot, dilute soup." Further chemical synthesis would take place, giving rise to the first primitive organisms, which would then feed on the rich, organic nutrients around them.

Pieces of the heterotrophs-first origin-of-life model were filled in. In 1945 Norman Horowitz offered an explanation of how intermediary metabolism might have arisen.[27] As nutrients (a particular amino acid, e.g.) were gradually depleted in the primordial soup, there would be a natural selection for the elaboration of biochemical synthetic pathways to make up the deficits, one by one, in reverse order of today's biochemical synthesis. Cornelis van Niel suggested that the aboriginal heterotrophic anaerobes would have resembled bacteria of the genus *Clostridium*.[28]

Life served up in a hot rich soup had become catechismal among biologists; it made the most sense for many microbiologists. The ability of an organism to synthesize all its cellular constituents using carbon dioxide as the only carbon source seemed to require a highly developed enzymatic apparatus, and microbiologists found it hard to imagine how such an apparatus could have originated by any mechanism in an inorganic world.

Certainly there were modifications made within the heterotrophs-first conceptual structure. For example, J.D. Bernal suggested that clays might have played a role in concentrating organic molecules.[29] In the earliest models, Earth's conditions were very hot because Earth was believed to have arisen as a fragment of the sun. Later, in the 1950s when there was the acceptance of the cold-accretion theory of Earth's formation, the primordial ocean soup was thought to have been cool. The primordial soup was heated up again in the 1970s when lunar exploration revealed that the Moon's crust may have been partially or completely molten during that body's initial 100 million years. Meteorites also suggested that there was a comparable heating 4.5 billion years ago. Consequently, it was determined that Earth's surface was initially hot if not actually molten.[30]

The possibility that organic compounds were first created abiotically won great support from experimental demonstrations that amino acids and a variety of other organic compounds could be created by an electrical spark in an anaerobic atmosphere of methane, ammonia, water, and hydrogen. Still, there was a long way from an aqueous mixture of amino acids, purines, and pyrimidines to an organism. Nothing was said about the origin of the genetic code, as Horowitz emphasized when he discussed "the Garden of Eden" concept embedded in Oparin's theory at a symposium on "Evolving Genes and Proteins," in 1965.[31] Whereas writers such as Margaret Dayhoff and Richard Eck believed that life on Earth emerged inevitably as cosmic process, others saw it as a singular event (see chapter 10). "Whether the origin of life was virtually inevitable, or whether the origin of life was an event of vanishingly low probability—almost an unrepeatable accident—is impossible to say at the present time," Horowitz wrote. "This is one of the major scientific questions that the exploration of Mars might answer, since there is reason to believe that the early development of Mars was similar to that of the Earth."[32]

The imagery in the origin-of-life stories was sometimes drawn from the Book of Genesis (see chapter 9). Oparin's account was a story of biblical proportions—of repeated crises and resolutions. The aboriginal organisms lived effortlessly in an ocean paradise, a place of plenty, what David Hawkins referred to as "an aquatic Garden of Eden."[33] But that happy lifestyle was doomed as the primitive organisms overreproduced and unwittingly depleted the great oceanic stores of nutrients. The further life progressed, the fewer nutrients were available, and the more strongly and bitterly a struggle for existence was waged. It was "a fight to the death," Oparin said.[34]

To survive the acute food shortage, organisms would have to either acquire nutrients by "eating their weaker comrades," or evolve intermediary metabolism and eventually learn to transform other forms of energy (chemical and then light) into biochemical energy to produce essential products.[35] Evolution thus turned in the direction of autotrophs that utilize inorganic substances such as ammonia, hydrogen sulfide, and ferrous iron. The present-day chemoautotrophs were relics of this epoch and represented what Oparin said was "an insignificant rivulet in the main evolutionary stream."[36] Another epoch was entered when the supply of inorganic substances was gradually exhausted, and photosynthetic organisms evolved.[37] The intermediate dark ages of the chemotrophs were over.[38]

No Time for the Soup

By 1979, Woese could point to data that directly conflicted with the heterotrophs-first model. Oparin had supposed that Earth was sterile for most of its history; it took longer to make the organic molecules than it did for all the species of organisms that evolved afterward. Although Earth was 4.5 billion years old, life in his scheme emerged only 1.5 to 2 billion years ago. Indeed, Oparin emphasized a long "sterile, lifeless period in the history of our planet as a necessary condition for the primary origin of life."[39]

That view of Earth's long sterile history no longer held up, as paleobiologists of the 1970s pushed the fossil record back billions of years. The fossil record of eukaryotes was extended back to 1.5 to 2 billion years. Discrete organic microfossils in the sediments of the Archaen Swaziland system of South Africa that dated to 3.4 billion years were confirmed to be relics of microbial life in 1977 when Andrew Knoll and Elso Barghoorn reported evidence of various stages of cell division.[40] Some of the oldest known limestone contained microbial fossils of what seemed to be blue-green algae (cyanobacteria), which have essentially the same photosynthetic apparatus as plants. They use energy from the sun to synthesize glucose from carbon dioxide and water, giving off oxygen. Also like plants and animals, all cyanobacteria can utilize oxygen (by aerobic respiration).

These are complex ways to live. Simpler anaerobic microbes, Woese reasoned, must have arisen far earlier than ever had been imagined.[41] The last common universal ancestor of all of life, the progenotes, would then be at least four billion years old:

> Thus, we are faced with the possibility that the origin of life on this planet virtually coincided with the origin of the planet itself. If so, then the origin of life can no longer be perceived as an improbable happening— requiring a series of unlikely and so protracted events, or the slow accumulation of compounds in a primeval ocean.[42]

Not only was there not enough earthly time for the primordial organic soup, but also Woese pointed to results from experiments in prebiotic chemistry that were at odds with colloidal droplets in a hot ocean broth.[43] Basic biochemical reactions were generally dehydrations. Chemists had to invoke ever increasingly water-free (or water trapped) primitive environments in order to effect primitive syntheses. "These Ptolemaic revisions of Oparinism should be recognized for what they are, and the question put squarely, Woese wrote, "It is not a matter of how to modify Oparinism, but whether to replace it."[44]

Mother Earth

The Oparin thesis had simply ceased to be a productive paradigm in Woese's view.[45] Indeed, there was another facet of the paradigm shift he called for that was more a matter of aesthetics. In the Oparin model, Woese argued, life evolved in a non-biological way: the life forms that emerge are only peripherally connected to the process giving rise to them.[46] It was also non-dynamic: there were no biocycles in its system of elements, compounds, and organisms. It did not have an ecosystem feel. There was a disconnection between the proposed geochemical context and life itself that Woese thought ought to create the conditions for life, just as it does today. "To the biologist, accustomed as he is to rapid biocycles of elements, compounds, and organisms," Woese commented,

> such a scenario feels alien....Most of all, perhaps, this scheme is unappealing because the life that arises is basically destructive of the organization that preceded it; it does not in itself contribute to the build-up of chemical complexity on the planet (until photosynthetic organisms finally evolve).[47]

In the autotrophs-first model that Woese proposed, life arose from the geochemical processes themselves. Biology would be modeled on non-biological chemical pathways. Carbon dioxide and hydrogen in the atmosphere would form methane without organisms; the emergence of methanogenic microbes paralleled an existing process. There was no dichotomy and opposition between life and the environment. Primitive autotrophs generated energy and complexity; they did not consume it as did heterotrophs. Organisms did not arise as adaptations to Earth; life played an active role in its own evolution.

The first cells developed "from sources of, not from the sinks for, prebiotic biochemistry; the earliest organisms would, therefore, be autotrophic and photosynthetic."[48] There was no fundamental discontinuity between the way life originated and the way it maintained itself afterward. As Woese put it,

> Prebiotic evolution is not a collection of special conditions, a peculiar dynamics whose essence is discarded and replaced by another dynamics, other conditions, once life arises. Preliving states must possess the basic attributes of living ones, for these attributes are not properties of "living organisms" per se; they are characteristics of a general process of transformation of energy into organization.[49]

Woese thus contrasted his own philosophically holistic outlook, with the reductionist heterotrophs-first model. In the Oparin heterotroph scenario, the origin of life was a two step-process in colloidal coacervates—spherical droplets of organic molecules—in the atmosphere and then in the oceans. In the scenario Woese envisaged, life had evolved simultaneously with Earth itself, emerged in one location, and made the conditions for oceans. Life began in the clouds of the primitive atmosphere, at a time when the planet's surface was too hot to sustain liquid water. Woese pointed to Venus to show how Earth might have been surrounded by vast cloud banks. Severe weather conditions and volcanic eruptions would have caused large quantities of minerals (dust), from the very dry surface, to be swept into the atmosphere. Atmospheric water vapor then condensed on the dust, partially dissolving it. The droplets in those clouds would contain minerals and accumulate organic compounds produced by interactions among atmospheric gases and other constituents.

Those droplets would be the precursors of cells, their surfaces coated with mixtures of the larger organic compounds, their interiors solutions of reactive compounds. Droplets (and hydrated dust) offered enormous amounts of surface, so surface chemistry would become all important in life's origins. "The droplet phase serves as a natural definition of the proto-cell.... In such an atmosphere the primary chemistry is 'membrane' (interface-associated) chemistry. Solution chemistry would be the by-product of 'membrane' chemistry, not the reverse."[50] Woese thus contrasted his one-theater model of life's origins to the "two-theater" Oparin model in which the initial reactive compounds produced in the atmosphere are subsequently quenched and protected in the ocean, where they accumulate and ultimately produced organisms.

Methanogens as ancient organisms seemed to fit the context of a primordial earthly system that was poised to react chemically in any case; organisms would simply facilitate, "catalyze" the process, and develop thereby.[51] CO_2 (and CO) and H_2 are thermodynamically unstable at normal temperatures, so as Earth cooled they would naturally convert to methane (CH_4). But methanogenic bacteria would emerge in concert, and Woese suggested that they would further increase the rates of conversion, depleting CO_2 and CO, known greenhouse gases. Thus, methanogens, he thought, might have broken a runaway greenhouse effect, and helped Earth to cool to its present state with liquid water. Life itself would play a role in the production of the first oceans, and as the production of H_2 dropped, Earth was on course toward its present atmosphere.[52] What was important, Woese said, was not whether his scheme was true or complete. "Its main function is to force a realization that there may exist genuine alternatives to Oparinism."[53]

Woese would later team up with the chemist and patent lawyer Günter Wächtershäuser.[54] Beginning in the 1980s, Wächtershäuser imagined "precursor organisms" that were acellular and lacked a mechanism for division, enzymes, and nucleic acids.[55] Life at this stage would consist of an autocatalytic metabolism in what was essentially a two-dimensional monomolecular organic layer. The negatively charged "surface organisms" were bonded to positively charged surfaces of pyrite at the interface of hot water. The energy for carbon fixation would be provided by converting ferrous ions and hydrogen sulfide into pyrite, which provided the all-important binding surface for the organic constituents. At a later stage, when a lipid membrane was grown, with an internal broth of detached constituents, the pyrite support would be abandoned and true cellular organisms arise.[56]

| Big Tree

A revolution is occurring in bacterial taxonomy. What had been a dry, esoteric, and uncertain discipline—where the accepted relationships were no more than officially sanctioned speculation—is becoming a field fresh with the excitement of the experimental harvest. For the most part the transition reflects the realization that molecular sequencing techniques permit a direct measurement of genealogical relationships.

—George Fox et al., "The Phylogeny of
Prokaryotes" (1980)

IN THE SUMMER of 1979, Woese, Fox, Stackebrandt, and Kandler were focused on "Big Tree," a universal phylogenetic tree of life. Kandler fixed on mapping the diverse wall types onto the 16S rRNA phylogenies that Woese's laboratory shared with him. "Hopefully, George sent you what you need in terms of SAB's [the similarity coefficient S_{AB}; see chapter 11]," Woese wrote him in September. "Our 'Big Tree' mss. is near completion; the figures are at the artist's."[1] That paper integrated the past decade of empirical and conceptual work on 16S rRNA phylogenies into an outline of the foundational lineages of life on Earth. Several matters had piled up. To date, 170 species of bacteria had been cataloged by the 16S rRNA method; the archaebacterial group and its defining characteristics had grown, an outline of the relationships among its members had been discerned, and many of the published taxonomic groupings of eubacteria based on morphology and physiology were reclassified in terms of ribosomal RNA phylogenies. Yet several critical methodological issues remained unresolved.

Phylogeny at the Crossroads

Great skepticism still prevailed about a phylogenetic approach to bacterial taxonomy—not only among those unfamiliar with the molecular techniques and concepts, but also among some of those who had employed them. The ability to distinguish a convergent bacterial character from a conserved character was held by some to be a problem as much at the molecular level as at the morphological level for microbial classification. Molecular phylogenies of animals were generally consistent with both the fossil record and morphological data. To support the inference, for example, that such shared patterns as the skeletal and muscular pattern of four-limbed vertebrates were homologous—that is, inherited from a common ancestor that also possessed that pattern—the fossil record would provide independent evidence. For bacteria, there was little fossil record, and classifications based on molecular methods frequently did not match the classifications based on morphology and biochemistry.

Molecular sequences of different species could be homologous for a number of reasons without actually being a phylogenetic measure of relatedness: lateral gene transfer, different rates of change, convergence, and gene duplication. Homology as applied to proteins and nucleic acids rested on the ability to demonstrate that the two sequences being compared represent identical regions. Correct comparisons relied on their correct alignment. But because of gene duplication, some argued, it would simply be impossible to conclude with certainty that any two proteins one compared from two species were actually homologous in the sense of common ancestry when there was no fossil record. Gene duplication was understood to be an evolutionary mechanism for increasing genomic size and for developing new gene functions. Once they are duplicated, identical genes can diverge to create two different genes.

Molecular phylogeneticists since the time of Zuckerkandl's and Pauling's first papers in the mid-1960s maintained that the probability of a group of structurally related genes arising independently was so small that those genes must have evolved from a common evolutionary progenitor. Hans Neurath and colleagues at the University of Washington disagreed.[2] Considering that the ability to fly, for example, had evolved independently several times over the eons, as in the case of insects, birds, and bats, they said, it was not at all unlikely that a single structural gene could have had several independent points of origin.[3] "The evolutionary biochemist has not and cannot have any independent experimental evidence relating to the question of ancestral genes."[4] To talk about the distinction between homology and analogy at the molecular level without a detailed fossil record, they said, was comparable to Lewis Carroll's Alice thinking of cats and bats as she fell down the rabbit hole:

> Down, down, down. There was nothing else to do, so Alice soon began talking again.... "Dinah my dear! I wish you were down here with me!

There are no mice in the air, I'm afraid, but you might catch a bat, and that's very like a mouse, you know. But do cats eat bats, I wonder?" And here Alice began to get rather sleepy, and went on saying to herself, in a dreamy sort of way, "Do cats eat bats?" And sometimes "Do bats eat cats? For you see, as she couldn't answer either question, it didn't much matter which way she put it."[5]

Recognizing that molecular homology did not necessarily imply common ancestry, in 1970 Walter Fitch distinguished between what he called "orthologous" genes and "paralogous" (Greek *para*, in parallel) genes:

> Where the homology is the result of gene duplication so that both copies have descended side by side during the history of an organism (e.g., α and β hemoglobin) the genes should be called *paralogous* (para = in parallel). Where the homology is the result of speciation so that the history of the gene reflects the history of the species (for example a hemoglobin in man and mouse) the genes should be called *orthologous* (ortho = exact). Phylogenies require orthologous, not paralogous, genes.[6]

For Neurath and colleagues, such a distinction was experimentally meaningless, but Fitch maintained that "within limits of reasonable alternatives" one could indeed determine whether two sequences had a common or independent origin, that is, homologous or analogous in the classical sense.[7] To determine whether two proteins, say, cytochrome *c* of an animal and of a fungus, were the result of divergence, not of convergence, one needed "only to show that the ancestral cytochrome *c* sequences were more alike than are the present day representatives of these two groups."[8]

The problem of distinguishing convergence from divergence, analogy from homology, arose again in 1979 when Robert Schwartz and Margaret Dayhoff published in *Science* a paper titled "Origins of Prokaryotes, Eukaryotes, Mitochondria, and Chloroplasts."[9] They gathered together the published amino acid sequence data to construct four trees. One was based on the iron-containing protein ferredoxin that functions in electron transport in chemical processes in green plants and certain types of anaerobic bacteria. A second was based on 5S rRNA; a third, on c-type cytochromes that function in electron transport in mitochondria, the chloroplasts of plants, and many aerobic bacteria (see figure 17.1). The fourth was a composite tree based on those three molecular genealogies.[10]

They interpreted their data in accordance with the classical Oparin–Haldane theory of life's origins. Following van Niel, they considered the anaerobic heterotroph *Clostridium* to be the most ancient of the bacteria.[11] And they provided phylogenetic data. Indeed, the most innovative aspect of their phylogenetic scheme was in their method for "rooting" their tree, that is, identifying its earliest branching by molecular methods (figure 17.2). It was later considered to

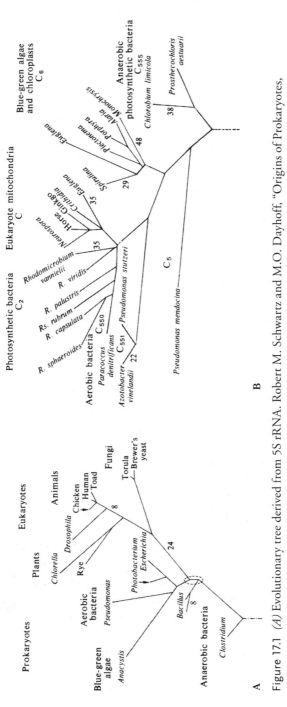

Figure 17.1 (A) Evolutionary tree derived from 5S rRNA. Robert M. Schwartz and M.O. Dayhoff, "Origins of Prokaryotes, Eukaryotes, Mitochondria, and Chloroplasts," *Science* 199 (1978): 395–403, at 399. With permission. (B) Evolutionary tree derived from c-type cytochromes. Robert M. Schwartz and M.O. Dayhoff, "Origins of Prokaryotes, Eukaryotes, Mitochondria, and Chloroplasts," *Science* 199 (1978): 395–403, at 399. With permission.

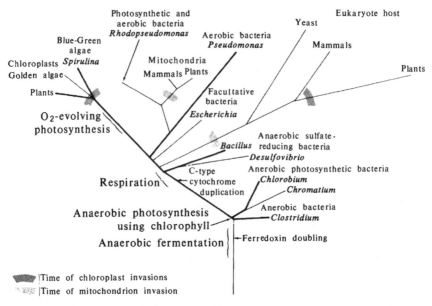

Figure 17.2 Composite evolutionary tree. This tree presents an overview of early evolution based on ferredoxin, c-type cytochromes, and 5S rRNA sequences. Robert M. Schwartz and M.O. Dayhoff, "Origins of Prokaryotes, Eukaryotes, Mitochondria, and Chloroplasts," *Science* 199 (1978): 395–403, at 400. With permission.

be ingenious. To root a tree, classical phylogeneticists added to the data set an "outgroup"—a species that is known to have branched off before all other taxa of the "ingroup" one is investigating. For vertebrates, for example, the fossil record could be used to fix the location of the earliest branching and its approximate date. In molecular phylogenetics, one could use sequences of nucleic acids or amino acids from organisms that are closely related to but not part of the group of organisms one is studying to root groupings within a tree. For example, if one were studying the evolution of birds, one could use crocodiles as the outgroup. Provided the position of its root can be fixed, molecular sequencing would be decisive. Those sequences that are closest to that of the common ancestral version are necessarily the most primitive.

But rooting the base of the universal tree was a different matter. By definition, there is no living species that could be used as an outgroup. Conventional wisdom among phylogeneticists was that the root of the universal tree could not be determined because no outgroup exists by which to position it. Schwartz and Dayhoff showed how it could be done. There was a gene doubling that had produced a highly conserved sequence, one that was shared by all the ferredoxin sequences they examined. Therefore, that gene doubling had to have occurred before all the species that contained the conserved sequence. This made it possible to deduce the point of earliest time in these trees. If one made phylogenetic trees using both ferredoxin molecules, then the branch that joins the clusters of the two molecules together identified the root. The organism whose sequence was closest to that of the common ancestral version was necessarily the most primitive. The ferredoxins from *Clostridium* were similar to the ancient protein.

Schwartz and Dayhoff used sequence data from different c-type cytochromes to test the symbiotic hypothesis for the origin of the chloroplasts and mitochondria. They used 5S rRNA data to infer the origin of the eukaryotic cytoplasm. Mitochondria branched from the purple photosynthetic bacteria; chloroplasts were similar to the blue-green algae; the eukaryotic nucleocytoplasm, they said, diverged from the branch leading to the aerobic bacteria and the blue-green algae.[12]

Their trees drew immediate criticism in the letters of *Science*. Vincent Demoulin at the University of Liège noted that their trees were sketchy and based on very few organisms: they were based on data cobbled together from partial sequences of diverse molecules, and there was very little overlap in the organisms used on each of their trees. In fact, there was not even one species present on all three of the individual trees. Without being able to compare the same organisms on different molecular trees, one could not know if the individual protein trees were actually compatible with one another, or whether the molecules one compared were orthologous and derived from a common parent, or had been duplicated in the course of evolution, resulting in false genealogies.[13]

Demoulin maintained that Schwartz and Dayhoff had provided no real answer to the question of the origin of chloroplasts, mitochondria, and eukaryotic cytoplasm because they had used different molecules for each of them. The only valid way to test the symbiotic theory, he reasoned, was to compare one kind of molecule, such as ribosomal RNA, which is coded for by genes in the nucleus, chloroplast, and mitochondria. A resolution of the question of how eukaryotes originated, he said,

> will have to wait for perfectly comparable data coming from the three eukaryotic cell compartments (for example, partial sequences for the large rRNAs). If they show clearly incompatible cladistic patterns, the symbiotic theory will then really be favored; if not, the autogenous theory.[14]

Schwartz and Dayhoff replied that Demoulin's comments regarding gene duplication were beyond scientific reason. It was simply impossible to refute his claims. "No information," they said, "is perfect. A tree based on partial RNA sequences might well suffer all the criticisms Demoulin has made."[15] They said that their data did not "prove" the symbiotic origin of organelles, but it did "support" that interpretation. And they took their composite tree to be "an excellent working hypothesis with which to organize new sequence data and our ideas."[16]

Proving Symbiosis Theory

The new era in the study of eukaryotic organelle evolution had already begun in 1975 when Woese's erstwhile technician Linda Bonen transferred the 16S rRNA cataloging technology from Woese's laboratory to Canada, where it was developed in the laboratories of Ford Doolittle and Michael Gray at Dalhousie University in Halifax, Nova Scotia. After completing her master's degree in biophysics at the University of Illinois, Bonen worked as a technician in Woese's laboratory for two years beginning in 1972, helping to develop the procedures and working on 16S rRNA cataloging. "I totally enjoyed it from day one," she recalled. "Carl was very generous in including me in all aspects of the work (even though I was a novice), and that helped greatly in transferring the technology to Halifax later on."[17] When she moved to Halifax, where her husband took up a faculty position in exercise physiology at Dalhousie, Woese suggested that she see Doolittle.[18]

Doolittle had grown up in Urbana but completed an undergraduate degree in biochemistry at Harvard before moving to Stanford to work with famed molecular biologist Charles Yanofsky.[19] After completing his Ph.D. in 1968, he returned to his hometown to begin what was to be a short postdoctoral fellowship with Sol Spiegelman. Doolittle socialized with Woese:

> Spiegelman was a fairly intimidating and unapproachable person whereas Woese was quite approachable and would like to go drinking beer with the boys and stuff like that. So many of Sol's graduate students socialized with Woese more than they did with Sol, who I think they were all afraid of. Woese was not intimidating and so I got to know him quite well.[20]

When Spiegelman moved to Columbia University (see chapter 11), Doolittle went to the National Jewish Hospital and Research Center in Denver, Colorado, for two years to work with Norman Pace, also a former student of Spiegelman's who had become a close associate of Woese.[21] Pace would revolutionize the study of bacterial diversity in the mid-1980s by developing the 16S rRNA methods for its study without the need for culturing (see chapter 20).[22]

Doolittle arrived at Dalhousie in 1972 and began to work on ribosomal RNA processing pathways in blue-green algae, but with Bonen's arrival two years later, he turned to evolutionary biology with the idea that he could test the hypothesis about organellar origins. At that time, the arguments for the symbiotic origin of chloroplasts rested on structural resemblances between chloroplasts and prokaryotic cells, strong similarities in structure and function between chloroplast and prokaryotic ribosomes, near identity of the pathways of photosynthetic electron flow and CO_2 fixation in chloroplasts and blue-green algae, and analogous cases of symbionts in protists. Still, the endosymbiotic theory was underdetermined by the data. Those who favored an endogenous origin accounted for those similarities by invoking the possibility that cytoplasmic organellar systems evolved much more slowly than did the nuclear genetic system (see chapter 9).

The notion that cyanobacteria (blue-green algae) represented the missing link between eukaryotes and prokaryotes was also still in the air. Indeed, at that time, Doolittle recalled, it was "still 50/50 whether you believed that cyanobacteria were the ancestors of plastids or whether you believed they were the ancestors of the entire eukaryotic."[23] If the endosymbiont hypothesis were not correct, that would leave the cyanobacteria as the most advanced prokaryote, which would have given rise to the most primitive eukaryotes, the red algae. Doolittle knew that there was also a strong antimolecular sentiment on the part of some microbiologists, such as Lynn Margulis, who were not prepared to accept a molecular phylogenetics.[24] The view that evolutionary arguments about cell origins belonged to the realm of metascience, as Stanier had advocated, was widespread at this time (see chapter 9).

The 16S rRNA method offered a quantitative analysis and measure of phylogenetic distance between chloroplasts and bacteria.[25] Bonen and Doolittle began cataloging 16S rRNA signatures from the chloroplasts of the red algae *Porphyridium*. They compared those signatures to those of the 18S RNA encoded in its nuclear genes, to the 16S rRNA of the blue-green algae *Anacystis nidulans*, and to 31 conserved sequence stretches that Woese's laboratory had identified in *B. subtilis* and *E. coli*.[26] The results were clear-cut: 25 of those conserved oligonucleotides were found in the 16S rRNA of the red alga chloroplasts, and only seven were present in the 18S rRNA encoded in the nuclear genes of the red alga.[27] The chloroplast ribosomal RNA shared little homology with the comparable ribosomal RNA derived from nuclear genes. Woese's laboratory simultaneously compared the 16S RNA of *Euglena gracilis* chloroplasts to the 16S rRNA of the blue-green algae *Anacystis nidulans*.[28] In both cases, there was an 80–90% homology to blue-green alga.

These results favored the symbiotic origin of the chloroplasts, but they did not prove it. As Bonen and Doolittle concluded, the results did not in themselves definitively refute the alternative possibility that chloroplast ribosomal RNA genes had evolved more slowly than their nuclear homologs.[29] The means for effectively proving the endosymbiotic origin of chloroplasts came the next year when mitochondrial 16S rRNA was compared.

The arguments that mitochondria arose from symbiosis were not as strong as they had been for chloroplasts. The shared properties of mitochondrial and bacterial ribosomes were fewer, and they seemed more superficial than those common to chloroplasts and bacteria. Mitochondrial ribosomes (at least those of animals) also differed substantially from those of bacteria (and among themselves) in size and protein content, and their RNAs differed from those of bacteria in size, base compositions, and transcriptional processes. These considerations had favored the view that, unlike chloroplasts, mitochondria most likely arose autogenously.[30]

Deploying the quantitative methods of ribosomal RNA cataloging to determine mitochondrial origins was also orders of magnitude more difficult than it was for chloroplasts.[31] Because chloroplasts are continuously photosynthesizing in the light, a plant cell has much more chloroplast rRNA than mitochondrial rRNA. The Dalhousie group worked on wheat embryos (wheat germ), and that made all the difference. Wheat germ contained large quantities of RNA from which to extract mitochondrial RNA. And as a by-product of the flour industry, wheat germ was also easily obtained at no cost. When wheat seeds are crushed, the embryos are discarded. Biochemists could get a sack of wheat germ from a flour mill. Gray, who was from western Canada, had been a graduate student in biochemistry at the University of Alberta in the mid-1960s, where he worked on the structure of RNA obtained from wheat embryos.[32]

In 1976, Doolittle and Bonen teamed up with Gray and his Ph.D. student Scott Cunningham. Their results were unambiguous. The SSU ribosomal RNA of mitochondria resembled that of four types of bacteria and had little resemblance to that encoded by the nuclear DNA of wheat. Bonen, Cunningham, Gray, and Doolittle argued in 1977 that the strong homology of mitochondrial and bacterial ribosomal RNAs and the lack of homology with nuclear-encoded ribosomal RNA "clearly supports the endosymbiont hypothesis for the origin of mitochondria."[33]

Still, they realized that when considered individually, those experiments, no more than those pertaining to chloroplasts, did not actually "prove" that mitochondria arose as symbionts. One could still argue that the nucleus and the mitochondria evolved from a common proto-eukaryotic ancestor: if one assumed that mitochondrial genomes evolved more slowly than nuclear genomes, then the similarities between mitochondria and bacteria would be expected, just as they would be for chloroplasts.[34] However, when the ribosomal RNA data from chloroplasts and mitochondria were considered together, the evidence was strong. If the chloroplast originated from cyanobacteria and the mitochondrion originated from another kind of bacteria, then the nucleus could not be both of those things.[35] Doolittle commented years later, "So at least one of the endosymbiont hypotheses was proven by those data...and probably both."[36]

Can Only God Make a Tree?

At the time the "Big Tree" paper was written, some molecular phylogeneticists were already abandoning bacterial taxonomy on the grounds that lateral gene transfer between taxa made the whole approach insolvent. Lateral gene transfer had been postulated since 1970 to be a stumbling block for bacterial phylogenetics (see chapter 10). Though it was still not known how prevalent it actually was, in principle it could cause havoc in tracing ancestries; it would mimic convergence. Woese was certainly apprehensive about lateral gene transfer, as Fox recalled:

> Carl was very concerned that the HGT argument would make people ignore SSU before enough data to truly test it could be obtained and I agreed. We therefore were very prompt in fighting this idea as soon as it came up. In a sense, the whole SSU cataloging thing was an "engineering" approach in that assumptions were made in order to make progress with the hope the answer would ultimately validate the assumptions.[37]

There were arguments against the view that ribosomal RNA genes would be readily transferred laterally. Recall that genes that were laterally transferred, such as those for antibiotic resistance, were grouped into clusters of functionally related genes in bacterial genomes (chapter 10). The complexity of the ribosomal system required the interaction of many genes that were not known to be grouped into clusters. Woese and Fox had argued in 1977 that "the large number of ribosomal components whose genes are not all necessarily contiguous argues that the ribosomal system would not be readily transferred genetically from one organism to another."[38] In 1978 Dayhoff and Schwartz similarly claimed that lateral gene transfer was infrequent enough to be ignored in large scale bacterial phylogenetics because it was based on molecules "involved in basic metabolic functions," not with adaptive traits.[39]

Still, the peril of lateral gene transfer for bacterial phylogenetics was raised the following year by Martin Kamen and his students Terrance Meyer and Richard Ambler at the University of California–San Diego. Kamen is well known today as the co-discoverer in 1940 of carbon-14 and its use in biochemistry as a tracer to follow such chemical reactions as those involved in photosynthesis, and for his studies of the biochemistry of cytochromes and their role in energy conversion.[40] His laboratory turned to cytochrome c to explore bacterial phylogenies. Cytochrome c comparison had been shown to correlate well with what was known of the phylogeny of animals based on morphological and paleontological evidence, but not so with the published classifications of bacteria.

Kamen and his students reported in *Nature* that cytochrome c–based classification of *Rhodospirillaceae* (purple nonsulfur photosynthetic bacteria) did not match their classification in *Bergey's Manual*; instead, it grouped them with the nonphotosynthetic bacteria.[41] They suggested that a single lateral gene transfer could account for such strange results. Because so little was known about evolutionary mechanisms in bacteria, molecular phylogenetics simply could not be trusted.[42]

There were two swift replies in the letters of *Nature*. Richard Dickerson, at the California Institute of Technology, had been working on cytochrome c phylogenetics since the late 1960s, and he had considerable interest in the phylogeny of the purple photosynthetic bacteria.[43] He emphasized that the disagreement with *Bergey's Manual* was not an issue because the classification therein did not necessarily reflect genealogies. If one were to claim that the determinative categories in *Bergey's* had phylogenetic meaning, it would have "to be proven." He also noted that the classification of *Rhodospirillaceae* by cytochrome *c* was in agreement both with the type of bacterial membrane and with the 16S rRNA sequence data. Such agreement would be difficult to explain by lateral transfer of genes, unless one assumed that genes for all three components were closely linked and jointly transferable. "It may be that lateral gene transfer and blurring of the evolutionary record will be a serious problem for some proteins, especially those that are intrinsically useful to a bacterium in their own right, without an attendant metabolic setting," he said.[44]

Woese, Jane Gibson, and Fox wrote the other reply. Like Dickerson, they argued that "if comparative analysis of several unrelated macromolecules yields essentially the same phylogenetic tree then that pattern is extremely unlikely to reflect the lateral transfer of genes."[45] Meyer and Kamen remained unconvinced. Later, they argued that convergent mutations and back mutations would blur the molecular evolutionary record. Maintaining "a wary optimism that a natural classification for bacteria will eventually emerge," in 1986, they quoted the poet Joyce Kilmer and suggested that "only God can make a tree."[46]

The Primary Lines of Descent

The "Big Tree" paper was formally titled "The Phylogeny of Prokaryotes" and published in *Science* in July 1980, authored by Fox, Stackebrandt, and 16 others who had collaborated with Woese over the previous decade. Cherished evolutionary dictums required reconsideration, beginning with the statement that microbial phylogenetics was impossible. "For the first time," they said, "a single experimental approach, SSU ribosomal RNA sequence characterization, has been used to develop an overview of phylogenetic relationships in the bacterial world. The technique permits the tracing of relationships back to the common ancestor of all extant life."[47] Far from being "an idiosyncrasy of rRNA," they emphasized that the phylogenetic patterns they discerned agreed with other molecular methods, insofar as they could be compared. Such congruency of phylogenies "effectively rules out the possibility that the interspecific transfer of genes can obscure evolutionary relationships in the bacterial world. Bacterial phylogenies can be determined experimentally!"[48]

The concept that life was divided into two basic phylogenetic categories, prokaryotes and eukaryotes, was to be replaced with three primary lines of descent, three urkingdoms.[49] "The tripartite division of extant life," they declared, "is incompatible with a view of two phylogenetic categories, prokaryotes and eukaryotes."[50] In each urkingdom there were important differences that, they said, implied that the three lines of descent had diverged before the level of complexity of the prokaryotic cell was attained: "Genetic control mechanisms seem to differ; RNA polymerase subunit structure differs; ribosomal RNA's and transfer RNA's differ in patterns of post transcriptional modification; cell walls differ in composition, as do lipids, and so on."[51] Prokaryotes, like progenotes and eukaryotes, were three levels of organizational complexity only. In accordance with the progenote concept, they argued that "those features of the cell that have to do with refining molecular functions, coping with a large genome, and so on, are evolved independently in the three primary lines of descent, as each reaches a more complex (that is, prokaryotic) level of organization."[52]

Thus, they presented a startling new schematic representation of the three fundamental lines of evolutionary descent, three urkingdoms: eubacteria (or true bacteria), archaebacteria, and eukaryotes, defined by their cytoplasmic ribosomal RNA, ascending from a common ancestral state (figure 17.3). A host of common characters set the methanogens, extreme halophiles, and certain thermoacidophiles apart from other bacteria.

> *Archaebacteriae* should be considered a separate kingdom of prokaryotes
> that possess (i) a variety of cell walls, none of which contain muramic acid

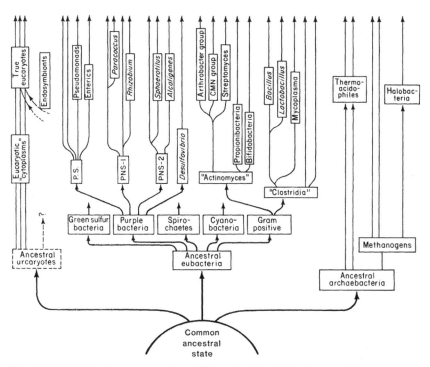

Figure 17.3 Schematic representation of the major lines of prokaryotic descent. G.E. Fox, E. Stackebrandt, R.B. Hespell, J. Gibson, J. Maniloff, T.A. Dyer, R.S. Wolfe, W.E. Balch, R.S. Tanner, L.J. Magrum, L.B. Zablen, R. Blakemore, R. Gupta, L. Bonen, B.J. Lewis, D.A. Stahl, K.R. Luerhsen, K.N. Chen, and C.R. Woese, "The Phylogeny of Prokaryotes," *Science* 209 (1980): 457–463, at 459. With permission.

(the hallmark of eubacterial walls), (ii) membranes whose major component is a branched chain (phytanyl), ether-linked lipid, (iii) transfer RNA's (tRNA) devoid of ribothymidine in the TΨC loop (T, thymine; Ψ, pseudouridine; C, cytidine), (iv) distinctive RNA polymerases subunit structures, and perhaps (v) an unusual but still not fully elaborated spectrum of coenzymes. (All archaebacteria demonstrate the first four of these properties; the fifth so far is confined largely to methanogens.)[53]

They further separated the methanogens and extreme halophiles into one group, and the thermoacidophiles into another. The diversity and phylogenetic depth of the archaebacteria were indicated by their four major cell wall types, compared to only one among all the eubacteria, and by their GC content, which ranged from 27% to 68%. Maintaining that the archaebacteria were indeed an ancient group, they suspected that their defining phenotype may reflect an ancient habitat and thus not reflect adaptations to present-day earthly conditions. "If so, basic archaebacterial metabolism and control mechanisms could have evolved to reflect conditions that no longer prevail."[54]

The eubacteria kingdom was reorganized into eight groups, some of them strikingly odd from the point of view of morphology and physiology. They insisted that the classical tendency to separate photosynthetic from the nonphotosynthetic eubacteria did not hold up. The purple photosynthetic bacteria and various genera of nonphotosynthetic bacteria were grouped together. The emphasis traditionally placed on morphological characteristic was also not justified: "Spherical shape is a principal offender," they said. "All spherical bacteria so far examined fall into phylogenetic categories defined in terms of nonspherical organisms."[55] Mode of cell division and lack of a cell wall were also deceptive characters for classification. Spore formation was a relatively good phylogenetic indicator, but lack of spore-forming capacity was not.[56]

The various taxonomic levels defined by traditional criteria also bore little relationship to the phylogenetic levels: older groups were distinguished by a greater range in S_{AB} values, indicating greater diversity. The first organisms were not heterotrophs (e.g., *Clostridium*), but autotrophs: the first eubacteria may have been photosynthetic, and the first archaebacteria may have been methanogenic.

The mitochondria and chloroplasts of eukaryotes originated exogenously from free-living bacteria: the chloroplast was derived from cyanobacteria, and the mitochondrion from the purple photosynthetic bacteria. Fox et al. were less certain of the origin of other structures. But it could not be assumed that the bulk of nuclear genes had a common ancestry. Descent with modification need not apply for the evolution of the eukaryotic cell. "The collection of genes constituting the nucleus," they said "could just as a well have arisen through myriad gene or gene cluster captures from all manner of sources. Until this matter is settled—and molecular phylogenetic studies will ultimately do so—it is unwise to speak of an ancestral eukaryotic line of descent."[57]

Nothing Left to Chance

When Woese sent the "Big Tree" paper to *Science* at the end of January 1980, he suggested several possible reviewers: Zuckerkandl, Doolittle, John Oró, Dickerson, and Fitch. The two reviewers remained anonymous. Their comments were unequivocal: "This is a *very* important paper, and it *belongs* in *Science*. It represents the summation—long awaited in evolutionary circles—of hundreds of analyses of ribosomal RNA sequences performed during the last decade by Woese and his collaborators," said one reviewer. "Some of these analyses have already drastically affected phylogenetic thinking.... The resulting tree has implications (a few of which they discuss) which can be debated and subjected to experimental test for years to come."[58] The second reviewer commented,

> A remarkable paper, of exceptional interest. It appears to the present reviewer that Carl Woese and his associates have made the greatest contribution yet achieved to our knowledge of the phylogeny of primitive unicellular organisms and therefore of the origins of the cell.... Publication of this paper is warmly recommended.[59]

When Woese sent a draft to several colleagues, their responses were equally admiring. Dickerson wrote to Woese congratulating him on such "a great article" that clearly defined the issues. He was convinced that Woese must "be right in principle" and that molecular data would provide the solution to bacterial phylogeny—"if an answer is ever to be found," he added.[60] Woese dedicated the paper to van Niel, then long retired, who was delighted by the honor.[61]

Stanier saw Woese's work as revolutionary. He was in the midst of writing an autobiography essay for *Annual Reviews of Microbiology*.[62] He wrote to Ralph Wolfe about Woese's success while comparing it to the "Main Outlines of Bacterial Classification" that he and van Niel published in 1941.[63] "Historically speaking, I think it's the most worthless paper I've written," he said.

> I think it represents the last gasp of speculative system-building, whereas Carl Woese has the signal merit of conceiving a new taxonomic treatment, based on the données of molecular biology. You contributed an essential component, in considering the methanogens phenotypically: the ester-linked lipids, the wall composition, the unique coenzymes, all contributed to establishing the solidity of the methanogens as a very ancient biological group, far antedating the traditional prokaryotes.[64]

Francis Crick was fascinated by "The Phylogeny of Prokaryotes," but he had a different view of life's origins on Earth. He and Leslie Orgel had proposed in 1973 that the first organisms on Earth were deliberately planted on Earth by intelligent beings from another planet.[65] It was a modification of the concept of panspermia (see chapter 4).[66] Such concepts were reconsidered at a meeting on extraterrestrial intelligence in 1971. Carl Sagan considered it unlikely that bacteria could have reached Earth as spores because of the lethal dose of radiation they would receive in interstellar space.[67] The probability of successful seeding would be greatly increased, Crick and Orgel argued, if the seeding of Earth were carried out deliberately by an existing technological civilization. As for the means of dispensation:

> The spaceship would carry large samples of a number of microorganisms, each having different but simple nutritional requirements, for example, blue-green algae, which could grow on CO_2 and water in "sunlight". A payload of 1,000 kg might be made up of 10 samples each containing 10^{16} microorganisms, or 100 samples of 10^{15} microorganisms.[68]

In 1980 Crick was in the midst of writing his book *Life Itself*, in which he elaborated his argument about "directed panspermia" that Earth may have been seeded by intelligent beings. He wrote to Woese:

> Leslie passed on to me your review on "The Phylogeny of Procaryotes" which I read with great pleasure and interest. I could not help wondering just why you find two (or perhaps three) distinct kingdoms and not any intermediate forms. Of course many explanations are possible but I was amused to notice that one of them would be Directed Panspermia! If you had to send bacteria to another earth-like planet at a stage in its development (i.e., without oxygen) you would probably send:
>
> 1) several different and distinct types of organisms
> which were 2) photosynthetic
> and also 3) autotrophic.
>
> I shall be interested to see how this rather unusual point of view stands up as further data come out.[69]

No question that "The Phylogeny of Prokaryotes" was landmark work for molecular phylogenetics. But it did not convince those who doubted the whole molecular approach to phylogeny. Two critical letters were immediately sent to *Science*. For some, the claim that the photosynthetic bacteria were related to the nonphotosynthetic bacteria was simply too much to accept, just as it had been for Ambler, Meyer, and Kamen. It clearly showed the failure of the 16S rRNA approach for determining evolutionary relationships. Others argued that the 16S rRNA approach was clearly "inappropriate," and because they were looking at a "single" character, their conclusions are not representative of the whole organism. Some pointed to the small "sample of bacterial species used," and the lack of details about the similarity coefficient S_{AB}.

The associate editor of *Science* forwarded the letters to Woese and Fox for their responses before accepting them for publication. Both recommended that they not be published. "The creation of a public elusion [*sic*] that the work is highly suspect (which it isn't!)," Fox replied, "would not be in the best interest of the field."[70] Woese responded in more detail:

> It is one thing to present useful criticisms of the work of others; it is another thing to create obfuscation. Unfortunately, both presentations are of the latter type. There is no way these articles if published by Science (i.e. sanctioned as valid criticism) can do anything but generate confusion among their readers....
>
> The criticisms appear to be "text-book" taxonomic criticisms, applied without their author's really caring to look at our approach in any depth.[71]

Regarding the criticism that 16S rRNA did not represent the phylogeny of the whole organism, Woese drew a distinction between "superficial" characters and fundamental or what he called "deep characters." Such deep characters he said differed from superficial ones "by how tightly they are coupled, integrated into the fabric of the cell as a whole. The latter tend not to vary rapidly over evolutionary time and (if the cytochrome c vs rRNA comparison be representative) are not subject to interspecific gene transfer." This conception would later be called "the complexity hypothesis" (see chapter 22). "The 'right' single character can be representative of the history of the whole genome," Woese said. For *Science* to publish those two letters, he concluded, would "merely tend to destroy (in the eyes of readers not sufficiently familiar with the system) a beautiful and correct construct."[72]

These issues aside, there were indeed methodological problems: the similarity coefficient S_{AB} sometimes did bias results. It was based on the concept that nucleotide changes in an organism occurred at the same rate over evolutionary time. Therefore, the extent to which two oligonucleotide sequences differed would be a measure of the time since the organism diverged from a common ancestor. However, some organisms changed at a faster rate than others—their mutational clocks were fast.[73] Woese referred to such rapidly evolving organisms as being "tachytelic" (Greek *takhus*, rapid, and *telos*, purpose) in keeping with the terminology coined by paleontologist George Gaylord Simpson in his *Tempo and Mode of Evolution* of 1944.[74]

Paleontologists had long inferred from the fossil record that rates of evolution were not constant among the lineages (or within a given lineage at various times). What Simpson noted was that when the rate (tempo) of evolution increased, the nature (mode) of the process qualitatively changed—many new lineages of varying stabilities and unexpected diversity suddenly would come into existence. Fossil lineages showed periods of very rapid evolution—evolutionary bursts—characterized by numerous drastic and unusual phenotypic changes. Comparing rapid and gradual evolutionary changes left the impression that no matter how long the slow evolutionary process was to continue, the changes accumulated would never come to resemble in kind those produced by the rapid ones.

Woese looked for this "tempo–mode" effect on the molecular level and found it in the wall-less mycoplasmas. When their RNA sequences diverged rapidly, they also showed a qualitative shift: a high fraction of the changes involved positions in the sequence whose composition had usually been very stable and invariant.[75] "The idea puts on the molecular level (and so makes compelling) the notion that fast-clock organisms must (evolve to) have unusual bizarre phenotypes," Woese wrote Joseph Felsenstein. "The idea begins to transfer the 'macro-evolution' phenomenon to the molecular level, and so potentially better defines it."[76]

| The Dawn Cell Controversy

They just told me I've received the Bergey Award for 1983....Any recognition for the Archaes and bacterial phylogeny is for the good. Perhaps, I enjoy most seeing how independent the archaebacteria have become of my lab and the initial work—a true sign of their validity and importance.

—Carl Woese to Wolfram Zillig, September 28, 1982

Whatever the final outcome of Carl Woese's work will be, he opened a door which nobody had expected to exist.

—Otto Kandler, June 26, 1988

THE 1980S WERE HEADY TIMES for bacterial phylogenetics. Comparative studies of 16S rRNA led to a breakthrough in deep phylogeny, and they provided crucial evidence for the bacterial descendents of mitochondria and chloroplasts. The archaebacterial concept illustrated how the direction of biological research in biochemistry, molecular genetics, and natural history could be profoundly affected by phylogenetic classification. It triggered the discovery of many organisms living in extreme habitats: new genera and even orders, unique lipids, walls, and variants of the genetic machinery.[1]

More than 50 researchers attended the "First International Workshop on the Archaebacteria" in Munich in the summer of 1981. Funded by the Volkswagen Foundation and the Max Planck Society, it focused on all facets of archaebacterial research: molecular and biochemical research and geochemical, paleontological, and taxonomic aspects. Kandler wrote in the introduction to the conference proceedings:

> The sequence similarity of ribosomal SSU RNA has provided us with startling new insights into the genealogical relationship between organisms. It must be considered a break-through in the search for a natural system of bacteria, the final goal of bacterial systematics. Moreover, the concept of the archaebacteria is exerting a marked stimulatory effect on molecular genetics, comparative biochemistry and physiology and has far reaching implications for ecological and applied microbiology.[2]

In recognition of the new phylogenetics and the expansion of bacterial taxonomy in 1984, *Bergey's Manual* began the first edition of a new series of volumes titled *Systematic Bacteriology.*[3]

Big technological change was also occurring. New methods for whole gene sequencing had become available in the late 1970s and 1980s.[4] In 1977, Frederick Sanger and Walter Gilbert independently developed methods for sequencing DNA. Three years later they shared the Nobel Prize in Chemistry with Paul Berg, who in 1972 had constructed the first DNA molecule recombined from different organisms, a crucial step in the development of genetic engineering. Then, in 1983 the polymerase chain reaction (PCR) was invented by Kary Mullis working at Cetus Corporation in Berkeley, California. It provided the means to make many copies of sequences from minute samples—to clone DNA to compare nucleotide sequences.[5] PCR transformed many aspects of biology.

Two back-to-back international workshops on archaebacteria were held in June 1985: one on the "Molecular Genetics of Archaebacteria" at the Max Planck Institute in Martinsried, and the other on the "Biology and Biochemistry of the Archaebacteria" at the Botanical Institute of the University of Munich. Some 65 papers and 38 poster abstracts documented the progress in archaebacterial research.[6] Comparisons of complete 16S rRNA sequences confirmed, refined, and extended the earlier concepts of archaebacterial phylogeny.[7] Studies of the diversity of the sulfur-dependent archaebacteria were extended, and Wolfram Zillig pioneered the archaebacterial virus–host system. The origin of viruses remained unknown: they could have arisen as escaped genes from organisms, or from a primordial soup. Zillig argued that studies of the novel structure of archaebacterial virus-like entities took on special importance in view of "the probably primitive (close to primeval) nature of the archaebacteria, and thus possibly of their viruses."[8]

Are the Archaebacteria Proto-eukaryotes?

Since the late 1970s, Woese, Zillig, and Kandler repeatedly discussed the question of whether or not the archaebacteria were more closely related to eukaryotes than to eubacteria. Zillig and Karl Stetter's research showed that they had several complex RNA polymerases that were similar to those found otherwise only in eukaryotes.[9] Those enzymes were resistant to the bacterial antibiotics rifampicin and streptolydigin just as were the polymerases of eukaryotes.[10] Some members of the archaebacteria also seemed to be more closely related to the eukaryotes than were others. Zillig wrote to Woese in the summer of 1979 that archaebacteria could be divided into two groups based on comparisons of their RNA polymerases, "namely the Methanogens Halophiles group and the Sulfolobus Thermoplasma group."[11] Zillig informed Woese in August 1982 that based on SSU rRNA–DNA hybridization experiments, the thermoacidophilic branch of the archaebacteria urkingdom consisted of two orders, the Sulfolobales with various *Sulfolobus* species and the Thermoproteales with the families within them.[12]

It seemed to Zillig, at least initially, in the early 1980s, that the thermoacidophiles might be more closely related to the archaebacteria than were the methanogens and extreme halophiles. Woese was reluctant to tease out any groups from the archaebacteria as being more closely related to eukaryotes than any others. He favored the hypothesis that all three lineages emerged from the progenotes. Similarities between the primary lineages could be accounted for by lateral gene transfer (see chapter 17). "Archeas have contributed to eucaryotes; how much and by what mechanisms? We have a long way to go, all exciting," he wrote to Zillig in January 1982.[13]

Zillig had no doubt that the archaebacterial lineage was ancient. "I am convinced that the term which you have chosen is fully justified," he wrote to Woese in December 1982:

> The immense phylogenetic depth of some features e.g. the 5S rRNA which reaches from eubacterial until more than eukaryotic [*sic*], the ribosome and the initiator tRNA on the one hand, the very pronounced eukaryotic character of some features like EFII, ribosomal A protein and RNA polymerase on the other hand but also the increasing evidence that *Sulfolobus* and the *Thermoproteales* might be "more eukaryotic" than the methanogens + extreme halophiles suggests that your urkaryote might have been something resembling the recent probably atavistic sulfur metabolizing archaebacteria whereas the first narrow offshoot leading to the eubacteria could have occurred somewhat earlier from the other corner of the urarchaebacteria, the urmethanogens. The eubacteria then would have evolved with high velocity before dividing into many phyla.
>
> The only grave argument against this assumption seems to be the special nature of the archaebacterial lipids.[14]

Zillig and Woese also considered the idea that archaebacteria might be at the base of the other two urkingdoms. "Although I am not yet willing to commit to the idea that the archaekingdom underlies the other two (i.e., methanogens et al gave rise to eubacteria, thermoproteus et al to the urcaryotes), I am most taken by it," Woese replied the next month.[15] Then, another striking molecular characteristic was found that seemed to link the archaebacteria closely to the eukaryotes. Archaebacteria possessed noncoding regions in their genome; such regions were thought to be characteristic of eukaryotes alone.

In the late 1970s, Phillip Sharp and his collaborators at MIT and Richard Roberts at New England Biolabs showed that eukaryotic genes had long intervening sequences, "introns," between coding regions of a given gene, "exons."[16] Before such "split genes" are expressed, the corresponding messenger RNA (mRNA) is "edited"; that is, the intervening sequences are removed and then spliced back together again. A small body called a "spliceosome," a protein–RNA complex, cut out various segments along the mRNA. The remaining pieces of RNA could be spliced together into a number of alternative combinations and these used as messages (mRNA). A single DNA gene sequence then could give rise to a number of different proteins.[17] Alternative RNA splicing was recognized to be one of several important mechanisms of genetic regulation in the differentiation of eukaryotic cells during the embryonic development.

Woese informed Zillig that his laboratory found suggestive evidence of introns in *Sulfolobus* transfer RNA genes.[18] "I am absolutely struck by your findings of introns in Sulfolobus tRNA genes and of the non coding of the CCA terminals of the corresponding tRNAs," he wrote to Woese in July 1983. "This adds very strong arguments to the 'urkaryotic' nature of the sulfur-dependent archaebacteria."[19] The following year, Woese also considered the idea that they might well be closest relatives to the eukaryotes.[20] Two years later introns were also found in extreme halophiles and other archaebacteria.[21] Still, the idea that the thermoacidophiles might be the mother of the eukaryote persisted.

The Eocyte Concept

Heated controversy broke out in the 1980s when James Lake at the University of California–Los Angeles went one step further and suggested that the thermoacidophiles be granted an urkingdom of their own as a group closely related to the eukaryotes. He named the proto-eukaryotes "eocytes" (dawn cells) in 1984.[22] Lake's views would change repeatedly. The following year he proposed another new urkingdom, "Photocyta," for the extreme halophiles and the eubacteria. Three years later, he would drop the kingdom Photocytes and propose instead two superkingdoms, "Parkaryotes" and "Karyotes." A few years later, he would drop those superkingdoms, too.

The eocyte concept caught the scientific limelight and was discussed in editorials and letters throughout the 1980s and 1990s. But the methods and the conclusions based on them attracted severe criticisms from various points. Lake's proposals were based on morphology—on the structural differences in ribosomes: specific bumps they exhibited on their surfaces—as determined by electron microscopy. Lake was well known for his electron microscopic studies of ribosomes. He maintained that the morphology of ribosomes was a phylogenetic characteristic, which offered a rapid method for classifying organisms. At first, he and his collaborators endorsed the three-urkingdoms model.[23] Then they reported that the ribosomes of the sulfur-dependent archaebacteria had a bump that was different from that found on the ribosomes of methanogens and halophiles.

The controversy began in the fall of 1982 when, after receiving cell preparations from Zillig, Lake reported the smaller ribosomal subunits of *Thermoproteus* and *Sulfolobus* had a distinctive ducklike "bill" that was lacking in eubacteria but present in eukaryotes.[24] The smaller subunit also possessed "lobes" that he saw to be similar to those of eukaryotes. The larger ribosomal subunits of those thermoacidophiles also possessed a lobe and a bulge that resembled those of eukaryotic large subunits. At first, Zillig and Woese took Lake's data to support their own.[25] But then, as Zillig explained to Woese in his letter of July 22, 1983, Lake went too far:

> About two weeks ago Jim Lake was here presenting his comparative analysis of the shapes of ribosomal subunits....I would take this as another [piece] of evidence, besides your recent findings and all the older arguments, that the sulfur metabolizers are indeed primitive and close to the urkaryote.
>
> Jim Lake, however, wants to make them a fourth urkingdom, the "eocytes," and he wants me to be a co-author in a paper on this subject....
>
> What do you think? I would either like to convince him that the same data should be presented in a less pretentious manner and remain a co-author or get off and ask him to just acknowledge our giving him the "eocytes."[26]

The following month Woese and Zillig independently wrote to Lake. "I disagree with your proposal," Woese wrote:

> I feel it is not sufficiently supported by facts. You cannot reasonably rule out convergence with your type of data. And you are defining the split by criteria different from those used to define the kingdom in the first place—i.e. genotypic evidence. By rRNA and tRNA sequence archaebacteria *are* a coherent grouping, in being closer to one another than any are to members of the other kingdoms. Therefore, all your proposal does is muddy the waters. There is an issue here, but I don't feel you're addressing it in a constructive way.[27]

Zillig replied to Lake similarly:

> My opinion is that more arguments speak against than for a further splitting though I believe it important to stress the deep division between

the two branches of archaebacteria....Another reason for not following your suggestion is that it is based on but one phenotypic and non-quantitative feature, the shape of the ribosome....I had contacted Carl in this matter...and he shares my opinion. But as you are free to have yours, go ahead and publish it if you like, without me being co-author, and acknowledge me furnishing the cells.[28]

Lake sent a second version of his paper introducing the eocyte kingdom to Woese, who countered in February 1984 that Lake's arguments were "not scientifically sound," that it was ludicrous to base a "kingdom on one or a few-ill defined phenotypic characters," and that when one considers any phenotypic characteristic individually or a few at a time, one could construct any tree at all. He suggested that Lake consult with an appropriate taxonomist such as Joseph Felsenstein. "Your discussion of this evidence," Woese wrote, "is altogether too cursory and, I think, misleading":

> You downplay or purposely ignore evidence that links all archaebacteria to one another, or is otherwise counter to your argument. Every sequence measure, be it rRNA catalogs, protein sequences, or whatever, shows all archaebacterial examples to be far closer to one another than to any sequence from an outgroup organism.
>
> Your apparent need to have there be a new kingdom detracts from your presentation of a solid contribution....Workers in archaebacteria have been aware for some time, as you know, that the division between the two sides of the kingdom is unprecedently [*sic*] deep phenotypically, and especially in Zillig's case...that the sulfur-dependent archaes are on the whole closer phenotypically to eucaryotes than are the methanogens....In my opinion, you can only do your reputation harm by this sort of innocent pronouncement you seem so intent on making.[29]

Woese suggested that Lake write two papers: one a "more technical report" in which both Zillig, who supplied the materials for study, and Alastair Matheson, who had helped in the preparation of the ribosomes for electron microscopy, were included as authors; and the other "your proposal to create a new kingdom, under your own name."[30] "I shall agree if he deletes the term 'eocyte,'" Zillig wrote to Woese the following month, "which subcutaneously anticipates the fourth urkingdom."[31] Lake published two papers, the first in the summer of 1984 with Zillig and Matheson among the authors. Although there was no mention of the fourth urkingdom in the body of the paper, Lake wrote in the reference notes "that the sulfur-dependent bacteria constitute a group (the eocytes) with a phylogenetic importance equal to that of the eubacteria, archaebacteria, and eukaryotes."[32] The paper on the eocyte kingdom appeared later that year.[33]

Then in 1985, Lake and collaborators introduced the new kingdom of "photocytes" (extreme halophiles and eubacteria).[34] Photosynthesis as it occurred in some of the eubacteria and all plants is based on chlorophyll, whereas photosynthesis in extreme halophiles was well known to be based on rhodopsin. Nonetheless, Lake and his students suggested that photosynthesis had emerged only once: "The eubacteria and halobacteria compose a monophyletic group, for which we propose the name 'photocytes.' "[35] "Phylogenetically" they said, "it is clear that the halophiles do not belong in the archaebacteria."[36]

The eocyte and the photocyte concepts found support from molecular phylogeneticist Allan Wilson at the University of California–Berkeley, well known for his studies of human evolution. Wilson's work first attracted great attention when in 1967 he and Vincent Sarich argued that humans and apes had diverged from a common ancestor only around five million years ago.[37] That time line flew in the face of the date favored by anthropologists of around 25 million years. Wilson shocked science again in 1975 he and Mary-Claire King reported that the average human protein is more than 99% identical to its chimpanzee counterpart.[38] He confronted traditional anthropological thinking again in the early 1980s with the "mitochondrial Eve" hypothesis.[39] Using mitochondrial DNA to track human evolutionary history, he and his collaborators concluded that modern humans had diverged from a single population in Africa about 150,000 years ago.

In a 1985 *Scientific American* article titled "The Molecular Basis of Evolution," Wilson included a phylogenetic tree based on Lake's proposals. He wrote that cells evolved nearly three billion years ago and led to "four main groups of bacteria: eubacteria (the major current form) and halobacteria, methanogens and eocytes (sulfur bacteria).... Photosynthesis probably originated in the common ancestor of chloroplasts, eubacteria and halobacteria."[40] He said that this phylogenetic tree "was inferred from comparing the base sequence of ribosomal RNA found in the various organisms."[41] Woese spoke with him about his paper after it appeared. As he explained to Zillig,

> Wilson said he was impressed with Lake's "cladistic analysis". He also says Lake told him that I accept the eocyte-photocyte concept (and by implication that others in the field do as well). I think we now have to do formal battle with this growing tissue of half-truths, misrepresentations, propaganda, etc.[42]

Conflicts in Nature

The battle of the kingdom keepers was waged in the letters of *Nature*. It began when Roger Garrett wrote an account of the Archaebacteria workshop on "The Molecular Genetics of Archaebacteria" held in Martinsried in June 1985.[43] Garrett was a former student of Maurice Wilkins, who had shared the Nobel Prize in Physiology or Medicine with Watson and Crick in 1962. Zillig met him at a meeting held in Denmark in the spring of 1984, and he was quickly integrated into the archaebacterial research program.[44] Garrett was not going to mention Lake's eocyte concept and its reception, but he was requested by the deputy editor of *Nature*, Peter Newmark, to include a discussion of it.[45]

Garrett wrote of the molecular evidence for the uniqueness of the archaebacteria with its three main orders: the extreme halophiles, the methanogens, and the sulfur-dependent extreme thermophiles. Each order, he said, had some special characteristics that reflect adaptation to a particular niche, but other characteristics were common to all archaebacteria. He noted exciting developments in the characterization of virus–host systems by Zillig, and that introns had been discovered not only in *Sulfolobus* but also more recently in *Thermoproteales* by his own group and in extreme halophiles by Ford Doolittle's group.[46] Garrett suspected that the possession of introns would be another fundamental characteristic distinguishing the archaebacteria as a whole from typical bacteria.

Then Garrett came to Lake's eocyte proposal "that the sulphur-dependent *Thermoproteales* and *Sulfurlobales* should be considered as a separate kingdom." That idea, he said, "received an extremely negative reaction," just as did Lake's contention "that the extreme halophiles should be grouped with eubacteria because their ribosome shapes are comparable and because it is unlikely that photosynthesis evolved in two separate events."[47] Garrett concluded on a positive note:

> There were optimistic messages for all participants. The biotechnologists were promised heat-stable enzymes from the extreme thermophiles and genetically engineered methanogens for improving industrial processes. The molecular geneticists were offered a wealth of novel and exciting problems. And the evolutionists were shown how to refute the age-old prejudice that first there were prokaryotes and then eukaryotes, and how to infer the nature of early forms of life on Earth.[48]

Lake was furious. He wrote to Garrett suggesting that he had deliberately tried to suppress "a major controversy," that his report was "grossly inaccurate and damaging" to his work, and that it amounted to scientific censorship in an attempt to protect archaebacterial orthodoxy.[49] He explained that controversy was good when based on competing techniques and results; it helped to crystallize ideas.[50] When Lake replied in the letters of *Nature*, he depicted his struggle as one between "a scientifically just alternative" and "archaebacterial dogma." "Our tree," he said, is based on "analyses of three dimensional ribosome structure," whereas "the archaebacterial tree is an artifact produced by greatly unequal rates in different arms of the tree." Because of different rates of change in fast clock and slow clock organisms, the "primary sequence data, as *presently analysed*, neither support nor disprove the archaebacterial tree or any other tree."[51]

As Zillig and Woese saw it, Lake was merely aiming for notoriety and disturbing the rational development of the field. Zillig aimed to stop him in his tracks before matters got out of hand. He wrote a letter to *Nature* asserting that that there was simply no factual basis for Lake's new kingdom, nor that the three-dimensional structure of ribosomes was a phylogenetically useful character.[52] He pointed out that he and Stetter had already published a paper in 1982 in which they placed methanogens and halophiles in one group of archaebacteria and the sulfur-dependent archaebacteria in another; he and Woese had emphasized the same again in 1984.[53] He concluded: "The kingdom of the archaebacteria remains a solid entity in our incomplete understanding of the early phase of biotic evolution."[54]

Zillig also noted electron microscopic results that contradicted Lake's claims. Georg Stöffler and coworkers at the University of Innsbruck tested Lake's hypothesis about the phylogenetic significance of the bulge in the ribosome, which, he said, was unique to the sulfur-dependent archaebacteria (the eocytes) and the eukaryotes. They reported that the ribosomes of some methanogenic archaebacteria also had that same shape.[55] Woese, Norman Pace, and Mitchell Sogin's former student Gary Olsen also sent a letter to *Nature* emphasizing that there was no evidence that the shapes of ribosomes were actually phylogenetic characters.[56] Zillig's colleague Hermann Lederer sent a letter to *Nature* contradicting Lake's claim that the archaebacterial tree was distorted because of unequal clock rates.[57] Lake replied in *Nature*, "We see no support for Woese's tree and taxonomic proposal from any molecular properties whereas deep and fundamental properties support our eocyte and photocyte trees and classification."[58]

Lake dropped the Photocyta urkingdom without explanation the next year in *Nature*. This time, he replaced the classical prokaryote–eukaryote bifurcation of life with a new bifurcation of two superkingdoms: parkaryotes ("essentially bacterial" organisms) and karyotes ("a proto-eukaryotic group") (figure 18.1).[59] He employed a "new rate-invariant treeing algorithm" that he devised for analyzing the 16S rRNA. He called it an "evolutionary parsimony" method and concluded "that the last common ancestor of all organisms lacked nuclei, metabolized sulphur and lived at near-boiling temperatures."[60] Lake abandoned parkaryotes and karyotes and reverted to the prokaryote–eukaryote dichotomy a few years later, but he maintained his view of the eocyte.[61]

Zillig's continued comparisons of RNA polymerase sequence data led him to reject his previous suggestion that sulfur-dependent archaebacteria were more similar to eukaryotes than were the others. It seemed to him that the publicity that Lake's ideas were getting in high-visibility journals was having a negative impact on his own experiment work. "I must admit that I have greatly underestimated the stupidity of the scientific community and/or the persuasive power of Lake," Zillig wrote to Woese in February 1988, after reading the latest reports on the eocyte concept in *Nature*.

> Though I am convinced that in the long run this will all be forgotten or judged as ridiculous intermezzo....I see the impact of these pseudo-ideas,

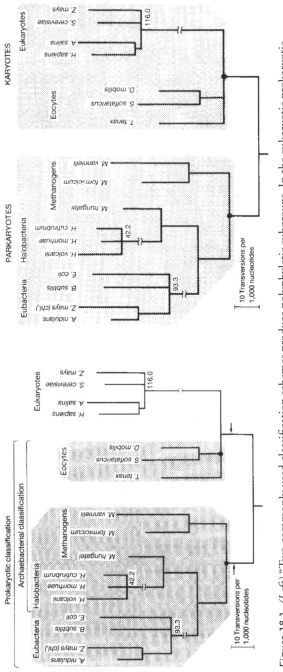

Figure 18.1 (*Left*) "Two commonly used classification schemes produce polyphyletic subgroups. In the eukaryotic prokaryotic scheme the prokaryotes the noneukaryotic organisms are polyphyletic. In the archaebacterial-eubacterial-eukaryotic scheme, the archaebacteria (the non eubacterial prokaryotes) are polyphyletic." (*Right*) "The parkaryotic-karyotic classification proposed in this paper defines two balanced monophyletic groups." James A. Lake, "Origin of the Eukaryotic Nucleus Determined by Rate-Invariant Analysis of rRNA Sequences," *Nature* 331 (1988), 184–186, at 186. With permission.

especially because of their publication in Nature and because of their treatment by an editorial....

...Our RNA polymerase manuscript has been rejected by Nature and later on also by Science...It seems that it's a bad time for publishing papers on archaebacteria in the science evening press. But this will change again. I'm sure.

As you know we ourselves had once nourished ideas similar to those of Lake but convinced ourselves especially from the rRNA data, but know also now from our polymerase sequence data that our interpretation of a list of features was wrong.[62]

The controversy over Lake's methods continued. In 1989 Olsen and Woese argued that the eocyte tree held up only when a particular alignment of sequences was used and when particular sequences were used for tree reconstruction.[63] On the other hand, phylogenetic trees based on rRNA, the RNA polymerases, as well as a host of phenotypic characters, supported the monophyletic conception of the Archaebacteria. In short, no method other than Lake's, they said, gave the results he reported.

Others tested Lake's rate-invariant algorithm ("evolutionary parsimony") method on more obvious phylogenetic relationships. When Manolo Gouy from Lyon and Wen-Hsiung Li at the University of Texas used it to compare the 16S rRNA from humans, *Drosophila*, rice, and the slime mold *Physarum*, humans and rice came out in one clade. On the other hand, when they used other methods (neighbor-joining and maximum parsimony methods) on the ribosomal RNA data, the results agreed with the archaebacterial tree.[64]

Gouy and Li published their paper in *Nature*, and Lake replied in *Nature*'s letters, suggesting that they had used only part of the relevant data set to reach their conclusions.[65] Li informed the assistant editor of *Nature* that Lake had declined his request for the source code for his algorithm, so it was difficult for him and others to understand his computational procedure.[66] When Li and Gouy replied in *Nature*, they said that theoreticians did not understand Lake's derivation because that part of the method was added at the proofs stage of the paper in which he described the method, and no detail of the derivation was given.[67] Lin Jin and Masatoshi Nei subsequently examined the theoretical basis of Lake's "evolutionary parsimony" method, concluding that it depended "on a number of unrealistic assumptions."[68]

Rooting the Tree

Woese wrote an extensive, landmark overview titled "Bacterial Evolution" in 1987.[69] In it, he discussed how, stymied by technical difficulties, microbiologists had given up on phylogenetics and then discounted evolution as unnecessary to the advancement of their field, just as did molecular biologists. He elucidated why ribosomal RNA sequences were the ultimate "molecular chronometers" and could be used to measure both close and the most distant genealogical relationships. He explained that the sequences in different positions in ribosomal RNA could vary from one taxon to another at vastly different rates. He pictured this as a chronometer with different dials/hands, which measured seconds, minutes, hours, and so on. The "second hand" would be most useful in measuring the distance between the most closely related taxa; the "minute hand," less closely related taxa; the "hour hand," distantly related taxa.[70] Sequencing information, he said, had thus extended the scope of evolutionary knowledge nearly tenfold. "It shows the evolutionist an intimacy between the evolution of the planet and the life forms thereon that he has never before experienced."[71]

The phylogenetic relationships between the three primary kingdoms remained unresolved (figure 18.2). "Our present, rather limited understanding," Woese wrote, "would suggest that the overall phenotypic resemblance is greatest between archaebacteria and eukaryotes."[72] The conventional wisdom held that the root of the universal tree could not be determined because no outgroup exists by which to position it. Recall that in cladistic analysis, branching order was determined by using an "outgroup." That is, organisms known to be

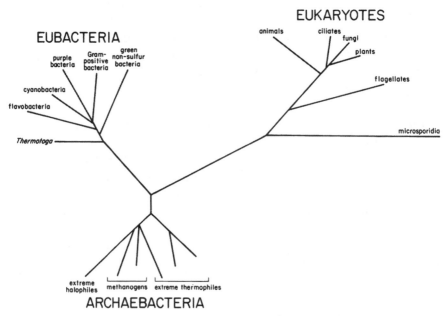

Figure 18.2 "Universal Phylogenetic Tree Determined from rRNA Sequence Comparisons. A matrix of evolutionary distances was calculated from an alignment of representative 16S rRNA sequences from each of the three urkingdoms." C.R. Woese, "Bacterial Evolution," *Microbiological Reviews* 51 (1987): 221–271, at 231. With permission.

phylogenetically outside the "ingroup" were used as a rooting point of a phylogenetic tree (see chapter 17). The same method was used in molecular phylogenetics. One could use several eubacterial outgroup sequences to root the purple bacteria, for example. But establishing a root for the universal tree of life, the branching order among the primary urkingdoms, was another matter entirely.

Woese pointed to the ingenious method of Margaret Dayhoff and Robert Schwartz based on gene duplication (chapter 17). "What is required," he said, "is a gene that has duplicated in the common ancestor state (as pointed out by M. Dayhoff long ago)."[73] A gene doubling for the protein ferredoxin made it possible to deduce the point of earliest branching in Dayhoff and Schwartz's trees. But ferredoxin was not present in all life forms so could not be used to root a universal tree.[74]

In 1989, two independent groups used more ancient gene doublings to root a universal tree. Both reached the same conclusion: archaebacteria was a sister group to eukaryotes. The first report was by Peter Gogarten, then a postdoctoral fellow at the University of California–Santa Cruz, and several other botanists. They used genes for two forms of H^+ ATPases, enzymes that catalyze a process involving the hydrolysis of ATP, the molecule that transports chemical energy in cells. The proteins encoded in each gene possessed different subunits, α and β. Since both types of subunits were homologous, they would have arisen by a gene duplication that occurred before the common ancestor of the major lineages diverged. The phylogenetic tree of the subunits could be rooted at the point where the gene duplication occurred (figure 18.3).[75] Gogarten and colleagues tested those genes on both *Sulfolobus* and *Methanococcus thermolithotrophicus*. They thus concluded that their results were "consistent with the monophyletic origin of the Archaebacteria."[76]

The second study was by Takashi Miyata's group at Kyushu University in Fukuoka, Japan. They compared the amino acid sequences of α and β subunits

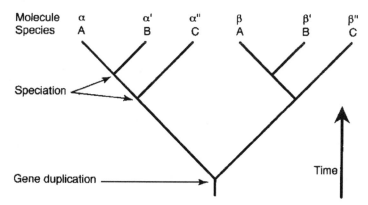

Figure 18.3 "Rooting Molecular Phylogenies." Peter Gogarten, "The Early Evolution of Cellular Life," *Trends in Ecology and Evolution* 10 (1995): 147–150, at 149. With permission.

of ATPase as well as another homologous pair of proteins resulting from gene duplication: the elongation factors, EF-Tu and EF-G, involved in the process of translation. They also concluded that "the archaebacteria are more closely related to eukaryotes than they are to eubacteria."[77] They also noted that their evidence that bacteria branched on one side and archaebacteria and eukaryotes on the other was in agreement with the evidence that 5S rRNA, RNA polymerases, and several ribosomal proteins from archaebacteria were similar to those in eukaryotes.

| # Three Domains

I have been searching for words that convey the aboriginal nature of these highest taxa, with only partial success. Hopefully, between us something implying primacy, aboriginal, archetypal, etc. will emerge. (It seems important somehow to convey that there are no higher groups than these).

—Carl Woese to Otto Kandler, December 8, 1989

THERE WAS GREAT CONFUSION over kingdoms in the 1980s. Woese and Fox had referred to Archaebacteria, Eubacteria, and Eucaryota informally as "urkingdoms"; later they were sometimes called "primary kingdoms," or just "kingdoms." Some referred to the Prokaryotae as a superkingdom; others referred to it as a kingdom. James Lake had put forth Eocyte and Photocyte urkingdoms, as well as Parkaryote and Karyote superkingdoms. Lynn Margulis continued to champion the five-kingdom model of the 1960s, in which bacteria were placed in the kingdom Monera, and eukaryote was not recognized as a taxon.[1]

New eukaryotic kingdoms were also added. In 1981 Tom Cavalier-Smith proposed the kingdom "Chromista" as distinct from the plant kingdom because its members did not acquire their chloroplasts by "primary symbiosis" (i.e., from cyanobacteria) but secondarily from the engulfment of eukaryotic algae.[2] Two years later, he proposed a new subkingdom, later kingdom, which he named the Archezoa and held to be a primitive form of eukaryotic life, lacking mitochondria (see chapter 22).[3] He placed archaebacteria and eubacteria as kingdoms within the superkingdom Prokaryotae.[4] "The subdivision of the living world into two major structural cell types (eukaryote and prokaryote) is in no way invalidated by recognition of the Archaebacteria and Eubacteria," he wrote in 1986.[5]

"Not Just Prokaryotes"

In the fall of 1989, Woese set out to make a formal taxonomic proposal for three primary lineages of life and, in so doing, to move microbiological evolution and deep phylogenetics to the center of biology.[6] He had long abandoned the concept that eukaryotes might be wholly chimeric. At the center of concern was a name for the archaebacteria that would divest them from the connotation that they were just "prokaryotes."

Prokaryotes were generally defined negatively as lacking a membrane-bound nucleus, organelles, and sexuality comparable to eukaryotes—they were noneukaryotes. The general biological conception of them was of more or less ancient relics, entities that had not progressed to the eukaryotic state. Molecular biologists had assumed that bacteria were more-or-less the same at the molecular genetic level. They treated *E. coli* as the model organism, and they were generally no more interested in the natural history, taxonomy, or evolution of bacteria than were *Drosophila* geneticists in entomology. Bacteria were raised as laboratory domesticates when used to probe gene structure and function; they served as little workers, harnessed to perform sundry tasks in industry; and when pathological they were understood as germs to be killed. Then there was the sentiment among those biologists who focused on plants and animals that bacterial diversity was no more interesting (and, for most, less so) than discovering an unknown bird species.

The rootings of the tree, published in the summer of 1989, indicating that archaebacteria were more closely related to eukaryotes than to bacteria were congruent with the evidence that various components of the transcription and translation apparatus of the archaebacteria resembled those of eukaryotes. The traits defining the archaebacteria as a group had piled up: the unusual ether-linked lipids in the membranes, the diversity of their cell walls, which lacked peptidoglycan, the structure of their transfer RNAs, and their unique transcription enzymes, viruses, and introns. Eubacteria were found to have introns, but the mechanisms for RNA splicing were different in all three groups. Archaebacterial introns were found to have an archaebacterial-specific splicing enzyme.[7]

Woese's aim was for a name that would avert the insidious inferences embedded in the prokaryote as a monophyletic group that preceded and gave rise to eukaryotes. The deliberations were considerable, measuring the meanings of words and their possible impact on the direction of biology. As he saw it, a formal taxonomic proposal at the highest taxonomic ranks would turn the heads of biologists to evolution's deep roots: the origins of the genetic code, the evolution of complexity at the molecular level. The direction of biological research, publishing, and funding for microbial phylogenetics was at stake. He carried a deep-seated regret for what he considered to be the indifference of the biological community to the discovery of the archaebacteria. He encountered that indifference again in the fall of 1989 when trying to get funding for genomic research on the archaebacteria.

The Human Genome Project was getting under way, coordinated by the U.S. Department of Energy and the National Institutes of Health. Five years later, the U.S. Department of Energy instituted a "Microbial Genome Initiative" (MGI) as an offshoot of the Human Genome Project. The Human Genome Project was rationalized in terms of its medical benefits, and the Microbial Genomics Initiative had been justified similarly: it was to provide data on selected microorganisms, each microbe for a specific practical purpose: medical, agricultural, or industrial. In its first five years, it focused on industrially important microbes, emphasizing those that live under extreme conditions, including the deep subsurface, geothermal environments, and toxic waste sites.

But Woese would argue for a deeper and more fundamental rationale.[8] Humans were stressing the biosphere, and there would soon come a day when a deep knowledge of the biosphere and its capacity to adapt will be critical. Bacteria are largely responsible for the overall state of the biosphere: our oxygen atmosphere exists (directly or indirectly) because of them, and they are vital to the regulation of the planet's surface temperature through their roles in carbon dioxide turnover and methane production and utilization. Genomics was needed to explore the unknown diversity of microbes and their interactions with each other and with their physical environments.

At the inception of that new genomics era, Woese and Gary Olsen were gearing up to sequence the entire genome of an archaebacterium when their grant application was turned down by the National Institutes of Health, partly on the basis that there was already funding for the genome sequencing of a bacterium. Woese recalled:

> We here had formed the CPGA [Center for Prokaryote Genome Analysis] to do genome sequencing of bacteria, motivated by the need to bring our knowledge of the archaes up to speed with that of the bacteria. We had a little initial funding from Amoco, but that stopped. . . . In searching for other funding we decided to try the Human Genome Project.
>
> Gary wrote the body of the grant and was PI [principal investigator]. The methodology was ahead of its time, but proved precisely to be the one used for the first bacterial genome(s): cloning small pieces of sheared DNA into a plasmid; i.e. the "shot gun" approach that Craig Venter so highly touted later.
>
> We were rejected at the first cut. . . . That's when I picked up the scuttle-butt that one of the reviewers said something like "We're already sequencing the *E. coli* genome. That's enough bacteria." A similar sentiment was voiced later by Eric Lander when the *E. coli* genome was finally finished. . . . It went something like "Now that the *E. coli* genome is done,

there is no need for further bacterial genomes" or perhaps "*E. coli* is the most important of all bacterial genomes"; etc.[9]

Woese's semantic strategy was to rename the archeabacteria in such a way that would do away with the impression that they were related to bacteria. He wrote to Kandler and Zillig in October 1989 asking if they were interested in collaborating, and tentatively suggesting "Archaeocytae" (the old cells) as an alternative to archaebacteria.[10] "As time goes by it becomes more and more obvious that I made a major mistake in naming the archaes," he wrote:

> Their relationship to eukaryotes... is becoming increasingly clear. The recent work on the rooting of the universal tree using the Dayhoff strategy (genes that duplicate in the common ancestor state, before the three primary lineages diverged from one another) definitely puts the root in the eubacterial branch, making the archaes specific (albeit distant) relatives of the eukaryotes. Both ATPase subunits and EF-Tu vs EF-G give the same result.
>
> Also there is now a *true* histone gene found in methanogens; the amino acid sequence is as close to the consensus H2 sequence as any eukaryotic sequence is. Undoubtedly many more truly eukaryotic genes will be found in the archaes.
>
> Unfortunately, with a name like archae*bacteria* the majority of biologists still say, "Oh, they are only prokaryotes," and *act* accordingly. This affects not only ideas, but career decisions, funding decisions, the structure of courses and text books, departmental organization, etc. (We just had a grant rejected and one of the panelists in rejecting it said "We're already funding the sequencing of the *E. coli* genome; isn't that enough bacteria?") If this unfortunate influence of the prokaryote-eukaryote dichotomy is not countered, archaebacteria (the study of them), and all of biology will suffer.
>
> May I have your opinion on the above?
>
> Also would you be willing to join me (and hopefully others) in the following partial solution to the problem: A group of major figures in the archae field should publish a joint paper proposing a formal name for the archaes at the kingdom *i.e.* highest) level (and argue why there should be a kingdom distinction made). The only name I can think of that would remove the stigma of their being seen as "bacteria" and retain some continuity of naming is "Archaeocytae." In this way archaebacteria can continue as the common name, and the formal name will draw attention to the fact that archaes are not "just bacteria." The contrast between trivial and formal names will also force some biologists to think, as well.
>
> What do you say about this?

The response was mixed. Zillig was flatly not interested (at least not initially). Kandler was receptive, but not with the proposed Archaeocytae. "I understand your worries," Kandler replied; "however, I am not sure if it is necessary and wise to switch from Archaebacteria to Archaecyta which resembles Lake's Eocyta."[11] Kandler agreed that the prefix "archae" should be kept to signify the presumed archaic physiology of the group, such as methanogenesis and sulfur metabolism. He suggested "two kingdoms of prokaryotic organized organisms. The Bacteriota (former trivial name Eubacteria) and the Archaeobacteriota or Archaeobiota (former trivial name archaebacteria; the 'o' is inserted due to latinization)." The term "Archaeobiota," he wrote, "would make it clear, also for a layman, that these organisms are not just 'bacteria.'" He proposed the name Eucaryota, as "an early offspring of the Archaebiota" characterized by its highly chimeric genome.[12]

The "Old Ones"

Formal taxonomic naming is complicated. "Archaebacteria" had slipped into the lexicon of microbiology with relative ease; it was a causal "trivial name," a "nickname." In the formal christening, the code of nomenclature had to be considered.[13] The bacteriological code of nomenclature had been approved at the Fourth International Congress for Microbiology in 1947 but was later discarded. The official "nomenclatural starting date" for the International Code of Nomenclature of Bacteria was January 1, 1980.[14] The name of all taxa above the species was to be of a single word of Latin or Greek origin or a latinized word of any origin. That taxon must stand for a group of organisms living today or for an organism that could be proven to have lived in the past by fossil evidence. One could not, for example, give a formal name for the hypothetical "progenotes."

There were two ways of thinking about new taxa at the highest level. One was to make them kingdoms and demote Animalia and Plantae to subkingdoms. The other was to keep Animalia and Plantae as kingdoms but make new taxa at a higher level. For whatever reasons, in the fall of 1989 Woese and Kandler were thinking only of kingdoms and no rank above them. After consulting with taxonomists in Germany and Austria, Kandler suggested Bacteriota and Archeota "for the two prokaryote kingdoms" and said that they found the distinction of "the two prokaryotic kingdoms necessary and the names (Bacteriota, Archaeobiota) adequate."[15] "The name you have come up with sounds perfect," Woese replied; "everyone who has heard the name *Archaeota* has a positive reaction to it."[16]

Then Woese discussed his draft manuscript with Mark Wheelis, one of the authors with Stanier and Adelberg of the 1979 and 1986 editions of *The Microbial World*.[17] Wheelis was a former student of Stanier's; he was a postdoctoral fellow at the University of Illinois with I.C. Gunsalus before his appointment at the University of California–Davis in 1971. He had planned to devote part of his upcoming sabbatical leave to working on a taxonomic proposal, and he suggested that Woese and Kandler raise the level of the taxa they were proposing above that of the kingdom. In effect, he reactivated the original idea about three urkingdoms. Wheelis subsequently wrote to Woese requesting that he be considered as an author.[18] He was brought on board and helped with writing the paper.[19]

On January 5, 1990, Kandler recommended shortening the names of the three primary taxa to Archaea, Bacteria, and Eucarya. As he explained to Woese, "Although I agreed with the ending -tida...there is a common tendency among taxonomists to use the most simple endings for the highest ranks."[20] "I'll go along with your change in ending," Woese replied; "however, I had become fond of '-tida'—having said it so many times now. And I'd like to try out the two alternatives on a few biologists to get their reactions before making the final decision. (Let's stay with Eucarya, because it is traditional.)"[21] Three days later it was unanimous. Woese wrote to Kandler, "'Archaea' et al. plays well in Peoria—which is American for the people I've asked like the newest names better than the previous '-tida' names. Hopefully your German colleagues approve of the names as well."[22] They did. "The idea to create a new level of taxons and the names Archaea, Bacteria and Eucarya are well accepted," Kandler rejoined.[23]

Anthropomorphisms

What to call the taxonomic level was troublesome; "urkingdom," "domain," "realm," "empire," were considered. All were geopolitical words carrying anthropomorphic and particularly militaristic connotations of power and conquest—which Woese sought to avoid. He and Fox had written about the "prokaryotic domain" in their paper introducing "the archaebacteria" in 1977.[24] And he suggested the term "domain" for the three primary lineages in November 1989 because it seemed to be more neutral than the others. Kandler offered other possibilities:

> Your *domain* for the taxon higher than the kingdom is not bad, on the contrary. However, if you wish to have something more clearly and specifically terminological, then you have the following candidates:
>
> (a) *realm* is logically a higher taxon than kingdom
> (b) *syntagm* or *syntagma* is the Greek word for "group"; the fact that is it used in, e.g., linguistics makes it easier to accept, but people usually have trouble deciding how to pronounce it
> (c) *fons* would probably come closest to your intention, because it is "fountain" in Latin. Problem: the Latin plural is *fontes*; will people know it? Another problem: what if someone one day discovers a yet higher taxon?[25]

Realm was a good candidate until Kandler learned that the Latin translation only returned its meaning to kingdom.[26] "In any case," Woese replied, "'realm' does not convey the meaning that should be conveyed here. We should stay away from all names that have military connotations in naming the primary taxon; don't you agree? This would rule out 'empire', 'realm' and perhaps even 'domain.'"[27]

When Woese suggested "urtype" and "protoform" to depict the fundamental organizational differences of the three groups, Kandler explained that both words were unacceptable because "the names of all the other ranks (genus to kingdom) comprise a quantity."[28] Finally on March 5, 1990, Woese returned to "domain"—which, with the right Latin translation, might be able to escape political connotations.[29] "I agree with 'domain' which may be translated into Latin as 'regio'," Kandler replied.[30] Thus they proposed three domains.[31]

They distinguished two kingdoms within the Archaea. One comprised what were variously called the "thermoacidophiles," "sulfur-dependent archaebacteria," or "extreme thermophiles." The other was phenotypically heterogeneous and comprised the methanogens and their relatives, the extreme halophiles, sulfate-reducing species, and two types of thermophiles.[32] Their names were to reflect the fact that the first group had stayed more-or-less constant and ancestral phenotypically, while the other group had evolved into new niches (including mesophilic ones) and generated a spectrum of metabolic diversity. Woese wrote to Kandler,

> The sulfur archaes have retained their ancestral (archae) phenotype (all being thermophilic and heavily centered on sulfur metabolism), while the methanogens and their relatives ostensibly have started from this same ancestral phenotype, but have generated a great deal of biochemical, ecological, etc. variety.[33]

Kandler and Woese had no difficulty in agreeing on Euryarchaeota (the wide old ones)—primitives adapted to a wide range of habitats—for the methanogens and halophiles.[34] But Kandler favored the adjective "sten" (narrow or contracted) for the thermophilic sulfur-dependent archaes, so as to carry an ecological connotation in reference to their apparently narrow ecological distribution in extreme hot environments. Woese preferred the adjective "cren" (spring or font) to carry an evolutionary connotation, as resembling the ancestral archaeal phenotype.[35] He explained his reasoning to Kandler, beginning with "the probable assumption" that the archaes were the specific relatives of the eukaryotes. In other words, the root of the universal tree lay between the eubacteria and everything else.[36] Since thermophily was the only general phenotype that occurred on both major branches of the Archaea, he supposed that it was the ancestral phenotype of the Archaea. They settled on the kingdom Crenarchaeota.

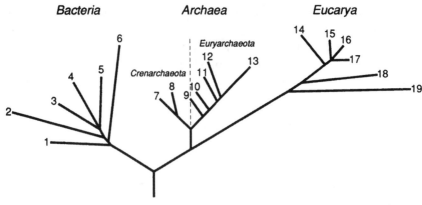

Figure 19.1 "Universal phylogenetic tree in rooted form, showing the three domains. Branching order and branch lengths are based upon rRNA sequence comparisons.... The numbers on the branch tips correspond to the following groups of organisms. Bacteria: 1, the Thermotogales; 2, the flavobacteria and relatives; 3, the cyanobacteria; 4, the purple bacteria; 5, the Gram-positive bacteria; and 6, the green nonsulfur bacteria. Archae: the kingdom Crenarchaeota: 7, the genus *Pyrodictium*; and 8, the genus *Thermoproteus*; and the kingdom Euryarchaeota: 9, the Thermococcales; 10, the Methanococcales; 11, the Methanobacteriales; 12, the Methanomicrobiales; and 13, the extreme halophiles. Eucarya: 14, the animals; 15, the ciliates; 16, the green plants; 17, the fungi; 18, the flagellates; and 19, the microsporidia." Carl Woese, Otto Kandler, and Mark Wheelis, "Towards a Natural System of Organisms: Proposal for the Domains Archaea, Bacteria, and Eucarya," *Proceedings of the National Academy of Sciences* 87 (1990): 4576–4579, at 4576. With permission.

Their paper was submitted to the *Proceedings of the National Academy of Sciences* that January. Both Woese and Kandler knew it was a keystone in the research literature on microbial phylogeny. "I still remember when we met first, almost exactly 14 years ago, in January 1977," Kandler wrote to Woese after reading the final draft.

> At that time you had just submitted the first paper on archaebacteria, and I had the first, although incomplete, results on the non-murien nature of cell walls of methanogens and halococci. I feel very honored by your invitation to join the authorship of the "final" paper which documents our continuous multiform cooperation in this exciting period.[37]

Published in June, "Towards a Natural System of Organisms" offered formal definitions of the three domains, the kingdoms of the Archaea, and a figure of the universal phylogenetic tree in rooted form (figure 19.1).[38]

1. Domain Eucarya (Greek, true nucleus) "captured its defining cytological characteristic—i.e., cells with well-defined encapsulated nuclei": "cells eukaryotic; cell membrane lipids predominantly glycerol fatty acyl diesters; ribosomes containing a eukaryotic type of rRNA."[39]

2. Domain Bacteria (Greek, small rod or staff) "cells prokaryotic; membrane lipids predominantly diacyl glycerol diesters; ribosomes containing a (eu)bacterial type of rRNA."

3. Domain Archaea (Greek, the old or very primitive ones) "cells prokaryotic; membrane lipids predominantly isoprenoid glycerol diethers or diglycerol tetraethers; ribosomes containing an archaeal type of rRNA."[40] The Archaea, denoted "the apparent primitive nature of the archaebacteria (vis-à-vis the eukaryotes in particular).... We recommend abandonment of the term 'archaebacteria' since it incorrectly suggests a specific relationship between the Archaea and the Bacteria."[41]

Woese, Kandler, and Wheelis presented themselves as following the lost and abandoned tradition of the Delft School (see chapter 7), seeking a natural classification of microbes from a phylogenetic point of view. Shifting from the morphological to the molecular allowed science to achieve what Kluyver, van Niel, and Stanier could not. In presenting their "natural system," they confronted the five-kingdom scheme of Whittaker and Margulis.[42] They said that the kingdoms Protista and Fungi were "artificial: "It is generally accepted that the metaphyta and metazoa evolved from unicellular eukaryotic ancestors; the extant groups of eukaryotic microorganisms, therefore, comprise a series of lineages some (or many) of which greatly antedate the emergence of the Plantae and Animalia."[43]

There were two critical faults with the five kingdom scheme, as Woese, Kandler, and Wheelis saw it. First, even in its own terms, it failed to recognize that the difference between Monera and the other four kingdoms was far more significant than the differences among those four.[44] Second, it failed to recognize that Monera was an ill-defined cytological distinction. While eukaryotes were defined positively, prokaryotes were defined negatively "by their lack of characteristics that define the eukaryotic cell," a definition, they said, that was "empty of meaningful internal phylogenetic information."[45] They pointed to the long history of concerns about whether the bacteria were a monophyletic grouping.

> On the cytological level, archaebacteria are indeed prokaryotes (they show none of the defining eukaryotic characteristics), but on the molecular level they resemble other prokaryotes, the eubacteria, no more (probably less) than they do the eukaryotes. *Prokaryotae* (and its synonym *Monera*) cannot be a phylogenetically valid taxon.[46]

They maintained that the molecular differences in the transcription and translation machinery of the eubacteria, archaebacteria, and eukaryotes, were "of a more profound nature than those that distinguish traditional kingdoms, such as animals and plants, from one another. This is most clearly seen in the functions that must have evolved early in the cell's history and are basic to its workings."[47]

The announcement of the three domains heralded the scale of the microbial phylogenetic revolution. The word spread quickly.[48] Woese had no regrets about the new naming, except in using "Bacteria" for the eubacterial domain. "In retrospect that was not the happiest of choices for a domain name, given the many connotations of bacterium that now exist," he wrote to Zillig in March 1992.[49] Fox and many other molecular phylogeneticists welcomed the name change because it made it clear that Archaea were something different. "Towards a Natural System of Organisms" brought deep evolution to the fore of biology. But whether called "domain," "empire," or "superkingdom," few biologists were willing to accept the Archaea on par with the eukaryotes; few were willing to reject the venerable prokaryote–eukaryote dualism. Classical evolutionists and microscopists were fiercely opposed to any comparison of the diversity they observed around them with that at the molecular and biochemical levels.

| Disputed Territories

The classification supported by me is based on the traditional principles of classification, which biology shares with all fields in which items are classified, as are books in a library or goods in a warehouse.

—Ernst Mayr, "More Natural Classification" (1991)

A global classification should reflect both principal dimensions of the evolutionary process: genealogical relationship and quality and extent of divergence within a group. The ultimate purpose of a global classification is not simply information storage and retrieval.

—Mark Wheelis, Otto Kandler, and Carl Woese, "On the Nature of Global Classification" (1992)

THE PROPOSAL OF FORMAL NAMES for the three primary lineages earned the ire of those who insisted that the Archaea were nothing other than bacteria, properly classified as prokaryotes. The fiercest opposition came from classical evolutionists. They wanted the term "Archaea" banished as misleading and erroneous. At the surface it might have appeared to be just another taxonomic debate, but underneath was a clash of world views. The rift was broad and deep—as great as one could imagine between biologists examining the two fundamentally different levels of organization: the morphological and the molecular.

The Clash

Theodosius Dobzhansky, George Gaylord Simpson, and Ernst Mayr were opposed to the field of molecular phylogenetics from its inception in the 1960s.[1] There were several aspects to their antagonism. To begin with, they resented the molecularists' ascendancy. Molecular biology, triumphant and confident with its claim to having discovered "the secret of life" in DNA, was well funded, driven by technical innovations, and rooted in the industrial biomedical complex, far removed from evolutionary biology. In 1963, at the time the genetic code was being cracked, Dobzhansky wrote a note to *Science* reminding readers that there was still stimulating research going on in "organismic as well as molecular genetics."[2] His note was followed by a commentary from Mayr requesting "more financial and moral support for the classical areas."[3]

Then, when the principles of molecular biology were applied to evolution (and vice versa), molecular evolutionists were perceived as interlopers by some. Evolutionary biology was supposed to be more or less a matter of applying the hard-won Darwinian principles established by those architects of the "grand synthesis." Classical evolutionists were ticked off by what they considered to be the erroneous methods and concepts of a "molecular clock." Evolution, they said, was an "affair of phenotypes." Attempts to stop the clocks occurred the moment that molecular evolutionists argued that the amino acid sequences of proteins could evolve without adaptation and natural selection. The nature of change at the molecular level of nucleic acids and proteins was different from that presumed to be the case for most morphological traits. Many random changes seemed to have no effect on protein function and were therefore of no adaptive value. Recall that Motoo Kimura had referred to it as "the neutral theory" and Jack King and Thomas Jukes called it "non-Darwinian evolution" (see chapter 10).[4]

Change constantly occurred, even if most of it is not acted upon by selection. To be sure, the pace of molecular changes was not linear. Different DNA sequences (or proteins) or even different parts of the same gene (or protein) evolve at markedly different rates. But neutral or not, those changes occurred at a rate that was independent of morphological evolution. As Woese phrased it in 1987, "Evolution has a tempo that is quasi-independent of its mode (the selected changes occurring in the phenotype). An analogy to a car and its motor is apt: a car does not go unless its motor is running, but the motor can run without the car moving."[5] Mayr and Simpson found it incredible that molecular and morphological evolution could be different in mechanism and rate. In any case, they argued, the only evolution that mattered operated by natural selection.[6] Changes at the molecular level that did not affect the phenotype, Simpson said, were really of "no interest for organismal biologists as they are not involved in the evolution of whole organisms."[7]

Classical evolutionary biology was not concerned with microorganisms, the origin of cells, the role of symbiosis in evolution: sharing resources among taxa through forms of lateral gene transfer, giving rise to new kinds of microorganisms. The "modern synthesis" of the first half of the twentieth century was "germ-free"; it was not about the origin of kingdoms, but about the origin of species of plants and animals. The microbial world seemed small, and the kind of diversity therein seemed insignificant to the fabulous morphological diversity displayed by plants and animals.

Matters of Taste

The vast difference between classical evolutionary biology and molecular phylogenetics manifested in a heated debate between Mayr and Woese, a feud that fumed in public and private correspondence during the last decade of the twentieth century. It began in December 1990, when Mayr published a letter in *Nature* asserting that to separate the Archaea from Bacteria as one of three "domains" on par with eukaryotes was grossly misleading. He called for the restoration of the traditional prokaryote–eukaryote dichotomy.[8] A professor at the Museum of Comparative Zoology at Harvard University, Mayr was the grand old man of evolutionary biology. In his late 80s and 90s, he was a force to be reckoned with, and well respected by all biologists, including Woese, 24 years his junior.

Mayr was the gatekeeper of the "evolutionary synthesis" and its historical reconstructions. He took historical scholarship seriously as providing a foundation for understanding of the basis of science. In 1980, he co-edited *The Evolutionary Synthesis*, a book dealing with the individuals, disciplines, and nations that contributed to formulating modern Darwinian theory. Two years later, he published his opus, *The Growth of Biological Thought*, a mix of his own biological views embedded in a historical narrative.[9] But while those books were written, evolutionary biology was changing under his feet, moving toward primordial processes in the deep recesses of evolution on Earth.[10]

Mayr's books made scant reference to microbes, and his discussions of kingdoms were haphazard and inconsistent.[11] In his *Growth of Biological Thought of 1982*, an opus of almost 1,000 pages, Mayr mentioned bacteria on two pages, and protozoa on two others. He referred to Woese and Fox's proposal to recognize "two kingdoms," but he was not then willing to even consider the idea. Because bacteria had so many characteristics in common, he said, they should be grouped into one kingdom, Monera, as prokaryotes, which he defined as lacking "an organized cell nucleus and complex chromosomes." He also recognized the fungi as a fourth kingdom separate from plants because of a lack of photosynthesis and haploidy. A fifth kingdom for protists, he suggested, was not really a scientific matter. Essentially, they were "one celled animals and plants." To recognize them a kingdom he said, was "a matter of taste."[12] It was also a matter of utility. "Since the literature on protozoans and one-celled algae is rather separate from that of metazoans and metaphytes," he commented, "such a separation might facilitate information retrieval."[13]

Mayr changed his mind dramatically about primary groupings eight years later when the three-domain proposal was launched. He no longer considered Monera to be a kingdom on par with the others. Echoing previous statements that the difference in cell structure between prokaryotes and eukaryotes is the "most drastic change in the whole history of the organic world," he assigned prokaryotes and eukaryotes to the rank above the kingdom—as two "domains" (later "empires"). He changed his views about protists, too. They had long been recognized to be too diverse to be contained within one kingdom (chapter 8). And 16S rRNA sequence comparisons in the late 1980s indicated that they were even more diverse than previously imagined. The ciliated protists alone were as genetically diverse as plants and animals.[14] Mayr now recommended that Protists be ranked above kingdom—as a "subdomain."[15] And although he was previously unwilling to give archaebacteria kingdom status, in 1990 he assigned them to the rank of a subdomain of the Prokaryota.

Nothing separated Woese and Mayr more than the importance of phylogeny. Woese's aim to understand life's course, the primordial genomes, and how the genetic apparatus evolved could only be deductions of phylogeny. Genealogy was primary, and phenotypic differences could *at best* corroborate taxonomy based on phylogenetic, that is, molecular characteristics. Because it was a polyphyletic grouping, the prokaryote was anathema to what he considered to be "a natural classification." In Mayr's view, utility, the organization of an effective storage system, should be the main aim of taxonomy. In that endeavor, phenotypic comparison was important; phylogenetic evidence, much less so. The degree of modification that occurred with the emergence of the eukaryote was what mattered in a "natural classification." "A basic classification of the living world will be used not only by specialists but also by non taxonomists and lay people," he said. "They will be more comfortable with the classical concept of classification."[16] It did not matter that "the prokaryote" was a polyphyletic group. It did not matter that archaea were more closely allied phylogenetically to eukaryotes than to bacteria. It was sufficient to classify based on degree of (morphological) modification, no matter how the genealogies played out.

Convergence

Mayr's and Woese's views converged only regarding their belief that evolution was fundamental to biological understanding. They had met only once, at a symposium in Cambridge in 1982 honoring the 100th anniversary of Darwin's death. Many of the luminaries in molecular biology were there. Woese was full of enthusiasm for what the molecular signals were telling him. Bacteria were at the fore; protists and full-genome sequencing was on the horizon. He was hopeful that there remained a sufficient number of ancient genes to reconstruct the aboriginal gene families: those of the "universal ancestor." Still, he found it piercingly ironic that the icons of molecular biology were front and center at the symposium honoring Darwin, considering how little they cared about evolution (see chapter 11). It was entirely fitting that Horace Judson's landmark book, *The Eighth Day of Creation: Makers in the Revolution in Biology*, published three years earlier, contained virtually no reference to evolution.[17] There was indeed little about evolution in accounts of what has been called *The Century of the Gene*.[18]

What was more disgraceful was that Fredrick Sanger, the two-time Nobel-Prize–winning scientist who led the way with the new sequencing technology for the field of molecular evolution, was absent, yet he lived right there in Cambridge. In Woese's view, the Cambridge symposium was a mockery of the importance of evolutionary biology. He found himself alone when he raised a glass in honor of Sanger. As he recalled in a letter to Mayr five years later when Mayr asked him for a copy of his overview "Bacterial Evolution": "I remember toasting Sanger, the absent spirit whose impact on evolution would come to rival that of Darwin," Woese wrote to Mayr; "embarrassingly, my glass was the only one raised."[19]

Woese told Mayr how he considered evolutionary understanding to be in a nascent state with new fundamental principles to be discovered beyond a Darwinian framework of selection and adaptation. Evolution, he said, belonged beside other rules in the universe. A proper understanding of evolution, he thought, had to be joined somehow with a new understanding of the second law of thermodynamics. Spontaneous change increasing entropy was not evolution's way. "When both are properly appreciated," he wrote Mayr, "the former will become an obvious and beautiful manifestation of the latter. Variation and selection are essential to the evolutionary process; yet they are somehow not its essence. The tempo-mode problem seems the key to understanding evolution— yet I don't know why I say this."[20] He continued:

> Evolution could be the key to the future of all science, ... it could even-
> tually emerge as the primary science, around which all others are struc-
> tured. Life will then become central to our world view, rather than being
> an nth-order-derived characteristic of the universe.
>
> I hope this scientific mysticism hasn't put you off too much.
>
> If so, please forgive me; I'm not always this way. Tomorrow is
> sequence-gel-reading day; this always brings one back to "reality."[21]

Classical evolutionists typically saw phylogenetics to be something wholly apart from the study of evolutionary *process*.[22] The latter was to be discerned by studies of the nature of gene mutations and recombination, by studying the changes in the frequencies of alleles in populations, and mechanisms for speciation. Phylogeny was thus limited to the construction of the evolutionary history of a group. At the core of the evolutionary synthesis was the principle that mechanisms of microevolution were the same as those of macroevolution. There was no reason to expect phylogeny, even the broadest and deepest, to reveal new mechanism of evolution.

Mayr was of the opinion that phylogeny could not address the most important "why" questions in evolution. Woese countered that the phylogenetics actually dealt with all three classes of questions: what, when, and why. The archaebacterial tree, for example, provided a probable answer to the question, "What is the nature of the ancestor of all archaebacteria?" The answer for Woese was that it resembles the extreme thermophiles. Then one would ask, "Since methanogenesis is not ancestral, how did methanogenesis arise from the sulfur-reducing metabolism of the extreme thermophiles?" This question would lead to, "Why did sulfur-reducing metabolism evolve into one that produces methane from carbon dioxide?" The strength of phylogeny was in posing the question that did not exist before, in opening questions for investigation. "I feel we are only at the beginning of the study of evolution," he replied in January 4, 1988,

> (and, incidentally, that reconstruction of the evolutionary history of this planet is a far more important thing to do than most scientists think). Therefore, we are still accumulating phenomena, finding out what questions to ask....
>
> Without a phylogenetic tree, we will be unable to reconstruct the evolution of biochemistry, and tie biological evolution into the evolutionary history of the planet....
>
> Evolutionists are winning their battle, but the battle is not yet won. Biology teaching has to be totally revamped; there is still too much medical rationale to biology, biologists have to take a bigger view of biology (in keeping with the move away from reductionism), etc.[23]

Woese later learned that Mayr supported his election to the National Academy of Sciences in 1988. But their relations became strained two years later when Mayr's letter against the three-domains proposal appeared in *Nature*.

Would the Real Darwinian Please Stand Up?

Like many disputes, part of this one was in defining what exactly it was about. Mayr portrayed it as being chiefly about competing kinds of taxonomy. Initially, at least, he described it in terms of a traditional approach that he saw as dating back to Linnaeus's *Systema Naturae*, which classified organisms on the basis of their overall phenotypic similarity, versus a cladistic approach dating to the

English translation of Willi Hennig's *Phylogeneic Systematics* of 1966, which classified strictly on the basis of genealogy or branching points.[24] Woese, in his view, was a neophyte in taxonomy who seemed to be following cladism but was ignorant of the fundamental issues that Mayr believed he himself had resolved.

Mayr's newly revised book *Principles of Systematic Zoology* with Peter Ashlock had just appeared, and he had just returned from a lecturing tour in the United States and Europe telling his audience of the perils of cladistic classification, dismissing it as a waste of time and counterproductive.[25] On January 14, 1991, he sent Woese a copy of his note to *Nature*, informing him of his deep disagreement about domains. "Somebody seems to have persuaded you to accept the principles of cladistic classification," he wrote.[26] Mayr explained that he had already had "an avalanche of supporting mail" for his side. Judging from the responses he had received, he said that "many if not the vast majority of biologists" would side with him.

Woese conceived of the conflict differently. He was not a follower of cladism. The debate was over differences in the traditional methods of morphologists versus those of molecular phylogeneticists. The issue for him was plain: Should classification be based on molecular phylogeny or on an amalgam of presumed relationships, morphology, and utility? Recall that cladists completely disregarded the amount of evolutionary change that had occurred between lineages (see chapter 13). In that way, they could avoid qualitative judgments when classifying. They would not accept genealogically incomplete (paraphyletic) groups and therefore would not exclude birds and mammals from reptiles. Woese, Kandler, and Wheelis did not ignore that evolutionary change. They recognized that classification had to reflect genealogy (branching order) and "degree of modification" (divergence). They accepted paraphyletic groups—those that contained its most recent common ancestor but that did not contain *all* the descendants of that ancestor.

But to put organisms with separate origins into a single group was anathema to any evolutionary taxonomic understanding. As Kandler emphasized in a letter to Woese years earlier, "One has to make compromises between phylogenetic age (time since the separation of the gene pools) and the extent of phenetic differences, which depends on selective pressure. However, taxa with different lines of phylogeny may never be mixed within one taxon."[27] Taxa *had* to be monophyletic in the sense of a shared common ancestor (parent) that itself is also a member of the group. It was axiomatic that different phylogenetic lines could not be crossed. The prokaryote was a polyphyletic grouping—an egregious violation of a "natural classification" after Darwin.[28]

The three-domains proposal was not based on cladism, as Mayr assumed. It signified that the archaebacteria were phylogenetically as distinct from bacteria as eukaryotes were. And as Kandler and Woese saw it, Mayr ignored the great biochemical and molecular differences between the Archaea and Bacteria: the differences in their walls and the lipids in their membranes, and the radical differences in their DNA replication, transcription, and translation. He had erroneously reported that the three-domains proposal was "based entirely on the amount of differences in ribosomal RNA."[29]

Nature's Bias

When Woese, Kandler, and Wheelis sent a reply to *Nature* in January 1991, they pointed to the rooting of the tree, which placed bacteria on one branch and archaebacteria and eukaryotes on the other. And they countered Mayr's remarks about "relatively small differences" between prokaryotes:

> Many biologists tend to see morphological complexity and diversity as being more significant than other kinds of complexity.... In terms of their coding capacity, genetic organization and gene regulation, prokaryotes are no simpler than eukaryotes, and in terms of their metabolisms they are considerably more complex and diverse.[30]

Indeed, if they inverted the comparison and compared the eukaryotes by the metabolic and other processes and patterns then "eucaryotes would seem relatively dull and very much all of a kind." Since Mayr's "default system" could in no way be considered "a natural system," they pointed to the aims of Mayr's taxonomy, which emphasized utility, "being more comfortable," familiarity for non-specialists, and the conspicuous phenotypic similarity.[31]

A reply was not difficult to write, but having it published was a different matter. Mayr did his best to prevent it. In January, he wrote to Woese explaining how there has to be "a subjective element in classification."[32] He also gave Woese instruction on aspects of molecular evolution, asserting that the three-domains classification was based on only "one or a few [molecular] characters," which, he asserted, could not be used owing to different rates of evolution of those molecules. Ribosomes, he said, reflected changes in prokaryotes, but not the dramatic evolutionary events that took place in evolution of eukaryotes. The differences between himself and Woese were such that they could not be properly dealt with in the correspondence of *Nature*. He recommended that Woese read the relevant chapters of the revised edition of his book *Principles of Systematic Zoology*. Mayr "obviously has not tried to understand our side of the issue," Woese wrote to Wheelis.

> He is schooling me because I "don't understand evolution" (and I don't, and don't want to, in the petty scholastic sense). And, I think he is sitting on our reply to Nature, hoping to win the day by not showing up for battle.

...What really disturbs me however, especially having read "Origin" on the matter of taxonomy...is that Darwin eschewed all but natural classification, going on at great length and in detail as to how that is the only meaningful classification; and E. Mayr appears not to see, among other things, that he is insisting on a non-Darwin (polyphyletic) grouping of "prokaryotes" (if one accepts the current root of the universal tree). His is the most egregious of all possible Darwinian transgressions.[33]

On February 20, Woese received notice that their letter was rejected. *Nature* had sent it to one reviewer, who recommended that it not be published, on the grounds that "Mayr did a reasonable job of making clear their views in his note."[34] Woese faxed a note to Mayr, asking him if he was that reviewer; he responded in the affirmative. It was obvious, he said, that Woese was "quite unaware of the *real* issues," and he suggested that any answer should be delayed until Woese read his book. He also sent Woese a reprint of a recently published paper he wrote in German.[35]

After Woese protested to the editor of *Nature*, their letter was published—six months after it was first submitted.[36] Woese sent a preprint to Mayr, who replied that he had tried to save Woese "from possible embarrassment," because as a "relative new comer" to the field Woese was "still unfamiliar with some of the basic principles." It was obvious, Mayr insisted, that the three-domain argument was based on cladistic principles.[37] He replied in *Nature* later that summer, reasserting that the "arrangement adopted by Woese et al. is Hennig's phylogenetic reference system"[38] Insisting again that there was very little difference between groups of prokaryotes, he wrote: "To give each of the two subdivisions of the prokaryotes the same rank (domain) as the eukaryotes, which differ by the possession of a nucleus, cytoplasmic organelles and many other drastic characteristics, violates all principles of hierarchical classification." It would not be until 1998 that Mayr admitted that the three-domain proposal was not based on cladism. Still, he rejected it.

Wheelis, Kandler, and Woese tried to reply to *Nature*, re-emphasizing that theirs was not a cladistic approach, but their letter was rejected. Ford Doolittle also sent a note to *Nature* asserting that the three-domain proposal was not based on cladism and that it was a perfectly sound proposal based on the molecular rooting data then available.[39] If the rooting had split Bacteria and Archaea on one side and Eukarya on the other, then, he argued, taxonomic rank would be "a matter of taste" and Mayr's position would be supportable. It did not. The stem split Archaea and Eukarya on one side and Bacteria on the other. Based on that rooting, in his view, Archaea was a paraphyletic group inasmuch as it excludes eukaryotes. It made more sense to recognize it as a domain of its own. Doolittle's letter was also rejected. But other letters supporting the prokaryote–eukaryote dichotomy were accepted.

In 1992, Patrick Forterre wrote to *Nature* to argue for the standard dichotomous classification, but he added a twist: that the last common universal ancestor (LUCA) was a primitive eukaryote and that prokaryotes evolved from them by reductive evolution as an adaptation to high temperature.[40] Tom Cavalier-Smith replied in *Nature* that Forterre's eukaryotes-first suggestion was preposterous.[41] He was adamantly opposed to the three-domain proposal. Like Mayr, he saw the fundamental difference of prokaryotes and eukaryotes to be one of progressive grades. He recognized archaebacteria and eubacteria as kingdoms within the "Empire Bacteria," and he preferred the kingdom name "neobacteria," or "metabacteria" that Horishi Hori continued to use, instead of archaebacteria.[42]

Wheelis, Kandler, and Woese had prepared a manuscript for *Proceedings of the National Academy of Sciences* in 1992 in which they offered a more detailed discussion of the issues.[43] Emphasizing that Mayr defined prokaryotes negatively, they commented: "It is formally equivalent to defining the reptilian grade only as lacking the characteristic avian features, a definition that would almost certainly include amphibians, fish, invertebrates, etc., among the reptiles."[44] They then explained how foreign molecular phylogenetics was to classical systematists, who had little appreciation for their methods and theory.

Some saw nucleotide and amino acid sequence data as being no more important than other things about organisms that one might use to demarcate taxa. Lynn Margulis and Ricardo Guerrero adopted this point of view when they voiced their objection to the three-domain proposal in *New Scientist* in 1992. They said that "the meaningfulness of any phylogenetic tree as a guide to evolutionary history depends critically on what, and how many characteristics were used to construct it." They then listed differences and similarities in "appearance, anatomical organization, development, mode of nutrition, metabolic pathways, gas emissions, pigments" as well as other characteristics.[45]

For molecular phylogeneticists, there could be no middle ground in this regard. Only nucleic acids and proteins kept the genealogical record. One might quarrel about which one of those molecules were to be used, the size of the sequences, or that there are not enough of them, but one could not argue that molecular data were only one type among other kinds of phylogenetic information. The world of phenotypes could be used as indicators of phylogenetic relationships only insofar as they corresponded to the molecular phylogenetic signals. Margulis continued to maintain the five -kingdom model of Animals, Plants, Fungi, Protists, and Monera, each of equal rank, until 1993. In the second edition of *Symbiosis in Cell Evolution*, she sometimes referred to "Prokaryotae or Monera" as a kingdom, and sometimes to "Prokaryota" and "Eukaryota" as superkingdoms.[46] While Margulis did not recognize archaebacteria as a kingdom, others referred to it as a kingdom within the superkingdom Prokaryotae.[47]

Of Birds and Bacteria

To gain a better understanding of the ideas of others about the prokaryote concept, Woese reflected more and more on the history of microbial phylogeny.[48] He looked back on the 1950s and 1960s as "the Dark Age" of microbiology when phylogenetics was disavowed, represented by Roger Stanier's declarations about it being "as a waste of time" and when in the "brave new molecular world evolutionary relationships counted for naught."[49] Yet, the monophyly of prokaryotes had become unquestioningly accepted. The prokaryote–eukaryote dichotomy fostered the notion that to understand bacteria one only had to determine how *E. coli* differed from eukaryotes.[50]

Microbiology, like molecular biology, had developed without the evolutionary dimension essential to defining and understanding any biological system. Lacking this unifying framework, without a central core, microbiology was largely a descriptive science, a collection of facts given shape by practical considerations. The prokaryote–eukaryote dichotomy, he said, only served to obscure profound differences among bacteria and to hide from view microbiologists' near total ignorance of the relationships among them. This was "not the unifying principle that we all once believed it to be. Quite the opposite: it is a wall, not a bridge." Woese decried in 1994: "Biology has been divided more than united, confused more than enlightened, by it … Biological thinking, teaching, experimentation, and funding have all been structured in a false and counterproductive and dichotomous way."[51]

The debate over the three domains between Woese and Mayr came to a head in the summer of 1998. In one of his visits to Woese's summer home on Martha's Vineyard, Norman Pace told Woese of a discussion he had had with Nicholas Cozzarelli, a molecular biologist at Berkeley and the editor of *Proceedings of the National Academy of Sciences*, over a contribution from Mayr titled "Two Empires or Three?" Reluctant to publish only one perspective on the issue, Cozzarelli asked Pace to approach Woese about writing a companion article. Woese jumped at the chance to bring his and Mayr's feud out into the open.[52] His paper, "Default Taxonomy; Ernst Mayr's View of the Microbial World," was published a month after Mayr's.

Mayr explained that among discoveries in biodiversity of the previous 100 years, none surpassed "Carl Woese's discovery of the archaebacteria," which, he said, "was like the discovery of a new continent."[53] The question was, "Where should one place this new group of microorganisms?" Not being a microbiologist, Mayr acknowledged that he had received considerable help from several specialists, including Margulis and James Lake. But then again, he erroneously asserted that Woese had had no training in biology: "Here it must be remembered that Woese was not trained as a biologist and quite naturally does not have an extensive familiarity with the principles of classification."[54]

Mayr reiterated his view of the purpose of taxonomy as "an information storage and retrieval system. Its aim is to permit you to locate an item with a minimum of effort and loss of time. This is as true for a classification of books in a library or goods in a store as for taxa of organisms."[55] But this time he claimed that his classification was Darwinian, whereas Woese's views were those of an inconsistent cladist. Woese recognized paraphyletic taxa; cladists did not. On the existing rooting, archaebacteria and eukaryotes were derived from the same stem. A cladist would then combine them into a single clade. Woese did not do that.[56]

Still, Mayr insisted that it was preposterous to compare the molecular genetic differences between eubacteria and archaebacteria to the huge morphological differences between eukaryotes and prokaryotes. Evolution, he reminded readers, was "an affair of phenotypes," and on that basis, he asserted that "all archaebacteria are nearly indistinguishable." Even if one took prokaryotes as a whole, he claimed it "does not reach anywhere the size and diversity of eukaryotes."[57] Microbial phylogeneticists had so far described only about 175 archaebacterial species and about 10,000 eubacterial species, whereas Mayr suspected that within eukaryotes there were more than 30 million species. There were 10,000 species of birds alone, and hundreds of thousands of insect species.[58] Protistologists counted 200,000 species of protists.[59]

Mayr was not concerned with the negative definition of the prokaryote: "The nonpossession of a character is as positive a character in any traditional classification as is its possession (except in cases when the loss of a character can be determined with certainty)."[60] He also denied that the three-domain classification was in keeping with a Darwinian classification system, in which he asserted "as many characters are to be used as are available."

Woese had based his classification solely on the highly important part of the genome that archaebacteria share with eukaryotes, the translation apparatus, and the information-processing genes. These, as he had long argued, were highly conserved, non-adaptive, universal "essential" characteristics at the core of all organisms and thus the appropriate probe for revealing deep phylogenetic relations. But in Mayr's view, the great difference between the prokaryotes and eukaryotes lay elsewhere:

> The eukaryote genome is larger than the prokaryote genome by several orders of magnitude. And it is precisely this part of the eukaryote genome that is most characteristic for the eukaryotes. This includes not only the genetic program for the nucleus and mitosis, but the capacity for sexual reproduction, meiosis, and the ability to produce the wonderful organic diversity represented by jellyfish, butterflies, dinosaurs, hummingbirds, yeasts, giant kelp, and giant sequoias. To sweep all this under the rug and claim that difference between the two kinds of bacteria is of the same weight as the difference between the

prokaryotes and the extraordinary world of the eukaryotes strikes me as incomprehensible.[61]

Woese responded that the difference between himself and Mayr was nothing less than a "pronouncement on the nature of biology":

> Mayr's biology is the biology of visual experience, of direct observation. Mine cannot be directly seen or touched; it is the biology of molecules, of genes and their inferred histories. Evolution for Dr. Mayr is an "affair of phenotypes." For me, evolution is primarily the evolutionary process, not its outcomes. The science of biology is very different from these two perspectives and its future even more so.[62]

Ever since Darwin, the primary aim of taxonomy was "to encapsulate organismal descent," and in this it would automatically provide the most utility as an "information storage and retrieving system." Above all, he said biological classification is theory,

> a de facto theory, exhibiting the three main characteristics of any good theory: A biological classification has *explanatory power*—i.e., it aids in and enriches the interpretation of findings, integrating them into a deeper, more meaningful context. A biological classification *makes testable predictions*, which lead to the design of experiments. And finally, like any overarching theory, a biological classification has *conceptual power*; it influences the focus of a discipline, steering it in certain directions and away from others.[63]

Viewing classification as "an overarching evolutionary theory," the prokaryote–eukaryote dichotomy was simply "a failed taxonomic theory."[64]

Comparing the diversity of eukaryotes to prokaryotes in the way that Mayr did made no sense because the great diversity in the bacterial world was not manifested in form. Mayr's "eye of the beholder" view of diversity, Woese argued, would not hold in the microbial world, any more than would his species count comparisons. Whether or not the concept of species could be applied to the bacterial world had long been recognized to be a thorny issue (see chapters 6 and 10). Furthermore, the diversity of microbes was hardly explored, as those who studied microbial diversity lamented in the middle of the century. "The human body is, as a matter of fact, practically the only habitat that has been comprehensively studied as a source of bacteria," the editors of *Bergey's* wrote in 1957. "Even in this case it is the bacteria that cause the disease that are best known."[65]

Ever since Pasteur, "microbiology as a science has always suffered from its eminent practical implications," Albert Jan Kluyver said in his Harvard lectures of 1956. "By far the majority of the microbiological studies were undertaken to answer questions either directly or indirectly connected with the well-being of mankind."[66] Given the microbe's role in cycling organic matter, Kluyver estimated that "the total weight of microbial protoplasm on Earth exceeds that of animal protoplasm by many times." It was "perhaps one half-of the living protoplasm on Earth."[67]

Woese pointed to estimates published in 1998 by William Whitman and colleagues at the University of Georgia, Athens, that collectively bacteria had a biomass comparable to that of all plants on Earth.[68] That data was a prima facie evidence for a microbial world that contains the bulk of the planet's biodiversity. Microbial diversity would have to be assessed at the molecular and biochemical levels. And a "genetic measure of diversity," he asserted, "it surely is: over 90% of the biodiversity on this planet is microbial."[69]

Pace and his students at the University of Colorado extended the 16S rRNA technology as a probe to study microbial diversity, opening a new era in microbial natural history.[70] Most microorganisms lived in a complex, codependent way. The great majority of bacteria could not be cultured in isolation and characterized by classical techniques. Molecular methods—using rRNA probe—circumvented the need for culturing when measuring their diversity. Pace and collaborators extracted rRNA genes and cloned them directly from a specific habitat. From those sequences, "phylogenetic stains" (specific probes) could then be designed, which would permit those organisms to be identified as "phylotypes" and counted.[71] The vast majority of microbial species-level groupings still remained undetected. Most of Earth's microbes live in the open ocean, in soil, and in oceanic and terrestrial subsurfaces. Fertile soil contains billions of them.[72]

That immense diversity was illustrated forcefully a few years later when the genome of closely related strains of *Escherichia coli* (O157:H7 and K12) were compared. The former, a nasty pathogen, bears some 1,387 genes not found in the latter, a benign laboratory strain that possess 528 genes not included in the pathogen.[73] Those two genomes differed in 26–12% of their genes. As Doolittle and colleagues commented, "There is probably more variation in genetic information between these two strains of what is considered a single species than within, say, all the mammals, maybe even all the vertebrates!"[74]

Woese and Mayr corresponded again when, in 2003, Mayr congratulated Woese on being awarded the Crafoord Prize for his discovery of the Archaea as a "third main group" of organisms.[75] While emphasizing that he still completely disagreed with the three-domain arrangement, Mayr said he deeply admired Woese's "reconstruction of the earliest steps of life after its origin" and his "earlier work on the classification of the eubacteria."[76] The next spring, when Woese wrote to Mayr saluting his 100th birthday, he said that while he had enjoyed their "public discussions, as surely many, many biologists have," both of them had made an error: "Both of us, I think, have made the mistake of confounding evolution on the multicellular level with that on the cellular level. The two are qualitatively distinct and must be treated accordingly."[77]

While Mayr and Woese were debating and reflecting on their different viewpoints, the whole prospect of microbial phylogenetics and of a universal tree of life had once again come under a swirling cloud of doubt. There was dissention among the ranks of molecular evolutionists.

TWENTY-ONE | Grappling with a Worldwide Web

In classifying bacteria microbiologists make two implicit assumptions: (i) that bacteria have a phylogeny, and (ii) that the taxonomic system that works well for the metazoa is actually applicable to, i.e., meaningful in, the microbial world. These two points require explication and discussion, for they are far from self-evident.

—Carl Woese, "Bacterial Evolution" (1987)

If instances of LGT can no longer be dismissed as "exceptions that prove the rule," it must be admitted (i) that it is not logical to equate gene phylogeny and organismal phylogeny and (ii) that, unless organisms are construed as either less or more than the sum of their genes, there is no unique organismal phylogeny. Thus, there is a problem with the very conceptual basis of phylogenetic classification.

—W. Ford Doolittle, "Phylogenetic Classification and the Universal Tree" (1999)

DEEP CONCERNS WERE RAISED ANEW before the turn of the century. Were the phylogenetic patterns discerned by 16S rRNA reflections of the one true course of organismal evolution? Were the three domains and the taxonomy of the universal tree of life discoveries of *natural kinds*? Or were they merely arti-facts of the techniques used to discern them, and prejudices in the interpre-tation thereof? These doubts arose in the context of genomics with evidence indicating that lateral gene transfer (LGT) among bacteria was far more extensive than phylogeneticists had generally imagined.

All microbial evolutionists came to agree that gene transfer across the taxonomic spectrum was a major source of evolutionary innovation. Its scope and significance became hotly contested beginning in the mid-1990s. Were all genes passed around through all bacteria in a haphazard manner so as to leave no trace of the genes that defined the organization of the cell? Was there an essential genetic core that was passed on vertically from cell generation to the next? Could one ever reconstruct an organismal genealogy? Essentially, there were two viewpoints.

Most phylogeneticists maintained that while LGT was indeed a powerful integrative evolutionary force, it did not erase an aboriginal organismal genealogical trace. They drew a distinction between two classes of genes: *informational genes* involved in transcription and translation and related processes, and *operational* genes involved in metabolism. This view came to be called the "complexity hypothesis."[1] Transcription and translation were complex systems involving the coevolution and interaction of hundreds of gene products; they were ancient and essential processes at the core of the organism. That complexity would restrict the lateral transfer of genes for the components of those systems. Representing the "essence of the organism," genes that were integrated into the "fundamental" system would seldom be transferred laterally. Operational genes, on the other hand would have fewer functional constraints because the corresponding enzymes tended to be involved in individual metabolic pathways such as the synthesis of amino acids, nucleotides, cofactors, cell envelope, and lipid synthesis. They would be transferred between phylogenetic groups.

Others came to see LGT as representing the basis of a wholly new paradigm that supplanted the aims of microbial phylogenetics. In denying an organismal phylogeny for bacteria, they rejected the concept of an organizational core, as an unproven and unwarranted assumption. Like numerical taxonomists before them, they argued that there were no weighted genes or characters that could be privileged over others. At best, all genes had to be considered in an essentially nonphylogenetic taxonomy based on "majority rule." According to this view, the natural classification sought by evolutionists from Darwin to Woese was impossible, not because of methodological impasse, but inherently. Descent with modification simply did not apply to the bacterial world because there was no organismal history to be discerned. Bacteria could not be arranged in a hierarchal manner of groups within groups: species within genera, genera within families, and so on. Lateral gene transfer was the essence of the "prokaryotic" evolutionary process; most, if not all genes, had at one time or another been transferred laterally. Bacterial taxa were no more real than their supposed genealogies; the only reality was in the individual histories of genes. Representations in terms of bifurcating trees were inadequate if not totally misleading. The "tree of life" was not really a tree at all but rather a thoroughly reticulated "world wide web."

Do Bacteria Really Have Organismal Histories?

None of the fin de siècle ideas about LGT were novel. Known since the early 1960s to play a role in widespread and rapid antibiotic resistance (see chapter 10), LGT had become a problem of great medical concern. The pharmaceutical industry's continued synthesis of large numbers of antibiotics, and their expanding use by the medical community, had led to crisis in antibiotic resistance in the 1990s.[2] Lateral gene transfer not only could confer antibiotic resistance, but could also transform bacteria from benign to pathogenic in a single saltational step: from the incorporation of a DNA fragment conferring virulence characteristics.[3] Genetic engineers were long familiar with manufacturing transgenic bacteria by isolating and amplifying DNA segments and inserting them into a foreign cell. Still, the fact of LGT in medicine and manufacture said nothing of its prevalence in the wild; it said nothing of how bacterial genomes evolve naturally.

Those few bacteriologists of the 1960s and 1970s who speculated on LGT's evolutionary effects had noted how the species concept, so basic to biologists' understanding of animals, broke down with gene transfer between taxa (see chapter 10). At the University of Montreal in the 1980s, Sorin Sonea's theorizing on LGT and "evolution without speciation" led him to conceive of the entire bacterial world as a superorganism.[4] Recall that Norman Anderson had suggested in 1970 that evolution depended largely on LGT by viral transduction across species and phylum barriers and that such virus infection might explain the universality of the genetic code itself.[5] In 1985, Michael Syvanen at the University of California–Davis continued this line of argument—that LGT was a mechanism of macroevolutionary change among all organisms.[6] He suggested that it could help explain many observations that puzzled evolutionists, such as rapid bursts in evolution and the widespread occurrence of parallelism in the fossil record. Such speculations were far from the mainstream of bacteriological theory and practice.

The implications of LGT for bacterial phylogenetics had been a simmering issue. Peter Sneath had discussed the problems it potentially posed since 1970, when he and Dorothy Jones argued that, because LGT could mimic convergence, all molecular characters should be considered to be unweighted phenetic traits.[7] Stanier had also noted the problem of LGT for molecular-based phylogenetics when he commented in 1971 that "a bacterium may be a genetic chimera."[8] In New Zealand, Darryl Reanney argued similarly in 1977 when pondering the effects of LGT among bacteria. "This communicability," he said, "may render suspect any phylogenetic scheme based on the assumption that two related species have evolved independently since their divergence from a common node."[9]

Those microbiologists in the 1960s who had employed ribosomal RNA hybridization methods to classify bacteria emphasized that ribosomes were highly conserved at the core of the organism. That argument was extended by Woese for 16S rRNA as "the ultimate molecular chronometer." Nested deep in the center of essential cellular functions, interacting with more than 100 coevolved proteins, made it least liable of all genes to experience LGT between taxa.[10] The capabilities of that ribosomal RNA chronometer seemed to be well confirmed by its predictive power and explanatory capacity. When evolutionists knew what to expect (as in the case of mitochondria and chloroplasts), ribosomal RNA phylogenies confirmed it. When that technology surprised microbiologists by predicting unexpected relationships, it was corroborated by other data. Nothing, of course, was more striking than the grouping together of the phenotypically diverse organisms that constitute the archaebacteria. That group was shown to possess unique common characteristics: the great diversity of their walls, all lacking peptidoglycan; the ether-linked lipids in their cell membranes; and the distinctive idiosyncrasies in their transfer RNAs, in their unique transcription enzymes, introns, viruses, and so on. The predictive success of the 16S rRNA signatures thus engendered new experimental programs in molecular biology, biochemistry, and ecology.

Still, the rRNA-defined groupings were not supported by many phenotypic traits. Some of them seemed to be strikingly at odds with common biochemistry, none more so than the purple bacteria, which contained both photosynthetic and nonphotosynthetic organisms (see chapter 17). In his well-read review on "Bacterial Evolution" in 1987, Woese drew on that example when discussing two of the most fundamental assumptions underlying 16S rRNA phylogenetics: that bacteria exhibited organismal genealogies and that grouping based on 16S rRNA would have common phenotypic properties.[11] Both suppositions would become the focal points of dissension a decade later.

Did bacteria actually possess organismal genealogies? Or was the bacterium merely "a collection of genes (or gene clusters), each with its own history" resulting from LGT? Pointing to the phenotypically diverse purple bacteria, he argued that because trees derived from both molecular chronometers (ribosomal RNA catalogs and cytochrome *c* sequences) had very similar topologies (branching order), it was likely that neither chronometer had been involved in LGT. "Although more extensive testing of the lateral transfer notion is highly desirable," he wrote, "it is now relatively safe to assume that bacteria do in principle have unique, characteristic evolutionary histories and that at least some of the cell's chronometers record them."[12]

Could bacterial phyla defined by 16S rRNA be partitioned into a natural hierarchy of taxa with phenotypic properties common to all members of the group? Woese recognized that the concept of species as an interbreeding group did not apply to bacteria because reproduction and gene transfer were not coupled as in animals and plants. The question was whether the higher taxa were natural or artificial constructs. "There is no compelling evidence to suggest that the bacteria fall into naturally defined taxa," he wrote. "In fact, existing evidence might even suggest the contrary."[13] Again he pointed to the many non-photosynthetic species grouped with the purple bacteria. Still, he was hopeful that the lack of common phenotypic resemblances in genealogical groups might simply reflect the lack of research.[14]

The Last Universal Common Ancestor

Woese had already envisaged a time in evolution when LGT was so pervasive that there were no genealogies as such: the progenote era, made up of precellular entities in the throes of developing translation, transcription, and replication, from which extant organisms emerged. Molecular phylogeneticists who dug into the roots of the universal tree of life typically searched for a universal single prokaryotic-like cell as the most recent common ancestor. Walter Fitch called this mother cell "the cenancestor."[15] Other microbial phylogeneticists called it LUCA (the last universal common ancestor). Many favored an autotrophic cenancestor. Autotrophic thermophilic bacteria and archaea using a hydrogen-based energy source seemed to be located on the deepest branches of the phylogenetic tree. Conditions like the environs of thermal vents on the ocean floor might have given rise to the mother cell, which was already organized in the way prokaryotes are today.

If one deduced the characteristics of the universal common ancestor based on shared characters of all organisms today, it was already complex. All organisms were cellular, and genetic information was stored in DNA transcribed into RNA and translated into proteins. All organisms use a very similar genetic code. Although there were differences in the transcription and translation machinery, the process is very similar in all cells. All cells use lipid membranes to separate their protoplasm from the environment, and they use the same energy-rich metabolites. Peter Gogarten commented in 1995:

> The last common ancestor was already a prokaryotic cell with ribosomes and energy-conserving membranes. This organism possessed ion-translocating ATPases that were already multisubunit enzymes. It had different elongation factors, two types of methionine transfer RNAs, and different dehydrogenases.[16]

Patrick Forterre and Hervé Philippe suggested that the belief that prokaryotes came before eukaryotes was sheer prejudice. Certainly prokaryotes predated modern eukaryotes, but the eukaryotic lineage itself could be older: "If characters present in both Archaea and Eukarya are primitive, LUCA might have been a complex organism, even harboring some features now present only in Eukaryotes."[17]

Woese rooted the tree of life deeper into an earlier hypothetical world of the progenote, which in 1982 he reconceived as representing "an ancestor *state*," "a genetic communion of precellular entities."[18] The proto-cell was a "less integrated, more ill-defined ephemeral entity at the progenote stage." The tempo of evolution would have been rapid, and driven by a high mutation rate of an error-prone genetic system and by extensive LGT of its still relatively loosely connected components.[19] There would be a "ready exchange, a flow, among sub-cellular entities—be they called genes, plasmids, viruses, selfish DNA" at the progenote stage giving rise to "a molecular mosaicism in its descendant lines."[20] According to this model, organisms with discernable lineages emerged, as information processing became more accurate, and the extreme levels of genetic exchange diminished.[21] The fast tempo in the progenote era would have given rise to major changes in the mode of evolution. As Woese commented in 1987:

> General differences in cell architecture among the three groups are remarkable, as are their differences in intermediary metabolism, and each kingdom seems to have its own unique version of every fundamental cellular function: translation, transcription, genome replication and control, and so on. The kind of variation that subsequently occurred within each of the kingdoms is minor by comparison. Thus the mode of evolution accompanying the transition from the universal ancestor is unusual; far more novelty arose during formation of the primary kingdoms than during the subsequent evolutionary course in any one of them.[22]

Though embraced by Otto Kandler in 1994, the conception of the progenote as a communal state was generally overlooked by phylogeneticists until Woese elaborated it again in 1998 in the context of new evidence for LGT.[23]

Flipping-Over Genomics

While none of the issues of the 1990s regarding LGT was new, the evidence was. Much of it came from genomics with the development of automated sequencing technology, the polymerase chain reaction, and new comparative methods. Genome sequencing and bioinformatics for medical and industrial purposes became big business as scientists constructed online data bases and computer programs. Bacterial and archaeal genome sequencing projects started loading data onto the Internet beginning in the mid-1990s. The first full genome sequence of a bacterium (that of *Haemophilus influenzae*) was determined in 1995 by researchers at The Institute for Genomic Research (TIGR), headed by Craig Venter, in Rockville, MD.[24] The following year, Woese and Gary Olsen together with researchers at TIGR published the first complete genome sequence of an archaeon: *Methanococcus jannaschi*.[25]

By 1998, the complete genomes of 12 microbes had been fully sequenced, promising the path to a whole new way of understanding how they cause disease, how they live, and how they evolve.[26] Many different approaches emerged to study genome evolution and the role of LGT.[27] Initially, there were essentially two: one involved comparing phylogenetic trees generated from different genes in the genome and assessing incongruities, and the other based on comparing differences in base (G+C) composition and codon usage.

Using the latter method, Jeffrey Lawrence of the University of Pittsburgh and Howard Ochman at the University of Rochester reported that in 1998, astonishingly, about 18% of the genes of *E. coli* were relatively recent acquisitions, and that all the genes that distinguished *E. coli* from *Salmonella* had occurred from LGT.[28] A recently transferred gene could be detected based on its anomalous G+C content when compared with other genes within a "species." They applied similar considerations to codon preferences within a "species." Bear in mind the redundancy of the genetic code: all but two amino acids are encoded by more than one codon (e.g., AAA and AAG for lysine). There were often species-specific preferences for one of the several codons that encode the same given amino acid. Codon preferences were thought to reflect a balance between mutational biases and natural selection for translational optimization: optimal codons help to achieve faster translation rates and higher accuracy. Such translational selection would be stronger in highly expressed genes.

Ancient gene transfers would be more difficult to detect because over time (through mutation in countless rounds of DNA replication and repair by the enzymes specific to a species) they would gradually become more similar in GC content to the rest of the genome. An acquired gene with an anomalous codon bias would over time undergo what they called "amelioration" and adaptation to the recipient's codon preferences.[29] Thus, Lawrence and Ochman suggested that "bacterial speciation" was not driven by point mutations, but rather from "a high rate of lateral transfer, which introduces novel genes, confers beneficial phenotypic capabilities, and permits the rapid exploitation of competitive environments."[30] They said nothing about the potential effect of LGT on phylogenetics.

No one advanced the idea that phylogenetics was moribund more than did Ford Doolittle. Lateral gene transfer, as he would come to understand it, was the rule, not the exception: the search for the universal tree of life was simply quixotic; there were no organismal genealogies in the bacterial world, only gene histories. Recall that when Linda Bonen exported the technology from Urbana to Halifax, Doolittle and his coworkers exploited it in a successful effort to demonstrate that mitochondria and chloroplasts had evolved as symbionts (chapter 17). Throughout the 1980s, he worked on diverse topics: 5S rRNA phylogenetics of fungi and the molecular genetics of cyanobacteria; he theorized on the origin of introns, and parasitic DNA, which he called "selfish DNA."[31] He had been Woese's champion in the 1980s, trumpeting what he called the "Woesian revolution": the remarkable success and predictive ability of ribosomal RNA phylogenies, and especially the evolutionary significance of the archaebacteria.[32]

Doolittle's phylogenetic doubts had emerged before the genomic data suggested that LGT was widespread. The problem for him was entangled in the roots: the search for the mother cell from which all life had sprung—the cenancestor. A number of disagreements between gene trees deep in the roots of the tree of life had been reported since the early 1990s.[33] Recall that two independent studies of ancient gene duplications based on elongation factors and on ATPases had placed Archaea and eukaryotes on one branch and Bacteria on the other (chapter 18). Eucarya and Archaea would then be sisters. That finding was consistent with the long-held belief that an archaebacterium or its ancestor had given rise to the eukaryotic nucleocytoplasm. However, the genes for other proteins seemed to place the archaea and bacteria together.[34] They had many shared genes. The highly conserved heat-shock protein Hsp70, a class of molecular chaperones involved in protein folding, grouped some of the archaebacteria together with Gram-positive bacteria.[35] Similarly, the gene for glutamine synthetase, a key enzyme in nitrogen metabolism in the archaeon *Pyrococcus woesei*, was closely related to *Clostridium*.[36] In 1993, Gogarten and Elena Hilario reported that the gene for ATPase used to root the tree four years earlier had been subjected to lateral transfer.[37] While recognizing that "the tree of life becomes a net of life," Gogarten was still confident in 1995 that "SSU-like rRNAs provided a solid backbone for future phylogenetic analyses concerning early evolution."[38]

By that time, Doolittle pondered abandoning the idea that 16S rRNA (or any genes) could provide a framework for understanding bacterial evolution. Symbiosis and LGT saturated the intellectual atmosphere of microbial evolution. Not only did the conflicting gene phylogenies have the ear marks of LGT, but there were suggestions that the eukaryote might be chimeric. Again, the idea was not new—Woese and Fox had once entertained the concept that eukaryotes were thoroughly chimeric, composed of genes of the other two primary lineages. Eukaryotes and archaea shared genes involved in transcription and translation, and there were increasing reports of bacterial-like metabolic genes in what were considered to be ancient amitochondriate eukaryotes (see chapter 22). "Extensive gene transfer may have played such an important role in early cellular evolution as to jeopardize the very concept of cellular lineages," Doolittle and his students wrote in 1994.[39]

The following year, Doolittle and James Brown rooted the universal tree using another duplicated gene family, the aminoacyl-tRNA synthetases. Those proteins are vital to the translation process; they catalyze the attachment of a specific amino acid to its cognate transfer RNA: "We consider our result to be the strongest to date in support of the sisterhood of archaea and eukaryotes." They were emphatic: "the separate monophyly of all three domains," they said, was "strongly supported by this analysis."[40] Still, Doolittle's reservations persisted even when yet another ancient gene duplication suggested that the root of the universal tree lies somewhere in the bacteria, thus positioning archaea and eukaryotes as sister groups.[41] "There is still uncertainty about this rooting," he and Brown commented in 1997, "since each duplicated gene data set has its own particular, and significant, shortcomings. Furthermore, three or four genes spanning a few thousand base pairs may not be representative of entire genomes with thousands of genes and, at least, several million base pairs."[42]

That year, Norman Pace published a key paper in *Science*, "Molecular View of Microbial Diversity and the Biosphere." In it he depicted an unrooted tree (figure 21.1).[43] It was based on 16S rRNA phylogenies, which, he said, were largely congruent with phylogenies based on genes involved in the information processing system of cells, but not so with those involved in metabolism, adaptations to the environment, and possibly resulting from LGT: "The tree can be considered a rough map of the genetic core of the cellular lineages that led to the modern organisms (sequences) included in the tree."[44]

In Doolittle's view, Pace's unrooted figure showed clearly that ribosomal RNA phylogenies in themselves were simply unable to trace life's earliest evolution. This was not because the methods were faulty but because of the nature of evolution itself. As Doolittle saw it, there simply was no uniquely correct tree of organisms.[45] He had already made a major philosophical shift in outlook the previous year at which time Woese, Olsen, and collaborators published the first full genome sequence of an archaeon. Doolittle had come to consider all taxonomic categories to be artifacts, not realities of nature. Anticipating the tsunami of genomic data that was on its way, he laid out the conceptual issues regarding LGT, while addressing the assumptions that Woese had articulated nine years earlier. Whereas Woese had spoken about an organismal core, the "essence of the organism" (e.g., translation, transcription, genome replication, and control), and maintained the existence of natural *kinds* to be discovered, Doolittle denied both. The organism was no more than the sum of all of its genes. As he saw it, a conception of bacterial phylogenetics, whether in terms of a core of genes or in terms of the most shared genes with a common history (majority rule), embodied an erroneous concept of essentialism:

> The "majority-rule" and "core function" approaches both seem arbitrary, and tinged by the same sort of essentialism that colors our thinking about "eukaryotes" and "prokaryotes." We want to believe that organismal and species lineages do have discrete and definable histories that we can discover, and not that we are choosing, arbitrarily, genes whose phylogeny we will equate with that history.[46]

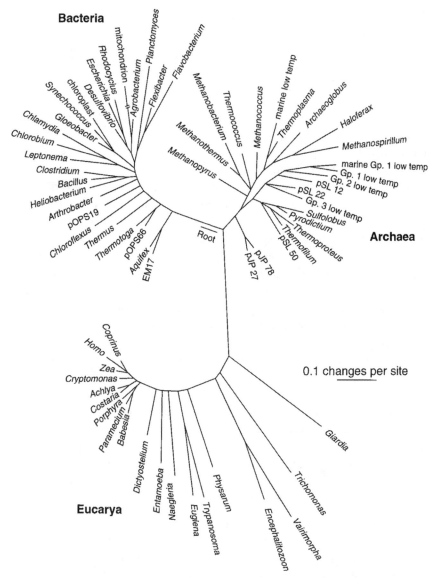

Figure 21.1 "Universal Phylogenetic Tree Based on SSU rRNA Sequences." Norman Pace, "A Molecular View of Microbial Diversity and the Biosphere," *Science* 276 (1997): 734–740, at 735. With permission.

As Doolittle saw it, the concept of a branching tree was being forced onto the data. The concept of a bifurcating tree resulting from common descent was embedded in the concept that inheritance and reproduction were joined at the hip. Descent with modification resulted from inheritance of characteristics from a common parent. Darwin emphasized the main issue in the concluding paragraph of *Origin*: "Inheritance is almost implied by reproduction."[47]

That seemed to be obviously true for animals and plants, but not so for the microbial world.

Trees without Organisms

To wholly reject the concept of a hierarchal tree, one would have to go further. From the time of Lamarck and Darwin to the present, the concept of a hierarchical progressive arrangement (exclusive or otherwise) was based on distinguishing essential or fundamental characteristics of the organism from relatively trivial adaptive characteristics. Woese referred to superficial versus deep characters; others referred to metabolic genes versus "the core" of informational genes. Pan-lateralism, the wholesale rejection of the treelike imagery for microbial evolution, relied on the assumption that such a distinction does not apply.

There were no "fundamental" organizational patterns that defined the major taxa of bacteria, as there were, for example, in the animal world, in Doolittle's view. Lateral gene transfer among bacteria, as he saw it, was analogous to genetic exchange between individuals of an animal species. To rest the phylogeny of bacteria on a small subset of "core" genes was similar in principle to showing the genealogy of two men and tracing them back several generations to a common ancestor based on the Y chromosome and their surnames.[48] In privileging the Y chromosome, one left out all the genetic mixing that had occurred through all the mothers along the way—other genes would give different results. Choosing the Y chromosome lineage would be quite arbitrary. Indeed, constructing such lineages within an interbreeding group would be arbitrary because of gene transfer between individuals resulting from sexual recombination. By analogy, he argued, there was no one universal "true history of genomic lineage" deep in the core of evolution. Gene histories were real, but organismal trees seemed arbitrary.

> [The] three-domain rRNA-based scheme and its many subsidiary branchings is not the only true representation of the phylogenetic relationships between their member species, and these domains (and the various kingdoms or phyla within the first two, at least) are not what Darwin (or for that matter, Linnaeus or Hennig) understood higher taxa to be.[49]

What microbial phylogeneticists were (unwittingly) constructing then were "gene trees, which for various periods of history and at various scales of resolution, have congruent topologies" (branching points).[50] Because the integrity of organismal lineages would be violated by gene transfers and endosymbioses, neither phenetics nor cladistics would provide the correct method to taxonomy. "No single philosophy of systematics will give us the 'right' answer about species history because there is no such right answer," Doolittle declared in 1996.[51] He opted for a "majority rule" classification of taxa, but stripped of phylogenetic meaning.[52]

Shaking the Tree of Life

Doolittle's prophesy of an imminent conceptual shift—that the entire tree of life hypothesis, including the three domains, would fall—was trumpeted widely in *Science* with editorials titled "Genome Data Shake Tree of Life" and "Is It Time to Uproot the Tree of Life?" by reporter Elizabeth Pennisi.[53] In May 1998, she explained how from gene-by-gene comparisons of newly sequenced genomes scientists had come across genes whose histories conflicted with trees determined by ribosomal RNA. When two staff scientists at Diversa Corporation in San Diego compared several of the genes of the thermophilic bacterium *Aquifex aeolicus* with their counterparts in a range of taxa, they got conflicting results.[54] The taxonomic place of the organisms differed depending on what gene they compared. Noting that some of the genes encoding metabolic enzymes in archaea were similar to those in bacteria, Pennisi suggested "that archaea might not be as coherent and distinct a group as the rRNA tree implies."[55] "Each gene has its own history... I think it's open whether the three domains will hold up," Robert Feldman from Diversa was quoted as saying.[56]

Pennisi also pointed to other kinds of evidence that she claimed further suggested that "archaea were not so distinct from true bacteria." Although archaeans had been "once considered to be limited to extreme environments," she said, they were also turning up in the milder surroundings. Edward DeLong, a former student of Pace's, extending the 16S rRNA phylotype approach to examine microbial diversity in the oceans, found archaea to be common and abundant components of aerobic marine plankton off the Pacific and Atlantic costs of North America.

DeLong did not question the coherence of the Archaea; he firmly supported the 16S rRNA approach, and genomics research in his view only further confirmed the tripartitite organization of life. Nor had the Archaea been limited to "extreme" habitats, as Pennisi had assumed. Mesophylic methanogens, for example, those that lived in temperate environments, had been part of the archaebacteria when that urkingdom was announced 21 years earlier (see chapters 12 and 14). What was important to DeLong was that there was so little basic information about the abundance, diversity, and distribution of bacteria on the planet.[57]

In 1992, he discovered cold crenarchaea, formerly known only to be extreme thermophiles (the sulfur-dependent archaea), on both coasts of North America. They were first detected by ribosomal RNA gene amplification and sequencing, and quantitative ribosomal RNA hybridization.[58] Their great distribution in oceans in a broad range of temperatures and depths, he said, made them "one of the most abundant prokaryotic cell types on earth."[59] The presence of archaea in moderate environments was a wonderful example of adaptive radiation from hot to cold environments as DeLong, Pace, and others saw it.[60] Archaea were found everywhere—soils, sediments, lakes, oceans. Their lipid structures were also determined; they were found to have the classical archaeal ether-linked tetraether lipids (chapter 14). In 2006, DeLong's group at MIT sequenced the genome of *Cenarchaeum symbiosum* (a symbiont of marine sponge).[61] Subsequently, the genome of another moderate crenarchaean, *Nitrosopumilus maritime* was sequenced. As DeLong put it, those "genomes have archaeon written all over them, although, as with other genomes, there are many gene homologues that match only bacteria."[62]

The Dispute at the "Core"

The new evidence for LGT did not in itself call for a wholesale change to a nonphylogenetic paradigm. Indeed, Doolittle's revolt against phylogeny cannot be reduced to a question of data. Though he and Woese had seen eye to eye early on when the discovery of the archaea signaled the evolutionary turn to microbiology, they became separated as deep philosophical differences between them surfaced. Neither of them was Darwinian. Their differences belonged to a wholly new evolutionary structure. Doolittle was an organismal reductionist; Woese a holist. Doolittle was a functionalist; for him, LGT and adaptation defined the organism and its evolution. From his nominalist perspective, there was no such thing as an archaeon—only gene histories were real. Woese was a structuralist; for him, the primary kingdoms were real, and laterally transferred genes were adaptations—LGT did not define the organisms, the organisms defined the dynamic of LGT.

While Doolittle speculated about the end of cellular/organismal phylogenetics, Woese saw just the opposite. The three domains depicted fundamental cellular organizations. There was no question for Woese that the new genomic era provided evidence of extensive LGT, no question that LGT was a major mechanism of evolution. But organisms could never be conceived simply in terms of bags of enzymes and of modular genes trafficked at large within the microbiosphere. The data on genome evolution would sort out those fundamental organismal properties from the auxiliary characteristics that lent themselves to LGT.

Woese replied to the apparent phylogenetic crisis in 1998 in a paper on "The Universal Ancestor," in which he rearticulated his conception of the progressive development of the translation apparatus and the organismal core. True organismal lineages would emerge from the communal progenote state of precellular entities with underdeveloped and error-prone replication and translation machinery.[63] Before the development of the modern translation apparatus, evolution would be driven by a different mode and tempo. There would be no individual entities as such because of the intensity of mutation and LGT. These processes would generate enormous diversity very quickly. Primitive systems would be modular and exchange parts freely. But, as the molecular genetic system evolved, becoming refractory to LGT, so too did definable lineages.

Woese likened the emergence of the three domains to physical annealing: there would first be a period of intense genetic "heat" (high mutation rates) and intense LGT when cellular entities were simple and information systems were inaccurate. This would be followed by genetic "cooling" with the development of the modern cell with a sophisticated translation apparatus, resulting in the emergence of genealogically recognized domains and taxa. He called this the "Darwinian transition."[64] To be sure, Woese recognized that the universal tree was "no conventional organismal tree." "Its primary branchings," he said, "reflect the common history of central components of the ribosome, components of the transcription apparatus, and a few other genes. But that is all. In its deep branches, this tree is merely a gene tree."[65]

Doolittle in effect replied to Woese with two feature articles, one in *Science* and another in *Scientific American* in 1999 and 2000.[66] He found Woese's progenote concept persuasive, and he abandoned the mother-cell concept, but he did not accept the idea of a progressively evolving translation apparatus resilient to LGT. He did not accept the "Darwinian transition" (see figures 21.2 and 21.3). He insisted that there "is no compelling reason other than pride of place to choose ribosomal RNA as the more reliable molecular chronometer."[67] He pointed to experimental evidence of Catherine Squires and colleagues indicating that 16S rRNA could be exchanged at least between closely related species.[68] Even if ribosomal genes were rarely transferred laterally, Doolittle maintained, their history would still not be the one true history. If the three domains could stand, they would do so only on a phenetic basis, devoid of phylogenetic meaning.

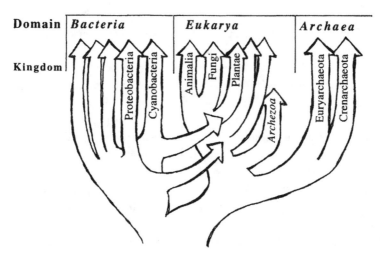

Figure 21.2 "The Current Consensus or Standard Model." W. Ford Doolittle, "Phylogenetic Classification and the Universal Tree," *Science* 284 (1999): 2124–2128, at 2125. With permission.

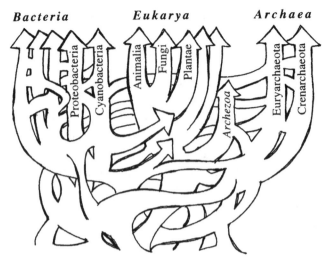

Figure 21.3 "A reticulated tree, or net, which might more appropriately represent life's history." W. Ford Doolittle, "Phylogenetic Classification and the Universal Tree," *Science* 284 (1999): 2124–2128, at 2127. With permission.

It might still be the case that there will be more genes that support a division of living things into Bacteria, Archaea and Eukarya than support any other single trifurcation (or other simple division) of all known taxa. Nevertheless, such a "majority rule" classification is not the "natural" scheme that Darwin, Zuckerkandl and Pauling, or Woese first had in mind. Inclusive organismal hierarchy may just not be a biological reality.[69]

In 2002, Doolittle teamed up with Gogarten and Lawrence to write a manifesto of pan-lateralism, mapping the parameters of a new "paradigm" far outside of the Darwinian models of a bifurcating tree, hierarchical classification, and organismal histories.[70] Their core tenets were unequivocal: (1) "tree like phylogenies are inadequate to represent the pattern of prokaryotic evolution at any level," (2) the quest for unraveling of "the complex histories of genes and genomes supersede the quest for one true 'organismal phylogeny,'" (3) "embracing gene transfer promises a broad and radical revision of the prokaryotic evolutionary paradigm," and (4) LGT would be the basis of a new paradigm forging a synthesis of once distantly related specialties: "a fusion of population genetics, molecular genetics, epidemiological and environmental genomics, microbial ecology, and molecular phylogeny."[71]

Studying the ramifications of LGT was not necessarily at odds with phylogenetics based on vertical inheritance. "Although we have presented the new view as if it were antithetical to traditional understandings of prokaryotic evolution," they wrote, "in the long run we endorse a synthesis that will acknowledge gene exchange and clonality, weblike and treelike behavior, and adaptation and the evolution of new function by many modes."[72] One paradigm did not completely supplant or preclude the other. There was no data to indicate that all genes were laterally transferred. There could still be gene sets that defined a hierarchy of higher taxa in the bacterial world just as in that of animals and plants.

At the level of species, at least for some bacteria, it was clear that there were no stable core of genes that could faithfully record that population's bifurcation back to that group's common ancestor. The species concept just did not seem to apply to bacteria.[73] But at higher taxonomic ranks, there could be a stable core of genes that "would truly give the same tree."[74] There could be a stable core for phyla, divisions, domain, or even a universal core. The clustering of genes into patches in prokaryotic genomes was taken to be evidence of genomic evolution by lateral transfer (see chapter 10). By 2003, ribosomal proteins were known to be often highly clustered.[75] However, ribosomal RNA genes were not clustered with them. In short, Doolittle and collaborators were convinced that *most* genes had been laterally transferred at some point in their history, but he was uncertain whether *all* genes were laterally transferred. "The community, and we ourselves, are divided on the issue."[76]

In 2004, Doolittle and Robert Charlebois reported that out of 147 "prokaryotic" genomes, there were about 50 genes that were shared by all.[77] That universal "core" was composed of informational genes, a high fraction of which were translational components. That core represented about 5% of the average prokaryotic genome. At first glance, their analysis might seem to support the theory that informational genes are less frequently transferred. However, by "core" they meant that these genes were shared by all other prokaryotes, not that they were resilient to LGT. In other words, these genes were core because they were universal; they were not universal because they were core.

Most microbial phylogeneticists maintained that, while LGT was a powerful force in prokaryotic evolution, there was a set of vertically inherited conserved genes that provided a strong phylogenetic signal.[78] For them, the genomic crisis regarding LGT was an illusion merely reflecting the failure to recognize the vital distinction between the fundamental core and relatively trivial or auxiliary characteristics. Some argued further that talk of a radical paradigm change was based on empty rhetoric and hyperbole propped up by editorials in leading journals.

If Doolittle's pan-lateralism represented one extreme, Charles Kurland's views that LGT was little more than a "nuisance" represented the other. As he saw it in 2005, a robust universal tree was indeed attainable, and the claims of Doolittle and his coauthors to the contrary was more "ideology" than biology. The most rigorous of their speculations about pan-LGT, he observed, tended to fall in the "what if" category. The impression of an imminent paradigm shift was merely a social construction, a "social phenomenon...a product of a breakdown of the referee system, particularly in the offices of the most widely read journals" such as *Nature* and *Science*.[79] It was also propped up by one-sided editorials against the universal ribosomal RNA tree in major journals such as *Science* "without inviting the publication of a single defense."

The error-ridden methods for identifying laterally transferred genes was another issue that Kurland said had contributed to the "inflated role" assigned to LGT.[80] Gene transfers between close relatives of bacteria were the least controversial. Transfers between distantly related taxa were thought to be much less likely to succeed for two reasons: (1) because conjugation or viral transduction of genes between different species would be less common and (2) because foreign DNA would have to be integrated in the recipient genome via nonhomologous recombination. Still, with a constant "rain" of DNA, even those rare events might become fixed in a lineage if they conferred selective advantage.

Theory aside, obtaining consistent data, and even a ballpark estimate about the percentage of genes transferred between taxonomic groups seemed unattainable during the first decade of the century. Methodological difficulties abounded, as complementary approaches often led to conflicting results.[81] One of the major difficulties was in the distinguishing genes that had been laterally transferred from those whose mutation rates differed, and from situations where genes had been lost. While gene loss could sometimes explain presence and absence of a gene in closely related taxa, a faster evolving gene might give the false reading of being much older than it was.[82] There were many estimates of laterally transferred genes, using a variety of "complementary approaches" yielding a broad range in the percentage of laterally transferred genes: from 1.5% to 66%.[83]

By 2007, it was becoming clear to Doolittle that neither the claims for or against a stable core were falsifiable.[84] As he saw it, if one were to insist on there being an organismal genealogical tree of life based on a small set of core genes, one could do so, but it was certainly not what Darwin had in mind as the explanation for our ability to classify "groups subordinate to groups."[85] Whether one accepted the genetic core or not, the scope and depth of LGT in the microbial world made it clear that Lamarck's and Darwin's concept of common descent did not apply to the microbial world.[86] As Doolittle put it, "we molecular phylogeneticists can claim to have found such a tree of life only if we admit that we have radically redefined (and weakened) what it is we were looking for."[87] The controversy over pan-LGT among bacteria remains essentially in stasis.

LGT was not expected to be common or at least to play the same role in the evolution of plants and animals that possess sex cells distinct from somatic cells. Certainly hybridization was well known among plants, and in some animals.[88] Hereditary symbiosis was known to be ubiquitous among protists, insects, and nematodes, and in some cases massive LGT was found to occur between symbiont and host.[89] Although viral transfer of genes among protists remained largely unexplored, integrated retroviruses (RNA-based viruses), relics of ancient germ-cell infections, were held to comprise about 9% of the human genome.[90] The eukaryotic "symbiome" thus typically comprised chromosomal genes, organellar genes, genes of viral ancestry, and symbionts.[91]

Molecular phylogenetics of bacterial viruses indicated that they are also chimeric, their genome evolution occurring through gene mutation and recombination with each other as well as by gene acquisition from bacteria. Indeed, "natural phage communities," which seem to be driving so much of bacterial evolution, may contain the greatest genetic diversity on Earth.[92] Some phylogeneticists suggests that viruses may well have preceded the emergence of the three domains, emerging at the progenote era of evolution. Whereas d'Herelle saw phages as the most primitive of organisms at the dawn of life, from which bacteria coalesced, and others following Muller and Haldane have seen them as having originated within bacteria as "genes that got loose" (chapter 5), some phylogeneticists suggest a third alternative: phage and bacteria were partners in their coupled evolution.[93]

| Entangled Roots and Braided Lives

The progenote is today the end of an evolutionary trail that starts with fact, progresses through inference, and fades into fancy. However, in science endings tend to be beginnings. Within a decade we will have before us at least an order of magnitude more evolutionary information than we now possess and will be able to infer a great deal more with a great deal more assurance than we now can. The root of the universal tree will probably have been determined.

—Carl Woese, "Bacterial Evolution" (1987)

MICROBIAL EVOLUTIONARY BIOLOGY crossed the twenty-first century in a tumultuous state of competing ideas over the three-domains concept and its ramifications for the structure of microbial evolutionary theory. The Archaea, Bacteria, and Eucarya trifurcation was widely taught in textbooks as representing fundamental phylogenetic lineages. *Bergey's Manual of Systematic Bacteriology* was based on the phylogenetic framework deduced from 16S rRNA data, which by 2003 was the largest data set in the world for a gene or gene product. Still, there was no agreement regarding the relationships among the three domains. Their roots were discovered to be intertwined with many shared genes, and it was not clear which was the best way to disentangle them.

Perplexed by the Prokaryote

The concept of three primary lineages contradicted the prokaryote–eukaryote dichotomy regarding phylogeny, as Woese and Fox had argued when they introduced the archaebacteria in 1977. Thirteen years later, Woese made a further step to correct the misinterpretation of the prokaryote, when he, Kandler, and Wheelis recommended abandoning the term "archaebacteria" (see chapter 19). Still, prokaryote continued to be used as a taxon comprising the Bacteria and the Archaea. Biologists generally insisted, just as Ernst Mayr had, that the prokaryote–eukaryote dualism be maintained because it represented the greatest schism in the evolution of biological organization (see chapter 20). Indeed, regarding their organismal organization, Woese, Kandler, and Wheelis had also defined the Bacteria and the Archaea as "prokaryotic cells" when they formally proposed the three domains in 1990. But by the mid-1990s, Woese would question the reality of anything "prokaryotic."[1]

What exactly constituted prokaryotic cell structure had become murkier as the molecular and biochemical differences grew between the Archaea and the Bacteria. Although the "Bacteria and the Archaea are both prokaryotic in cell type (whatever that now means)," Woese commented in 1994, "the members within each domain share many common molecular characteristics, making each of the three as distinct an entity for the biologist, as elephants, ants, and flowers are for the layman."[2] The prokaryote concept, he argued, had fostered the notion among molecular biologists that to understand bacteria, one had only to determine how *E. coli* differed from eukaryotes. In short, it obscured the profound differences among the bacteria, and it had concealed microbiologists' nearly total ignorance of the relationships in the first place.[3]

Those who supported the dichotomy insisted that the "archaebacteria" met every criterion of the prokaryote as "classically defined." To further embed the dichotomy into the foundations of biology, they constructed myths of its origins: how in the 1920s or 1930s with "singular prescience" Edouard Chatton had articulated the distinction as the basis for two taxa at the highest levels. Like the tale of Gregor Mendel's discovery of the laws of inheritance, lost then found three decades later, so too the truth of Chatton's discovery was not widely recognized until decades later.[4] No one advanced that myth more than did Mayr, and his authority as a historian led others to follow his account.[5] "Although foreshadowed by suggestions made by earlier authors," he wrote in 1998,

> by far the most important advance made in our understanding of the living world as a whole was the realization by Chatton (1937) [*sic*] that there are two major groups of organisms, the prokaryotes (bacteria) and the eukaryotes (organisms with nucleated cells). This classification was confirmed and made more widely known by Stanier and van Niel, and it was universally accepted by biologists until recently.[6]

The power of the myth notwithstanding, Chatton did not articulate the prokaryote–eukaryote dichotomy (chapter 7), and Stanier and van Niel did not conceive of the prokaryote as a taxon; it was a type of cellular organization. Like their predecessors, they had defined the prokaryotes largely in negative terms as lacking a nuclear membrane, lacking organelles, and lacking sex. A negative definition was really no definition at all. The only positive feature they included was the presence of peptidoglycan in prokaryotic walls, a claim vitiated by its absence in archaebacterial walls.

The prokaryote–eukaryote dichotomy had taken on an essentialist meaning just as had the plant and animal dualism of old. But it had slipped into a new age of evolutionary biology in which it could no longer be taken for granted. "It is as if we believed that the words 'eukaryote' and 'prokaryote' named natural kinds whose properties we need only to discover," Ford Doolittle commented in 1996.

> But in fact they are categories we ourselves invented 30–40 years ago (when our understanding of cell and molecular biology was pretty rudimentary) to define organizational grades or identify evolutionary clades. We have not only the right but the obligation to change them now and in future, as our knowledge grows.[7]

Few would agree that the prokaryote–eukaryote dichotomy was merely a human-made construction. That dualism was generally assumed to be as real and natural as the evolutionary discontinuity it reflected. Mayr was, however, especially critical of essentialist-typological thinking, which he claimed had stood in the way of the Darwinian concept of evolving populations. "To preclude misunderstandings," he commented in 1998, "let me emphasize that I support the dichotomy prokaryotes vs eukaryotes not owing to a philosophical preference for a dichotomous division but because this is where the great break is in the living world."[8] In the years that followed, several molecular phylogeneticists followed him in assuming the reality of the prokaryotic type, and they sought molecular properties to define it.

When Mayr first wrote against the concept of three domains in the letters of *Nature* in 1990, Doolittle had tried to support Woese by denouncing the prokaryote concept (see chapter 20). As he then saw it, to group archaea and bacteria together as prokaryotes was an egregious error because the rootings of the universal tree at that time indicated that bacteria branched on one side and archaea and eukaryotes on the other, which was congruent with the similarities and differences in the translation and transcription mechanisms of the three lineages. But Doolittle's views changed radically as that rooting became entangled in lateral gene transfers (see chapter 21). He rejected the reality of taxa, but he believed that the word "prokaryote" was an adequate descriptor for the Archaea and the Bacteria. In 2005, he pointed to some molecular features to conjoin them as "prokaryotic domains": they possessed "a typically (but not always) circular chromosome(s); absence of spliceosomal introns; organization of many genes into operons (sometimes with homologous genes in the same order)."[9] Still, these criteria were insufficient for a phylogenetic classification, because all could be the result of convergent evolution.

Back to Nature

The whole question about what a prokaryote was, and how and if it ever could be defined evolutionarily and organizationally, came to a head when Norman Pace wrote the essay "Time for a Change," published in *Nature* in 2006, in which he recommended that the word be expunged from the lexicon of biology because of its misleading phylogenetic connotation and negative definition. He provided a figure of a trifurcated universal tree as microbial phylogeneticists generally perceived it (figure 22.1). "No one can define what is a prokaryote, only what it is not," he said:

> I believe it is critical to shake loose from the prokaryote/eukaryote concept. It is outdated, a guesswork solution to an articulation of biological diversity and an incorrect model for the course of evolution. Because it has long been used by all texts of biology, it is hard to stop using the word prokaryote. But the next time you are inclined to do so, think what you teach your students: a wrong idea.[10]

Pace's announcement got the attention he sought. Lines were drawn, as microbiologists outside the archaeal and ribosomal RNA research programs rallied to defend the traditional dichotomy. Among the most adamant were those who had abandoned the concept of microbial phylogenetics altogether and who

A. Three-Domain Model

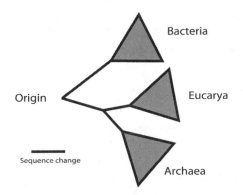

Bacteria

Origin

Eucarya

Sequence change

Archaea

Figure 22.1
"Two models for phylogenetic organization and the course of evolution. The wedges represent relatedness groups of organisms." Norman Pace, "The Molecular Tree of Life Changes How We See, Teach Microbial Diversity," *Microbe* 3 (2008): 15–20, at 18. With permission.

B. Procaryote-Eucaryote Model

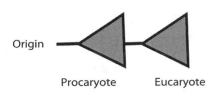

Origin

Procaryote   Eucaryote

denied the three-domain concept. A series of letters were subsequently published in *Nature* under the headings "A Positive Definition of Prokaryotes," "Advances in Biology Reveal Truth about Prokaryotes," and "Concept of a Bacterium Still Valid in Prokaryote Debate." The first was by William Martin and Eugene Koonin, who proposed that the coupling of transcription and translation was the key positive character of the prokaryote: DNA messages are translated into the amino acids of proteins at the same time that they are being transcribed to RNA. Prokaryotes, they said, are cells "with co-transcriptional translation on their main chromosomes; they translate nascent messenger RNAs into protein. The presence of this character distinguishes them from cells that possess a nucleus and do not translate nascent transcripts on their main chromosomes."[11] Dismissing the trifurcated tree as a question of "belief" they wrote: "Regardless of what any gene tree might suggest and regardless of what anyone might believe about early evolution, modern cells lacking spliceosomal introns and spliceosomes, a nucleus, and mitochondria do possess transcriptionally coupled translation—they are prokaryotes."[12]

The second letter was from Michael Dolan and Lynn Margulis, who offered a definition of eukaryotes based on a mixture of feeding behavior, symbiosis, sex, and intracellular motility (the prokaryote was implicitly defined in terms of what eukaryotic features they lacked):

> Eukaryotes—whether protoctists, fungi, animals or plants—routinely open their membranes to take in (or let out) nuclear genomes, whole cells or other large particles, in processes such as ingestion, fertilization and hybridization. They reseal their membranes and live happily ever after. All eukaryotic sexuality requires cell fusion. Nearly all eukaryotic cell phenomena involve microscopically visible intracellular motility that never happens in prokaryotes.[13]

The third letter was from Tom Cavalier-Smith.[14] He rejected the progenote concept and instead favored the evolution of the three cellular types, beginning with eubacteria. He maintained that the odd walls of the archaebacteria diverged after the invention of peptidoglycan, not before it. The absence of peptidoglycan in the archaebacteria resulted from the loss of the ability to make it.[15] In 2002, he maintained that the archaebacteria and eukaryotes were sisters forming a clade, which he called the "neomura" (new walls). He also postulated a very late divergence of the neomura from bacteria—about 850 million years ago.[16]

To exemplify the need for the term prokaryote, Cavalier-Smith pointed to an essay titled "Biology's Next Revolution," in which Nigel Goldenfeld and Woese argued for a revision of concepts of organisms, species, and evolution on the basis of "the emerging picture of microbes as gene-swapping collectives."[17] "The uselessness of the species concept," they said, was "inherent in the recent forays into metagenomics"—the study of genomes taken directly from environments as opposed to clonal cultures. Many genes were "cosmopolitan," wandering within communities of microbes in accordance with environmental conditions. Early life evolved in a "Lamarckian way" by lateral gene transfer. Microbes, they said, could not be regarded as individuals with discrete genomes but rather as cooperative populations.

Goldenfeld and Woese also pointed to "quorum sensing," which Bonnie Bassler and colleagues had recently shown to be a widespread phenomenon of cell–cell communication among bacteria that collectively coordinate gene expression according to the local density of their population.[18] "In fact, their communication by genetic or quorum sensing channels indicate that micro-bial behaviour must be understood as predominantly cooperative." Addressing the prokaryote question, they commented, "Sometimes, language expresses ignorance rather than knowledge as in the case of the word 'prokaryote', now superseded by the terms archaea and bacteria."[19] As Cavalier-Smith saw it, their persistent use of the word "microbe" when speaking of lateral gene transfer and their failure to distinguish bacteria from eukaryotic protists amply illustrated the need for the term "prokaryote," because the concept of species applied per-fectly well for sexual protists.[20]

Nature also received several letters from microbiologists who asserted that Martin and Koonin had made critical errors. Pace sent a reply in which he noted that they ignored the polyphyletic nature of the prokaryote, but also that there was no actual evidence that transcriptionally coupled translation was a common characteristic of the Archaea.[21] Patrick Forterre and David Pranishvilli submit-ted a letter asserting that although there was *some* evidence that cotranscrip-tional translation occurred in *some* archaea (though none had yet been pub-lished), there was no evidence that coupled transcription and translation was a primitive ancestral trait.[22] Only a shared derived trait could be used to define a valid phylogenetic group. John Fuerst noted still another error in defining the prokaryote as a cell that lacked a nucleus (i.e., a nuclear membrane). He pointed to Planctomycetes, a bacterial phylum, which he had shown to possess intracel-lular membrane compartmentalization including a membrane-bound nucleus.[23]

Astonishingly, *Nature* chose not to publish any of these letters; only those supporting the traditional dichotomy saw the light of day. Pace and Woese were flabbergasted by what they saw to be *Nature*'s lack of impartiality. In their view, the correspondence editors of *Nature* had arrogated to themselves a censorship role, as they had in debates with Mayr in the early 1990s (see chapter 20).[24] In April 2007, Woese and Goldenfeld sent a letter to *Nature* summarizing the points in the other rejected letters, and emphasizing that there was a general consensus that Archaea, Batceria, and Eucarya were distinct phylogenetic lin-eages. Their letter was also rejected by *Nature*.

Pace published a feature essay in *Microbe* the following year pointing to the false phylogenetic connotation of the word "prokaryote" as a monophyletic group of organisms that preceded and gave rise to eukaryotes. "Fundamentally," he said, "there are three phylogenetic kinds of organisms, representative of the three primary domains. Moreover, none of those primary domains is derived from another."[25] John Ingraham, from the University of California, Davis, replied, accepting the tripartite division of life but defending the prokaryote–eukaryote dichotomy as an organizational distinction between fundamentally distinct cell types.[26] The word "prokaryote" did not prevent one from understanding diversity, he said. Microbiology students were "fully capable of comprehending that microbes are distributed among the three domains: eucaryotes, bacteria, and archaea, two of which, because of their dramatically distinct cell structure, are called procaryotes." Nor did the prokaryote necessarily imply that they came before eukaryotes. "Pro" did not necessarily mean "before"; it could also mean "in place of" or "substituting for." So Ingraham insisted that "Procaryote is an accurate and useful word that in no way contradicts emerging ideas about the phylogenetic connections among life's three domains. How could a modern textbook be written without it?"[27]

Though the prokaryote held no phylogenetic meaning, many biologists agreed that no degree of molecular difference between Archaea and Bacteria could compare with the structural organization of eukaryotes with its complex and highly regulated process of nuclear and cell division and elaborate choreography of chromosomes. Those were the features that underlie the development and evolution of all life forms we see around us. Indeed, while zoologists spoke of "the Cambrian explosion" in animal diversity some 550 million years ago, those who studied the origins of the eukaryote argued that biology's real big bang had occurred much earlier.

Who Is the Eukaryotic Mother?

In 1997, Russell Doolittle and colleagues determined divergence time for the major kingdoms based on amino acid sequence comparisons of proteins. According to their protein clock, plants and animals shared a common ancestor about 1,200 million years ago (Mya), fungi diverged from either of those groups at about 1,275 Mya, protists diverged from other eukaryotes at about 1,550 Mya, and eukaryotes diverged from eubacteria 2.5 billion years ago. According to their analysis, archaebacteria did not cluster together with eukaryotes, but diverged from eubacteria between 3,000 and 4,000 Mya.[28] Still, there were many sources of error arising from variations in the rate of change within sequences, and corruption of data by lateral gene transfer and symbiosis.[29] All claims about the divergence times and relations of the Bacteria, archaea, and eukaryotes remained widely disputed.

The evolutionary relations among the three primary lineages remained enigmatic. All the old questions from the 1970s and 1980s persisted: Were the Archaea as ancient as their name implied, or did they diverge relatively recently from Bacteria? Did eukaryotes evolve from Archaea? Did they evolve shortly after life began, or were they relatively modern? Did eukaryotes actually represent a third fundamental lineage of descent, in the same sense as Archaea and Bacteria? Or were they wholly chimeric constructs?

That the nucleus arose by a symbiosis of some kind is an old idea. It dates back to the nineteenth century when cytological and chemical characteristics of the nucleus and cytoplasm left the impression on some that the nucleus was suspended in the body of a different organism (see chapter 9). Shôsaburô Watasé explored that concept in 1893. Eleven years later, Theodor Boveri suggested that chromosomes were symbionts living in symbiosis with the cytoplasm of cells. In 1909, Constantin Merezhkowsky proposed that nucleus was formed when bacteria, composed of "mycoplasm," found a home inside another kind of organism composed of "amoeboplasm."

Woese and Fox had considered a chimeric origin of eukaryotes when "Big Tree" was published in 1980 (see chapter 17): "The question of eukaryotic evolution," they said, "then becomes the manner in and the extent to which eubacterial and archaebacterial genes, as well as genes from other sources, are represented in the eukaryotic chimera."[30] Under the progenote concept, the eukaryotes would be expected to share various genes with the other two domains, before they emerged as a distinct lineage. "Fundamentally," Wheelis, Woese, and Kandler commented in 1992, the eukaryote "is an anaerobic cell with but a single (nuclear) genome."[31] They pointed to anaerobic protists living today that lacked mitochondria. "It is these phenotypes, if any," they said, "that would most closely resemble the ancestral eukaryotic phenotype."[32] More than a thousand amitochondriate protists were discovered by that time.[33] Cavalier Smith had grouped them into a kingdom called the "Archezoa."[34]

The idea that the eukaryotic cell arose by symbiosis became popular again when such amitochondriate cells were shown to contain both bacterial and archaeal genes. There was no consensual account. In fact, there were almost as many versions as proponents. Each scheme differed in the molecular data it emphasized, what it was that needed to be explained, and why and how it happened. By the early twenty-first century, symbiosis theories could be grouped into two main categories: (1) those that proposed that the eukaryote (E) arose as a symbiosis (or fusion) between archaea (A) and bacteria (B): E = A + B; and (2) those that proposed that the nucleus arose from the engulfment of a representative of both archaea (A) and bacteria (B) by an extinct third cell type (C): E = A + B + C.[35]

E = A + B

In 1991, Wolfram Zillig proposed that eukaryotes arose as an ancient symbiosis of some kind, when he discussed the evidence that eukaryotic proteins involved in the genetic machinery of the cell (e.g., ribosomal proteins, translation factors, and RNA polymerases) resembled those of the archaea whereas eukaryotic enzymes involved in glycolysis and central metabolism resembled those of eubacteria. As a possible explanation, he suggested that perhaps "the ancestor of the *Eucarya*, arose by some sort of fusion event between an archaeal and one or several (possibly bacterial) ancestors."[36] To support this suggestion, Zillig pointed to results published two years earlier of a "bacterial-like protein found in an ancient eukaryote devoid of mitochondria, *Giardia lamblia*.[37]

That *Giardia* arose from a symbiosis found further support beginning in 1994, when Radhey Gupta and colleagues reported that the heat-shock protein Hsp70 of *Giardia lamblia* resembled that of bacteria.[38] Like Zillig, they also proposed that the eukaryote arose from a fusion of an archaean and a Gram-negative bacterium.[39] Gupta proposed that the fusion event would have taken place in an aerobic environment, predominated by antibiotic-producing bacteria. Those two selective forces, oxygen and antibiotic warfare, would have led to a primitive eukaryote that was antibiotic resistant and oxygen tolerant. An oxygen-tolerant eubacterium (related to proteobacteria) provided protection against oxygen, and the other partner, an archaebacterium, provided antibiotic protection.[40]

Gupta fixed on the Hsp70 family of proteins, which, among other functions, play a vital role in folding proteins into their correct three-dimensional forms. His comparative analysis of their amino acid sequences did not support a clear separation of the domains Archaea and Bacteria. Instead, he divided the bacteria into two different fundamental groups: Gram-positive (the earliest group) and Gram-negative bacteria. He renamed them, respectively, as "monoderms" (those with a single membrane) and "diderms" (those with a double membrane). In Gupta's view, archaebacteria evolved from Gram-positive bacteria as adaptations—but not in response to extreme environments, but rather to antibiotic selection pressure.

Gupta heralded his data and concepts as representing a "paradigm change." But his critics argued that his conclusions far exceeded his data. The Hsp70 trees conflicted not only with the ribosomal RNA results, but also with the phylogeny of many other proteins, including that of another heat-shock family protein, Hsp60, which supported the ribosomal RNA tree.[41] After comparing the congruence of many proteins with the ribosomal RNA trees in 2005, Wolfgang Ludwig and Karl-Heinz Schleifer suspected that the anomalous Hsp70 sequences resulted from limitations of the techniques he used to compare proteins, from lateral gene transfer, or from gene duplication or losses.[42]

James Lake also became a strong advocate for the view that the eukaryote arose as a symbiosis, not by a fusion event, but rather as an engulfment of an archaebacterium by a bacterium.[43] In his scheme, the host would have been anaerobic photosynthetic purple sulfur bacteria, and the symbiont anaerobic eocyte that generated energy from the reduction of sulfur with hydrogen.[44]

Such an engulfment, he said, would account for the double membrane of the nucleus, just as it did for that of mitochondria and chloroplasts.

Margulis also favored a chimeric origin of eukaryotes. But, for decades she had argued that the nucleus had emerged autogenously after the acquisition of mitochondria.[45] The next symbiotic step resulted in the centrioles/kinetosomes, and related cell structures (cytoskeleton, the mitotic spindle, and motility features of the cell). They would have evolved from symbiotic spirochetes.[46] She shifted her views after 1995, by which time it had become clear that DNA was not located in centrioles themselves.[47] Perhaps the ancient bacterial genes in the nucleus of *Giardia* were actually those of the spirochetes. Mitochondria would originate later.[48] Thus, she hypothesized that the eukaryotic cell "evolved from a symbiotic consortium of Spirochaeta-like eubacteria with archaebacteria that resembled extant *Thermoplasma*."[49] She invited Darwin to have the last word: "Anyone whose disposition leads him to attach more weight to unexplained difficulties than to the explanation of a certain number of facts will certainly reject my theory."[50]

There were difficulties. By that time, the archezoa kingdom, composed of putatively ancient amitochondriate eukaryotes, was collapsing.[51] Like the prokaryote concept, the Archezoa had been defined negatively in terms of what its members lacked. As it turned out, the kingdom was polyphyletic: the similarities within and between the four phyla said to be in the kingdom were an evolutionary illusion arising from convergence. It appeared to be likely that many, if not all, of those organisms had once possessed mitochondria but lost them. Some of those organisms were also shown not to be deep in the tree of life as had been claimed.[52] During the second half of the 1990s, there were several reports by Ford Doolittle and by several of his former students of evidence of "mitochondria genes" (alpha-proteobacterial homologs) in the nucleus of those amitochondriate protists.[53]

Perhaps there never was a eukaryote that lacked mitochondria. Perhaps mitochondria were acquired first by an archaeal host. Several models of this kind had been suggested earlier. In 1978, Denis Searcy had proposed that the wall-less thermoacidophilic *Thermoplasma* was the mitochondrial host; Margulis supported that suggestion in 1981.[54] That mitochiondria were at the basis of eukaryotic origins received further support with the new evidence of what appeared to be 'mitochondrial genes' in amitochondriate protists. One of the most prominent mitochondria-first models was proposed in 1998 by William Martin and Miklós Müller. They called it "the hydrogen hypothesis for the first eukaryote."[55] It offered a radically different scheme for why the symbiosis occurred.

In the textbook narrative, the proto-mitochondrion had been acquired by an anaerobic eukaryotic host for its present-day function of aerobic respiration and energy.[56] Accordingly, as the primitive Earth's atmosphere changed from anaerobic to aerobic around 1.5 billion years ago, as a result of photosynthetic oxygen production, anaerobic prokaryotes were forced either to adapt to aerobic conditions or to become restricted to anaerobic environments. In the complementarity of symbiont and host, the proto-mitochondrion was in charge of aerobic respiration and energy (ATP), while the host was responsible for the breakdown of organic substances and protection.[57] Margulis commented in 1993 that "the release of ATP to its host would be analogous to throwing cash into the streets."[58]

The symbiotic scenario for the mitochondrion–host relationship was typically depicted in terms of "slavery" (chapter 9). John Maynard Smith and Eörs Szathmáry suggested that "host cells may have kept protomitochondria as humans keep pigs." They switched to other similes to explain their integration into the cell system, and compared the transfer of mitochondrial genes to the nucleus to "paying taxes to a central government."[59] Others accounted for the gene transfer to the nucleus in terms of decreasing mutation load and the importance of increasing efficiency and regulation.[60]

Martin and Müller rejected all of it. In their scheme, the proto-mitochondrion was acquired by an archaeal host in an anaerobic environment; the nucleus evolved afterward. The initial advantage to the host was not ATP, but rather the excretion of H_2. The clue was in the hydrogen-producing organelles called hydrogenosomes found in some protists and fungi. Hydrogenosomes were postulated to have evolved from the same symbiont as modern-day mitochondria.[61] In support of their model, Martin and collaborators denounced the kingdom Archezoa, asserting that there was no evidence of any protists that did not at one time possess mitochondria.[62]

While that was true, there were also weaknesses in their model. First of all, it was far from conclusive that those bacterial genes in amitochondriate protists were actually "mitochondrial genes." The evidence for mitochondrial genes in the nucleus of a species typically depended on finding those same genes present in mitochondria of another species. Yet, the putative mitochondrial genes in amitochondriate protists had not been found in any mitochondria.[63] Those bacterial genes could have been acquired from other symbionts that were subsequently lost, or by routine ingestion of food bacteria by protists.[64] Indeed, gene acquisitions from eating bacteria may be pervasive among protists as they were among prokaryotes.

Although classical evolutionists have long considered mitochondria and chloroplasts to be exceptions, representing "the quirky and incidental side" of evolution, the symbiotic origin of mitochondria and chloroplasts was actually the exception that proved the rule.[65] Hereditary symbiosis was known to be pervasive among eukaryotic protists. Among animals, lateral gene transfer was hypothesized to be rare in part because they possessed a sequestered germline. Maynard Smith and Szathmáry asserted in 1999 that "transmission of symbionts though the host egg is unusual."[66] But that statement was also proving false as molecular phylogenetic techniques for screening revealed that bacteria of the genus *Wolbachia* are inherited through the cytoplasm of the eggs of a great majority of all insects and nematodes.[67] Far from being slaves, they manipulate the reproduction and development of their hosts. There was also growing evidence of widespread and massive lateral gene transfer of intracellular bacteria to their multicellular hosts.[68]

Given the near omnipresence of hereditary symbiosis among eukaryotic protists, it was difficult to prove that those mitochondrial-like genes in the nucleus of amitochondriate protists were derived from mitochondria. And even if all the eukaryotes known today did once possess mitochondria, critics reasoned, it was still possible that one might find a real archezoan hidden in the great diversity of protists.[69] Or perhaps, after all, the hypothetical amitochondriate proto-eukaryote left no survivors.[70]

These issues aside, there seemed to be a fundamental overarching weakness in all of the above models that proposed a symbiosis between bacteria and archaea: it was widely assumed that only eukaryotic cells with a cytoskeleton could engulf symbionts, accomplishing it by phagocytosis.[71] To eat, bacteria generally absorb nutrients from their surroundings, sometimes secreting digestive enzymes to break down more complex environmental substances. Bacteria and archaea have none of the complex cellular apparatus that makes it possible for a cell to wrap part of itself around another cell. Only eukaryotes could eat in that way.

The idea was an old one. In 1915, Edward Minchin proposed that two different cell types evolved from a prebacterial cell that was formed from a synthesis of scattered and independent chromatin granules of diverse genetic constitution.[72] One cell type specialized in the vegetative mode of life, while the other developed a predatory existence. The chief event in the evolution to the predatory type, he said, was the formation of a surrounding matrix of protoplasm. The streaming movements of the protoplasm would enable it to flow round and engulf other creatures.

> Thus arose in the beginning the brand of Cain, the prototype of the animal, that is to say, a class of organism, which was no longer able to build up its substance from inorganic materials in the former peaceful manner, but which nourished itself by capturing, devouring and digesting other living organisms.[73]

The next stage in its evolution was the organization of the chromatin grains into a definite cell nucleus. The subsequent perfection of mitosis conditioned the possibility of large multicellular forms.[74]

In 1970, Stanier suggested that the predatory ability to engulf was at the root of the great morphological diversity of eukaryotes vis-à-vis the enormous biochemical diversity of prokaryotes (see chapter 9).[75] While prokaryotes evolved toward new and diverse modes of energy-yielding metabolism, the elaboration of the cytoskeleton and the ability to engulf prey freed the eukaryotic lineage from the need to evolve the diversity of energy metabolisms characteristic of bacteria. If phagocytosis was limited to eukaryotes, then the cytoskeletal system would have had to evolve without symbiosis, and the host for mitochondria would already have been some kind of eukaryote.

Indeed, the assumption that only eukaryotes were capable of phagocytosis had been a chief reason for postulating the archezoan host in the first place. As Cavalier-Smith asserted in 1987, "it is only the existence of such fully eukaryotic phagotrophs that makes a symbiotic origin of mitochondria mechanistically plausible."[76] Those who argued for a primordial symbiosis at the origin of eukaryotes pointed to the existence of predatory bacteria that penetrate their bacterial prey.[77] But Cavalier-Smith was confident that "no bacteria, not even predatory ones, can take up or harbor other living cells in their cytoplasm, and to suppose that any ever did is to stray into the realms of science fiction."[78] Actually, at least one case was then known of bacteria harboring bacteria within them.[79] And even if such cases were exceptional, one could argue that the symbiogenesis of the nucleus would have been a rare event, in which case the rarity of the phenomenon would strengthen the hypothesis, not weaken it. The hypothesis that the eukaryotic nucleus was derived from a symbiosis of archaea and bacteria continued to be tested by searching for homologies between proteins shared among the three domains.[80]

E = A + B + C

The hypothesis that the proto-eukaryote represented a third line of descent that subsequently engulfed archaea and bacteria was also a testable possibility. In 1984, Hyman Hartman proposed that the nucleated cell emerged when an RNA-based proto-eukaryote, which possessed a cytoskeleton, engulfed an archaeal lineage that had invented DNA.[81] That hypothesis could explain why transcription occurs in the nucleus and translation occurs in the cytoplasm. The separation between transcription and translation of eukaryotic cells today would then be the result of the communication setup between the DNA-based engulfed cell and the RNA-based host cell. Messenger RNA made in the endosymbiont would be transported and translated in the "cytoplasm" of the host.[82]

Mitchell Sogin made a similar proposal in 1991.[83] His molecular phylogenetic data at the Center for Molecular Evolution in Woods Hole, Cape Cod, supported the view that the eukaryotic lineage was much more ancient than had been commonly assumed. 16S rRNA sequence comparisons indicated that protists were even more diverse than previously imagined. The ciliated protists alone were as genetically diverse as plants and animals.[84] Given the diversity of the protists, he suspected that that "the eukaryotic lineage could be at least as ancient as its eubacterial and archaebacterial counterparts."[85] As he envisaged it, the chimeric organism resulting from the engulfment would have derived most of its protein-coding DNA from an archaebacterium, and its translation apparatus and cytoskeleton from the protoeukaryote.[86]

In 2002, Hartman and Alexei Fedorov reported further molecular phylogenetic evidence for an ancient third cell type that they called a "chronocyte." Instead of searching for eukaryotic proteins that were shared with the other domains, they searched for proteins that were unique to eukaryotes. They found 347 "eukaryotic signature proteins," among them several associated with the cytoskeleton, which they reasoned implied that there was a third cellular domain. They proposed that the nucleus would have been formed when a chronocyte engulfed several archaea as well as bacteria. "This formation of the nucleus," they said, "would restore the three cellular domains, as the Chronocyte was not a cell that belonged to the Archaea or to the Bacteria."[87]

The concept of an ancient eukaryotic lineage of anaerobic amitochondriate microbes found support from other perspectives.[88] In terms of their evolutionary affinities, eukaryotes would thus have three kinds of genes: archaeal-like genes for the translation machinery, bacterial genes for many metabolic functions, and chronocyte genes for proteins that give eukaryotes their distinctive cellular character (actin, tubulin, the main components of the microtubules of their cytoskeleton, proteins of the nuclear pore complex, and many other structural complexities) which are not clearly either bacterial or archaeal.[89]

Concluding Remarks

MOLECULAR PHYLOGENETICS HAS BROUGHT about a profound revolution in biology—in making taxonomy an experimental, quantitative, evolutionary science; in distinguishing the primary lineages of life; and in bringing microbes to the fore. Concepts of domains and microbial kingdoms are now at biology's center. And investigations of microbial diversity and relationships have led to a fundamentally new understanding of the evolutionary process.

From the nineteenth century to the present, there have been repeated claims that bacteria defy an evolutionary understanding and classification comparable to that for animals and plants. But the reasons for that have changed. We can distinguish four phases in the study of bacterial diversity, each characterized by a range of possibilities defined by the current theories and beliefs about them, the techniques for studying them, and the way in which they are discussed.

The first phase began with the rise of germ theory. The evolutionary history of microbes was of little interest to pathologists of Pasteur's day. A genealogical classification was considered neither possible, nor necessarily beneficial. When a morphological classification first confronted concepts of bacterial pleomorphism, it was uncertain whether difficulties reflected the limitations of technique, or the reality of nature itself. Subsequent to the development of pure culture techniques, discussions of a natural bacterial classification centered on whether morphological and/or physiological traits should be the primary consideration used in a phylogenetically oriented classification. An unweighted classification based on as many kinds of traits as possible was considered to be both most practical and to most closely approximate a natural hierarchical order. The concept of species was difficult to apply because bacteria reproduced by fission and there was no discernable sexual recombination. Bacteria also seemed to evolve by a "Lamarckian" process of environmentally induced adaptive changes.

Bacteria were generally considered to be plants, and speculative phylogenetic trees were rooted in assumptions about the nature of the primary organisms, the starting point from which an ever increasing complexity would have evolved. Most microbiologists considered the first organisms to be autotrophs. There was indirect evidence that life existed some 1.5 billion years ago. Debates over the definition of bacteria centered over their anatomy: whether or not they lacked a membrane-bound nucleus. A kingdom of Monera was proposed, but whether the bacteria represented a group derived from common ancestry or whether they were only polyphyletically related was as uncertain as was their structure and that of the viruses and the blue-green algae.

In a second phase, viruses were distinguished as entities unto themselves on the basis of electron microscopy, whereas blue-green algae and bacteria were confirmed to possess basically the same cell anatomy. Bacteria were said to have a prokaryotic organization because they lacked a membrane-bound nucleus, organelles, and sexuality comparable to other microbes. The prokaryote–eukaryote dichotomy was confirmed as a central organizing concept of modern biology.

Though a phylogenetic understanding of the bacteria was reaffirmed to be unachievable, bacteria were generally presumed to have originated from one common ancestor, as inferred from their common prokaryotic structure and supported by the concept of the unity of life. A conceptual consensus was formed and elaborated that life's universal ancestors were not autotrophs as formerly supposed, but rather heterotrophs that fed on, and separated from, a preexisting primordial soup that had evolved over billions of years on a sterile earth. A five-kingdom scheme of monera, protists, fungi, plants, and animals was formulated on the basis that bacteria represented a taxon on a par with plants and animals. Others proposed Prokaryota and Eukaryota as superkingdoms on the basis that the gross difference in their cellular organizations represented the greatest discontinuity in the evolution of life on Earth.

Escherichia coli was the model prokaryotic organism of molecular biology. Lamarckian mechanisms of bacterial evolution were rejected, but some adaptive characteristics were shown to be acquired through lateral gene transfer. A concept of infective heredity was constructed. Lateral gene transfer was exploited in molecular genetics, and biotechnology, and it was ultimately recognized to be the basis of widespread antibiotic resistance among bacteria. Still, leading microbiologists insisted that bacterial hybridization through conjugation, transformations, and viral transductions were rare in nature. The idea that mitochondria and chloroplasts were of symbiotic origin came to the fore when these organelles were discovered to have their own DNA. But competing theories of organellar origins were considered to be irresolvable by objective science. The evolution of cells was considered to be outside the boundaries of empirical science.

In a third phase, molecular sequences of amino acids, ribosomal RNA, and DNA were used to erect microbial phylogenetic trees, because classifications of bacteria based on morphology and/or physiology were often contradicted.

Protests against the new phylogenetics centered on whether its classifications were truly phylogenetic or resulted from faulty analyses or conclusions. While microbiologists who supported numerical taxonomy warned of the hazards of lateral gene transfer, those who advocated molecular phylogenetic methods adopted the concept of a genetic core of highly integrated conserved genes resistant to lateral gene transfer.

Ribosomal RNA emerged as the universal evolutionary chronometer. The conjecture that mitochondria and chloroplasts had originated from bacterial symbionts was rigorously confirmed on that basis. A radically new bacterial systematics unfolded, while a startling new outline of a universal tree of life was constructed with three primary lineages, archaebacteria, eubacteria, and eukaryotes, emerging from a simpler common hypothetical ancestral state composed of ancient precellular entities in the throes of evolving the modern molecular genetic apparatus. That fundamental trifurcation of life was posited in direct conflict with the conception of the prokaryote as a monophylogenetic grouping that preceded and gave rise to eukaryotes. There were three, not two, primary lineages of life, as previously assumed. As the microbial fossil record was extended back some 3.5 billion years for putative cyanobacteria, the idea reemerged that the first organisms were (thermophilic) autotrophs. Discussions among molecular phylogeneticists centered over the relationships among the three primary lineages.

Rootings of the universal tree of life were congruent with other molecular and phylogenetic evidence that distinguished bacteria, on one hand, from eukaryotes and archaebacteria, on the other. None of the three primary lineages was derived from the other. A clash with classical morphologists ensued when a formal taxonomic proposal of three domains was put forward: Archaea, Bacteria, and Eucarya. Classical biologists insisted that classification on the basis of gross morphological differences (gradistic classification) took precedence over genealogy in the case of the primary groupings of life.

A fourth phase is defined by the emergence of genomics and startling new phylogenetic evidence of extensive lateral gene transfer among bacteria. Skepticism was renewed regarding the ability to reconstruct a universal tree of life. The species concept was now considered of doubtful application to bacteria—not because of a lack of genetic recombination, but because there seemed to be so little barrier to it. The concept of the organismal core was retained, and so too were the three primary lineages. Debates over the prokaryote concept continued over its phylogenetic connotation, and its organismal meaning as a group negatively defined.

From the perspective of the three-domain concept, the prokaryote was a phylogenetic and organismal illusion, a specter from a bygone era when evolutionary biology was abandoned by both microbiology and molecular biology alike. The very criteria by which the prokaryote was defined and distinguished concealed and prevented understanding of bacterial evolution. Still, those who upheld the prokaryote concept insisted that the molecular and biochemical differences between the Archaea and the Bacteria could not compare to the organizational differences at the supramolecular level that distinguished the eukaryotic form.

While the relations among the three primary lineages remained uncertain and the common ancestor of life remained unknown, chimeric conceptions of present-day eukaryotes arose as genomic studies revealed nuclear genomes that comprised genes from all three domains: informational genes like those of the Archaea, metabolic genes like those of the Bacteria, and cytoskeletal genes of strictly eukaryotic descent. Symbiosis in the origin of the eukaryotic cell was considered to be only the tip of the iceberg of the inheritance of acquired genes and genomes as a broader conceptual framework of evolution emerged that entailed profound modifications in the old.

Thus, molecular phylogeneticists have transcended Darwinian explanation in revealing evolution's complementary processes of genetic divergence and integration. We also see these complementary forces in the process of scientific change: the emergence of microbial phylogenetics is an interdisciplinary fusion of microbiology and molecular biology, natural history and ecology. It is a stunning illustration of the saltational effects resulting from the lateral transfer of concepts, techniques, and individuals among fields. Indeed, the new evolutionary biology was far from being a simple refinement of or addition to classical evolutionary biology. The new microbial phylogenetics was not a matter of completing evolutionary theory with new methods through which to investigate the emergence of organisms and a universal tree of life.

The new phylogenetics marks a new beginning with concepts that fundamentally contradict the principal tenets and assumptions of Darwinian theory: (1) The concept of species as discrete objective entities with more or less isolated gene pools is confronted with evidence of extensive lateral gene transfer between taxa. (2) The notion that gene mutation and recombination within species are the principal fuels of evolutionary change is denied. (3) That evolutionary change is gradual—that evolution does not take leaps—is refuted by evidence of saltational changes resulting from the inheritance of acquired genes and genomes. Finally, (4) the fundamental conception of evolution itself as descent with modification, an ordering of group within group based on characters inherited from a common parent, is contradicted by evidence that most genes are transferred laterally.

The tree of life today is not the same kind of tree that Darwin and his followers had imaged. Still, to go further, to adopt pan-lateralism and wholesale chimericism and completely abandon a hierarchical organization of microbial life is to reject the distinction between essential characteristics of organisms that transcended adaptive function. The search for such essential characteristics had been led by Lamarck and his followers of the early nineteenth century and upheld by Darwin and his followers in the second half of that century. But a tree embracing all of life was possible only with the emergence of molecular phylogenetics. Those genes by which one could trace the universal tree of life were, at the "core," highly integrated and far removed from the interactions between the organism and its environment. They functioned in the replication of DNA, in its transcription into RNA, and in the translation of that RNA into the amino acid sequences of proteins.

Therein lay another fundamental difference between microbial phylogenetics and the evolutionary biology that preceded it. For the champions of "the modern synthesis" of the 1940s, species were real; higher taxa were matters of convenience. The inverse was true for many microbial phylogeneticists. Kingdoms and domains reached deep within the concept of organism. Nothing could be more real than the body plans that defined the primary groupings.

NOTES

CHAPTER 1

1. Several myths about Leeuwenhoek were generated: that he was a "physician," "a surgeon," "a professor," and that he was the inventor of the microscope. See Elmer Bendiner, "The man who did not invent the microscope," *Hospital Practice* 139 (1984): 144–160, 165–174. No two accounts agreed until Clifford Dobell's important biography, *Antony van Leeuwenhoek and his "little animals"* (New York: Russell and Russell, 1958), 2. See also Brian J. Ford, *The Leeuwenhoek legacy* (London: Biopress, 1991).

2. Brian J. Ford, *Single lens: The story of the simple microscope* (London: Harper and Row, 1985). See also Ford, "From dilettante to diligent experimenter, a reappraisal of Leeuwenhoek as microscopist and investigator," *Biology History* 5 (1992): 35–39; Ford, "First steps in experimental microscopy, Leeuwenhoek as practical scientist," *Microscope* 43 (1995): 47–57.

3. Robert D. Huerta, *Giants of Delft: Johannes Vermeer and the natural philosophers: The parallel search for knowledge during the age of discovery* (Lewisburg: Bucknell University Press, 2003).

4. Dobbell, *Antony Leeuwenhoek*, 317.

5. Robert Hooke, *Micrographia: Or, Some physiological descriptions of minute bodies made by magnifying glasses* (London: J. Martyn and J. Allestry, 1665).

6. Dobell, *Antony Leeuwenhoek*, 313–338.

7. Ibid., 330.

8. Ibid., 132–133.

9. Ibid., 179.

10. Ibid., 133.

11. Ibid., 372–376.

12. Ibid., 373.

13. Ibid., 52–53.

14. Dobell, *Antony Leeuwenhoek*, 375.

15. On the origins of the herbarium cabinet, see Staffan Müller-Wille, "Linnaeus' herbarium cabinet: A piece of furniture and its function," *Endeavour* 30 (2006): 60–64.

16. Quoted in Antoine Lavoisier, *Traite élémentaire de chimie* (Paris: Cuchet, 1789), vi. See also Antoine Lavoisier, *Elements of chemistry* (1789), translation by Robert Kerr (Edinburgh, 1790).

17. Lavoisier, *Elements of chemistry*, vi.

18. Ibid., vii.

19. Ibid., vi.

20. See J.L. Larson, *Reason and experience: The representation of natural order in the work of Carl Linnaeus* (Berkeley: University of California Press, 1971); T. Frängsmyr, ed., *Linnaeus—the man and his work* (Berkeley: University of California Press, 1983); L. Koerner, *Linnaeus: Nature and nation* (Cambridge: Harvard University Press, 1999).

21. Quoted in Staffan Müller-Wille, "Gardens of paradise," *Endeavour* 25 (2001): 48–54, at 53.

22. See Michel Foucault, *The order of things: An archaeology of the human sciences* (London: Tavistock, 1974); Francois Jacob, *The logic of life: A history of heredity* (New York: Vintage books, 1976), 19–32.

23. Alec Panchen, *Classification, evolution, and the nature of biology* (Cambridge: Cambridge University Press, 1992), 16.

24. Quoted in E. Mayr, *The growth of biological thought* (Cambridge: Harvard University Press, 1982), 177.

25. Quoted in. Thomas B. Wilson and John Cassin, "On a Third Kingdom of Organised Beings," *Proceedings of the Academy of Natural Sciences, Philadelphia* 15 (1863): 113–121, at 113.

26. Andreas Cesalpino, *De plantis libri*, XVI, I, 14 (Florence, 1583), 28, quoted in Jacob, *The logic of life*, 23.

27. Quoted in J.-B. Marie Georges Bory de Saint-Vincent, "Dictionnaire classique d'histoire naturelle; rédige par une société de naturalistes, avec une nouvelle distribution des corps naturels en cinq règnes," *Revue Encyclopédique*, 17th année, 2nd series (1825): 46–61, at 49.

28. Carolus Linnaeus, *Systema naturae*, 10th ed., 2 vols. (1758–1759), Vol. 1: 880, 821.

29. Carolus Linnaeus, *Systema naturae*, 12th ed.; Dobell, *Antony Leeuwenhoek*, 377.

30. William Bulloch, *The history of bacteriology* (London: Oxford University Press, 1938), 36; Dobell, *Antony Leeuwenhoek*, 378.

31. Bulloch, *The history of bacteriology*, 37.

32. Giulio Barsanti, "Linné et Buffon: Deux visions différentes de la nature et de l'histoire naturelle," in *Les fondements de la botanique: Linné et la classification des plantes* (Paris: Viubert, 2005), 101–129.

33. Georges-Louis Leclerc Buffon, *Premier discours, histoire naturelle generale et particuliere avec la description du cabinet du roy*, 44 vols. (Paris: Imprimerie Royale 1749–1804), Vol. 1: 12.

34. Georges-Louis Leclerc Buffon, *Histoire naturelle, De la manière de traiter et d'étudier l'histoire naturelle*, 44 vols. (Paris: Imprimerie Royale, 1749–1804), Vol. 1: 54.

35. Georges-Louis Leclerc Buffon, *Histoire naturelle, Générale et particuliére, avec la description du cabinet du roy*, 44 vols. (Paris: Imprimerie Royale, 1749–1804), Vol. 4: 8.

36. Ibid., 252.

37. Jacob, *The logic of life*, 136–143; Mayr, *The growth of biological thought*, 260–263; John H. Eddy, Jr., "Buffon's *Histoire Naturelle*: A critique of recent interpretations," *Isis* 85 (1994): 644–661.

38. Georges-Louis Leclerc Buffon, *Oeuvres complètes*, Vol. 3: *Vue de la nature*, 418, quoted in Jacob, *The logic of life*, 138.

39. L.J.-M. Daubenton, *Séances des ecoles normales*, 9 vols. (Paris: L. Reynier, 1795), Vol. 5: 277.

40. J.B. Lamarck, *Zoological philosophy* (Chicago: University of Chicago Press, 1984), 44.

41. Richard Burkhardt, *The spirit of the system: Lamarck and evolutionary biology* (Cambridge: Harvard University Press, 1977); Madeleine Barthélemy-Madaule, *Lamarck: Ou le mythe du précursor* (Paris: Seuil, 1979); Pietro Corsi, *The age of Lamarck: Evolutionary theories in France, 1790–1830* (Berkeley: University of California Press, 1988).

42. See Conway Zirkle, "The early history of the idea of the inheritance of acquired characters and of pangenesis," *Transactions of the American Philosophical Society* 35 (1946): 91–151.

43. Quoted in Robert Young, *Darwin's metaphor: Nature's place in Victorian culture* (Cambridge: Cambridge University Press, 1995), 116.

44. C.R. Darwin, *The variation of animals and plants under domestication* (London: John Murray, 1868).

45. J.B. Lamarck, *Zoological philosophy* (Chicago: University of Chicago Press, 1984), 56.

46. Michel Foucault, *The order of things*; François Jacob, *The logic of living systems: A history of heredity*, translation by Betty Spillmann (London: Allen Lane, 1974).

47. A. Lovejoy, *The great chain of being: A study of the history of an idea* (Cambridge: Harvard University Press, 1936).

48. See M.P. Winsor, *Starfish, jellyfish, and the order of life: Issues in nineteenth century science* (New Haven: Yale University Press, 1976); Panchen, *Classification, evolution, and the nature of biology*.

49. Lamarck, *Zoological philosophy*, 58.

50. Ibid., 58.

51. Ibid., 57.

52. Ibid.

53. Ibid., 58.

54. Ibid., 35.

55. Ibid., 236.

56. Ibid., 66.

57. Lamarck, *Zoological philosophy*, 62.

58. Ibid., 64.

59. Ibid., 131.

60. Ibid., 135.

61. Ibid.

62. Ibid., 238.

63. Ibid., 236.

64. William Coleman, *Georges Cuvier, zoologist* (Cambridge: Harvard University Press, 1964); Dorinda Outram, *Georges Cuvier: Vocation, science, and authority in post-revolutionary France* (Dover, N.H.: Manchester University Press, 1984); Martin Rudwick, *Georges Cuvier* (Chicago: University of Chicago Press, 1997).

65. Toby Appel, *The Cuvier-Geoffroy debate: French biology in the decades before Darwin* (New York: Oxford University Press, 1987).

66. Lamarck, *Zoological philosophy*, 60.

67. Georges Cuvier, "Sur un nouveau rapprochement à établir entre les classes qui composent le règne animal," *Annales Muséum d'Histoire Naturelle* 19 (1812): 73–84, at 83.

68. J.-B. Marie Georges Bory de Saint-Vincent, "Psychodiaire. Règne," *Encyclopédie méthodique* (Paris: Agasse, 1824), 657–663; Bory de Saint-Vincent, "Dictionnaire classique d'histoire naturelle," 51. He was well known as the editor of *Dictionnaire classique d'histoire naturelle* (1822–1831), to which he made many entries on genera of microscopic algae. M.A. Ragan and R.R. Gutell, "Are red algae plants?" *Botanical Journal of the Linnean Society* 118 (1995): 81–105.

69. Ibid., 659.

70. Bory de Saint-Vincent, "Dictionnaire classique d'histoire naturelle," 48.

71. Ibid., 51.

72. Ibid., 60.

73. Bory de Saint-Vincent, "Psychodiaire. Règne," 660.

74. Ibid., 662.

75. Ibid., 659.

76. Ibid., 658.

77. Ibid., 659.

CHAPTER 2

1. Ferdinand Cohn, *Bacteria: The smallest of living organisms* (1881), translation by Charles S. Dolley (Baltimore: Johns Hopkins Press, 1939), 14.

2. C.G. Ehrenberg, *Sylvae mycologicae berolinensis* (Berlin: Bruschke, 1818).

3. C.G. Ehrenberg, *Naturgeschichtliche Reisen durch Nord-Afrika und West-Asien in den Jahren 1820 bis 1825 von Dr. W. F. Hemprich und Dr. C. G. Ehrenberg, Historischer Theil* (Berlin: Mittler, 1828).

4. Richard Owen, *Paleontology, Or a systematic summary of extinct animals and their geological relations* (1860), reprint edition (New York: Arno Press, 1980), 16–17.

5. Owen, *Paleontology*, 17.

6. See W.A.S. Sarjeant, "Hundredth year memorium: Christian Gottfried Ehrenberg 1795–1877," *Palynology* 2 (1978): 209–221.

7. A. Hughes, *A history of cytology* (London: Abelard-Schuman, 1959).

8. Lynn Gamwell, "Beyond the visible—microscopy, nature, and art," *Science* 299 (2003): 49–50.

9. On *Naturphilosophie*, see, e.g., Robert Richards, *The meaning of evolution. The morphological construction and ideological reconstruction of Darwin's theory* (Chicago: University of Chicago Press, 1992); Timothy Lenoir, *The strategy of life: Teleology and mechanics in nineteenth century German biology* (Dordrecht: Reidel, 1982).

10. Peter Bowler, *Fossils and progress: Paleontology and the idea of progressive evolution in the nineteenth century* (New York: Science History Publications, 1976), 47–53.

11. See Adrian Desmond, *Archetypes and ancestors: Paleontology in Victorian London, 1850–1875* (London: Blond and Briggs, 1982).

12. Erik Nordensköild, *The history of biology* (New York: Tudor, 1946), 426–429; William Bulloch, *The history of bacteriology* (London: Oxford University Press, 1938), 174.

13. F.J.R. Taylor, "The collapse of the two-kingdom system, the rise of protistology and the founding of the International Society for Evolutionary Protistology (ISEP)," International Journal of Systematic and Evolutionary Microbiology 53 (2003): 1707–1714.

14. Nordensköild, *History of biology*, 428.

15. Ibid., 428–429.

16. See Bulloch, *History of bacteriology*, 175.

17. Matthias Schleiden, "Beitrage zur Phyto- genesis," *Archiv für Anatomie, Physiologie, und wissenschaftliche Medizin* (1838): 137–176, at 140.

18. R. Virchow, *Die Cellularpathologie in ihrer Begrundung auf physiologische und pathologische Gewebelehre* (1858), published in English as *Cellular pathology* (New York: Dover, 1971), 12.

19. E.W. Brücke, "Die Elementarorganismen," *Sitzungsberichte der Kaierlichen Akademie Wien* 44 (1861): 381–406.

20. Herbert Spencer, "Professor Weismann's theories," *Contemporary Review* 63 (1893): 743–760.

21. Edmund B. Wilson, *The cell in development and inheritance* (New York: MacMillan, 1896).

22. Owen, *Paleontology*.

23. G.A. Goldfuss, *Uber die Entwicklungsstufen des Thieres* (Nuremberg: Leonard Schrag, 1817).

24. C.T.E. von Siebold and F.H. Stannius, *Lehrbuch der vergleichenden Anatomie der Wirbellosen Thiere* (Berlin: Von Veit, 1845).

25. See Desmond, *Archetypes and ancestors*; Evelyn Richards, "A question of property rights: Richard Owen's evolutionism reassessed," *British Journal for the History of Science* 20 (1987): 129–171; Michael Ruse, "Were Owen and Darwin Naturphilosophen?" *Annals of Science* 50 (1993): 383–388; Lynn Nyhart, *Biology takes form* (Chicago: University of Chicago Press, 1995); Mario Di Gregorio, "A wolf in sheep's clothing: Carl Gegenbaur, Ernst Haeckel, the vertebral theory of the skull, and the survival of Richard Owen," *Journal of*

the *History of Biology* 28 (1995): 247–280; Ron Amundson, "Typology reconsidered: Two doctrines on the history of evolutionary biology," *Biology and Philosophy* 13 (1998): 153–177.

26. Desmond, *Archetypes and ancestors*, 71.

27. Owen, *Paleontology*, 17.

28. John Lindley, *An introduction to the natural system of botany, Or, A systematic view of the organization, natural affinities, and geographical distribution of the whole vegetable kingdom* (New York: G. & C. & H. Carvill, 1831), 15.

29. Jan van der Hoeven, *Handbook of zoology*, translation by W. Clark (Cambridge: Longman, Brown, Green, Longmans, and Roberts, 1856–1858).

30. Richard Owen, *Lectures on the comparative anatomy and physiology of the invertebrate animals*, 2nd ed. (London: Longman, 1855), 2.

31. Ibid., 4.

32. Ibid.

33. Ibid., 8.

34. *Owen, Paleontology,* 4; see also L.J. Rothschild, "Protozoa, Protista, Protoctista: What's in a name?" *Journal of the History of Biology* 22 (1989): 277–305; M. Regan, "A third kingdom of eukaryotic life: History of an idea," *Archive für Protistenkund* 148 (1997): 225–243.

35. Owen, *Paleontology*, 5–8.

36. John Hogg, "On distinctions of a plant and an animal, and on a fourth kingdom of nature," *Edinburgh New Philosophical Journal* 12 (1860): 216–225.

37. Owen changed the name of his kingdom *Protozoa* to *Acrita* in the second edition of *Paleontology* (Edinburgh: Adam and Charles Block, 1861), 6.

38. Hogg, "On distinctions of a plant and an animal," 222.

39. Ibid.

40. Ibid., 219.

41. Ibid., 222.

42. Ibid., 218–219.

43. Ibid., 221.

44. Ibid., 223.

45. Ibid., 218–219.

46. Ibid., 220.

47. J.M. Scamardella, "Not plant or animals: A brief history of the origin of the kingdoms Protozoa, Protoctista and Protista," *International Microbiology* 2 (1999): 207–216.

48. See Desmond, *Archetypes and ancestors*; Nicolaas Rupke, *Richard Owen: Victorian naturalist* (New Haven: Yale University Press, 1994); Peter Bowler, *Life's splendid drama* (Chicago: University of Chicago Press, 1996).

49. Owen, *Paleontology*, 413–414.

50. Ibid., 413.

51. Scamardella, "Not plant or animals," 208.

52. Ibid.

53. Hogg, "On distinctions of a plant and an animal," 224.

54. Ibid.

55. Ibid.

56. Thomas Wilson and John Cassin, "On a third kingdom of organized beings," *Proceedings of the Academy of Natural Sciences of Philadelphia* 15 (1863): 113–121, at 118.

57. Ibid., 117.

58. Ibid., 113.

59. Ibid.

60. Ibid., 115.

61. Ibid., 116.

62. Ibid., 116–117.

CHAPTER 3

1. Ernst Haeckel, *The history of creation*, 6th ed., 2 vols. (New York: Appleton, 1892), Vol. 1: xiii.

2. See Adrian Desmond, *Archetypes and ancestors: Paleontology in Victorian London, 1850–1875* (London: Blond and Briggs, 1982), 155.

3. See Haeckel, *History of creation*, Vol. 1: xiii.

4. Erik Nordenskiöld, *The history of biology* (New York: Tudor, 1929), 505. 506.

5. Ernst Haeckel, *Die Radioloarian. (Rhizopoda Radiaria.) Eine Monographie*, 2 vols. (Berlin: Georg Reimer, 1862).

6. See Desmond, *Archetypes and ancestors*.

7. Ernst Haeckel, *The riddle of the universe* (1899), translation by Joseph McCabe (New York: Prometheus Books, 1992), 78–79.

8. Charles Darwin, *On the origin of species*, facsimile of 1859 edition (Cambridge: Harvard University Press, 1969), 411.

9. Ibid., 420.

10. Ibid., 129.

11. On the relations between the concept of division of labor in the social sciences and biology, see Camille Limoges, "Milne-Edwards, Darwin, Durkheim and division of labor: A case study in reciprocal conceptual exchanges between the social and natural sciences," in I.B. Cohen, ed., *The relations between the natural sciences and the social sciences* (Princeton: Princeton University Press, 1994), 317–343; Limoges and Claude Ménard, "Organization and division of labor: Biological metaphors at work in Alfred Marshall's principles of economics," in Philip Mirovski, ed., *Natural images in economics* (Cambridge: Cambridge University Press, 1994), 336–359.

12. See David Kohn, "Darwin's principle of divergence as internal dialogue," in Kohn, ed., *The Darwinian heritage* (Princeton: Princeton University Press, 1988), 245–263.

13. Darwin, *Origin of species*, 471.

14. Ibid., 479.

15. Charles Darwin to Thomas Henry Huxley, September 26, 1857, *Darwin correspondence project* [online database], available at www.darwinproject.ac.uk/.

16. Darwin, *Origin of species*, 414.

17. Charles Darwin to Joseph Hooker, January 11, 1844, *Darwin correspondence project* [online database], available at www.darwinproject.ac.uk/.

18. Darwin, *Origin of species*, 414.

19. Ibid., 479.

20. See Stephen Jay Gould, *Ontogeny and phylogeny* (Cambridge: Harvard University Press, 1977).

21. See Peter Bowler, *Evolution: The History of an Idea* (Berkeley: University of California Press, 1984), 247; Robert Richards, *The meaning of evolution: The morphological construction and ideological reconstruction of Darwin's theory* (Chicago: University of Chicago Press, 1992), 40; Timothy Lenoir, *The strategy of life: Teleology and mechanics in nineteenth-century German biology* (Dordrecht: D. Reidel, 1978).

22. Richards, *The meaning of evolution*.

23. Darwin, *Origin of species*, 338.

24. Ibid., 440–442.

25. Haeckel, *The history of creation*, Vol. 1: 10–11.

26. See Brian K. Hall, "Balfour, Garstang and de Beer: The first century of evolutionary embryology," *American Zoologist* 40 (2000): 718–728; Hall, *Evolutionary developmental biology*, 2nd ed. (Dordrecht: Kluwer, 1998); Gould, *Ontogeny and phylogeny*; Gavin de Beer, *Embryology and evolution* (Oxford: Oxford University Press, 1930); de Beer, *Embryology and ancestors* (Oxford: Oxford University Press, 1940).

27. Haeckel was known for his sometimes questionable methods of blending fact with hypothesis in the depictions of genealogies based on comparative embryology. In order to support his recapitulation theory, he published various drawings to illustrate the striking similarities among different types of embryos. When he was later attacked by anti-evolutionists for his "frauds and forgeries," Haeckel admitted that about 6–8 percent of his numerous drawings were falsified. As he saw them, they were merely hypothetical genealogies to be replaced by more complete and more accurate phylogenies. See Jan Sapp, *Where the truth lies: Franz Moewus and the origins of molecular biology* (New York: Cambridge University Press, 1990).

28. Francis Darwin, ed., *The life and letters of Charles Darwin*, 3 vols. (London: Murray, 1887; New York: Johnson Reprint Corp., 1969), Vol. 3: 105.

29. Oscar Schmidt, *The doctrine of descent and Darwinism* (New York: Appleton, 1874), 250.

30. Desmond, *Archetypes and ancestors*, 157.

31. T.H. Huxley, "*The natural history of creation*," *Academy* 1 (1869): 13–14, 40–43, at 41.

32. Darwin, *Origin of species*, 488.

33. Ibid., 484.

34. Ibid.

35. Ibid., 490.

36. Haeckel, *Die Radiolarian*, 231–232, quoted in John Farley, *The spontaneous generation controversy from Descartes to Oparin* (Baltimore: Johns Hopkins University Press, 1977), 75.

37. C. Darwin to J.D. Hooker, 29 March 1863, in Francis Darwin, ed., *The life and letters of Charles Darwin*, 3 vols. (London: Murray, 1887; New York: Johnson Reprint Corp., 1969), Vol. 3: 17.

38. Charles Darwin, *The origin of species*, 6th ed. (London: John Murray, 1872), 429.

39. Charles Darwin to Mr. J. Fordyce, May 7, 1879, in Francis Darwin, ed., *The life and letters of Charles Darwin*, 2 vols. (New York: D. Appleton, 1959), Vol. 1: 304.

40. C.R. Darwin to J.D. Hooker, February 1, 1871, in Francis Darwin, ed., *The life and letters of Charles Darwin*, 2 vols. (New York: D. Appleton, 1959), Vol. 2: 202–203.

41. Gould, *Ontogeny and phylogeny*; M. Ruse, *The Darwinian revolution* (Chicago: University of Chicago Press, 1979); Desmond, *Archetypes and ancestors*; Peter Bowler, *Life's splendid drama* (Chicago: University of Chicago Press, 1996); Lynn Nyhart, *Biology takes form* (Chicago: University of Chicago Press, 1995); Daniel Gasman, *The scientific origins of national socialism* (New York: Peter Lang, 1998).

42. See Farley, *The spontaneous generation controversy*; James Strick, *Sparks of life* (Cambridge: Harvard University Press, 2000).

43. Quoted in Farley, *The spontaneous generation controversy*, 108.

44. See Gerald L. Geison, "The protoplasmic theory of life and the vital-mechanist debate," *Isis* 60 (1969): 273–292.

45. Ernst Haeckel, *The wonders of life: A popular study of biological philosophy* (New York: Harper and Brothers, 1904), 352.

46. Ibid.

47. T.H. Huxley, "On the physical basis of life" (1868), *Lay sermons, addresses and reviews*, in *Collected Essays I*, 11 vols. (London: Macmillan, 1893), Vol. 1: 130–165, 142.

48. Haeckel, *Wonders of life*, 353.

49. He distinguished his monist philosophy from theoretical materialism, on the one hand, which "dissolves the world into a heap of dead atoms," and from theoretical spiritualism, on the other, which "considers the world to be a specially arranged group of 'energies' or immaterial natural forces." Haeckel, *Riddle of the universe*, 20.

50. E. Haeckel, *Generelle Morphologie der Organismen*, 2 vols. (Berlin: Reimer, 1866), Vol. 2: 451, translated in Farley, *The spontaneous generation controversy*, 76.

51. Haeckel, *Generelle Morphologie*, Vol. 1: 164, quoted in Farley, *The spontaneous generation controversy*, 76.

52. Haeckel, *History of creation*, Vol. 1: 180.

53. Haeckel *Wonders of life*, 35.

54. Haeckel, *History of creation*, Vol. 2: 64.

55. Ibid., 49.

56. Ibid.

57. Ibid., 58, 75.

58. Ibid., 67.

59. Ibid., 72.

60. Ibid.

61. Ibid., 73.

62. Ibid., 46.

63. Ibid.

64. T.H. Huxley, "On the classification of the animal kingdom," *Nature* 11 (1874): 101–102, at 101.

65. William Saville-Kent, *A manual of the Infusoria: Including a description of all known flagellate, ciliate, and tentaculiferous protozoa, British and foreign, and an account of the organization and affinities of the sponges* (London: David Bogue, 1880–1882).

66. See discussion in Herbert Copeland, "Progress report on basic classification," *American Naturalist* 81 (1947): 340–361, at 341.

67. T.H. Huxley, "On some organisms living at a great depths in the North Atlantic Ocean," *Quarterly Journal of Microscopic Science* 8 (1868), 203–210, at 210.

68. Philip F. Rehbock, "Huxley, Haeckel, and the oceanographers: The case of *Bathybius haeckelii*," *Isis* 66 (1976): 504–553. Haeckel considered the discovery of *Bathybius* as evidence for spontaneous generation (*History of creation*, Vol. 1: 344).

69. Haeckel, *Wonders of life*, 206.

70. Haeckel, *History of creation*, Vol. 2: 73–74.

71. Haeckel, *Wonders of life*, 205.

72. Ibid.

73. Ibid., 32.

74. Ibid., 195–196.

75. Ibid., 208.

76. Ibid., 356.

77. Ibid., 357.

78. Bacteria were mentioned, e.g., on three pages of some 1,200 in the third edition of E.B. Wilson, *The cell in development and heredity* (New York: Macmillan, 1925).

79. Wilson, *The cell*, 642.

80. E.B. Wilson, *The cell in development and inheritance*, 2nd ed. (New York: Columbia University Press, 1900), 40.

81. E.A. Minchin, "The evolution of the cell," *Report of the Eighty-Fifth Meeting of the British Association for the Advancement of Science* (September 7–11, 1915): 437–464, at 442.

82. Ibid., 442. See also Wilson, *The cell*, 2nd ed., 32.

83. See William Coleman, "Cell, nucleus, and inheritance: An historical study," *Proceedings of the American Philosophical Society* 109 (1965): 124–158.

84. Wilson, *The cell*, 3rd ed., 210; Lester B. Sharp, *Fundamentals of cytology* (New York: McGraw Hill, 1943), 25, 162.

85. See René Dubos, *The bacterial cell* (Cambridge: Harvard University Press, 1945), 22–25; I.M. Lewis, "The cytology of bacteria," *Bacteriological Reviews* 5 (1941): 181–230.

86. Minchin, "The evolution of the cell," 447.

87. Wilson, *The cell*, 3rd ed., 25.

88. Wilson, *The cell*, 2nd ed., 434.

CHAPTER 4

1. P. De Kruif, *The microbe hunters* (London: Jonathan Cape, 1927); W.W. Ford, *Bacteriology* (New York: Hafner, 1939); W. Bulloch, *The history of bacteriology* (New York: Oxford University Press, 1938); W.D. Foster, *A history of*

medical bacteriology and immunology (London: Heinemann, 1970); K. Codell Carter, "The development of Pasteur's concept of disease causation and the emergence of specific causes in nineteenth-century medicine," *Bulletin of the History of Medicine* 65 (1991): 528–548. See also N.J. Tomes and J.H. Warner, "Introduction to special issue on rethinking the reception of the germ theory of disease: Comparative perspectives," *Journal of the History of Medicine and Allied Sciences* 52 (1997): 7–16.

2. J. Farley, *The spontaneous generation controversy from Descartes to Oparin* (Princeton: Princeton University Press, 1974); Farley and G.L. Geison, *The private science of Louis Pasteur* (Princeton: Princeton University Press, 1995); J. Strick, *Sparks of life: Darwinism and the Victorian debate over spontaneous generation* (Chicago: University of Chicago Press, 2000); H. Harris, *Things come to life: Spontaneous generation revisited* (Oxford: Oxford University Press, 2002).

3. See Natasha Jacobs, "From unity to unity: Protozoology, cell theory, and the new concept of life," *Journal of the History of Biology* 22 (1989): 215–242.

4. K. Vernon, "Pus, sewage, beer and milk: Microbiology in Britain, 1870–1940," *History of Science* 28 (1990): 290–325.

5. Lister referred to the "theory of germs" in a letter in 1874 to Pasteur, who subsequently used it instead of his expression "theory of organized ferments." See Carter, "Development of Pasteur's concept," 530.

6. The term "microbe" was introduced by Charles Sédillot in 1876.

7. See, e.g., G.S. Woodhead, *Bacteria and their products* (London: Walter Scott, 1891), 24.

8. A.T. Henrici, *The biology of bacteria: An introduction to general microbiology*, 2nd ed. (Boston: D.C. Heath, 1939), 82.

9. Carl von Nägeli, "Bericht über die Verhandlungen der Botanischen Section d. 33. Versammlung deutscher Naturforscher und Aertzte," *Botanische Zeitung* 15 (1857): 760–761.

10. See Bulloch, *History of bacteriology*, 200–202; Milton Wainwright, "Extreme pleomorphism and the bacterial life cycle: A forgotten controversy," *Perspectives in Biology and Medicine* 40 (1997): 407–414.

11. Carl von Nägeli, *Die niederen Pilze in ihren. Bezihungen zu den Infectionskrankheiten und der gesundheitspflege* (München: Oldenburg, 1877), 20, quoted in Christoph Gradmann, "Isolation, contamination, and pure culture: Monomorphism of pathogenic micro-organisms as research problem 1860–1880," *Perspectives on Science* 9 (2000): 147–171, at 151; Bulloch, *History of bacteriology*, 200.

12. Bulloch, *History of bacteriology*, 192.

13. Jean Théodoridès, "Casimir Davaine (1812–1882): A precursor of Pasteur," *Medical History* 10 (1966): 155–165.

14. Bulloch, *History of bacteriology*, 207–211; R. Koch, "Die Aetologie der Milzbrand-krankeheit begründet auf die Entwicklungsgeschichte des Bacillus anthracis," *Beiträge zur Biologie der Pflanzen*, hrsg v. F. Cohn, 2 (1876): 277–310.

15. Morris Leikind, "Introduction," in Ferdinand Cohn, *Bacteria: The smallest of living organisms* (1872), translation by Charles S. Dolley (1881) (Baltimore: Johns Hopkins Press, 1939).

16. Cohn, *Bacteria*, 5.

17. Ibid., 25.

18. Ibid., 35–36.

19. Ibid., 36.

20. See Svante Arrhenius, "Panspermy: The transmission of life from star to star," *Scientific American* 196 (1907): 196; Arrhenius, *Worlds in the making: The evolution of the universe* (New York: Harper and Row, 1908); Arrhenius, *The destinies of stars*, translation by J.E. Fries (New York: G.P. Putnam's Sons, 1918).

21. Arrhenius, *The destinies of stars*, 19.

22. Ibid., 37.

23. Cohn, *Bacteria*, 16–17.

24. Ibid., 18.

25. Ibid., 17–18.

26. Ferdinand Cohn, "Untersuchungen fiber die Entwicklungsgeschichte mikroskopische Algen and Pilze," *Nova Acta Academiae Caesareae Leopoldino—Carolinae Germanine naturae cursrosorum* 24 (1853): 103–256.

27. Ferdinand Cohn, "Untersuchungen über Bacterien II," *Beiträge zur Biologie der Pflanzen* 1/3 (1875): 141–207, at 201, quoted in E.G. Pringsheim, "Relationship between bacteria and myxophyceae," *Bacteriological Reviews* 13 (1949): 47–98, at 48.

28. Ibid.

29. Morris Leikind, "Introduction," 3.

30. Cohn, *Bacteria*, 15.

31. See Otto Rahn, *Microbes of merit* (Lancaster, PA: Jaques Cattell Press, 1945), 9.

32. Ferdinand Cohn, "Untersuchungen über Bacterien," *Beiträge zur Biologie der Pflanzen* 1/2 (1872): 126–224.

33. Cohn, "Untersuchungen über Bacterien," 135–136.

34. Cohn, *Bacteria*, 16.

35. See Bulloch, *History of bacteriology*, 200–202.

36. Ibid.

37. Ibid.

38. Cohn, "Untersuchungen über Bacterien"; Cohn, "Untersuchungen über Bacterien II."

39. Arthur Parker Hitchens and Morris C. Leikind, "The introduction of agar-agar into bacteriology," *Bacteriology* 37 (1939): 485–493; Wolfgang Hesse, Walther Hesse, and Fanny Angelina Hesse, "Early contributors to microbiology," translation by D.H.M. Gröschel, *ASM News* 58 (1992): 425–428.

40. Bulloch, *History of bacteriology*, 230.

41. See Jan Sapp, *Evolution by association: A history of symbiosis* (New York: Oxford University Press, 1994).

42. Selman A. Waksman, *Sergie N. Winogradsky: His life and work* (New Brunswick: Rutgers University Press, 1953); Lloyd Ackert, "The role of

microbes in agriculture: Sergei Vinogradskii's discovery and investigation of chemosynthesis, 1880–1910," *Journal of the History of Biology* 39 (2006): 373–406; Serge Winogradsky, "Recherches physiologiques sur les sulphobacteries," *Annales de l'Institut Pasteur* 3 (1889): 49–60; Winogradsky, *Microbiologie du Sol: Problems et Methods, Cinquante ans de Recherches* (Paris: Masson et Cie Editeurs, 1949).

43. See Sapp, *Evolution by association*.

44. Anton de Bary, "Die Erscheinung der Symbiose," *Vortrag auf der Versammlung der Naturforsher und Aertze zu Cassel* (Strassburg: von Karl J. Trubner, 1879): 1–30.

45. Waksman, *Winogradsky*, 14–15.

46. Ibid., 17.

47. Ibid., 20–21.

48. C.B. van Niel, "The 'Delft School' and the rise of general microbiology," *Bacteriological Reviews* 13 (1949): 161–174, at 163–164; A.F. Kamp, J.W.M. La Rivière, and W. Verhoeven, eds., *Albert Jan Kluyver: His life and work* (Amsterdam: North Holland, 1959); Susan Spath, *C.B. van Niel and the culture of microbiology, 1920–1965* (Ph.D. diss., University of California–Berkeley, 1999).

49. G. Van Iverson, L.E. den Dooren de Jong, and A.J. Kluyver, *Martinus Beijerinck: His life and work* (The Hague: Martinus Nijhoff, 1940).

50. M.J. Beijerinck, "Concerning a *contagium vivum fluidum* as cause of the spot disease of tobacco leaves," *Verhandelingen der Koninkyke akademie Wettenschapppen te Amsterdam* 65 (1898): 3–21, translation published in English as *Phytopathological Classics* no. 7 (St. Paul: American Phytopathological Society Press, 1942); L. Bos, "The embryonic beginning of virology: Unbiased thinking and dogmatic stagnation," *Archives of Virology* 140 (1995): 613–619; H. Fraenkel-Conrat, "The history of tobacco mosaic virus and the evolution of molecular biology," in M.H.V. Van Regenmortel and H. Fraenkel-Conrat, eds., *The plant viruses*, 2 vols. (New York: Plenum Press, 1986), Vol. 2: 5–17.

51. M.W. Beijerinck, "Die Bacterien der Papilionaceen Knollchen," *Botanische Zeitung* No. 48 (1888): 769–771.

52. Quoted in van Niel, "The 'Delft School,'" 163–164.

53. See Woodhead, *Bacteria and their products*.

CHAPTER 5

1. W.T. Sedgwick, "The genesis of a new science," *Bacteriology Journal of Bacteriology* 1 (1918): 1–4.

2. Ibid., 4.

3. C.E.A. Winslow, "The characterization and classification of bacterial types," *Science* 39 (1914): 77–91, at 77.

4. See R.E. Buchanan, *General systematic bacteriology* (Baltimore: Williams and Wilkins, 1925), 11.

5. Ibid.

6. S. Orla-Jensen, "Die Hauptlinien des natürlichen Bakteriensystems nebst einer Uebersicht der Gärungsphenomene," *Zentralblatt für Bacteriologie Parasitenkunds Infectionskrankheiten und Hygiene Abteilung* II Bd 22

(1909): 305–346; Orla-Jensen, "The main outlines of the natural bacterial system," *Journal of Bacteriology* 6 (1921): 263–273.

7. Winslow, "The characterization and classification of bacterial types."

8. Ibid., 78.

9. Ibid., 84.

10. Walter Migula, *System der Bakterien. Handbuch der Morphologie, Entwickelungsgeschichte und Systematik der Bakterien*, 2 vols. (Jena: G. Fischer, 1897–1900).

11. Winslow, "The characterization and classification of bacterial types," at 84.

12. Ibid., 85.

13. Ibid.

14. Ibid., 80. See also C.N. Hinshelwood, "Bacterial Growth," *Biological Reviews* 19 (1944): 150–163.

15. Winslow, "The characterization and classification of bacterial types," at 80.

16. Ibid., 85.

17. Ibid., 87.

18. Ibid., 84–85.

19. C.-E.A. Winslow, Jean Broadhurst, R.E. Buchanan, Charles Krumwiede, L.A. Rogers, and George Smith, "The families and genera of bacteria," *Journal of Bacteriology* 2 (1917): 505–566, at 525.

20. Ibid., 537.

21. Ibid., 548–549.

22. Ibid., 549.

23. C.-E.A. Winslow, Jean Broadhurst, R.E. Buchanan, Charles Krumwiede, L.A. Rogers, and George Smith, "The families and genera of bacteria: Final report of the Committee of the Society of American Bacteriologists on Characterization and Classification of Bacterial Types," *Journal of Bacteriology* 5 (1920): 191–229, at 192.

24. R.S. Breed, H.J. Conn, and J.C. Baker, "Comments on the evolution and classification of bacteria," *Journal of Bacteriology* (1918) 3: 445–459.

25. Winslow et al., "The families and genera of bacteria" (1917), 542. For criticisms on this point, see Breed et al., "Comments," 450.

26. Henry Fairfield Osborn, *The origin and evolution of life* (New York: Charles Scribner's Sons, 1917), 68.

27. Ibid., 84.

28. I.J. Kligler, "The evolution and relationships of the great groups of bacteria," *Journal of Bacteriology* 2 (1917): 165–176.

29. Ibid., 169.

30. T.C. Chamberlin, *The origin of the earth* (Chicago: University of Chicago Press, 1916); T.C. Chamberlin and R.T. Chamberlin, "Early terrestrial conditions that may have favored organic synthesis," *Science* 28 (1908): 897–911.

31. R.L. Moore and T.A. Webster, "Synthesis by sunlight in relationship to the origin of life," *Proceedings of the Royal Society of London Series B* 87 (1913): 163–176, at 164.

32. Benjamin Moore, *The origin and nature of life* (London: Williams and Norgate, 1912), 181–193.

33. Breed et al., "Comments," 451, 449–450.

34. Ibid., 448.

35. Ibid., 445.

36. Ibid., 451.

37. See F.W. Twort, "An investigation on the nature of ultra-microscopic viruses," *Lancet* 2 (1915): 1241–1243.

38. Félix d'Herelle, "Sur un microbe invisible antagoniste des bacilles dysénteriques," *Comptes Rendus de l'Academie des Sciences* 165 (1917): 373–375. See William Summers, *Félix d'Herelle and the origins of molecular biology* (New Haven: Yale University Press, 1999), 56.

39. Summers, *Félix d'Herelle*, 98. See also S.S. Hughes, *The virus: A history of the concept.* London: Heinemann, 1977); Scott Podoslsky, "The Role of the Virus in Origin-of-Life Theorizing," *Journal of the History of Biology* 29 (1996): 79–126.

40. Summers, *Félix d'Herelle*, 101.

41. H.J. Muller, "Variation due to change in the individual gene," *American Naturalist* 56 (1922): 32–50, at 48.

42. J.B.S. Haldane, "The origin of life," *Rationalist Annual* (1929): 148–169, reprinted in J.D. Bernal, *The origin of life* (London: Weidenfeld and Nicolson, 1967), 242–249, at 245.

43. Winslow et al., "The families and genera of bacteria" (1920), 192.

44. D.H. Bergey, *Handbook of practical hygiene* (Philadelphia: Saunders, 1899); Bergey, *The principles of hygiene; a practical manual for students, physicians, and health-officers* (Philadelphia: Saunders, 1901).

45. Arranged by a committee of the Society of American Bacteriologists: David H. Bergey, Francis C. Harrison, Robert S. Breed, Bernard W. Hammer, and Frank M. Huntoon, *Bergey's manual of determinative bacteriology* (Baltimore: Williams and Wilkins, 1923), iii.

46. Quoted in Arthur Henrici, *The biology of bacteria: An introduction to general microbiology* (Boston: D.C. Heath, 1939), 317. After the publication of the fourth edition of 1934, the Society of American Bacteriologists relinquished sponsorship for *Bergey's Manual*. It transferred all of the rights and interests in the manual to Bergey, who would create an educational trust (designating himself, Robert S. Reed, and E.G.D. Murray as trustees) for the purpose of preparing, editing, and publishing revisions and successive editions of the manual.

47. See Buchanan, *General systematic bacteriology*, 11.

48. See discussion in Robert Breed, "The present status of systematic bacteriology," *Journal of Bacteriology* 15 (1928): 143–163, at 144.

49. Ibid.

50. See Olga Amsterdamska, "Stabilizing instability: The controversy over cyclogenic theories of bacterial variation during the interwar period," *Journal of the History of Biology* 24 (1991): 191–222. See also C.-E.A. Winslow, "The changing bacteria," *Science* 75 (1932): 121–123; W.H. Manwaring, "Environmental transformation of bacteria," *Science* 79 (1934): 466–470; W. Braun, "Bacterial dissociation: A critical review of a phenomenon of bacterial variation," *Bacteriological Reviews* 11 (1947): 75–114.

51. See O. Rahn, "Contributions to the classification of bacteria, I–IV," *Zentralblatt für Bakteriologie II* 78 (1929): 1–21 and 79 (1929): 321–343; Rahn, "New principles for the classification of bacteria," *Zentralblatt für Bakteriologie II* 96 (1937): 273–286; J. Kluyver and C.B. van Niel, "Prospects for a natural system of classification of bacteria," *Zentralblatt für Bakteriologie II* 94 (1936): 369–402. See also R.S. Breed, "Historical survey of classifications of bacteria, with emphasis on outlines proposed since 1923," in David H. Bergey, Robert S. Breed, E.G.D. Murray, and A. Parker Hitchens, *Bergey's Manual of Determinative Bacteriology*, 5th ed. (Baltimore: Williams and Wilkins, 1939), 4–38, at 14.
52. See Henrici, *The biology of bacteria*, 318.
53. Robert Breed, "The present status of systematic bacteriology," at 143.
54. Ibid., 144.
55. Ibid., 145.
56. Ibid.
57. Ibid.
58. E.A. Minchin, "The evolution of the cell," *Report of the Eighty-Fifth Meeting of the British Association for the Advancement of Science* (September 7–11, 1915): 437–464.
59. Ibid., 445.
60. F.A. Bather, "Biological classification: Past and future," *Proceedings of the Geological Society of London* 83 (1927): lxii–civ, at cii.
61. Ibid., c.
62. Ibid., ci.

CHAPTER 6

1. The Society of Protozoologists was founded in the United States in 1947, and the *Journal of Protozoology* was first published in 1954.
2. Adolf Engler and E. Gilg, *Syllabus der Pflanzenfamilien*, 10th ed. (Berlin: Bornträger, 1924).
3. In his *Fresh-water algae of the United States* (New York: McGraw-Hill, 1933), Gilbert Smith at Stanford listed the divisions Schizophyta, Rhodophyta, Chrysophyta, Phaeophyta, Chlorophyta, and Pyrophyta as comprising a "natural classification of the algal members of the plant Kingdom." "In reality," he commented, "the six divisions listed above represent six kingdoms, all plant-like in nature" (9–10). He said nothing more.
4. See Herbert Copeland, *The classification of lower organisms* (Palo Alto: Pacific Books, 1956), 4; W.H. Wagner, Jr., "Edwin Bingham Copeland, 1873–1964," *Taxon* 14/2 (February 1965): 33–41.
5. E.B. Copeland, "What is a plant?" *Science* 65 (1927): 388–390, at 388.
6. Ibid., 390.
7. Ibid.
8. Herbert F. Copeland, "The kingdoms of organisms," *Quarterly Review of Biology* 13 (1938): 383–420, at 384–385.
9. Ibid., 386.
10. Ibid., 392.

11. Ibid.
12. C.C. Dobell, "Contributions to the cytology of the bacteria," *Quarterly Journal of Microscopical Sciences* 56 (1911): 395–506, at 488, 489.
13. B.F. Lutman, *Microbiology* (New York: McGraw-Hill, 1929), 416.
14. Ibid., v.
15. William Park, Anna Williams, and Charles Krumwiede, *Pathogenic microorganisms: A practical manual for students, physicians and health officers*, 9th ed. (Philadelphia: Lea and Febiger, 1929), 28.
16. Copeland, "Kingdoms of organisms," 416.
17. Herbert F. Copeland, "Progress report on basic classification," *American Naturalist* 81 (1947): 340–361, at 342.
18. Ibid., 386.
19. Ibid.
20. Ibid.
21. Ibid., 388.
22. Ibid., 345.
23. Ibid.
24. Ibid.
25. Ibid.
26. Ibid., 343.
27. Ibid., 344.
28. Ibid., 343.
29. Louis Agassiz, *Contributions to the natural history of the United States of America*, 4 vols. (Boston: Little Brown, 1857–1862).
30. H.F. Copeland, "Progress report on basic classification," 346.
31. See Mary P. Winsor, "Non-essentialist methods in pre-Darwinian taxonomy," *Biology and Philosophy* 18 (2003): 387–400.
32. Copeland, "Progress report on basic classification," 349. See also T.A. Sprague, "Standard-species," *Kew Bulletin* (1926): 96–100.
33. Copeland, "Progress report on basic classification," 351.
34. Günter Enderlein, *Bakterien-Cyclogenie. Prolegomena zu Untersuchungen über Bau, geschlechtliche und ungeschlechtliche Fortpflanzung und Entwicklung der Bakterien* (Berlin: Walter de Gruyter, 1925). Enderlein was notorious for promoting various unorthodox theories about the nature of bacteria. Best known among them was the theory of cyclogeny, according to which, under certain conditions, a particular microbial species can manifest itself in various forms and developmental stages. Such claims about bacterial life cycles were generally disregarded by the mid-1940s. C.B. van Niel, "The classification and natural relationships of bacteria," *Cold Spring Harbor Symposia on Quantitative Biology* 11 (1946): 285–301, at 290; see also W. Braun, "Bacterial dissociation: A critical review of a phenomenon of bacterial variation," *Bacteriological Reviews* 11 (1947): 75–114. Copeland observed that because Mychota originated in a notoriously heterodox system, there was "not under any nomenclatorial code a valid objection to it." "Progress report on basic classification," 351.
35. See Peter J. Bowler, *The eclipse of Darwinism: Anti-Darwinian evolution theories in the decades around 1900* (Baltimore: Johns Hopkins University Press,

1983); David Hull, *Darwin and his critics: The reception of Darwin's theory of evolution by the scientific community* (Cambridge: Harvard University Press, 1983); Ernst Mayr and William B. Provine, eds., *The evolutionary synthesis: Perspectives on the unification of biology* (Cambridge: Harvard University Press, 1980).

36. Julian Huxley, *The modern synthesis* (London: Allen and Unwin, 1942).

37. See Vassiliki B. Smocovitis, *Unifying biology: The evolutionary synthesis and evolutionary biology* (Princeton: Princeton University Press, 1996).

38. R.A. Fisher, *The genetical theory of natural selection* (Oxford: Clarendon Press, 1930); J.B.S. Haldane, *The cause of evolution* (London: Longmans, Green, 1932); Sewall Wright, "Evolution in Mendelian populations," *Genetics* 16 (1931): 97–159; William Provine, *The origins of theoretical population genetics* (Chicago: Chicago University Press, 1971); Provine, *Sewall Wright and evolutionary biology* (Chicago: University of Chicago Press, 1986).

39. Theodosius Dobzhansky, *Genetics and the origin of species*, 3rd ed. (New York: Columbia University Press, 1951), 16.

40. Embryology was also not included in the synthesis. As many embryologists saw it, Mendelian genetics was concerned with relatively trivial difference, not with fundamental organizational characteristics of organisms. See Jan Sapp, *Beyond the gene: Cytoplasmic inheritance and the struggle for authority in genetics* (New York: Oxford University Press, 1987).

41. W.H. Manwaring, "Environmental transformation of bacteria," *Science* 79 (1934): 466–470; 466–467.

42. C.D. Darlington, *The evolution of genetics systems* (Cambridge: Cambridge University Press, 1939), 70.

43. Huxley, *Evolution: The modern synthesis*, 131–132.

44. Julian Huxley, ed., *The new systematics* (Oxford: Oxford University Press, 1940).

45. Ibid., 1.

46. Ernst Mayr, *Systematics and the origin of species* (New York: Columbia University Press, 1942).

47. There is considerable debate among scholars about essentialism or typological thought in taxonomic practice before Darwin. See Elliott Sober, "Evolution, population thinking, and essentialism," *Philosophy of Science* 47 (1980): 350–383. Ernst Mayr argued that such essentialist typological taxonomic methods had been widespread before Darwin's *Origin*, which broke with the concept of the *eidos* (idea, type, or essence) that had been part of philosophy since Plato. Other scholars claim that taxonomic methods widely used before Darwin were not essentialist (Winsor, "Non-essentialist methods in pre-Darwinian taxonomy"). Still others argue that while higher taxonomic categories were not essentialist in conception before Darwin, species concepts generally were. See David Stamos, "Pre-Darwinian taxonomy and essentialism—a reply to Mary Winsor," *Biology and Philosophy* 10 (2005): 76–96.

48. M.J. Kottler, "Charles Darwin: Biological species concept and theory of geographic speciation: The transmutation notebooks," *Annals of Science* 35 (1978): 275–298.

49. Charles Darwin, *On the Origin of Species*, facsimile of 1859 edition (Cambridge: Harvard University Press, 1964), 47.

50. Mayr, *Systematics and the origin of species*, 10.

51. Julian Huxley, "Toward the new systematics," in Huxley, *New systematics*, 11–16.

52. Ibid., 16. In this Huxley disagreed with only one of the contributors to the volume, who maintained that species were as equally artificial with genus. See J.S.L. Gilmour, "Taxonomy and philosophy," in Huxley, *New systematics*, 461–474. See also discussion in Mary P. Winsor, "Species, demes, and the omega taxonomy: Gilmour and the new systematics," *Biology and Philosophy* 15 (2000): 349–388.

53. William Turrill, "Species," *Journal of Botany* 63 (1925): 359–366, at 362.

54. Quoted in Huxley, "Toward the new systematics," 17.

55. Quoted in Ernst Mayr, *Systematics and the origin of species* (reprint: New York: Dover Publications, 1964), 120.

56. Ibid.

57. George Gaylord Simpson acknowledged problems with the biological or genetic definition of species; he concluded that it was appropriate for time slices of sexual species, but a more inclusive definition of species needed to embrace asexual organisms: "An evolutionary species is a lineage (an ancestral-descendant sequence of populations) evolving separately from others and with its own unitary evolutionary role and tendencies." G.G. Simpson, *Principles of animal taxonomy* (New York: Columbia University Press 1961), 153.

58. See Theodosius Dobzhansky, *Genetics and the origin of species*, 1st ed. (New York: Columbia University Press, 1937), 316–21, and 3rd ed. (New York: Columbia University Press, 1951), 275.

59. See O. Rahn, "Contributions to the classification of bacteria, I–IV," *Zentralblatt fur Bakteriologie II* 78 (1929): 1–21 and 79 (1929): 321–343; Rahn, "New principles for the classification of bacteria," *Zentralblatt fur Bakteriologie II* 96 (1937): 273–286. See also van Niel, "Classification and natural relationships of bacteria," 297.

60. S. Winogradsky, "Sur la classification des bactéries," *Annales de l'Institut Pasteur* 82 (1952): 125–131, at 131. Winogradsky held the same to be true for viruses, which he believed would always be identified only by their destructive effects. C.B. van Niel, "Classification and taxonomy of the bacteria and blue green algae," in van Niel, *A century of progress in the natural sciences 1853–1953* (San Francisco: California Academy of Sciences, 1955), 89–114.

61. See Mayr, *Systematics and the origin of species* (reprint), 122.

62. Smocovitis, *Unifying biology*; Smocovitis, "G. Ledyard Stebbins, Jr. and the evolutionary synthesis (1924–1950)," *American Journal of Botany*, 84 (1997): 1625–1637.

63. G. Ledyard Stebbins, *Variation and evolution in plants* (New York: Columbia University Press, 1950), 252.

64. Huxley "Toward the new systematics," 19–20.

65. Mayr, *Systematics and the origin of species*, 280.

66. Ibid., 119.

67. Huxley, *New systematics*, 2.

68. T.H. Huxley, "On the classification of the animal kingdom," *Nature* 11 (1874): 101–102, at 101.

69. Huxley, "Toward the new systematics," 19.

70. W.T. Calman, "A museum zoologist's view of taxonomy," in Huxley, *New systematics*, 458.

> Wherever it is reticulate, whether by allotetraploidy in plants, by hybridization followed by apogamy, or by the meeting and crossing of subspecies originally differentiated in isolation from each other, as seems not infrequent in animals, our existing taxonomic methods inevitably fail to denote phylogeny. (Huxley, "Toward the new systematics," 20, 21)

71. Winsor, "Species, demes, and the omega taxonomy."

72. W.J. Arkell and J.A. Moy-Thomas, "Paleontology and the taxonomic problem," in Huxley, *New systematics*, 395–434, at 395.

73. Ibid., 396.

74. Calman, "A museum zoologist's view of taxonomy," 455.

75. Ibid., 457.

76. T.A. Sprague, "Taxonomic botany, with special reference to the angiosperms," in Huxley, *New systematics*, 435–454, at 435.

77. Gilmour, "Taxonomy and philosophy."

78. H.S. Conard, *Plants of Iowa* (Grinnell Flora 5th ed.), Iowa Academy of Sciences Biological Survey Publication 2, 1–95. See also Winona H. Welch and Fabius LeBlanc, "Henry S. Conard (1874–1971)," *Bryologist* 75 (1972): 558–565.

79. A.T. Henrici, *The biology of bacteria* (Boston: D.C. Heath, 1939), 82.

80. André Prévot, *Manuel de classification et de determination de bactéries anaérobies* (Masson: Paris, 1940), 10.

81. R.Y. Stanier and C.B. van Niel, "The main outlines of bacterial classification," *Journal of Bacteriology* 42 (1941): 437–466; W. Rothmaler, "Uber das natürliche System der Organismen," *Biologische Zentralblatt* 67 (1948): 242–250. See also Theodore Jahn and Frances Jahn, *How to know the Protozoa* (Dubuque: W.C. Brown, 1949), 7; Copeland, *Classification of lower organisms*, 37–38.

CHAPTER 7

1. On microbes for biochemical genetics, see Jan Sapp, *Where the truth lies: Franz Moewus and the origins of molecular biology* (Cambridge: Cambridge University Press, 1990).

2. C.B. van Niel, "The 'Delft School' and the rise of general microbiology," *Bacteriological Reviews* 13 (1949): 161–174; Susan Spath, *C. B. van Niel and the culture of microbiology, 1920–1965* (Ph.D. diss., University of California–Berkeley, 1999).

3. A.F. Kamp, J.W.M. La Rivière, and W. Verhoeven, eds., *Albert Jan Kluyver: His life and work* (Amsterdam: North-Holland, 1959); A.J. Kluyver, *The chemical activities of microorganisms* (London: London University Press, 1931); Kluyver and H.J.L. Donker, "The unity in the chemistry of the fermentative sugar dissimilation processes of microbes," *Koninklijke Akad Wetenschappen Amsterdam*

338 | NOTES TO PAGES 83–85

28 (1925): 297–313; Kluyver and C.B. van Niel, *The microbe's contribution to biology* (Cambridge: Harvard University Press, 1956).

4. See Spath, *C. B. van Niel*; Spath, "van Niel's course in general microbiology," *ASM News* 70/8 (2004): 360–363.

5. C.E.A. Winslow, Jean Broadhurst, R.E. Buchanan, Charles Krumwiede, L.A. Rogers, and George Smith, "The families and genera of the bacteria," *Journal of Bacteriology* 2 (1917): 505–565, at 537.

6. A.R. Prévot, "Etudes de systématique bactérienne, I, II," *Annales des sciences naturelles, botanique* 10e série 15 (1933): 223–261.

7. A.J. Kluyver and C.B. van Niel, "Prospects for a natural system of classification of bacteria," *Zentralblatt für Bakteriologie II* 94 (1936): 369–402, at 377.

8. R.S. Breed, "Historical survey of classifications of bacteria, with emphasis on outlines proposed since 1923," in David H. Bergey, Robert S. Breed, E.G.D. Murray, and A. Parker Hitchens, *Bergey's manual of determinative bacteriology*, 5th ed. (Baltimore: Williams and Wilkins, 1939), 4–38, at 38.

9. R.Y. Stanier and C.B. van Niel, "The main outlines of bacterial classification," *Journal of Bacteriology* 42 (1941): 437–466, at 438.

10. Ibid.

11. Ibid., 437–438.

12. Ibid., 440.

13. Ibid., 449.

14. Ibid., 439.

15. Ibid., 449.

16. Ibid., 443. See also Kluyver and van Niel, "Prospects for a natural system," 377.

17. A.I. Oparin, *Origin of life*, translation with annotations by Sergius Morgulis (New York: Dover, 1938).

18. J.B.S. Haldane, "The origin of life," *The Rationalist Annual* 3 (1939): 148–153, reprinted as an appendix in J.D. Bernal, *The origin of life* (London: Weidenfeld and Nicolson, 1967), 242–249.

19. Stanier and van Niel, "The main outlines of bacterial classification," 444.

20. C.B. van Niel, "The classification and natural relationships of bacteria," *Cold Spring Harbor Symposia on Quantitative Biology* 11 (1946): 285–301, at 289. Copeland also changed his previous autotrophs-first model of 1938 to heterotrophs first. See H.F. Copeland, "Progress report on basic classification," *American Naturalist* 81 (1947): 340–361.

21. Stanier and van Niel, "The main outlines of bacterial classification," 290, 444.

22. Ibid.

23. R.S. Breed, H.J. Conn, and J.C. Baker, "Comments on the evolution and classification of bacteria," *Journal of Bacteriology* 3 (1918): 445–459, at 451; Kluyver and van Niel, "Prospects for a natural system."

24. Kluyver and van Niel, "Prospects for a natural system," 387–388.

25. van Niel, "The classsifcation and natural relationships of bacteria," 293. For a critique of the system proposed by Kluyver and van Niel, see P.B. White, "Remarks on bacterial taxonomy," *Zentralblatt für Bakteriologie II* 96 (1937): 145–149. Se also E.G. Pringsheim, "The relationship between bacteria and myxophyceae," *Bacteriological Reviews* 13 (1949): 47–98.

26. Stanier and van Niel, "The main outlines of bacterial classification," 458.

27. Ibid., 444.

28. C.B. van Niel to R. Stanier, August 13, 1941 (Stanier Papers, National Archives of Canada).

29. van Niel, "The classification and natural relationships of bacteria," 289.

30. Ibid., 290.

31. S. Winogradsky, "Sur la classification des bactéries," *Annales de l'Institut Pasteur* 82 (1952): 125–131, at 131.

32. C.B. van Niel, "Classification and taxonomy of the bacteria and blue green algae," in Edward L. Kessel, ed., *A century of progress in the natural sciences 1853–1953* (San Francisco: California Academy of Sciences, 1955), 89–114.

33. Ibid.

34. Roger Stanier, Michael Doudoroff, and Edward Adelberg, *The microbial world* (Englewood Cliffs: Prentice-Hall, 1957), 297.

35. C.F. Robinow, "Nuclear apparatus and cell structure of rod-shaped bacteria," addendum in René Dubos, *The bacterial cell* (Cambridge: Harvard University Press, 1945), 355–377, at 357.

36. Ibid., at 362.

37. Dubos, *The bacterial cell,* 23.

38. van Niel, "Classification and taxonomy of the bacteria and blue green algae," 93.

39. Stanier et al., *Microbial world*, 101.

40. On the development of the electron microscopy, see Nicolas Rasmussen, *Picture control: The electron microscope and the transformation of biology in America, 1940–1960* (Stanford: Stanford University Press, 1997).

41. Dubos, *The bacterial cell*, 26–27, 356.

42. Stanier et al., *Microbial world*, 127.

43. H.K. Schachman, A.B. Pardee, and R.Y. Stanier, "Studies on the macromolecular organization of microbial cells," *Archives of Biochemistry and Biophysics* 38 (1952): 245–260.

44. Stanier et al., *Microbial world*, 133–134.

45. Ibid., 102.

46. M. Calvin and V. Lynch, "Grana-like structures of *Synechococcus cedorum*," *Nature* 169 (1952): 455–456.

47. See Sapp, *Where the truth lies.*

48. O.T. Avery, C.M. MacLeod, and M. McCarty, "Studies on the chemical nature of the substance inducing transformation of pneumococcal types," *Journal of Experimental Medicine* 79 (1944): 137–158.

49. J. Lederberg and E. Tatum, "Gene recombination in *Escherichia coli*," *Nature* 158 (1946): 558; Lederberg and Tatum, "Novel genotypes in mixed cultures of biochemical mutants of bacteria," *Cold Spring Harbor Symposia on Quantitative Biology* 11 (1946): 113–114.

50. William Hayes, "Recombination in bact. *coli* K12: Unidirectional transfer of genetic material," *Nature* 169 (1952): 118–119; J.D.Watson and W. Hayes, "Genetic exchange in *Escherichia Coli* k12: Evidence for three linkage groups," *Proceedings of the National Academy of Sciences* 39 (1953): 416–426.

51. Stanier et al., *Microbial world*, 401.

52. van Niel, "Classification and taxonomy of the bacteria and blue green algae," 92.

53. Stanier et al., *Microbial world*, 55–57.

54. Ibid., 105.

55. Ibid., 106.

56. R.Y. Stanier and C.B. van Niel, "The concept of a bacterium," *Archiv für Mikrobiologie* 42 (1962): 17–35, at 17.

57. Ibid.

58. Ibid., 18.

59. Ibid.

60. A.P. Waterson and Lise Wilkinson, *An introduction to the history of virology* (Cambridge: Cambridge University Press, 1978), 10. On the history of virology, see also Ton van Helvoort, "What is a virus? The case of tobacco mosaic disease," *Studies in History and Philosophy of Science* 22 (1991): 557–588; van Helvoort, "The construction of bacteriophage as bacterial virus: Linking endogenous and exogenous thought styles," *Journal of the History of Biology* 27 (1994): 91–139; van Helvoort, "History of virus research in the twentieth century: The problem of conceptual continuity," *History of Science* 32 (1994): 185–236; A.N.H. Craeger, *The life of a virus: Tobacco mosaic virus as an experimental model, 1930–1965* (Chicago: University of Chicago Press, 2002).

61. Stanier et al., *Microbial world*, 360.

62. R.S. Breed, E.G.D. Murray, and A. Parker Hitchens, *Bergey's manual of determinative bacteriology*, 6th ed. (Baltimore: Williams and Wilkins, 1957), 9.

63. Theodore Jahn and Frances Jahn, *How to know the Protozoa* (Dubuque: W.C. Brown, 1949), 7.

64. H.J. Muller, "Variation due to change in the individual gene," *American Naturalist* 56 (1922): 32–50.

65. André Lwoff, "Lysogeny," *Bacteriological Reviews* 17 (1953): 269–337.

66. J. Lederberg, "Cell genetics and hereditary symbiosis," *Physiological Reviews* 32 (1952): 403–430; Lederberg, "Genetic transduction," *American Scientist* 44 (1956): 264–280.

67. André Lwoff, "The concept of virus," *Journal of General Microbiology* 17 (1957): 239–253. See also A. N. Creager, G. J. Morgan, "After the double helix: Rosalind Franklin's research on tobacco mosaic virus," *Isis* 99 (2008): 239–272; Erling Norrby, "Nobel Prizes and the emerging virus concept," *Archives of Virology* 153 (2008): 1109–1123.

68. Lwoff, "The concept of virus," 252.

69. Stanier and van Niel, "The concept of a bacterium," 19.

70. E. Chatton, "Pansporella perplexa: Reflexions sur la biologie et la phylogenie des protozoaires," *Annales des sciences naturelles, zoologique* 10e serie, 7 (1925): 1–84. Though published in 1925, he wrote the paper in 1923 at the University of Strasbourg.

71. Chatton, "Pansporella perplexa," 76.

72. For a fuller account, see Jan Sapp, "The prokaryote-eukaryote dichotomy: Meanings and mythology," *Microbiology and Molecular Biology Reviews* 69 (2005): 292–305.

73. Edouard Chatton, *Titre and travaux scientifique (1906–1937) de Edouard Chatton* (Sottano, Italy: Sette, 1938), 50 [author's translation].
74. R.Y. Stanier, "La place des bactéries dans le monde vivant," *Annales de l'Institute Pasteur* 101 (1961): 297–312.
75. Stanier and van Niel, "The concept of a bacterium."
76. Stanier and van Niel, "The concept of a bacterium," 21.
77. Ibid., 22.
78. Ibid., 25.
79. Ibid., 24.
80. Ibid., 20.
81. Ibid., 33.
82. R. Stanier, M. Doudoroff, and E. Adelberg, *The microbial world*, 2nd ed. (Englewood Cliffs: Prentice-Hall, 1963), 65.
83. Ibid., 85 [emphasis original].

CHAPTER 8

1. See Roger Stanier, "Toward a definition of the bacteria," in I.C. Gunsalus and Roger Stanier, eds., *The bacteria: A treatise on structure and function*, 12 vols. (New York: Academic Press, 1964), Vol. 5: 445–464.
2. A. Pascher, "Systematische Übersicht über die mit Flagellaten in Zusammenhang stehenden Algenreihen und Versuch einer Einreihung dieser Algenstämme in die Stämme des Pflanzenreichs," *Beiheftezum Botanischen Centralblatt* 48 (1931): 317–332, at 330, quoted in E.G. Pringsheim, "Relationship between bacteria and myxophyceae," *Bacteriological Reviews* 13 (1949): 47–98, at 48.
3. René Dubos, *The bacterial cell* (Cambridge: Harvard University Press, 1945), 7.
4. Ibid., 6.
5. André Prévot, *Manuel de classification et de determination de bacteries anaerobies* (Masson: Paris, 1940), 9. He considered blue-green algae, like protozoa, to be "cousins of the bacteria."
6. E.G. Pringsheim, "Zur Kritik der Bakteriensystematik," *Lotos* 71 (1923): 357–377. For a biographical sketch, see Dieter Mollenhauer, "The protistologist Ernst Georg Pringsheim and his four lives," *Protist* 154 (2003): 157–171.
7. E.G. Springsheim, "Relationship between bacteria and myxophyceae," *Bacteriological Reviews* 13 (1949): 47–98, at 52.
8. Ibid., 53, 87–88.
9. Ibid., 91.
10. Ibid., 88.
11. Ibid., 48–49.
12. E.C. Dougherty, "Neologism needed for structures of primitive organisms. 1. Types of nuclei," *Journal of Protozoology* 4 (1957): 14.
13. E.C. Dougherty, "Comparative evolution and the origin of sexuality," *Systematic Zooloogy* 4 (1955): 145–169, at 149.
14. R. Stanier, M. Doudoroff, and E. Adelberg, *The microbial world*, 2nd ed. (Englewood Cliffs: Prentice-Hall, 1963), 105.
15. R.Y. Stanier and C.B. van Niel, "The concept of a bacterium," *Archiv für Mikrobiologie* 42 (1962): 17–35, at 33.

16. Stanier et al., *Microbial world*, 409.

17. Ibid., 85.

18. R.Y. Stanier, "Toward an evolutionary taxonomy of the bacteria," 595–604, in A. Perez-Miravete and D. Pelaéz, eds., *Recent advances in microbiology* (Mexicao City: Associación Mexicana de Microbiologia, 1971).

19. T.H. Huxley, "On the physical basis of life" (1868), *Lay sermons, addresses and reviews* (London: Macmillan, 1870), *Collection of Essays* 1: 130–165, at 142.

20. F.G. Hopkins, "The dynamic side of biochemistry," *Nature* 92 (1913): 213–23, at 220.

21. For an overview of the concept of the unity of biochemistry, see Herbert C. Friedmann, "From 'Butribacterium' to 'E. coli,'" *Perspectives in Biology and Medicine* 47/1 (2004): 47–66.

22. C.B. van Niel, "The 'Delft School' and the rise of general microbiology," *Bacteriological Reviews* 13 (1949): 161–174, at 172.

23. Jacques Monod and François Jacob, "General conclusions: Teleonomic mechanisms in cellular metabolism, growth and differentiation," *Cold Spring Harbor Symposia on Quantitative Biology* 26 (1961): 389–401, at 393.

24. François Jacob, *La logique du vivant* (Paris: Editions Gallimard, 1970), published in English as *The logic of life: A history of heredity*, translation by Betty E. Spillmann (New York: Vintage Books, 1976), 306.

25. Friedmann, "From 'Butribacterium' to 'E. coli.'"

26. R.G.E. Murray, "Fine structure and taxonomy of bacteria," *Symposia of the Society for General Microbiology* 12 (1962): 119–145.

27. R.G.E. Murray to R.Y. Stanier, May 15, 1962 (National Archives of Canada), MG 31, Accession J35, Vol. 6.

28. R.Y. Stanier to R.G.E. Murray, May 21, 1962 (National Archives of Canada), MG 31, Accession J35, Vol. 6.

29. R.G.E. Murray, "Microbial structure as an aid to microbial classification and taxonomy," *SPISY* 34 (1968): 249–252.

30. A. Allsopp, "Phylogenetic relationships of the Procaryota and the origin of the eucaryotic cell," *New Phytologist* 68 (1969): 591–612, at 592.

31. Allsopp, "Phylogenetic relationships," 599. See also S. Soriano and R.A. Lewin, "Gliding microbes: Some taxonomic reconsiderations," *Antonie van Leeuwenhoek. Journal of Microbiology and Serology* 31 (1965): 66–80.

32. For a more complete account, see Jan Sapp, "The prokaryote-eukaryote dichotomy. Meanings and mythology," *Microbiology and Molecular Biology Reviews* 69 (2005): 292–305.

33. Peter Raven to R.E. Buchanan, October 8, 1970 (National Archives of Canada), MG 31, Accession J35, Vol. 6. See also P. Raven and H. Curtis, *The biology of plants* (New York: Worth Publishers, 1970).

34. Peter Raven to Roger Stanier, November 3, 1970, National Archives of Canada, MG 31, Accession J35, Vol. 6; Roger Stanier to Peter Raven, November 5, 1970, National Archives of Canada, MG 31, Accession J35, Vol. 6.

35. R.G.E. Murray, "A place for bacteria in the living world," in R.E. Buchanan and N.E. Gibbons, eds., *Bergey's manual of determinative bacteriology*, 8th ed. (Baltimore: Williams and Wilkins, 1974), 4–10.

36. See Lynn Margulis, *Symbiosis in cell evolution* (New York: W.H. Freeman, 1981), 34.

37. See Gunther Stent, *Molecular genetics: An introductory narrative* (San Francisco: W.H. Freeman, 1971), 43.

38. R.H. Whittaker, "New concepts of kingdoms of organisms," *Science* 163 (1969): 150–163, at 151.

39. E.P. Odum, *Fundamentals of ecology* (Philadelphia: W.B. Saunders, 1959), 12.

40. Herbert Copeland, *The classification of lower organisms* (Palo Alto: Pacific Books, 1956).

41. R.H. Whittaker, "The kingdoms of the living world," *Ecology* 38 (1957): 536–538, at 536.

42. R.H. Whittaker, "On the broad classification of organisms," *Quarterly Review of Biology* 34 (1959): 210–226, at 217.

43. Theodore Jahn and Frances Jahn, *How to know the Protozoa* (Dubuque: W.C. Brown, 1949), 7.

44. Whittaker, "The kingdoms of the living world," 537.

45. Whittaker, "On the broad classification of organisms," 217.

46. Whittaker, "New concepts of kingdoms of organisms," 151.

47. Ibid., 154, 157.

48. Verne Grant, *The origin of adaptation* (New York: Columbia University Press, 1963), 40–44.

49. Grant, *Origin of adaptation*, 59.

50. Ibid., 58–59.

51. Ibid., 42.

52. Ibid., 86.

53. Charles Darwin, *On the origin of species*, with an introduction by Ernst Mayr, facsimile of 1859 edition (Cambridge: Harvard University Press, 1964), 307.

54. On the controversy over Precambrian fossil evidence, see C.F. O'Brien, "Eozoön canadense, 'the dawn animal of Canada,'" *Isis* 61 (1970): 206–223. The classificatioin of fossils found in 1946 in the Ediacara Hills of Australia to Cambrian animals is still uncertain. The Ediacaran biota were soft-bodied marine creatures, unusual fossils, originally interpreted as jellyfish, strange worms, and frondlike corals that lived 635–542 million years ago. See Stephen Jay Gould, *Wonderful life: The Burgess Shale and the nature of history* (London: Hutchison Radius, 1989), 59; J. William Schopf, *The Cradle of life: The discovery of Earth's earliest fossils* (Princeton: Princeton University Press, 1999), 35–70.

55. See Gould, *Wonderful life*, 25.

56. See, e.g., Brian K. Hall, *Evolutionary developmental biology*, 2nd ed. (Dordrecht: Kluwer, 1999).

57. See J. William Schopf, "Solution to Darwin's dilemma: Discovery of the missing Precambrian record of life," *Proceedings of the National Academy of Sciences* 97 (2000): 6947–6953. Cambridge botanist Albert C. Seward commented in 1931, "We can hardly expect to find in Pre-Cambrian rocks any actual proof of the existence of bacteria." *Plant life through the ages* (Cambridge: Cambridge University Press, 1931), 92, quoted in Schopf, *Cradle of life*, 34.

That pronouncement, it has been said, stifled the hunt for Precambrian fossils for decades (Schopf, *Cradle of life*, 34).

58. Schopf, *Cradle of life*, 23–34.

59. Elso S. Barghoorn and Stanley A. Tyler, "Microorganisms from the Gunflint Chert," *Science* 147 (1965): 563–577; Preston E. Cloud, "Significance of the Gunflint (Precambrian) microflora," *Science* 148 (1965): 27–35.

60. Albert Engel, Bartholomew Nagy, Lois Nagy, Celeste Engel, Gerhard Kremp, and Charles M. Drew, "Alga-like forms in Onverwacht Series, South Africa: Oldest recognized lifelike forms on Earth," *Science* 161 (1968): 1005–1008.

61. E. Barghoorn and A. Knoll, "Precambrian eukaryotic organisms: A reassessment of the evidence," *Science* 190 (1975): 52–54; J.W. Schopf and D.Z. Oehler, "How old are the eukaryotes?" *Science* 193 (1976): 47–49.

62. K.A. Bisset, "Do bacteria have a nuclear membrane?" *Nature* 241 (1973): 45; Bisset, "This 'prokaryote-eukaryotic' business," *New Scientist* (February 8, 1973): 296–297.

63. Bisset "This 'prokaryote-eukaryotic' business," 296.

64. Ibid., 297.

65. Stent, *Molecular genetics*.

66. Ibid., 298.

67. Ibid.

68. Bisset, "Do bacteria have a nuclear membrane?" 45.

69. R. Stanier, "Some aspects of the biology of cells and their possible evolutionary significance," *Symposia of the Society for General Microbiology* 20 (1970), 1–38, at 31.

CHAPTER 9

1. On the history of symbiosis theory and research, see Donna Mehos, "Ivan E. Wallin's theory of symbionticism," in L.N. Khakhina, L. Margulis, and M. McMenamin, eds., *Concepts of symbiogenesis: History of symbiosis as an evolutionary mechanism* (New Haven: Yale University Press, 1993), 149–163; Jan Sapp, *Evolution by association* (New York: Oxford University Press, 1994); Sapp, "Cell evolution and organelle origins: Metascience to science," *Archiv für Protistenkunde* 145 (1995): 263–275; Sapp, "Freewheeling centrioles," *History and Philosophy of the Life Sciences* 20 (1998): 255–290; Sapp, "Paul Buchner and hereditary symbiosis in insects," *International Microbiology* (September 5, 2002): 145–160; Sapp, "The dynamics of symbiosis. An historical overview," *Canadian Journal of Botany* 82 (2004): 1–11; Sapp, "Mitochondria and their host: Morphology to molecular phylogeny," in W. Martin and M. Müller, eds., *Mitochondria and hydrogenosomes* (Heidelberg: Springer, 2007), 57–84; Sapp, Francisco Carrapiço, and Mikhail Zolotonosov, "Symbiogenesis: The hidden face of Constantin Merezhkowsky," *History and Philosophy of the Life Sciences* 24 (2002): 413–440.

2. Anton de Bary, "Die Erscheinung der Symbiose," in *Vortrag auf der Versammlung der Naturforsher und Aertze zu Cassel* (Strassburg: von Karl J. Trubner, 1879), 1–30, at 5.

3. Ibid., 29–30.

4. A.F.W. Schimper, "Ueber die Entwicklung der Schlorophyllkörner und Farbkörper," *Botanische Zeitung* 41 (1883): 105–114, at 112–113.

5. Ernst Haeckel, *The wonders of life: A popular study of biological philosophy* (New York: Harper and Brothers, 1904), 195–196.

6. See also A. Famintsyn, "Die Symbiose als Mittel der Synthese von Organismen," *Biologisches Centralblatt* 27 (1907): 253–264.

7. S. Watasé, "On the nature of cell organization," *Woods Hole Biological Lectures* (1893): 83–103, at 86.

8. Ibid., 101.

9. T. Boveri, *Ergebnisse über die Konstitution der chromatischen Substanz des Zellkerns* (Jena: G. Fischer, 1904), 90, quoted in E.B. Wilson, *The cell in development and heredity* (New York: Macmillan, 1925), 1108.

10. See Sapp, "Freewheeling centrioles."

11. L.F. Henneguy, "Sur les rapports des cils vibratiles avec les centrosomes," *Archives d'Anatomie Microscopique et de Morphologie Expérimentale* 1 (1898): 481–496, at 481; M. von Lenhossék, "Ueber Flimmerzellen," *Verhandlungen der Anatomischen Gesellschaft* 12 (1898): 106–128.

12. For an account of Merezkowsky's life, see Sapp et al., "Symbiogenesis"; C. Mereschkowsky, "Theorie der zwei Plasmaarten als Grundlage der Symbiogenesis, einer neuen Lehre von der Entstehung der Organismen," *Biologisches Centralblatt* 30 (1910): 277–303, 321–347, 353–367. See C. Merezkowsky, "Über Natur und Ursprung der Chromatophoren im Pflanzenreiche," *Biologishes Centralblatt* 25 (1905): 593–604; for English translation, see W. Martin and K.V. Kawallik, "Annoted English translation of Merezkowsky's 1905 paper 'Über Natur und Ursprung der Chromatophoren im Pflanzenreiche,'" *European Journal of Phycology* 34 (1999): 287–295. See also C. Mérejkovsky, "La plante considérée comme un complexe symbiotique," *Bulletin de la Société Naturelles* 6 (1920): 17–98.

13. R. Altmann, *Die Elementarorganismen* (Leipzig: Veit and Company, 1890).

14. The authenticity of many other cell structures, including centrioles, spindle fibers, and the Golgi apparatus, was also doubted for many years. See P. Mazzerello and M. Bentivoglio, "The centenarian Golgi apparatus," Nature 392 (1998): 543–544; F. Schrader, *Mitosis: The movements of chromosomes in cell division* (New York: Columbia University Press, 1953); Sapp, "Freewheeling centrioles."

15. For a comprehensive early review, see E.V. Cowdry, "The mitochondrial constituents of protoplasm," *Carnegie Institution Publications, Contributions to Embryology* 8/25 (1918): 41–144; Sapp, "Mitochondria and their host."

16. See J. Sapp, *Beyond the gene: Cytoplasmic inheritance and the struggle for authority in genetics* (New York: Oxford University Press, 1987).

17. P. Portier, *Les symbiotes* (Paris: Masson, 1918).

18. I.E. Wallin, *Symbionticism and the origin of species* (Baltimore: Williams and Wilkins, 1927). For historical accounts of Wallin's theorizing, see Sapp, *Evolution by association.*

19. Wallin, *Symbionticism,* 147.

20. Félix d'Herelle, "Sur un microbe invisible antagoniste des bacilles dysentériques," *Comptes Rendus de l'Academie des Sciences* 165 (1917): 373–375. See

also William Summers, *Félix d'Herelle and the origins of molecular biology* (New Haven: Yale University Press, 1999).

21. Félix d'Herelle, *The bacteriophage and its behavior*, translation by George H. Smith (Baltimore: Williams and Wilkins, 1926), 320.

22. See Sapp, *Evolution by association*.

23. Portier, *Les symbiotes*, 294.

24. Wallin, *Symbionticism*, 8.

25. E.M. East, "The nucleus-plasma problem," *American Naturalist* 68 (1934): 289–303, 402–439, at 431.

26. Sapp, *Evolution by association*; J. Sapp, *Genesis: The evolution of biology* (New York: Oxford University Press, 2003).

27. Wilson, *The cell in development and heredity*, 739.

28. Sapp, *Beyond the gene*.

29. Ibid.

30. J. Lederberg, "Cell genetics and hereditary symbiosis," *Physiological Reviews* 32 (1952): 403–430. For a contemporary critique of cytoplasmic heredity, see H.J. Muller, "The development of the gene theory," in L.C. Dunn, ed., *Genetics in the twentieth century* (New York: Macmillan, 1951), 77–100.

31. T.M. Sonneborn, "The cytoplasm in heredity," *Heredity* 4 (1950): 11–36.

32. C.D. Darlington, "Mendel and the determinants," in Dunn, ed., *Genetics in the twentieth century*, 315–332, 331.

33. N. Zinder and J. Lederberg, "Genetic exchange in Salmonella," *Journal of Bacteriology* 64 (1952): 679–699.

34. Lederberg, "Cell genetics," 413.

35. J. Lederberg, "Genetic studies in bacteria," in Dunn, ed., *Genetics in the twentieth century*, 263–289; Lederberg, "Cell genetics."

36. Lederberg, "Cell genetics," 425.

37. R. Dubos, "Integrative and creative aspects of infection," in M. Pollard, ed., *Perspectives in virology*, 10 vols. (Minneapolis: Burgess, 1961), Vol. 2: 200–205.

38. Ibid., 204.

39. H. Ris, "Ultrastructure and molecular organization of genetic systems," *Canadian Journal of Genetics and Cytology* 3 (1961): 95–120; H. Ris and W. Plaut, "Ultrastructure of DNA-containing areas in the chloroplasts of *Chlamydomonas*," *Journal of Cell Biology* 12 (1962): 383–391; M.M.K. Nass and S. Nass, "Intramitochondrial fibers with DNA characteristics," *Journal of Cell Biology* 19 (1963): 613–628; S. Nass, "The significance of the structural and functional similarities of bacteria and mitochondria," *International Review of Cytology* 25 (1969): 55–129; M.M.K. Nass, "Mitochondrial DNA: Advances, problems and goals," *Science* 165 (1969): 25–35.

40. R. Sager, *Cytoplasmic genes and organelles* (New York: Academic Press, 1972); N. Gillham, *Organelle heredity* (New York: Raven Press, 1978).

41. H. Ris and R.N. Singh, "Electron microscope studies on blue-green algae," *Journal of Biophysical and Biochemical Cytology* 9 (1961): 63–80.

42. Ris and Plaut, "Ultrastructure of DNA-containing areas," 390.

43. Nass and Nass, "Intramitochondrial fibers," 621.

44. Katherine B. Warren, ed., *Formation and fate of cell organelles*. Symposia of the International Society for Cell Biology, Vol. 6 (New York: Academic Press, 1967).

45. A.L. Lehninger, *The mitochondrion* (New York: W.A. Benjamin, 1965).

46. Jostein Goksøyr, "Evolution of eucaryotic cells," *Nature* 214 (1967): 1161.

47. Lynn Sagan, "On the origin of mitosing cells," *Journal of Theoretical Biology* 14 (1967): 225–274; Sapp, "Freewheeling centrioles."

48. See, e.g., Kenneth Bissett, "This 'prokaryotic-eukaryotic' business," *New Scientist* 58 (1973): 296–298.

49. L.R. Cleveland, and A.V. Grimstone, "The fine structure of the flagellate *Mixotricha paradoxa* and its associated micro-organisms," *Proceedings of the Royal Society of London Series B* 159 (1964): 668–686.

50. L. Margulis, *Origin of eukaryotic cells* (New Haven: Yale University Press, 1970).

51. Ibid., 231.

52. Ibid., 247.

53. See Sapp, "Freewheeling centrioles."

54. A. Allsopp, "Phylogenetic relationships of the procaryota and the origin of the eucaryotic cell," *New Phytologist* 68 (1969): 591–612, at 599; T. Cavalier-Smith, "The origin of nuclei and of eukaryotic cells," *Nature* 256 (1975): 463–468, at 467.

55. P.H. Raven, "A multiple origin for plastids and mitochondria," *Science* 169 (1970): 641–646, at 645.

56. Seymour S. Cohen, "Are/were mitochondria and chloroplasts microorganisms?" *American Scientist* 58 (1970): 281–289, at 282.

57. Gunther Stent, *An introduction to molecular genetics* (San Francisco: W.H. Freeman, 1971), 622.

58. Lewis Thomas, *The lives of a cell: Notes of a biology watcher* (New York: Viking Press, 1974), 71.

59. Ibid., 73–74.

60. F.J.R. Taylor, "II. Implications and extensions of the serial endosymbiosis theory of the origin of eukaryotes," *Taxon* 23 (1974): 229–258.

61. T. Uzell and C. Spolsky, "Mitochondria and plastids as endosymbionts: A revival of special creation?" *American Scientist* 62 (1974): 334–343, at 343.

62. See F.J.R. Taylor, "Autogenous theories for the origin of eukaryotes," *Taxon* 25 (1976): 377–390.

63. Aharon Gibor, "Inheritance of cytoplasmic organelles," in Warren, *Formation and fate of cell organelles*, 305–316.

64. Ibid., 314.

65. Ruth Sager, "Mendelian and non-Mendelian heredity: A reappraisal," *Proceedings of the Royal Society of London Series B* 164 (1966): 290–297, at 296.

66. Philip John and F.R. Whatley, "*Paracoccus denitrificans* and the evolutionary origin of the mitochondrion," *Nature* 254 (1975): 495–498, at 498. See also Rudolf A. Raff and Henry R. Mahler, "The non symbiotic origin of

mitochondria," *Science* 177 (1972): 575–582; Raff and Mahler, "The symbiont that never was: An inquiry into the evolutionary origin of the mitochondrion," *Symposia of the Society for Experimental Biology* 29 (1975): 41–92.

67. Roger Stanier, "Toward a definition of the bacteria," in I.C. Gunsalus and Roger Stanier, eds., *The bacteria: A treatise on structure and function*, 5 vols. (New York: Academic Press, 1964), Vol. 5: 445–464, at 461–462.

68. R. Klein and A. Cronquist, "A consideration of the evolutionary and taxonomic significance of some biochemical, micromorphological and physiological characteristics in the thallophytes," *Quarterly Review of Biology* 42 (1967): 105–296, at 167.

69. Ibid., 118.

70. Karl Pearson, *The grammar of science* (London: Dent, 1937), 5.

71. Allsopp, "Phylogenetic relationships of the procaryota," 605.

72. Cavalier-Smith, "The origin of nuclei and of eukaryotic cells," 463.

73. Raven, "A multiple origin for plastids and mitochondria," 646.

74. W.H. Woolhouse, "A review of *The Plastids* by J.T.O. Kirk and R.A.E. Tilney-Bassett," *New Phytologist* 66 (1967): 832–833, at 833.

75. R. Stanier, "Some aspects of the biology of cells and their possible evolutionary significance," *Symposia of the Society for General Microbiology* 20 (1970), 1–38, at 31.

76. J.T.O. Kirk and R.A.E. Tilney-Bassett, *The plastids: Their chemistry, structure, growth and inheritance*, 2nd ed. (Amsterdam: Elsevier/North-Holland Biomedical Press, 1978), 247–248.

77. S.J. Gould and R. Lewontin, "The spandrels of San Marco and the Panglossian paradigm: A critique of the adaptationist programme," *Proceedings of the Royal Society of London Series B* 205 (1979): 581–598.

CHAPTER 10

1. E.G. Pringsheim, "The relationship between bacteria and myxophyceae," *Bacteriological Reviews* 13 (1949): 47–98, at 88.

2. François Jacob, *The logic of life*, translation by Betty E. Spillman (New York: Vintage Books, 1973), 236.

3. F.H.C. Crick, "The biological replication of macromolecules," *Symposia of the Society for Experimental Biology* 12 (1958): 138–163, at 142.

4. Jacob, *The logic of life*, 243.

5. For a summary of conceptual changes to classical molecular biology, see Jan Sapp, *Genesis: The evolution of biology* (New York: Oxford University Press, 2003).

6. See F. Sanger, "Chemistry of insulin," *Science* 129 (1959): 1340–1344.

7. Ibid., 1344.

8. L. Pauling and E. Zuckerkandl, "Chemical paleogenetics: Molecular 'restoration studies,' of extinct forms of life," *Acta Chemica Scandinavica* 17 (1963): 9–16.

9. E. Zuckerkandl and L. Pauling, "Evolutionary divergence and convergence in proteins," in V. Bryson and H. Vogel, eds., *Evolving genes and proteins* (New York: Academic Press, 1965), 97–166; E. Zuckerkandl and L. Pauling,

"Molecules as documents of evolutionary history," *Journal of Theoretical Biology* 8 (1965): 357–366. See also G.L. Morgan, "Emile Zuckerkandl, Linus Pauling, and the molecular evolutionary clock, 1959–1965," *Journal of the History of Biology* 31 (1998): 155–178; M. Dietrich, "The origins of the neutral theory of molecular evolution," *Journal of the History of Biology* 27 (1994): 21–59; Dietrich, "Paradox and persuasion: Negotiating the place of molecular evolution within evolutionary biology," *Journal of the History of Biology* 31 (1998): 85–111; W. Provine, "The neutral theory of molecular evolution in historical perspective," in B. Takahata and J. Crow, eds., *Population biology of genes and molecules* (Tokyo: Baifukan, 1990), 17–31.

10. Zuckerkandl and Pauling, "Evolutionary divergence and convergence in proteins," 148.

11. Zuckerkandl and Pauling, "Molecules as documents of evolutionary history."

12. See M. Kimura, "The rate of molecular evolution considered from the standpoint of population genetics," *Proceedings of the National Academy of Sciences* 63 (1969): 1181–1188; Jack King and Thomas Jukes, "Non-Darwinian evolution," *Science* 164 (1969), 788–798; M. Kimura, *The neutral theory of molecular evolution* (Cambridge: Cambridge University Press, 1983).

13. Zuckerkandl and Pauling, "Evolutionary divergence and convergence in proteins," 164.

14. R.F. Doolittle, "The comparative biochemistry of blood coagulation" (Ph.D. diss., Harvard University, 1961). For an overview, see R.F. Doolittle and D.F. Feng, "Reconstructing the history of vertebrate blood coagulation from a consideration of the amino acid sequences of clotting proteins," *Cold Spring Harbor Symposia on Quantitative Biology* 52 (1987): 869–874.

15. W.M. Fitch and E. Margoliash, "Construction of phylogenetic trees," *Science* 155 (1967): 279–284.

16. E. Margoliash, "Primary structure and evolution in cytochrome *c*," *Proceedings of the National Academy of Sciences* 50 (1963): 672–679.

17. Dayhoff's application of computers began with her Ph.D. thesis project in quantum chemistry, completed in 1948 at Columbia University, in which she devised a method of applying punched-card machines to calculate molecular resonance energies of several polycyclic organic molecules (those composed of many rings of bonds). M.O. Dayhoff and G.E. Kimball, "Punched card calculation of resonance energies," *Journal of Chemical Physics* 17 (1949): 706–717 (Ph.D. Thesis, Columbia University, Graduate School of Chemistry, 1949).

18. See Richard V. Eck and Margaret O. Dayhoff, *Atlas of protein sequence and structure*, 5 vols. (Silver Spring, MD: National Biomedical Research Foundation, 1969), Vol. 4: xxi.

19. Eck and Dayhoff, *Atlas of protein sequence and structure* (1966), Vol. 1: xxiv.

20. In 1980, Dayhoff began an online computer database, one of the largest in the world.

21. Eck and Dayhoff, *Atlas of protein sequence and structure* (1969), Vol. 4: xxi.

22. Ibid., 2.

23. Ibid., 4.

24. Ibid., 5.

25. B.J. McCarthy and E.T. Bolton, "An approach to the measurement of genetic relatedness among organisms," *Proceedings of the National Academy of Sciences* 50 (1963): 156–164.

26. J. de Ley, "Molecular biology and bacterial phylogeny," in T. Dobzhansky, M.K. Hects, and W.C. Steare, eds., *Evolutionary biology*, 2 vols. (Amsterdam: North Holland, 1968), Vol. 2: 103–156, at 105.

27. Ibid., 128.

28. Ibid., 123.

29. Ibid., 140–141.

30. Ibid., 123.

31. R.Y. Stanier, "Toward an evolutionary taxonomy of the bacteria," in A. Perez-Miravete and D. Peláez, eds., *Recent advances in microbiology* (Mexico City: Asociatión Mexicana de Microbiología, 1971), Vol. 7: 595–604, at 595.

32. On the dispute over cladism, see David Hull, "Certainty and circularity in evolutionary taxonomy," *Evolution* 21 (1967): 174–189; Hull, "Contemporary systematic philosophies," *Annual Review of Ecology and Systematics* 1 (1970): 19–54; Hull, *Science as process: An evolutionary account of the social and conceptual development of science* (Chicago: University of Chicago Press, 1988).

33. W. Hennig, *Phylogenetic systematics*, translation by D. Davis and R. Zangerl (Urbana: University of Illinois Press, 1966).

34. Sergius G. Kiriakoff, "Cladism and phylogeny," *Systematic Zoology* 15 (1966): 91–93, at 92.

35. R.R. Sokal and P.H.A. Sneath, *Principles of numerical taxonomy* (San Francisco: W.H. Freeman, 1963).

36. See P.H.A. Sneath, "Thirty years of numerical taxonomy," *Systematic Biology* 44 (1995): 281–298, at 283.

37. See J. Hagen, "The introduction of computers into systematic research in the United States during the 1960s," *Studies in Philosophy of Biological and Biomedical Sciences* 32 (2001): 291–314. See also M.T. MacDonnel and R.R. Colwell, "The contribution of numerical taxonomy to the systematics of Gram-negative bacteria," 107–128 in M. Goodfellow, D. Jones, and F.G. Priest, eds., *Computer assisted bacteria systematics* (Orlando: Academic Press, 1985).

38. P.H. Sneath and R.R. Sokal, *Numerical taxonomy* (San Francisco: W.H. Freeman, 1973).

39. Sokal and Sneath, *Principles of numerical taxonomy*, 97.

40. Ibid., 98–99.

41. Ernst Mayr, "Numerical phenetics and taxonomy," *Systematic Zoology* 14 (1965): 73–97, at 95.

42. Ernst Mayr, "Cladistic analysis or cladistic classification?" *Zeitschrift für Zoologische und Systematische Evolutionsforschung* 12 (1974): 94–128, at 104.

43. Ibid., 122.

44. Ibid.

45. Ibid.

46. Willi Hennig, "Cladistic analysis or cladistic classification? A reply to Ernst Mayr," *Systematic Zoology* 24 (1975): 244–256.

47. P.H.A. Sneath, "Phylogeny of micro-organisms," *Symposia of the Society for General Microbiology* 24 (1974): 1–39, at 6.

48. Ibid., 1.

49. Ibid., 2.

50. De Ley, "Molecular biology and bacterial phylogeny," 113.

51. Ibid., 123.

52. Ibid., 137.

53. J. Lederberg, and P.R. Edwards, "Serotypic recombination in *Salmonella*," *Journal of Immunology* 71 (1953): 232–240; P. Schaeffer and M.E. Ritz, "Transfer interspecifique d'un charactere héréditaire chez des bactéries du genre Hemophilus," *Comptes Rendus de l'Academie des Sciences* 240 (1955): 1491–1493; E.S. Lennox, "Transduction of linked genetic characters of the host by bacteriophage P1," *Virology* 1 (1955): 190–206.

54. S.E. Luria, and M. Delbrück, "Mutations of bacteria from virus sensitivity to virus resistance," *Genetics* 28 (1943): 491–511.

55. Ibid., 510.

56. M. Demerec, "Origin of bacterial resistance to antibiotics," *Journal of Bacteriology* 56 (1948): 63–74, at 63.

57. Joshua Lederberg and Esther Lederberg, "Replica plating and indirect selection of bacterial mutants," *Journal of Bacteriology* 63 (1952): 399–406, at 406.

58. B.A.D. Stocker, "Bacteriophages and bacterial classification," *Journal of General Microbiology* 12 (1955): 375–381, at 377–378.

59. W. Hayes, "Recombination in bact. coli K-12: Unidirectional transfer of genetic material," *Nature* 159 (1952): 118–119.

60. W. Hayes. "Observations on a transmissible agent determining sexual differentiation in *Bact. coli.*" *Journal of General Microbiology* 8 (1953): 72–88.

61. S.E. Luria and J. Burrous, "Hybridization between *Escherichia coli* and *Shigella*," *Journal of Bacteriology* 74 (1957): 461–476, at 461.

62. S.E. Luria, "Recent advances in bacterial genetics," *Bacteriological Reviews* 11 (1947): 1–40, at 32.

63. Luria and Burrous, "Hybridization between *Escherichia coli* and *Shigella*," 461.

64. Roger Stanier, Michael Douderoff, and Edward Adelberg, *The microbial world* (Englewood Cliffs, Prentice-Hall, 1957), 616.

65. Stanier, Doudoroff, and Adelberg, *The microbial world*, 2nd ed. (Englewood Cliffs: Prentice-Hall, 1962), 466.

66. Ibid., 509.

67. S.G. Bradley, "Applied significance of polyvalent bacteriophages," *Advances in Applied Microbiology* 8 (1968): 101–135.

68. Susan Wright, *Molecular politics: Developing American and British regulatory policy for genetic engineering, 1972–1982* (Chicago: University of Chicago Press, 1994).

69. See W.C. Summers, "A historical introduction," in K. Lewis, A.A. Salyers, H.W. Taber, and R.G. Wax, eds., *Bacterial resistance to antimicrobials* (New York: Marcel Dekker, 2002), xvii–xxiii.

70. Mary Barber, "Coagulase-positive staphylococci resistant to penicillin," *Journal of Pathology and Bacteriology* 59 (1947): 373–384.

71. Mary Barber, "Staphylococcal infection due to penicillin-resistant strains," *British Medical Journal* 1947: 863–865, at 863, 865.

72. Ibid., 864; M. Barber, "Hospital infection yesterday and today," *Journal of Clinical Pathology* (London) 14 (1961): 2–10.

73. Barber, "Hospital infection yesterday and today," 9.

74. Ibid., 4.

75. Angela Creager, "Adaptation or selection? Old issues and new stakes in the postwar debates over bacterial drug resistance," *Studies in the History and Philosophy of Biology and Biomedicine* 38 (2007): 159–190. Such issues about extrachromosomal hereditary factors and the inheritance of acquired characteristics took on additional significance in the midst of the cold war. See Jan Sapp, *Beyond the gene* (New York: Oxford University Press, 1987).

76. Ibid., 5.

77. T.D. Brock, *The emergence of bacterial genetics* (Cold Spring Harbor, NY: Cold Spring Harbor Laboratory Press, 1990), 107.

78. F. Jacob and E.L. Wollman, "Les épisomes, élements génétiques ajoutés," *Comptes Rendus de l'Académie des Sciences* 247 (1958): 75–92.

79. T. Watanabe, "Infective heredity of multiple drug resistance in bacteria," *Bacteriological Reviews* 27 (1963): 87–115, at 87.

80. Watanabe, "Infective heredity."

81. Ibid., 108.

82. R.P. Novick, and S.I. Morse, "*In vivo* transmission of drug resistance factors between strains of *Staphylococcus aureus*," *Journal of Experimental Medicine* 125 (1967): 45–59; E. Meynell, G.G. Meynell, and N. Datta, "Phylogenetic relationships of drug-resistances factors and other transmissible bacterial plasmids," *Bacteriological Reviews* 32 (1968): 55–83; H. Richmond, "Extrachromosomal elements and the spread of antibiotic resistance in bacteria," *Biochemistry Journal* 113 (1969): 225–234.

83. See H.C. Neu, "The crisis in antibiotic resistance," *Science* 257 (1992): 1064–1073.

84. Anonymous, "Antibiotics and animals," *Lancet* 2 (1969): 1173–1175, at 1173.

85. A. Williams-Smith, and S. Halls, "Transfer of antibiotic resistance from animal and human strains of *Escherichia coli* to resident *E. coli*," *Lancet* 1 (1969): 1174–1176.

86. Anonymous, "Antibiotics and animals."

87. Norman Anderson, "Evolutionary significance of virus infection," *Nature* 227 (1970): 1346–1347, at 1346.

88. Ibid.

89. Ibid.

90. Stanier, "Toward an evolutionary taxonomy of the bacteria," 597.

91. E.S. Anderson, "Possible importance of transfer factors in bacterial evolution," *Nature* 209 (1966): 637–638, at 637.

92. Ibid., 638.

93. Dorothy Jones and P.H.A. Sneath, "Genetic transfer and bacterial taxonomy," *Bacteriological Reviews* 34 (1970): 40–81, at 44–45.

94. Ibid., 69.

95. Ibid.

96. R.W. Hedges, "The pattern of evolutionary change in bacteria," *Heredity* 28 (1972): 39–48.

97. Ibid., 40. That the grouping of related genes was associated with the partial genetic transfer mechanisms of bacteria was suggested when the non-randomness of the genes on the bacterial chromosome was first observed. See M. Demerec, "A comparative study of certain gene loci in *Salmonella*," *Cold Spring Harbor Symposia on Quantitative Biology* 21 (1957): 113–121; R.C. Clowes, "Fine genetic structure as revealed by transduction," *Symposia of the Society for General Microbiology* 10 (1960): 92–114.

98. Hedges, "The pattern of evolutionary change in bacteria," 39.

99. See A. Lwoff, "La notion d'espèce bactérienne à la lumière des découvertes recentes. L'espèce bactérienne," *Annales de l'Institut Pasteur* 94 (1958): 137; P. Shaeffer, "La notion d'espèce bactérienne à la lumière des découvertes récentes. La notion d'espèce après les recherches récente de génétique," *Annales de l'Institut Pasteur* 94 (1958): 167–178.

100. S.T. Cowan, "The microbial species—a macromyth," *Symposia of the Society for General Microbiology* 12 (1962), 433–455, at 451.

101. Ibid., 445.

102. Ibid., 446.

103. D. Dubnau, I. Smith, P. Morell, and J. Marmur, "Gene conservation in *Bacillus* species. I. Conserved genetic and nucleic acid base sequence homologies," *Proceedings of the National Academy of Sciences* 54 (1965): 491–498, at 495.

104. Ibid., 497.

105. Ibid., 495.

106. Ibid., 491.

107. R.H. Doi and R.T. Igarashi, "Conservation of ribosomal and messenger ribonucleic acid cistrons in *Bacillus* species," *Journal of Bacteriology* 90 (1965): 384–390.

108. R.L. Moore and B.J. McCarthy, "Comparative study of ribosomal ribonucleic acid cistrons in enterobacteria and myxobacteria," *Journal of Bacteriology* 94 (1967): 1066. See also M. Takahashi, M. Saito, and Y. Ikeda, "Species specificity of the ribosomal RNA cistrons in bacteria," *Biochimica Biophysica Acta* 134 (1967): 124–133.

109. Sneath, "Phylogeny of micro-organisms," 22.

110. Ibid., 30.

111. Ibid., 17.

112. Ibid., 2–3.

113. Ibid., 32.

114. Ibid.

CHAPTER 11

1. Author interview with Carl Woese, University of Illinois–Urbana, November 18, 2001; C.R. Woese, "Thermal inactivation of animal viruses," *Proceedings of the New York Academy of Sciences* 83 (1960): 741–751; C.R. Woese, R. Langridge, and H.J. Morowitz, "Microsome distribution during germination of bacterial spores," *Journal of Bacteriology* 79 (1960): 777–782.

2. C.R. Woese, "Phage induction in germinating spores of *Bacillus megaterium*," *Radiation Research* 13 (1960): 871–878.

3. The Knowles Lab was equivalent to the Bell Labs at AT&T; its luminaries included William David Coolidge, the inventor of the X-ray tube, and Irving Langmuir, the surface chemist and cloud seeder who was awarded the Nobel Prize in 1932.

4. On the cracking of the genetic code, see Lily Kay, *Who wrote the book of life? A history of the genetic code* (Stanford: Stanford University Press, 2000), 156–192.

5. See F.H. Crick, J.S. Griffith, and L.E. Orgel, "Codes without commas," *Proceedings of the National Academy of Sciences* 43 (1957): 416–420.

6. George Gamow, "Possible relation between deoxyribonucleic acids and protein structure," *Nature* 173 (1954): 318; Gamow, "Information transfer in the living cell," *Scientific American* 193 (1955): 70–78.

7. Gamow, "Possible relation," 318.

8. See Kay, *Who wrote the book of life?* 131–155.

9. Sydney Brenner, "On the impossibility of all overlapping triplet codes in information transfer from nucleic acid to proteins," *Proceedings of the National Academy of Sciences* 43 (1957): 687–94. That same year, Francis Crick, John Griffith, and Leslie Orgel considered a commaless nonoverlapping triplet code. See Crick et al., "Codes without commas."

10. F.H.C. Crick, L. Barnett, S. Brenner, and R.J. Watts-Tobin, "General nature of the genetic code for proteins," *Nature* 192 (1961): 1227–1232.

11. See Kay, *Who wrote the book of life?* 156–192.

12. C.R. Woese, "Composition of various ribonucleic acid fractions from microorganisms of different deoxyribonucleic acid composition," *Nature* 189 (1961): 920–921.

13. C.R. Woese, "Coding ratio for the ribonucleic acid viruses," *Nature* 190 (1961): 697–698.

14. C.R. Woese, "Nature of the biological code," *Nature* 194 (1962): 1114–1115.

15. Marshall W. Nirenberg and J. Heinrich Matthaei, "The dependence of cell-free protein synthesis in *E. coli* upon naturally occurring or synthetic polyribonucleotides," *Proceedings of the National Academy of Sciences* 47 (1961): 1588–1602; J. Heinrich Matthaei and Marshall W. Nirenberg, "Characteristics and stabilization of DNA ase-sensitive protein synthesis in *E. coli* extracts," *Proceedings of the National Academy of Sciences* 47 (1961): 1580–1588, at 1588.

16. Philip Leder and Marshall Nirenberg, "RNA codewords and protein synthesis: II. Nucleotide sequence of a valine RNA codeword," *Proceedings of the National Academy of Sciences* 52 (1964): 420–427; M. Nirenberg, P. Leder, M. Bernfield, R. Brimacombe, J. Trupin, F. Rottman, and C. O'Neal, "RNA codewords and protein synthesis: VII. On the general nature of the RNA code," *Proceedings of the National Academy of Sciences* 53 (1965): 1161–1168. See Kay, *Who wrote the book of life?* 235–293.

17. Jacques Monod and François Jacob, "General conclusions: Teleonomic mechanisms in cellular metabolism, growth and differentiation," *Cold Spring Harbor Symposia on Quantitative Biology* 26 (1961): 389–401.

18. S. Brenner, F. Jacob, and M. Meselson, "An unstable intermediate carrying information from genes to ribosomes for protein synthesis," *Nature* 190 (1961): 576.

19. Francois Jacob, and Jacque Monod, "Genetic regulatory mechanisms in the synthesis of proteins," *Journal of Molecular Biology* 3 (1961): 318–356. See Kay, *Who wrote the book of life?* and Horace Judson, *The eighth day of creation: The makers of the revolution in biology* (New York: Simon and Schuster, 1979).

20. See Gunther Stent, *Molecular genetics; an introductory narrative* (San Francisco: W.H. Freeman, 1871): 499–506.

21. See C.R. Woese, "Universality in the genetic code," *Science* 144 (1964): 1030–1031.

22. S.L., Miller, "A production of amino acids under possible primitive earth conditions," *Science* 117 (1953): 528–529.

23. C.R. Woese, "On the evolution of the genetic code," *Proceedings of the National Academy of Sciences* 54 (1964): 1546–1552, at 1546.

24. On plasmagene theory, see Jan Sapp, *Beyond the gene: Cytoplasmic inheritance and the struggle for authority in genetics* (New York: Oxford University Press, 1987).

25. Hall, B.D. and S. Spiegelman, "Sequence complementarity of T2-DNA and T2-specific RNA," *Proceedings of the National Academy of Sciences* 47 (1961): 137–146, at 146; S. Spiegelman, B.D. Hall, and R. Storck, "The occurrence of natural DNA-RNA complexes in *E. coli* infected with T2," *Proceedings of the National Academy of Sciences* 47 (1961): 1135–1141.

26. Author interview with Carl Woese, Urbana, Illinois, November 19, 2001.

27. C.R. Woese, "Order in the genetic code," *Proceedings of the National Academy of Sciences* 54 (1965): 71–75, at 71.

28. C.R. Woese, *The genetic code: The molecular basis for genetic expression* (New York: Harper and Row, 1967), 194.

29. Carl Woese, "The problem of evolving a genetic code," *Bioscience* 20 (1970): 471–485, at 471.

30. Woese, "Order in the genetic code," 71.

31. Ibid.

32. F.H. Crick, "The origin of the genetic code," *Journal of Molecular Biology* 38 (1968): 367–379, at 370.

33. Francis Crick, *What mad pursuit: A personal view of scientific discovery* (New York: Basic Books, 1988), 96. Crick's unpublished note was titled "On degenerative templates and the adaptor hypothesis: A note to the RNA Club." See F.H.C. Crick, "On protein synthesis," *Symposia of the Society for Experimental Biology* 12 (1958): 138–185.

34. Crick, "On protein synthesis," 155.

35. Ibid., 156.

36. Ibid., 155.

37. Ibid.

38. Crick, *What mad pursuit*, 96.

39. See, e.g., Stent, *An introduction to molecular genetics*, 499–506.

40. See Mahlon B. Hoagland, "The present status of the adaptor hypothesis," *Brookhaven Symposia in Biology* 12 (1959): 40–46; Hans-Jorg Rheinberger,

"Experiment, difference, and writing: I. Tracing protein synthesis," *Studies in the History and Philosophy of Science* 23 (1991): 305–331; Rheinberger, "Experiment, difference, and writing: II. The laboratory production of transfer RNA," *Studies in the History and Philosophy of Science* 23 (1991): 389–422; Mahlon Hoagland, "Biochemistry or molecular biology? The discovery of 'soluble RNA,'" *Trends in Biological Sciences* 21 (1996): 77–80. See also Kay, *Who wrote the book of life?* 243–244.

41. M.B. Hoagland, M.L., Stephenson, J.F. Scott, L.I. Hecht, and P.C. Zamecnik, "A soluble ribonucleic acid intermediate in protein synthesis," *Journal of Biological Chemistry* 231 (1958): 241–257, at 255.

42. Ibid., 157–158.

43. Hoagland, "The present status of the adaptor hypothesis," 44.

44. Ibid.

45. Ibid., 45.

46. Decades later, Hoagland commented that, as biochemists and "lacking involvement with the coding problem," he and his colleagues "were not as uncomfortable with its size." Hoagland "Biochemistry or molecular biology?" 79. Crick also commented 30 years later that although "these transfer RNA molecules were considerably bigger than I expected....I soon saw that there were no grounds for my objection." Crick, *What mad pursuit*, 96.

47. C.R. Woese, "Translation: In retrospect and prospect," *RNA* 7 (2001): 1055–1067, at 1061.

48. Carl Woese, "The evolution of cellular tape reading processes and macromolecular complexity," *Brookhaven Symposia in Biology* 23 (1972): 326–365, at 328.

49. Woese, "Translation: In retrospect and prospect," 1064.

50. Woese, "The evolution of cellular tape reading processes," 327.

51. Ibid., 332.

52. Alfred North Whitehead, *Science and the modern world* (New York: Macmillan, 1925; reprint, New York: Free Press, 1967), 75.

53. Ibid., 57.

54. Carl Woese, "The evolution of cellular tape reading processes," 332–333.

55. Woese, *The genetic code*, 5.

56. Ibid., 114.

57. Ibid., 189. See also C.R. Woese, "Just so stories and Rube Goldberg machines: Speculations on the origin of the protein synthetic machinery," in G. Chambliss, G.R. Craven, J. Davies, K. Davis, L. Kaban, and M. Nomura, eds., *Ribosomes: Structure, function and genetics* (Baltimore: University Park Press, 1980), 357–373.

58. Woese, *The genetic code*, 190.

59. Ibid., 192.

60. Ibid.

61. Ibid., 193.

62. Carl Woese, "Molecular mechanics of translation: A reciprocating ratchet mechanism," *Nature* 226 (1970): 817–820.

63. Carl Woese, "The evolution of cellular tape reading processes," 364–365.

64. C.R. Woese, "The emergence of genetic organization," in C. Ponnamperuma, ed., *Exobiology* (Amsterdam: North-Holland, 1972), 301–341.

65. Woese, "Translation: In retrospect and prospect," 1063.

66. Woese, *The genetic code*, 188–194. No attempt was made to test the ratchet hypothesis. See F.H.C. Crick, S. Brenner, A Klug, and G. Pieczenik, "A speculation on the origin of protein synthesis," *Origins of Life* 7 (1976): 389–397.

67. T.R. Cech, "Self-splicing RNA: Implication for evolution," *International Review of Cytology* 93 (1985): 3–22; Cech, "RNA as enzyme," *Scientific American* 255 (1986): 64–74; Cech, "The chemistry of self-splicing RNA and RNA enzymes," *Science* 236 (1987): 1532–1539.

68. The expression "RNA world" was coined by Walter Gilbert, "The RNA world," *Nature* 319 (1986): 618.

69. James D. Watson, *Molecular biology of the gene* (New York: W.A. Benjamin, 1965), 1–2.

70. Theodosius Dobzhansky, "Nothing in biology makes sense except in the light of evolution," *American Biology Teacher* 35 (1973): 125–129.

71. See Gunther Stent, "DNA," *Daedalus* 99 (1970): 909–937.

72. T.M. Sonneborn, "Degeneracy of the genetic code: Extent, nature and genetic implications," in V. Bryson and H. Vogel, eds., *Evolving genes and proteins* (New York: Academic Press, 1965), 377–397; Woese, "Order in the genetic code"; Woese, *The genetic code*; A.L. Goldberg and R.E. Wittes, "Genetic code: Aspects of organization," *Science* 153 (1966): 420–424; F. Crick, "Origin of the genetic code," *Nature* 213 (1967): 119; Crick, "Origin of the genetic code," *Journal of Molecular Biology* 38 (1968): 367–379; L.E. Orgel, "Evolution of the genetic apparatus," *Journal of Molecular Biology* 38 (1968): 381–393.

73. François Jacob, *The logic of living systems: A history of heredity*, translation by Betty Spillmann (London: Allen Lane, 1974), 305.

74. Ibid.

75. Carl Woese to Francis Crick, June 24, 1969 (Woese Papers, University of Illinois).

76. C.R. Woese, "The genetic code in prokaryotes and eukaryotes," *Symposia of the Society for General Microbiology* 20 (1970), 39–54.

77. Ibid., 41.

78. Ibid., 49–50.

79. Ibid., p. 49. See also C.R. Woese, "Evolution of the genetic code," *Naturwissenschaften* 60 (1973): 447–459.

80. F. Sanger, G.G. Brownless, and B.G. Barrell, "A two-dimensional fractionation procedure for radioactive nucleotides," *Journal of Molecular Biology* 13 (1965): 373–398, at 396.

81. Ibid.

82. Sogin completed his dissertation in 1972: M.L. Sogin, *Relationships among precursor and mature ribosomal RNAs* (Ph.D. diss., University of Illinois, 1972).

83. Woese devised a technique he called T_1 overcutting, which was essential to resolving the order of bases in pyrimidine stretches. As George Fox commented, "Carl's lab was light-years ahead of the competition in fingerprinting technology. This is a tribute to Carl's innate ability as an experimentalist—he

was always thinking about ways to improve the methods, get more information." George Fox to the author, January 21, 2005.

84. George Fox, Kenneth Pechman, and Carl Woese, "Comparative cataloging of 16S ribosomal ribonucleic acid: Molecular approach to prokaryotic systematics," *International Journal of Systematic Bacteriology* 27 (1977): 44–57, at 44.

85. The first was the 16S rRNA gene from *E. coli*: J. Brosius, M.L. Palmer, P.J. Kennedy, and H.F. Noller "Complete nucleotide sequence of a 16S ribosomal RNA gene from *Escherichia coli*," *Proceedings of the National Academy of Sciences* 75 (1978): 4801–4805. Woese and his colleagues published the first complete sequence of archaebacterial 16S rRNA five years later: R. Gupta, J.M. Lanter, and C.R. Woese, "Sequence of the 16S ribosomal RNA from *Halobacterium volcani*, an archaebacterium," *Science* 221 (1983): 656–659.

86. George Fox to the author, January 21, 2005. His thesis adviser was Dr. Phillip A. Rice.

87. See G.E. Fox to J.D. Watson, July 5, 1979 (Fox Papers, University of Houston).

88. The other instructors in the course included Anatol Eberhard, Kenneth H. Nealson, and Jane Gibson, a leader in bacterial photosynthesis at Cornell University, as well as Ralph Wolfe.

89. George Fox to the author, January 21, 2005.

90. See, e.g., C.R. Woese, M.L. Sogin, and L. Sutton, "Procaryote phylogeny I: Concerning the relatedness of *Aerobacter aerogenes* to *Escherichia coli*," *Journal of Molecular Evolution* 3 (1974): 293–299; L.B. Zablen, L. Bonen, R. Meyer, and C.R. Woese, "Procaryote phylogeny II: The phylogenetic status of *Pasteurella pestis*," *Journal of Molecular Evolution* 4 (1975): 347–358; L.B. Zablen and C.R. Woese, "Procaryote phylogeny IV: Phylogenetic status of a photosynthetic bacterium," *Journal of Molecular Evolution* 5 (1975): 25–34; C.R. Woese, M.L. Sogin, D. Stahl, B.J. Lewis, and L. Bonen, "A comparison of the 16S ribosomal RNA from mesophilic and thermophilic bacilli: Some modifications in the Sanger method for RNA sequencing," *Journal of Molecular Evolution* 7 (1976): 197–213.

91. C. Woese, G.E. Fox, L. Zablen, T. Uchida, L. Bonen, K. Pechman, B.J. Lewis, and D. Stahl, "Conservation of primary structure in 16S ribosomal RNA," *Nature* 254 (1975): 83–86.

92. Carl R. Woese and George E. Fox, "Phylogenetic structure of the prokaryotic domain: The primary kingdoms," *Proceedings of the National Academy of Sciences* 74 (1977): 5088–5090.

CHAPTER 12

1. Ralph S. Wolfe, "My kind of biology," *Annual Reviews of Microbiology* 45 (1991): 1–35, at 4.

2. Ibid.

3. Ibid., 5.

4. Ibid.

5. E.A. Wolin, M.J. Wolin, and R. Wolfe, "ATP-dependent formation of methane from methylcobalamin by extracts of *Methanobacillus omelianskii*," *Biochemical and Biophysical Research Communications* 12 (1963): 464–468.

6. See R.L. Switzer, E.R. Stadtman, and T.C. Stadtman, "H.A. Barker 1907–2000," *Biographical Memoirs of the National Academy of Sciences* 84 (2003): 3–20.

7. H.A. Barker, S. Ruben, and M.D. Kamen, "The reduction of radioactive carbon dioxide by methane-producing bacteria," *Proceedings of the National Academy of Sciences* 26 (1940): 426–430, at 426.

8. L. Demain and Ralph F. Wolfe, "Marvin P. Bryant," *Biographical Memoirs of the National Academy of Sciences* 81 (2002): 66–69. In previous editions they had been classified according to their diverse morphologically—rods, spirals, cocci, and sarcina cells, and thus were placed into many groups and spread intermittently throughout the manual.

9. M.P. Bryant, B.C. McBride, and R.S. Wolfe, "*Methanobacillus omelianskii*, a symbiotic association of two species of bacteria," *Archiv für Mikrobiologie* 59 (1967): 20–31.

10. Ibid., 9.

11. B.C. McBride and R.S. Wolfe, "A new coenzyme of methyl transfer, coenzyme M," *Biochemistry* 10 (1971): 2317–2324.

12. C.D. Taylor, and R.S. Wolfe, "Structure and methylation of coenzyme M," *Journal of Biological Chemistry* 249 (1974): 4879–4885; C.D. Taylor, B.C. McBride, R.S. Wolfe, and M.P. Bryant, "Coenzyme M, essential for growth of a rumen strain of *Methanobacterium ruminantium*," *Journal of Bacteriology* 120 (1974): 974–975.

13. W.E. Balch, and R.S. Wolfe, "New approach to the cultivation of methanogenic bacteria: 2-Mercaptoethanesulfonic acid (HS-CoM)-dependent growth of *Methanobacterium ruminantium* in a pressurized atmosphere," *Applied and Environmental Microbiology* 32 (1976): 781–791.

14. A.J. Kluver and H.J.L. Donker, "Die Einheit in der Biochemie," *Chemie der Zelle und der Gewebe* 13 (1926): 134–190. See Wolfe, "My kind of biology," 11. See also C.B. van Niel, "The 'Delft School' and the rise of general microbiology," *Bacteriological Reviews* 13 (1949): 161–174, at 172.

15. Wolfe, "My kind of biology," 12.

16. Ralph Wolfe, "The Archaea: A personal overview of the formative years," *Prokaryotes* 3 (2006): 3–9, at 2.

17. Carl Woese to the author, January 21, 2005.

18. George Fox to the author, January 21, 2005.

19. Ibid.

20. Carl Woese, "The birth of the Archaea: A personal retrospective," introduction to Hans-Peter Klenk, *Archaea: Evolution, physiology, and molecular biology*, edited by Roger A. Garrett (Malden, MA: Blackwell, 2007), 1–15.

21. Wolfe, "The Archaea," 5; Wolfe, "My kind of biology," 13.

22. George Fox to the author, January 24, 2005.

23. Ibid.

24. George Fox to the author, January 21, 2005.

25. Carl Woese, "The birth of the Archaea."

26. Wolfe, "The Archaea," 2.

27. George Fox, comments to the author, October 10, 2008.

28. Ibid.

29. Ibid.

30. William Balch, Linda Magrum, George Fox, Ralph Wolfe, and Carl Woese, "An ancient divergence among the bacteria," *Journal of Molecular Evolution* 9 (1977): 305–311.

31. Ibid., 310.

32. Carl Woese, "A comment on methanogenic bacteria and the primitive ecology," *Journal of Molecular Evolution* 9 (1977): 369–371.

33. Barker et al., "The reduction of radioactive carbon dioxide."

34. C.R. Woese, "The genetic code in prokaryotes and eukaryotes," *Symposia of the Society for General Microbiology* 20 (1970): 39–54.

35. Carl R. Woese and George E. Fox, "The concept of cellular evolution," *Journal of Molecular Evolution* 10 (1977): 1–6, at 3.

36. Ibid. See also G.E. Fox and C.R. Woese, "5S RNA secondary structure," *Nature* 256 (1975): 505–507; C.R. Woese, G.E. Fox, L.B. Zablen, T. Uchida, L. Bonen, K.R. Pechman, B.J. Lewis, and D.A. Stahl, "Conservation of primary structure in SSU ribosomal RNA," *Nature* 254 (1975): 83–86.

37. Ibid., 85.

38. C.R. Woese and G.E. Fox, "Progenotes and the origin of the cytoplasm," manuscript submitted to the *Journal of Molecular Evolution*, March 1977 (Fox Papers, University of Houston).

39. Woese and Fox, "The concept of cellular evolution," 5.

40. L.B. Zablen, M.S. Kissle, C.R. Woese, and D.E. Buetow, "The phylogenetic origin of the chloroplast and the prokaryotic nature of its ribosomal RNA," *Proceedings of the National Academy of Sciences* 72 (1975): 2418–2422.

41. Carl R. Woese, "Endosymbionts and mitochondrial origins," *Journal of Molecular Evolution* 10 (1977): 93–96, at 93.

42. Anonymous peer review report for Carl Woese and George Fox, "Progenotes and the origin of the cytoplasm" (Woese Papers, University of Illinois).

43. Born in Vienna, Zuckerkandl's family moved to Paris in 1938 followed by Algiers to escape the Nazi persecution of Jews. At the end of the Second World War, he spent one year at the Sorbonne, and then earned a master's degree in biology at the University of Illinois in 1947 before returning to the Sorbonne for his Ph.D. He moved to the United States in 1959, as a postdoctoral fellow at the California Institute of Technology, where he and Linus Pauling began their collaborative research on the amino acid sequences of various hemoglobins and developed the concept of the molecular clock. In 1965, Zuckerkandl moved back to France as founding director of the Research Center for Macromolecular Biology of the Centre National de Recherche Scientifique. He returned to the United States as founding editor of the *JME*, which began in 1971. In the late 1970s he was president of the Linus Pauling Institute of Science and Medicine in Palo Alto. As late as 1978, he drew no salary from the Linus Pauling Institute; his salary was from France. See Emile Zuckerkandl to Carl Woese, February 28, 1978 (Woese Papers, University of Illinois).

44. Emile Zuckerkandl to Carl Woese, March 15, 1977 (Woese Papers, University of Illinois).

45. Carl Woese to Emile Zuckerkandl, March 29, 1977 (Woese Papers, University of Illinois).

46. Emile Zuckerkandl to Carl Woese, April 15, 1977 (Woese Papers, University of Illinois).

47. G.E., Fox, L.J. Magrum, W.E. Balch, R.S. Wolfe, and C.R. Woese, "Classification of methanogenic bacteria by SSU ribosomal RNA characterization," *Proceedings of the National Academy of Sciences* 74 (1977): 4537–4541.

48. Ibid., 4538–4539.

49. Ibid., 4540.

50. Ibid. See also R. Gupta, and C.R. Woese, "Unusual modification patterns in the transfers RNAs of archaebacteria," *Current Microbiology* 4 (1980): 245–249.

51. O. Kandler and H. Hippe, "Lack of peptoglycan in the cell walls of *Methanosarcina barkeri*," *Archives of Microbiology* 113 (1977): 57–60.

52. Carl R. Woese and George E. Fox, "Phylogenetic structure of the prokaryotic domain: The primary kingdoms," *Proceedings of the National Academy of Sciences* 74 (1977): 5088–5090, at 5089.

53. T.M. Sonneborn, "Degeneracy of the genetic code: Extent, nature, and genetic implications," in H. Vogel, ed., *Evolving genes and proteins* (New York: Academic Press, 1965), 377–401; C.R. Woese, "Evolution of macromolecular complexity," *Journal of Theoretical Biology* 33 (1971): 29–34. Sonneborn was known for challenging biological orthodoxy with his research on non-Mendelian genetics. See Jan Sapp, *Beyond the gene: Cytoplasmic inheritance and the struggle for authority in genetics* (New York: Oxford University Press, 1987).

54. The name "archaebacteria" was suggested by Woese's research assistant Linda Magrum and his colleague David Nanney.

55. Woese and Fox, "Phylogenetic structure of the prokaryotic domain," 5089.

56. Ibid.

57. Ibid., 5090.

58. Ibid.

59. Ibid., 5088.

60. Ibid.

61. Ibid., 5090.

62. Ibid., 5089. See Jean Whatley, "Bacteria and nuclei in *Pelomyxa Palustris*: Comments on the theory of serial endosymbiosis," *New Phytologist* 76 (1976): 111–120.

CHAPTER 13

1. Steven J. Dick and James Strick, *The living universe: NASA and the development of astrobiology* (New Brunswick: Rutgers University Press, 2004), 47.

2. Ibid.; Joshua Lederberg, "Exobiology, experimental approaches to life beyond the earth," in L.V. Berkner and H. Odishaw, eds., *Science in space* (New York: McGraw-Hill, 1961), 407–425; Berkner and H. Odishaw, "Signs of life: Criterion system of exobiology," *Nature* 207 (1965): 9–13.

3. NASA had become a significant player, along with the National science Foundation, the National Institutes of Health, and the Atomic Energy

Commission, in the funding of biological research, especially those on the origins of life. Dick and Strick, *The living universe*, 37.

4. Anonymous, "Dawn of life: Archae-bacteria or methanogens," *Time* 110 (November 14, 1977): 56; P. Gwynne, "Oldest life: Methanogens," *Newsweek* 90 (November 14, 1977): 81.

5. Richard D. Lyons, "Scientists discover a form of life that predates higher organisms," *New York Times* (November 3, 1977): 1 and A20. It overshadowed the U.S. Court of Appeals upholding of the conviction of Patty Hearst for robbing a bank with a group of criminals.

6. Lyons, "Scientists discover a form of life," A20.

7. Ralph Wolfe, "The Archaea: A personal overview of the formative years," in Martin Dworkin, Stanley Falkow, and Eugene Rosenberg, eds., *The prokaryotes: An evolving electronic resource for the microbiological community*, 3rd ed., release 3.7 (New York: Springer, November 2, 2001), available at http://link.springer-ny.com/link/service/books/10125/.

8. Ibid.

9. Ibid., 1.

10. Ibid.

11. Boyce Rensberger, "Earliest life forms are found in rocks," *New York Times* (October 24, 1977): 1.

12. C. Sysbesma to Carl Woese, November 7, 1977 (Woese Papers, University of Illinois).

13. Carl Woese to C. Sysbesma, November 15, 1977 (Woese Papers, University of Illinois).

14. Joshua Lederberg to Denis Smith, November 9, 1977 (Lederberg Archives, Profiles in Science, National Library of Medicine).

15. T.M. Sonneborn to Carl Woese, December 13, 1976 (Woese Papers, University of Illinois).

16. Anonymous peer reviewer report for Carl R. Woese and George E. Fox, "Phylogenetic structure of the prokaryotic domain: The primary kingdoms," *Proceedings of the National Academy of Sciences* 74 (1977): 5088–5090.

17. Anonymous peer reviewer report for Woese and Fox, "Phylogenetic structure of the prokaryotic domain" (Woese Papers University of Illinois); T.M. Sonneborn to Carl Woese, July 6, 1977, and August 15, 1977 (Woese Papers University of Illinois). Sonneborn sent the revised paper on to *Proceedings of the National Academy of Sciences* on August 15, 1977.

18. Anonymous peer reviewer report on manuscript submitted by G.E. Fox, L.J. Magrum, W.E. Balch, R.S. Wolfe, and C.R. Woese, "Phylogenetic analysis of methanogenic organisms," published as G.E. Fox, L.J. Magrum, W.E. Balch, R.S. Wolfe, and C.R. Woese, "Classification of methanogenic bacteria by SSU ribosomal RNA characterization," *Proceedings of the National Academy of Sciences* 74 (1977): 4537–4541 (Woese Papers, University of Illinois).

19. Charles Darwin, *On the origin of species*, facsimile of 1859 edition, with an introduction by Ernst Mayr (Cambridge: Harvard University Press, 1964), 417.

20. Roger Y. Stanier, Michael Doudoroff, and Edward A. Adelberg, *The microbial world*, 3rd ed. (Englewood Cliffs: Prentice-Hall, 1970), 528–529.

21. Roger Stanier to Carl Woese, February 9, 1977 (Woese Papers, University of Illinois).

22. Carl Woese to Roger Stanier, February 17, 1977 (Woese Papers, University of Illinois).

23. Carl Woese to R.G.E. Murray, July 18, 1977 (Woese Papers, University of Illinois).

24. R.G.E. Murray to Carl Woese, November 18, 1977 (Woese Papers, University of Illinois).

25. R.G.E. Murray to Carl Woese, December 6, 1977 (Woese Papers, University of Illinois).

26. George Fox, Kenneth R. Pechman, and Carl R. Woese, "Comparative cataloging of SSU ribosomal ribonucleic acid: Molecular approach to procaryotic systematics," *Internal Journal of Systematic Bacteriology* 27 (1977): 44–77, at 57.

27. Anonymous peer reviewer report on Fox et al., "Phylogenetic analysis of methanogenic organisms" (Woese Papers, University of Illinois).

28. Anonymous reviewer to H.A. Barker, "Re: Dr. Woese's comments of 28 June 1977 (Woese Papers, University of Illinois).

29. C.R. Woese to G.E. Fox, November 19, 1977 (Woese Papers, University of Illinois).

30. François Jacob to Carl Woese, September 21, 1977 (Woese Papers, University of Illinois).

31. Cyril Ponnamperuma to G.E. Fox, May 30, 1979 (Fox Papers, University of Houston).

32. R.D. Smith to G.E. Fox, November 15, 1977 (Fox Papers, University of Houston).

33. J.D. Watson to G.E. Fox, August 12, 1977 (Fox Papers, University of Houston).

34. Thomas H. Maugh II, "Phylogeny: Are methanogens a third class of life?" *Science* 198 (1977): 812; J.J. Wilkinson, "Methanogenic bacteria: A new primary kingdom?" *Nature* 271 (1978): 707.

35. Maugh, "Phylogeny."

36. Wilkinson, "Methanogenic bacteria."

CHAPTER 14

1. Stephen Gould and Richard Lewontin, "The spandrels of San Marco and the Panglossian paradigm: A critique of the adaptationist program," *Proceedings of the Royal Society Series B* 205 (1978): 581–598. On the effect of Gould and Lewontin's "Spandrels" paper, see Massimo Pigliucci and Jonathan Kaplan, "The fall and rise of Dr. Pangloss: Adaptationism and the Spandrels paper 20 years later," *Trends in Ecology and Evolution* 15 (2000): 66–69.

2. On insect flight, see, e.g., J.H. Marden and M.G. Kramer, "Locomotor performance of insects with rudimentary wings," *Nature* 377 (1995): 332–334; Marden, M.R. Wolf, and K.E. Weber. "Aerial performance of *Drosophila melanogaster* from populations selected for upwind flight ability," *Journal of Experimental Biology* 200 (1997): 2747–2755.

3. Voltaire, *Candide, or Optimism*, translation by John Butt (London: Penguin Books, 1947), 20. See also William Bateson, "Address of the President of the

British Association for the Advancement of Science," *Science* 40 (1914): 287–302, at 293.

4. Charles Darwin, *On the Origin of Species,* facsimile of 1859 edition, with an introduction by Ernst Mayr (Cambridge: Harvard University Press, 1964), 414.

5. Ibid.

6. See R.Y. Stanier, M. Doudoroff, and E.A. Adelberg, *The microbial world,* 2nd ed. (Englewood Cliffs: Prentice-Hall, 1963), 358.

7. A.D. Brown and K.Y. Cho, "The walls of extremely halophilic cocci. Gram-positive bacteria lacking muramic acid," *Journal of General Microbiology* 62 (1970): 267–270; R. Reistad, "Cell wall of an extremely halophilic coccus. Investigation of ninhydrin-positive compounds," *Archiv für Mikrobiologie* 82 (1972): 24–30.

8. R.S. Breed, E.G.D. Murray, N.R. Smith, et al., *Bergey's manual of determinative bacteriology* (Baltimore: Williams and Wilkins, 1957), 208.

9. Carl R. Woese and George E. Fox, "Phylogenetic structure of the prokaryotic domain: The primary kingdoms," *Proceedings of the National Academy of Sciences* 74 (1977): 5088–5090, at 5090.

10. See Carl Woese to Otto Kandler, August 11, 1977 (Woese Papers, University of Illinois).

11. Ibid.

12. S.T. Bayley, "Composition of ribosomes of an extremely halophilic bacterium," *Journal of Molecular Biology* 15 (1966): 420–427.

13. Stanley Bayley to Carl Woese, January 6, 1978 (Woese Papers, University of Illinois).

14. N. Smith, A.T. Matheson, M. Yaguchi, G.E. Willick, and R.N. Nazar, "The 5-S RNA–protein complex from an extreme halophile, *Halobacterium cutirubrum*: Purification and characterization," *European Journal of Biochemistry* 89 (1978): 501–509; 501; Matheson, Makoto Yaguchi, and L.P. Visentin, "The conservation of amino acids in the N-terminal position of ribosomal and cytosol proteins from *Escherichia coli, Bacillus stearothermophilus,* and *Halobacterium cutirubrum,*" *Canadian Journal of Biochemistry* 53 (1975): 1323–1327; P. Falkenberg, Matheson, and C.F. Rollin, "The properties of ribosomal proteins from a moderate halophile," *Biochimica et Biophysica Acta* 434 (1976): 474–482.

15. R.N. Nazar, A.T. Matheson, and G. Bellemare, "Nucleotide sequence of *Halobacterium cutirubrum* ribosomal 5S ribonucleic acid." *Journal of Biological Chemistry* 253 (1978): 5464–5469, at 5464.

16. R.N. Nazar to the author, March 15, 2005.

17. Ibid.

18. Carl Woese to Alistair Matheson, January 17, 1978 (Woese Papers, University of Illinois).

19. Carl Woese, personal communication, May 2, 2005.

20. C.R. Woese to Emile Zuckerkandl, January 13, 1978 (Woese Papers, University of Illinois).

21. Emile Zuckerkandl to C.R. Woese, January 17, 1978 (Woese Papers, University of Illinois).

22. L.J. Magrum, K.R. Luehrsen, and C.R. Woese. "Are extreme halophiles actually "bacteria"? *Journal of Molecular Evolution* 11 (1978): 1–8, at 1.

23. C.R. Woese to Emile Zuckerkandl, February 2, 1978 (Woese Papers, University of Illinois).

24. Stanier et al, *The microbial world*, 354–355.

25. Ibid., 354.

26. See L.L. Campbell and B. Pace, "Physiology of growth at high temperatures," *Journal of Applied Bacteriology* 31 (1968): 24–35; R.W. Castenholz, "Aggregation in a thermophilic oscillatoria," *Nature* 215 (1967): 1285–1286; T. Cross, "Thermophilic actinomycetes," *Journal of Applied Bacteriology* 31 (1968): 36–53.

27. See Lethe E. Morrison and Fred W. Tanner, "Studies on thermophilic bacteria," *Journal of Bacteriology* 7 (1922): 343–366, at 344.

28. T.D. Brock, "Life at high temperatures," *Science* 158 (1967): 1012–1019, at 1013.

29. See T.D. Brock, "The Road to Yellowstone—and beyond," *Annual Review of Microbiology* 49 (1995): 1–28, at 17–18.

30. Brock, "Life at high temperatures," 1018.

31. G. Darland, T.D. Brock, W. Samsonoff, and S.F. Conti, "A thermophilic acidophilic mycoplasm isolated from a coal refuse pile," *Science* 170 (1970): 1416–1418.

32. Stanier et al., *The microbial world*, 184–185.

33. Ibid.

34. Darland et al., "A thermophilic acidophilic mycoplasma," 1417.

35. D.C. Wallace and H.J. Morowitz, "Genome size and evolution," *Chromosoma* 40 (1973): 121–126.

36. T.D. Brock, K.M. Brock, T.R. Belly, and R.L. Weiss, "*Sulfolobus*: A new genus of sulfur-oxidizing bacteria living at low pH and high temperature," *Archiv für Mikrobiologie* 84 (1972): 54–68.

37. Ibid.

38. See T. Brock, *Thermophilic microorganisms and life at high temperatures* (New York: Springer, 1978), 107; M. deRosa A. Gambacorta, G. Millonig, and J.D. Bu'Lock, "Convergent characters of extremely thermophilic acidophilic bacteria," *Experientia* 30 (1974):866–868, at 866.

39. Brock et al., "*Sulfolobus*," 54.

40. Ibid., 55.

41. Ibid., 58.

42. Ibid.

43. C.R. Woese, J. Maniloff, and L.B. Zablen, "Phylogenetic analysis of the mycoplasmas," *Proceedings of the National Academy of Sciences* 77 (1980): 494–498.

44. Ibid.

45. *Sulfolobus acidocaldarius* was characterized by 16S rRNA analysis in November 1977; *Thermoplasma acidophilum* was characterized the next month.

46. Peter Newmark to Carl Woese, June 2, 1978 (Woese Papers, University of Illinois).

47. R. Gupta and C.R. Woese, "Unusual modification patterns in the transfer RNAs of Archaebacteria," *Current Microbiology* 4 (1980): 245–249.

48. S.N. Sehgal, M. Kates, and N.E. Gibbons, "Lipids of *Halobacterium cutiru-brum*," *Canada Journal of Biochemistry and Physiology* 40 (1962): 69–81.

49. T.A. Langworthy, P.F. Smith, and W.R. Mayberry, "Lipids of *Thermoplasma acidophilum*," *Journal of Bacteriology* 112 (1972): 1193–1200; Langworthy, Smith, and Mayberry, "Long-chain glycerol diether and polyol dialkyl glycerol triether lipids of *Sulfolobus acidocaldarius*," *Journal of Bacteriology* 119 (1974): 106–116.

50. T. Langworthy, "Comparative lipid composition of heterotrophically and autotrophically grown *Sulfolobus acidocaldarius*," *Journal of Bacteriology* 130 (1977): 1326–1332, at 1131. See Mario De Rosa, Salvador de Rosa, Agata Gambacorta, and John Bu'Lock, "Lipid structures in the Caldariella group of extreme thermoacidophile bacteria," *Journal of the Chemical Society Chemical Communications* 447 (1977): 514–515. See also deRosa et al., "Convergent characters of extremely thermophilic acidophilic bacteria."

51. Brock, *Thermophilic microorganisms*, 107. See also deRosa et al., "Convergent characters of extremely thermophilic acidophilic bacteria," 866.

52. Brock, *Thermophilic microorganisms*, 174.

53. Woese put it this way:

> The rRNA analysis (and all other sequence position-specific molecu-lar measures) reveals nothing directly about the nature of the "organ-ism as an entity." The lipids, every classical trait, are, in a sense, higher level properties, ones that have "emerged." ... Remember too that I have always referred to the rRNA analyses as providing a "skeleton" that needs to be "fleshed out." Actually the RNA tree is not even that; it's more like a table of contents in a book. It's a map, as [Norman] Pace says, that lets you go where you want, but doesn't tell you what is there. (Carl Woese to the author, August 6, 2005)

54. T.G. Tornabene to the author, January 25, 2005. Bryant said nothing of the 16S rRNA results. T.G. Tornabene to the author, August 19, 2005.

55. Thomas G. Tornabene, memorandum to the author, August 8, 2005, pp. 1–3.

56. John Oró, "Mechanism of synthesis of adenine from hydrogen cyanide under possible primitive earth conditions," *Nature* 191 (1961): 217–227.

57. T.G. Tornabene, M. Kates, E. Gelpi, and J. Oró, "Occurrence of squalene, di and tetrahydrosqualenes and vitamin MK8 in an extremely halophilic bacterium, *Halobacterium cutirubrum*," *Journal of Lipid Research* 10 (1969): 294–303. See also J.K. Lanyi, W.Z. Placy, and M. Kates, "Lipid interactions in membranes of extremely halophilic bacteria II. Modification of the bilayer structure by squalene," *Biochemistry* 13 (1974): 4914–4920.

58. T.G. Tornabene, memorandum to the author, August 8, 2005, p. 1.

59. Ibid.

60. Ibid.

61. Thomas G. Tornabene, "Non-aerated cultivation of *Halobacterium cutirubrum* and its effect on cellular squalenes," *Journal of Molecular Evolution* 11 (1978): 253–257, at 257.

62. Thomas Tornabene to the author, August 18, 2005.

63. T.G. Tornabene, R.S. Wolfe, W.E. Balch, G. Holzer, G.E. Fox, and J. Oro, "Phytanyl-glycerol ethers and squalenes in the archaebacterium *Methanobacterium thermoautotrophicum*," *Journal of Molecular Evolution* 11 (1978): 259–266.

64. C.R. Woese to John Oró, March 9, 1978 (Woese Papers, University of Illinois).

65. Langworthy et al., "Lipids of *Thermoplasma acidophilum*"; Langworthy et al., "Long-chain glycerol diether."

66. Thomas Langworthy, "Tom's recollection of the Archaebacteria history," memorandum to Carl Woese, July 28, 2005, 10 pp., at 2.

67. Ibid.

68. Langworthy et al., "Lipids of *Thermoplasma acidophilum*," 1199.

69. Langworthy et al., "Long-chain glycerol diether," 115.

70. Langworthy, "Tom's recollection of the Archaebacteria history," 3.

71. Ibid., 5.

72. As Langworthy put it, "That shot the ideas about how ideal tetraethers are for life at high temperature and in hot acid since most methanogens grow at normal temperatures." Ibid., 6.

73. T.G. Tornabene and T.A. Langworthy, "Diphytanyl and dibiphytanyl glycerol ether lipids of methanogenic archaebacteria," *Science* 203 (1979): 51–53, at 53.

74. C.R. Woese to J.D. Bu'Lock, March 30, 1978 (Woese Papers, University of Illinois).

75. R.H. Whittaker and L. Margulis, "Protist classification and the kingdoms of organisms," *BioSystems* 10 (1978): 3–18.

76. Lynn Margulis to C.R. Woese, January 15, 1978 (Woese Papers, University of Illinois).

77. Joan Steitz, "Methanogenic bacteria," *Nature* 273 (1978): 101; Carl Woese, "Methanogenic bacteria," *Nature* 273 (1978): 101.

CHAPTER 15

1. Otto Kandler, ed., *Archaebacteria* (Stuttgart: Gustav Fischer, 1982).

2. R.S. Wolfe to Otto Kandler, November 11, 1976 (Kandler Papers, Universität Münchin).

3. J. Steber and K.H. Schleifer, "*Halococcus morrhuae*: A sulfated heteropolysaccharide as the structural component of the bacterial cell wall," *Archiv für Mikrobiologie* 105 (1975): 173–177.

4. O. Kandler and H. Hippe, "Lack of peptoglycan in the cell walls of *Methanosarcina barkeri*," *Archiv für Microbiologie* 113 (1977): 57–60.

5. Bryant often visited with Norbert Pfennig at the Institut für Mikrobiologie of the University of Göttingen. He was spending some time (September 1976 to March 1977) with Pfennig, who was known for his studies of photosynthesis in the absence of oxygen (anoxygenic phototrophism) in green and purple bacteria.

6. Otto Kandler to Ralph Wolfe, November 17, 1976 (Kandler Papers, Universität Münchin).

7. Carl Woese to Otto and Trudy Kandler, March 5, 2001 (Woese Papers, University of Illinois).

8. Otto Kandler to Ralph Wolfe, November 17, 1976 (Kandler Papers, Universität München).

9. Otto Kandler to Carl Woese, December 15, 1997 (Kandler Papers, Universität München).

10. There were a few books in Kandler's house as he grew up: one on the life of animals, an encyclopedia, and one on the universe and mankind. But Kandler tells of how he had read about Darwin as a youth, and when he was about 12 years old, he shared the idea with a local priest, who gave him two strikes on his hands with a cane. Author interview with Otto Kandler and Trudy Kandler, Munich, July 28, 2005.

11. See, e.g., O. Kandler, "Versuche zur Kultur isolierten Pflanzengewebes *in vitro*," *Planta* 38 (1950): 564–585; Kandler, "Untersuchungen über den Zusammenhang zwichen Atmungsstoffwechsel und Wachstumsvorgänge bei *in vitro* kultivierten Maiswurzeln," *Zeitschrift für Naturforschung* 5b (1950): 203–211; Kandler, "Uber den 'Synthetischen Wirkungsgrad' in vitro kultivierter Embryonen, Wurzeln and Sprosse," *Zeitschrift für Naturforschung* 8b (1953): 109–117.

12. From 1949 until 1953, when he completed his *Habilitation*, Kandler was an assistant professor University of Munich; then, from 1953 to 1957, associate professor of plant physiology. He was the first to obtain experimental evidence of photophosphorylation. Subsequently, he was offered a Rockefeller Fellowship, from 1956 to 1957, to work in the Brookhaven National Laboratories in Upton, New York, and in the Radiation Laboratories at the University of California–Berkeley.

13. Kandler maintained his interest in applied science throughout his career. From 1983 until 1988, he was Chairman of the Scientific Advisory Board of the National Research Center of Biotechnology in Braunschweig.

14. Kandler's professional style contrasted sharply to that of Woese. While Woese remained aloof from administration, Kandler held many positions. He was dean of the Faculty of Science at the Technical University of Munich in 1964–1965 and dean of the Faculty of Science at the University of Munich from 1973 to 1975. He also served as a member of the Senate of the Deutsche Forschungsgemeinschaft (German Science Foundation) from 1969 until 1976.

15. Kandler was elected to the German Academy of Natural Sciences/Scientists Leopoldina in 1970 and to the Bavarian Academy of Sciences in 1982. He also received the Bergey Award (1982) and the Ferdinand Cohn medal of the German Society of Hygiene and Microbiology (1989).

16. See Otto Kandler to Carl Woese, April 12, 1978 (Kandler Papers, Universität München).

17. Otto Kandler to Carl Woese, April 12, 1978 (Kandler Papers, Universität München).

18. R.Y. Stanier and C.B. van Niel, "The concept of a bacterium," *Archiv für Mikrobiologie* 42 (1962): 17–35, at 21.

19. K.H. Schleifer and Otto Kandler, "Peptidoglycan types of bacterial cell walls and their taxonomic implications," *Bacteriological Reviews* 36 (1972): 407–477, at 408.

20. Ibid.

21. C.S. Cummins and H. Harris, "The chemical composition of the cell wall in some Gram-positive bacteria and its possible value as a taxonomic character," *Journal of General Microbiology* 14 (1956): 583–600, at 596.
22. Schleifer and Kandler, "Peptidoglycan types of bacterial cell walls," at 460.
23. Ibid.
24. Ibid., 462.
25. See M.R.J. Salton, *The bacterial cell wall* (New York: Elsevier, 1964); H. Larsen, "Biochemical aspects of extreme halophilism," *Advances in Microbial Physiology* 1 (1967): 97–132; W. Stoeckenius and R. Rowen, "A morphological study of *Halobacterium halobium* and its lysis in media of low salt concentrations," *Journal of Cell Biology* 34 (1967): 365–393; A.D. Brown and K.Y. Cho, "The walls of extremely halophilic cocci. Gram-positive bacteria lacking muramic acid," *Journal of General Microbiology* 62 (1970): 267–270; R. Reistad, "Cell wall of an extremely halophilic coccus. Investigation of ninhydrin-positive compounds," *Archiv für Mikrobiologie* 82 (1972): 24–30.
26. See Schleifer and Kandler, "Peptidoglycan types of bacterial cell walls," at 435.
27. T.D. Brock, K.M. Brock, T.R. Belly, and R.L. Weiss, "*Sulfolobus*: A new genus of sulfur-oxidizing bacteria living at low pH and high temperature," *Archive für Mikrobiologie* 84 (1972): 54–68, at 62.
28. Ibid., 67.
29. O. Kandler and H. Hippe, "Lack of peptidoglycan in the cell walls of *Methanosarcina barkeri*," *Archiv für Microbiologie* 113 (1977): 57–60, at 59.
30. "In the last sentence I mentioned your work as a personal communication," he wrote. "Would you please look it up and let me know if you agree or if we should change this remark?" Otto Kandler to Carl Woese, February 18, 1977 (Woese Papers, University of Illinois).
31. Carl Woese to Otto Kandler, August 11, 1977 (Woese Papers, University of Illinois).
32. O. Kandler and H. König, "Chemical composition of the peptidoglycan-free cell walls of methanogenic bacteria," *Archiv für Microbiologie* 118 (1978): 141–152, at 150.
33. Ibid., 151.
34. Ibid.
35. O. Kandler, "Cell wall biochemistry and the three domain concept of life," *Systematic and Applied Microbiology* 16 (1994): 501–509.
36. Ibid., 501.
37. E. Stackebrandt to C. Woese, May 24, 1977 (Woese Papers, University of Illinois).
38. O. Kandler to C. Woese, May 24, 1977 (Kandler Papers, Universität Münchin).
39. C.R. Woese to E. Stackebrandt, September 13, 1977 (Woese Papers, University of Illinois).
40. E. Stackebrandt to C.R. Woese, November 17, 1977 (Woese Papers, University of Illinois).
41. C.R. Woese to Otto Kandler, November, 1978 (Woese Papers, University of Illinois).

42. C.R. Woese to Otto Kandler, December 22, 1978 (Woese Papers, University of Illinois).

43. Schleifer's laboratory received a generous grant for six years from the Deutsche Forschungsgemeinschaft to launch a research program on "systematics of Gram-positive cocci and coryneform bacteria." A part of those funds was used to resolve the second problem of developing new techniques. Karl Schleifer to the author, February 22, 2005.

44. Stackebrandt did not have permission to work with ^{32}P at the Technical University of Munich. Instead, he worked at the Max Planck Institute in Martinsried. Erko Stackebrandt to Carl Woese, June 25, 1979 (Woese Papers, University of Illinois).

45. Erko Stackebrandt to Carl Woese, May 26, 1979 (Woese Papers, University of Illinois).

46. Stackebrandt was at Kiel from 1984 to 1990, and for the next 3 years headed the Department of Microbiology at the University of Queensland, Brisbane, Australia (1990–1993). There he and Werner Liesack, was the first to introduced molecular environmental ecology to soil samples. He is presently Director of the German Collection of Microorganisms and Cell Cultures, the world's largest microbial collection, and he has a chair in Systematics of Prokaryotes at the Technical University Braunschweig.

47. C.R. Woese, "Bacterial evolution," *Microbiological Reviews* 51 (1987): 221–271.

48. E. Stackebrandt to C.R. Woese, January 9, 1979 (Woese Papers, University of Illinois).

49. Carl Woese to Erko Stackebrandt, January 9, 1979 (Woese Papers, University of Illinois).

50. See, e.g., Joseph Ben-David, *The scientist's role in society* (Englewood Cliffs: Prentice-Hall, 1971).

51. Author interview with Karl Stetter, Munich, Germany, July 18, 2005.

52. "At least in Germany," Stetter remarked, "this 'revolution' was in reality a destruction of an excellent research structure by chaotic disgusting people with no sense of serious research and responsibility." Karl Stetter to the author, February 15, 2006.

53. Author interview with Karl Stetter, Munich, Germany, July 18, 2005.

54. K.O. Stetter and O. Kandler, "Manganese requirement of the transcription processes in *Lactobacillus curvatus*," *FEBS Letters* 36 (1973): 5–8.

55. Molecular biology in Germany emerged from interdisciplinary cooperative focusing on plant–virus research between Butenandt and the directors of the Kaiser-Wilhelm Institute for Biology headed by Alfred Kühn and Fritz von Wettstein. Butenandt had become director of the biochemistry institute in 1936, a year after the Nazis ousted Jewish biochemist Carl Neuberg. Similarly, Alfred Kühn had become one of the directors of the Kaiser-Wilhelm Institute for Biology, the next year, after his predecessor, Richard Goldschmidt was forced to leave. See Jeffrey Lewis, "From virus research to molecular biology: Tobacco mosaic virus in Germany, 1936–1956," Journal of the History of Biology 37 (2004): 259–301.

56. Lewis, "From virus research to molecular biology."

57. All the directors in Tübingen were exempt from the denazification courts. Lewis, "From virus research to molecular biology."

58. A. Butenandt, P. Karlson, and W. Zillig, "Occurrence of kynurenine in silkworm pupae," *Hoppe Seyler's Zeitschrift für Physiologische Chemie* 288 (1951): 125–129; Butenandt, Karlson, and Zillig, "Dioxo-piperazines from pupae of silkworm," *Hoppe Seyler's Zeitschrift für Physiologische Chemie* 288 (1951): 279–283.

59. J.D. Watson, "The structure of tobacco mosaic virus I: X-ray evidence of a helical arrangement of sub-units around the longitudinal axis," *Biochimica et Biophysica Acta* 13 (1954): 10–19; Lewis, "From virus research to molecular biology"; G. Schramm, G. Schumacher, and W. Zillig, "An infectious nucleoprotein from tobacco mosaic virus," *Nature* 175 (1955): 549–550; R.R. Rueckert and W. Zillig, "Biosynthesis of virus protein in *Escherichia coli* C in vivo following infection with bacteriophage phi-X-174," *Journal of Molecular Biology* 5 (1962): 1–9.

60. W. Zillig, "Biological function of ribonucleic acid: Recent results of virus research," *Deutsche Medizinische Wochenschrift* 82 (1957): 1925–1926; Zillig, "Intracellular protein synthesis," *Deutsche Medizinische Wochenschrift* 83 (1958): 980–982.

61. See, e.g., G. Walter, W. Zillig, P. Palm, and E. Fuchs, "Initiation of DNA-dependent RNA synthesis and the effect of heparin on RNA polymerase," *European Journal of Biochemistry* 3/2 (1967): 194–201; M. Schachner and W. Zillig, "Fingerprint maps of tryptic peptides from Subunits of *Escherichia coli* and T4-modified DNA-dependent RNA polymerases," *European Journal of Biochemie* 22/4 (1971): 513–519; R. Schafer, W. Zillig, and K. Zechel, "A model for the initiation of transcription by DNA-dependent RNA polymerases from *Escherichia coli*," *European Journal of Biochemistry* 33/2 (1973): 207–214.

62. Author interview with Karl Stetter, Munich, July 18, 2005.

63. Ibid.

64. Ibid.

65. K.O. Stetter and W. Zillig, "Transcription in *Lactobacillaceae*: DNA-dependent RNA polymerase from *Lactobacillus curvatus*," *European Journal of Biochemistry* 48 (1974): 527–540.

66. K.O. Stetter, "Transcription in *Lactobacillaceae*. DNA-dependent RNA polymerase from *Lactobacillus casei*. Isolation of transcription factor y," *Hoppe Seyler's Zeitschrift für Physiologische Chemie* 358 (1977): 1093–1104.

67. Author interview with Karl Stetter, Munich, Germany, July 18, 2005.

68. Ibid.

69. B. Gregory Louis and P.S. Fitt, "Nucleic acid enzymology of extremely halophilic bacteria *Halobacterium cutirubrum* deoxyribonucleic acid-dependent ribonucleic acid polymerase," *Biochemical Journal* 121 (1971): 621–627, at 626; Louis and Fitt, "Isolation and properties of highly purified Halobacterium cutirubrum deoxyribonucleic acid-dependent ribonucleic acid polymerase," *Biochemical Journal* 127 (1972): 69–80.

70. Author interview with Karl Stetter, Munich, Germany, July 18, 2005.

71. Ibid.

72. W. Zillig, K.O. Stetter, and M. Tobien, "DNA-dependent RNA polymerase from *Halobacterium halobium*," *European Journal of Biochemistry* 91 (1978): 193–199.

73. W. Zillig, K.O. Stetter, and D. Janekovic, "DNA-dependent RNA polymerase from the archaebacterium *Sulfolobus acidocaldarius*," *European Journal of Biochemistry* 96 (1979): 597–604.

74. Mario De Rosa, Salvador de Rosa, Agata Gambacorta, and John Bu'Lock, "Lipid structures in the Caldariella group of extreme thermoacidophile bacteria," *Journal of the Chemical Society Chemical Communications* 447 (1977): 514–515; De Rosa, De Rosa, Gambacorta, M. Carteni-Farina, and V. Zappia, "The biosynthetic pathway of new polyamines in *Caldariella acidophila*," *Biochemical Journal* 176 (1978): 1–7.

75. Brock et al., "*Sulfolobus.*"

76. "The subunit composition of the enzyme resembles that of other bacterial polymerases." M.G. Cacace, M. De Rosa, and A. Gambacorta, "DNA-dependent RNA Polymerase from the thermophilic bacterium *Caldariella acidophila*. Purification and basic properties of the enzyme," *Biochemistry* 15 (1976): 1692–1696, at 1695.

77. W. Zillig, K.O. Stetter, S. Wunderl, W. Schulz, H. Preiss, and I. Scholz, "The Sulfolobus-'Caldariella' group: Taxonomy on the basis of the structure of DNA-dependent RNA polymerases," *Archives of Microbiology* 125 (1980): 259–269.

78. K.O. Stetter, J. Winter, and R. Hartlieb, "DNA-dependent RNA polymerase of the archaebacterium *Methanobacterium thermoautotrophicum*," *Zentralblatt für Bakteriologie Mikrobiologie und Hygiene I Abteilung Originale c-Allgemeine Angewandte und Okologische Mikrobiologie* (1980): 201–214, at 268.

79. C.R. Woese to O. Kandler, February 10, 1978 (Woese Papers, University of Illinois).

80. C.R. Woese to Otto Kandler, November 20, 1978 (Woese Papers, University of Illinois).

81. Otto Kandler to Carl Woese, March 3, 1979 (Kandler Papers, Universität Münchin).

82. Otto Kandler to C.R. Woese, December 8, 1978 (Kandler Papers, Universität Münchin).

83. W. Zillig to Carl Woese, December 18, 1978 (Woese Papers, University of Illinois).

84. C.R. Woese to O. Kandler, April 5, 1979 (Woese Papers, University of Illinois).

85. O. Kandler to C.R. Woese, May 22, 1979 (Kandler Papers, Universität Münchin).

86. Carl Woese to W. Zillig, April 5, 1979 (Woese Papers, University of Illinois).

87. C.R. Woese to W. Zillig, August 6, 1979 (Woese Papers, University of Illinois).

88. W. Zillig, K.O. Stetter and D. Janekovic, "DNA-dependent RNA polymerase from the archaebacterium *Sulfolobus acidocaldarius*," *European Journal of*

Biochemistry 96 (1979): 597–604; Zillig et al., "The Sulfolobus-'Caldariella' group."

89. W. Zillig, R. Schnabel, J. Tu, and K.O. Stetter, "The phylogeny of Archaebacteria, including novel anaerobic thermoacidophiles in the light of RNA polymerase structure," *Naturwissenschaften* 69 (1982): 197–204.

90. R. Wolfe, "The Archaea: A personal overview of the formative years," in M. Dworkin, S. Falkow, and E. Rosenberg, eds., *The prokaryotes: An evolving electronic resource for the microbiological community*, 3rd ed., release 3.7 (New York: Springer, November 2, 2001), available at http://link.springer-ny.com/link/service/books/10125/.

91. See J. Tu, D. Prangishvilli, H. Huber, G. Wildgruber, W. Zillig, and K.O. Stetter, "Taxonomic relations between Archaebacteria including 6 novel genera examined by cross hybridization of DNAs and 16S rRNAs," *Journal of Molecular Evolution* 18 (1982): 109–114.

92. K.O. Stetter, M. Thomm, J. Winter, G. Wildgruber, H. Huber, W. Zillig, D. Janecovic, H. König, P. Palm, and S. Wunderl, "*Methanothermus fervidus*, sp. nov., a novel extremely thermophilic methanogen isolated from an Icelandic hot spring," *Zentralblatt für Mikrobiologie und Parasitenkunde Infektionskreicht Hygiene Abteilung. Originale C2* (1981): 166–178.

93. K.O. Stetter, "Ultrathin mycelia-forming organisms from submarine volcanic areas having an optimum growth temperature of 105°C," *Nature* 300 (1982): 258–260.

94. K.O. Stetter, H. König, and E. Stackebrandt, "*Pyrodictium* gen. nov., a new genus of submarine disc-shaped sulphur reducing Archaebacteria growing optimally at 105°C," *Systematic and Applied Microbiology* 4 (1983): 535–551.

95. W.D. Reiter, W. Zillig, and P. Palm, "Archaebacterial viruses," *Advances in Virus Research* 34 (1988): 143–188.

CHAPTER 16

1. Carl Woese to Otto Kandler, August 13, 1979 (Kandler Papers, Universität Münchin).

2. C.R. Woese to J.F. Wilkinson, March 27, 1978 (Woese Papers, University of Illinois).

3. C.R. Woese to Erko Stackebrandt, January 9, 1979 (Woese Papers, University of Illinois).

4. H. Hori and S. Osawa, "Evolutionary change in 5S RNA secondary structure and a phylogenic tree of 54 5S RNA species," *Proceedings of the National Academy of Sciences* 76 (1979): 381–385.

5. S. Osawa and T.H. Jukes, "Codon reassignment (codon capture) in evolution," *Journal of Molecular Evolution* 4 (1989): 271–278.

6. After completing his Ph.D. at Nagoya University in 1951, Osawa worked at the Rockefeller Institute for Medical Research in New York for three years. See V.G. Allfrey, A.E. Mirsky, and Syozo Osawa, "Protein synthesis in isolated cell nuclei," *Journal of General Physiology* 40 (1957): 451–490; Osawa, Allfrey, and Mirsky, "Mononucleotides of the cell nucleus," *Journal of General Physiology* 40 (1957): 491–513; Mirsky, Osawa, and Allfrey, "The nucleus as

a site of biochemical activity," *Cold Spring Harbor Symposium on Quantitative Biology* 21 (1956): 49–58.

7. In 1984 Hori moved to Nagoya University. He completed his Ph.D. in 1988 on the "Molecular phylogeny of living organisms by using 5S rRNA sequences."

8. O. Kandler to C.R. Woese, September 28, 1979 (Kandler Papers, Universität Münchin).

9. H. Hori, "Evolution of 5S RNA," *Journal of Molecular Evolution* 7 (1975): 75–86.

10. G. Fox and C. Woese, "5S RNA secondary structure," *Nature* 256 (1975): 505–507.

11. H. Hori, "Molecular evolution of 5S RNA," *Molecular and General Genetics* 145/2 (1976): 119–123; Hori, K. Higo, and S. Osawa, "The rates of evolution in some ribosomal components," *Journal of Molecular Evolution* 9 (1977): 191–201.

12. C.R. Woese to O. Kandler, October 4, 1979 (Woese Papers, University of Houston).

13. Hori and Osawa, "Evolutionary change in 5S RNA," 384.

14. L.J. Magrum, K.R. Luehrsen, and C.R. Woese, "Are extreme halophiles actually 'bacteria'?" *Journal of Molecular Evolution* 11 (1978): 1–8.

15. George Fox to the author, March 11, 2005.

16. H. Hori, T. Itoh, and S. Osawa, "The phylogenic structure of the metabacteria," in O. Kandler, ed., *Archaebacteria* (Stuttgart: Gustav Fischer, 1982), 18–30.

17. W. Zillig to Carl Woese, October 11, 1979 (Woese Papers, University of Illinois). Zillig's point in regard to the homology with 5.8S was well taken, and research results published in 1977 supported the view that 5.8S RNA evolved from bacterial 5S. Paul Wrede and A. Volker, "Erdmann *Escherichia coli* 5S RNA binding proteins L18 and L25 interact with 5.8S RNA but not with 5S RNA from yeast ribosomes," *Proceedings of the National Academy of Sciences* 74 (1977): 2706–2709.

18. C.R. Woese to W. Zillig, October 22, 1979 (Woese Papers, University of Houston).

19. Carl Woese to Otto Kandler, October 26, 1979 (Woese Papers, University of Houston).

20. See H. Hartmann, "Speculations on the origin and evolution of metabolism," *Journal of Molecular Evolution* 4 (1975): 359–370.

21. A.I. Oparin, *The origin of life* (1924), translation by Sergius Morgulis, 2nd ed. (New York: Dover, 1965), 203, 207.

22. Ibid., 248.

23. Ibid., 249.

24. Ibid., 193.

25. Ibid., 251.

26. J.B.S. Haldane, "The origin of life," *Rationalist Annual* (1929): 148–169, reprinted in J.D. Bernal, *The origin of life* (London: Weidenfeld and Nicolson, 1967), 242–249.

27. N.H. Horowitz, "On the evolution of biochemical syntheses," *Proceedings of the National Academy of Sciences* 31 (1945): 153–157, at 157.

28. C.B. van Niel, "The classification and natural relationships of bacteria," *Cold Spring Harbor Symposia on Quantitative Biology* 11 (1946): 285–301, at 289. The next year, Hebert Copeland changed his autotrophs-first views to heterotrophs-first. See H.F. Copeland, "Progress report on basic classification," *American Naturalist* 81 (1947): 340–361.

29. J.D. Bernal, "The problem of stages in biopoesis," in *Proceedings of the First International Symposium on the Origin of Life on Earth* (London: Pergamon Press, 1959), 40–52.

30. C.R. Woese, "A proposal concerning the origin of life on the planet Earth," *Journal of Molecular Evolution* 13 (1979): 95–101, at 98.

31. N.H. Horowitz, "The evolution of biochemical synthesis—retrospect and prospect," in Vernon Bryson and Henry J. Vogel, eds., *Evolving genes and proteins* (New York: Academic Press, 1965), 15–23, at 16.

32. Ibid., 17.

33. David Hawkins, *The language of nature* (San Francisco: W.H. Freeman, 1964).

34. Oparin, *Origin of life*, 231.

35. A.I. Oparin, *The origin of life* (1924), translation by Ann Synge, in Bernal, *The origin of life*, 199–241, at 231.

36. Oparin, *Origin of life* (1965), 231–237.

37. Ibid., 236.

38. Ibid., 233.

39. Ibid., 63. In the forward (p. viii), Oparin's translator, Sergius Morgulis, commented, "It is one of Oparin's great contributions to the theory of the origin of life that he postulated a long chemical evolution as a necessary preamble to the emergence of Life."

40. See J.W. Schopf, *Cradle of life: The discovery of earth's earliest fossils* (Princeton: Princeton University Press, 1999); A. Knoll and E.S. Barghoorn, "Archaean microfills showing cell division from the Swaziland system of South Africa," *Science* 198 (1977): 396–398.

41. See also H.D. Holland, "Evidence for life on Earth more than 3850 million years ago," *Science* 275 (1997): 38–39.

42. Woese, "A proposal concerning the origin of life on the planet Earth," 98.

43. Ibid., 96.

44. Ibid., 97.

45. Carl Woese, "An alternative to the Oparin view of the primeval sequence," in H.O. Halvorson and K.E. Van Holde, eds., *The origins of life and evolution* (New York: Alan R. Liss, 1980), 65–76.

46. Woese, "A proposal concerning the origin of life on the planet Earth," 96.

47. Ibid.

48. Ibid., 97.

49. Ibid.

50. Ibid., 99.

51. Ibid., 100.

52. Ibid.

53. Ibid. The concept that methanogens were among the most ancient of organisms and played an important role in the production of the atmosphere persists

to the present. But their role changed when it became obvious that methane, too, is a greenhouse gas. James F. Kasting, "When methane made climate," *Scientific American* (July 2004): 78–86.

54. C.R. Woese, and G. Wächtershäuser, "Origin of life," in E.G. Briggs and P.R. Crowther, eds., *Palaeobiology: A synthesis* (London: Blackwell Scientific 1990), 3–9.

55. Günter Wächtershäuser, "Before enzymes and templates: Theory of surface metabolism," *Microbiological Reviews* 52 (1988): 452–484, at 452; Wächtershäuser, "Evolution of the first metabolic cycles," *Proceedings of the National Academy of Sciences* 87 (1990): 200–204; Wächtershäuser, "The origin of life and its methodological challenge," *Journal of Theoretical Biology* 487 (1997): 483–494; Wächtershäuser, "From volcanic origins of chemoautotrophic life to Bacteria, Archaea and Eukarya," *Philosophical Transactions of the Royal Society Series B* 361 (2006): 1787–1808.

56. Wächtershäuser, "Before enzymes and templates," 453.

CHAPTER 17

1. Carl Woese to Otto Kandler, September 18, 1979 (Woese Papers, University of Illinois).

2. Hans Neurath, Kenneth A. Walsh, and William P. Winter, "Evolution of structure and function of proteases," *Science* 158 (1967): 1638–1644, at 1639. The word "homology," they said, should be taken "to connote the occurrence of a degree of structural similarity among proteins greater than might be anticipated by chance alone."

3. Ibid., 1639.

4. William P. Winter, Kenneth A. Walsh, and Hans Neurath, "Homology as applied to proteins," *Science* 162 (1968): 1433.

5. Lewis Carroll, *Alice's adventure in wonderland* (1865; reprint, London: Penguin Books, 1994), 4.

6. W.M. Fitch, "Distinguishing homologous from analogous proteins," *Systematic Zoology* 19 (1970): 99–113, at 113.

7. Ibid., 112.

8. Ibid., 100.

9. Robert M. Schwartz and M.O. Dayhoff, "Origins of prokaryotes, eukaryotes, mitochondria, and chloroplasts," *Science* 199 (1978): 395–403.

10. Ibid., 395.

11. Ibid., 396.

12. Ibid., 399.

13. Vincent Demoulin, "Protein and nucleic acid sequence data and phylogeny," *Science* 205 (1979): 1036–1038, at 1037.

14. Ibid.

15. R.M. Schwartz and M.O. Dayhoff, "Protein and nucleic acid sequence data and phylogeny," *Science* 205 (1979): 1038–1039, at 1039.

16. Ibid.

17. Linda Bonen to the author, January 18, 2006; T. Uchida, L. Bonen, H.W. Schaup, B.J. Lewis, L. Zablen, and C.R. Woese, "The use of ribonuclease U2

in RNA sequence determination," *Journal of Molecular Evolution* 3 (1974): 633–677; C.R. Woese, G.E. Fox, L. Zablen, L. Uchida, L. Bonen, L. Pechman, K. Lewis, B.J. Sthal, and D. Stahl, "Conservation of primary structure in SSU ribosomal RNA," *Nature* 254 (1975): 83–85; L. Zablen, L. Bonen, B. Meyer, and C.R. Woese, "The phylogenetic state of *Pasteurella pestis*," *Journal of Molecular Evolution* 4 (1975): 307–318.

18. Bonen completed her Ph.D. in 1981; at this writing she is a professor at the University of Ottawa.

19. See W.F. Doolittle and C. Yankofsky, "Mutants of *Escherichia coli* with an altered tryptophanyl-transfer ribonucleic acid synthetase," *Journal of Bacteriology* 95 (1968): 1283–1294.

20. Doolittle regarded Woese "as an intellectual mentor." Author interview with Ford Doolittle, Dalhousie University, July 16, 1999.

21. W.F. Doolittle and N.R. Pace, "Synthesis of 5S ribosomal RNA in *Escherichia coli* after rifampicin treatment," *Nature* 228 (1970): 125–129; W.F. Doolittle and N.R. Pace, "Transcriptional organization of the ribosomal RNA cistrons in *Escherichia coli*," *Proceedings of the National Academy of Sciences* 68 (1971): 1786–1790.

22. N.R., Pace, D.A. Stahl, D.J. Lane, and G.J. Olsen, "The analysis of natural microbial populations by ribosomal RNA sequences," *Advances in Microbial Ecology* 9 (1986): 1–55.

23. Author interview with Ford Doolittle, Dalhousie University, July 16, 1999.

24. Ibid.

25. L. Bonen and W.F. Doolittle, "On the prokaryotic nature of red algal chloroplasts," *Proceedings of the National Academy of Sciences* 72 (1975): 2310–2314, at 2310.

26. Woese et al., "Conservation of primary structure in SSU ribosomal RNA"; W.F. Doolittle, C.R. Woese, M.L. Sogin, L. Bonen, and D. Stahl, "Sequence studies on SSU ribosomal RNA from a blue-green alga," *Journal of Molecular Biology* 4 (1975): 307–315.

27. Bonen and Doolittle, "On the prokaryotic nature of red algal chloroplasts," 2314.

28. L.B. Zablen, M.S. Kissil, C.R. Woese, and D.E. Buetow, "Phylogenetic origin of the chloroplast and prokaryotic nature of its ribosomal RNA," *Proceedings of the National Academy of Sciences* 72 (1975): 2418–2422.

29. Bonen and Doolittle, "On the prokaryotic nature of red algal chloroplasts," 2314.

30. See R.A. Raff and H.R. Mahler, "The symbiont that never was: An inquiry into the evolutionary origin of the mitochondrion," *Symposia of the Society for Experimental Biology* 29 (1975): 41–92. The old arguments for the great differences between mitochondria and bacteria had been based on animals. See W.M. Brown, M. George, Jr., and A.C. Wilson, "Rapid evolution of animal mitochondrial DNA," *Proceedings of the National Academy of Sciences* 76 (1979): 1967–1971. Animal mitochondria, as it turned out, were highly divergent. Plant mitochondria were different: their ribosomes were similar to bacteria. L. Bonen, R.S. Cunningham, M.W. Gray,

and W.F. Doolittle, "Wheat embryo mitochondrial 18S ribosomal RNA: Evidence for its prokaryotic nature," *Nucleic Acids Research* 4 (1977): 663–671, at 664.

31. There is about 1/10th as much ribosomal RNA coded by chloroplast genes than ribosomal RNA coded by nuclear genes and then there is 1/10th as much again mitochondrial ribosomal RNA.

32. L. Hudson, M. Gray, and B.G. Lane, "The alkali-stable dinucleotide sequences and the chain termini in soluble ribonucleates from wheat germ," *Biochemistry* 4 (1965): 2009–2016; Gray and Lane, "Studies of the sequence distribution of 2'-O-methylribose in yeast soluble ribonucleates," *Biochimica et Biophysica Acta* 134 (1967): 243–257; Gray and Lane, "5-Carboxymethyluridine, a novel nucleoside derived from yeast and wheat embryo transfer ribonucleates," *Biochemistry* 7 (1968): 3441–3453.

33. Bonen et al., "Wheat embryo mitochondrial 18S ribosomal RNA," 670.

34. Ibid.

35. M.W. Gray and W.F. Doolittle, "Has the endosymbiont hypothesis been proven?" *Microbiological Reviews* 46 (1982): 1–42, at 3.

36. Author interview with Ford Doolittle, Halifax, Nova Scotia, July 16, 1999.

37. George Fox, comments to the author, August 18, 2005.

38. G. Fox, Kenneth R. Pechman, and Carl R. Woese, "Comparative cataloging of SSU ribosomal ribonucleic acid: Molecular approach to procaryotic systematics," *Internal Journal of Systematic Bacteriology* 27 (1977): 44–77, at 44.

39. Schwartz and Dayhoff, "Origins of prokaryotes, eukaryotes, mitochondria, and chloroplasts," 396.

40. Martin D. Kamen, *Radiant science, dark politics: A memoir of the nuclear age* (Berkeley: University of California Press, 1985); Kamen, *Primary process in photosyntheses* (New York: Academic Press, 1963).

41. T.E. Meyer, R.G. Bartsch, and M.D. Kamen, "Cytochrome *c2* sequence variation among the recognised species of purple nonsulphur photo synthetic bacteria," *Nature* 278 (1979): 659–660. R.P. Ambler, T.E. Meyer, and M.D. Kamen, "Anomalies in amino acid sequence of small cytochrome *c* and cytochromes *c'* from two species of purple photo synthetic bacteria," *Nature* 278 (1979): 661–662.

42. Meyer et al., "Cytochrome *c2* sequence variation"; Ambler et al., "Anomalies in Amino Acid Sequence."

43. E. Margoliash, W.M. Fitch, and R.E. Dickerson, "Molecular expression of evolutionary phenomena in the primary and tertiary structures of cytochrome c," *Brookhaven Symposia in Biology* 21 (1968): 259–305. Based on cytochrome *c* sequence data, Dickerson proposed that respiration arose repeatedly from photosynthesis and that the mitochondrion and related respiring bacteria had evolved from the purple photosynthetic bacteria. See Richard Dickerson, "Cytochrome c and the evolution of energy metabolism," *Scientific American* 242 (1980): 136–153.

44. Richard Dickerson, "Evolution and gene transfer in purple photosynthetic bacteria," *Nature* 283 (1980): 210–212, at 212.

45. C. R Woese, J. Gibson, and G.E. Fox, "Do genealogical patterns in purple photosynthetic bacteria reflect interspecific gene transfer?" *Nature* 283 (1980): 212–214.

46. T.E. Meyer, M.A. Cusanovich, and M.D. Kamen, "Evidence against use of bacterial amino acid sequence data for construction of all-inclusive phylogenetic trees," *Proceedings of the National Academy of Sciences* 83 (1986): 217–220, at 220.

47. G.E. Fox, E. Stackebrandt, R.B. Hespell, J. Gibson, J. Maniloff, T.A. Dyer, R.S. Wolfe, W.E. Balch, R.S. Tanner, L.J. Magrum, L.B. Zablen, R. Blakemore, R. Gupta, L. Bonen, B.J. Lewis, D.A. Stahl, K.R. Luerhsen, K.N. Chen, and C.R. Woese, "The phylogeny of prokaryotes," *Science* 209 (1980): 457–463, at 458.

48. Ibid., 460.

49. Ibid., 458, 462.

50. Ibid., 458.

51. Ibid., 463.

52. Ibid.

53. Ibid.

54. Ibid.

55. Ibid., 459.

56. Ibid.

57. Ibid., 462.

58. Anonymous peer review report on Fox et al., "The phylogeny of prokaryotes" (Woese Papers, University of Illinois).

59. Ibid.

60. Dick Dickerson to C.R. Woese, February 17, 1980 (Woese Papers, University of Illinois).

61. Woese sent van Niel a copy via van Niel's former student Robert Hungate at the University of California–Davis. C.R. Woese to Robert Hungate, October 3, 1980; R.E. Hungate to Carl Woese, December 17, 1980 (Woese Papers, University of Illinois).

62. Roger Y. Stanier, "The journey, not the arrival matters," *Annual Review of Microbiology* 34 (1980): 1–48, at 45.

63. R.Y. Stanier and C.B. van Niel, "The main outlines of bacterial classification," *Journal of Bacteriology* 42 (1941): 437–466.

64. Roger Stanier to Ralph Wolfe, February 5, 1980 (Woese Papers, University of Illinois).

65. F.H.C. Crick and L.E. Orgel, "Directed panspermia," *Icarus* 19 (1973): 341–346.

66. S. Arrhenius, *Worlds in the making* (New York: Harper and Row, 1908).

67. Crick and Orgel, "Directed panspermia," 341.

68. Ibid., 341.

69. Francis Crick to Carl Woese, February 25, 1980 (Woese Papers, University of Illinois).

70. G.E. Fox to Eleanore Butz, March 26, 1981 (Fox Papers, University of Houston).

71. C.R. Woese to Eleanore Butz, March 25, 1981 (Woese Papers, University of Illinois).

72. Ibid.

73. See C.R. Woese, J. Maniloff, and L.B. Zablen, "Phylogenetic analysis of the mycoplasmas," *Proceedings of the National Academy of Sciences* 77 (1980): 494–498.

74. George Gaylord Simpson, *Tempo and mode of evolution* (1944; reprint, New York: Columbia University Press, 1984).

75. C.R. Woese, E. Stackebrandt, and W. Ludwig, "What are mycoplasmas: The relationship of tempo and mode in bacterial evolution," *Journal of Molecular Evolution* 21 (1985): 305–316.

76. Carl Woese to Joe Felsenstein, August 30, 1983 (Woese Papers, University of Illinois).

CHAPTER 18

1. See also C.R. Woese, "Archaebacteria," *Scientific American* 244 (1981): 98–122.

2. Otto Kandler, ed., *Archaebacteria* (Stuttgart: Gustav Fischer, 1982), i.

3. Noel Krieg and John G. Holt, eds., *Bergey's Manual of Systematic Bacteriology*, Vol. 1 (Baltimore: Williams and Wilkins, 1984), xiii. See also Helmut Koign and Karl O. Stetter, "Archaeobacteria," in James T. Staley, Marvin Bryant, Norbert Pfenning, and John G. Holt, eds., *Bergey's Manual of Systematic Bacteriology*, Vol. 3 (Baltimore: Williams and Wilkins, 1989), 2171–2173.

4. F. Sanger, S. Nicklen, and A.R. Coulson, "DNA sequencing with chain-terminating inhibitors," *Proceedings of the National Academy of Sciences* 74 (1977): 5463–5467; A.M. Maxam and W. Gilbert, "A new method for sequencing DNA," *Proceedings of the National Academy of Sciences* 74 (1977): 1258.

5. K.F. Mullis, F. Faloona, S. Scharf, R. Saiki, G. Horn, and H. Erlich, "Specific enzymatic amplification of DNA in vitro: The polymerase chain reaction," *Cold Spring Harbor Symposia on Quantitative Biology* 51 (1987): 263–273; H.F. Judson, "A history of the science and technology behind gene mapping and sequencing," in D.J. Kevles and L. Hood, eds., *The code of codes: Scientific and social issues in the human genome project* (Cambridge: Harvard University Press, 1992), 37–80; Paul Rabinow, *Making PCR: A story of biotechnology* (Chicago: University of Chicago Press, 1996).

6. Otto Kandler and Wolfram Zillig, eds., *Archaebacteria '85: Proceedings of the EMBO Workshop on Molecular Genetics of Archaebacteria* (Stuttgart: Gustav Fischer, 1986).

7. The first full sequence of the rRNA of *E. coli* was published in the fall of 1978. J. Brosius, M.L. Palmer, P.J. Kennedy, and H.F. Noller, "Complete nucleotide sequence of a SSU ribosomal RNA gene from *Escherichia coli*," *Proceedings of the National Academy of Sciences* 75 (1978): 4801–4805. The first archaebacterial 16S rRNA sequence was published in 1983: J.R. Gupta, J.M. Lanter, and C.R. Woese, "Sequence of the SSU ribosomal RNA from *Halobacterium volcani*, an archaebacterium," *Science* 221 (1983): 656–659; Carl Woese and Gary Olsen, "Archaebacterial phylogeny: Perspectives on the urkingdoms," *Systematic and Applied Microbiology* 7 (1986): 161–177.

8. Wolfram Zillig, Felix Gropp, Agnes Henschen, Horst Neumann, Peter Palm, Wolf-Dieter Reiter, Michael Rettenberger, Heinke Schnabel, and

Siobhan Yeats, "Archaebacterial virus host systems," *Systematic and Applied Microbiology* 7 (1986): 58–66, at 58.

9. W. Zillig, R. Schnabel, J. Tu, and K.O. Stetter, "The phylogeny of archaebacteria, including novel anaerobic thermoacidophiles in the light of RNA polymerase structure," *Naturwissenschaften* 69 (1982): 197–204.

10. W. Zillig, K.O. Stetter, and D. Janekovic, "DNA-dependent RNA polymerase from the archaebacterium *Sulfolobus acidocaldarius*," *European Journal of Biochemistry* 96 (1979): 597–604; Zillig, Stetter, R. Schnabel, and M. Thomm, "DNA-dependent RNA polymerases of the Archaebacteria," in C.R. Woese and R.S. Wolfe, eds., *The bacteria* (New York: Academic Press, 1985), Vol. 8: 499–524.

11. W. Zillig to Carl Woese, July 27, 1979 (Woese Papers, University of Illinois).

12. Wolfram Zillig to Carl Woese, August 6, 1982 (Woese Papers, University of Illinois).

13. Carl Woese to Wolfram Zillig, January 5, 1982 (Woese Papers, University of Illinois). Woese believed that the relationship between the three urkingdoms would be resolved when "we understand the progenote state." Carl Woese to Wolfram Zillig, September 28, 1982 (Woese Papers, University of Illinois).

14. Wolfram Zillig to Carl Woese, December 17, 1982 (Woese Papers, University of Illinois); Zillig et al., "The phylogeny of archaebacteria."

15. Carl Woese to Wolfram Zillig, January 14, 1983 (Woese Papers, University of Illinois); C.R. Woese, R. Gupta, C.M. Hahn, W. Zillig, and J. Tu, "The phylogenetic relationships of three sulfur dependent archaebacteria," *Systematic and Applied Microbiology* 5 (1984): 97–105.

16. See Phillip A. Sharp, "Split genes and RNA splicing" [Nobel Lecture], *Cell* 77 (1994): 805–815; B. Lewin, "Alternatives for splicing: Recognizing the ends of introns," *Cell* 22 (1980): 324–326.

17. A. Andreadis, M.E. Gallego, and B. Nadal-Ginard, "Generation of protein isoform diversity by alternative splicing: Mechanistic and biological implications," *Annual Review of Cell Biology* 3 (1987): 207–242. See also Bruce Alberts, Dennis Bray, Julian Lewis, and Keith Roberts, *Molecular biology of the cell*, 2nd ed. (New York: Garland, 1989), 536, 588.

18. Carl Woese to Wolfram Zillig, February 10, 1983 (Woese Papers, University of Illinois). See also P.B. Kaine, R. Gupta, and C.R. Woese, "Putative introns in tRNA genes of procaryotes," *Proceedings of the National Academy of Sciences* 80 (1983): 3309–3312.

19. Wolfram Zillig to Carl Woese, July 22, 1983 (Woese Papers, University of Illinois).

20. Woese et al., "The phylogenetic relationships of three sulfur dependent archaebacteria."

21. B.J. Kjems and R.A. Garrett, "An intron in the 23S rRNA gene of the archaebacterium *Desulfurococcus mobile*," *Nature* 318 (1985): 675–677; C.J. Daniels, R. Gupta, and W.F. Doolittle, "Transcription and excision of a large intron in the tRNA Trp gene of an archaebacterium, *Halobacterium volcanii*," *Journal of Biological Chemistry* 260 (1985): 3132–3134.

22. J.A. Lake, E. Henderson, M. Oakes, and M.W. Clark, "Eocytes: A new ribosome structure indicates a kingdom with a close relationship to eukaryotes," *Proceedings of the National Academy of Sciences* 81 (1984): 3786–3790, at 3786. The term "eocytes" was suggested by William Schopf.

23. James lake, Eric Henderson, Michael Clark, and A.T. Matheson, "Mapping evolution with Ribosome Structure: Intralineage Constancy and Interlineage variation," *Proceedings of the National Academy of Sciences* 79 (1982): 5948–5952.

24. Carl Woese to Wolfram Zillig, September 28, 1982 (Woese Papers, University of Illinois).

25. Carl Woese to Wolfram Zillig, January 14, 1983 (Woese Papers, University of Illinois).

26. Wolfram Zillig to Carl Woese, July 22, 1983 (Woese Papers, University of Illinois).

27. Carl Woese to James Lake, August 10, 1983 (Woese Papers, University of Illinois).

28. Wolfram Zillig to James Lake, August 18, 1983 (Woese Papers, University of Illinois).

29. Carl Woese to James Lake, February 27, 1984 (Woese Papers, University of Illinois).

30. Carl Woese to James Lake, February 27, 1984 (Woese Papers, University of Illinois). Matheson also disagreed with Lake's eocyte proposal. Alastair Matheson, K. Andrea Louie, and George Henderson, "The evolution of the ribosomal "A" protein in archaebacteria," *Systematic and Applied Microbiology* 7 (1986): 147–150.

31. W. Zillig to C. Woese, March 22, 1984 (Woese Papers, University of Illinois).

32. E.H. Hendersen, M. Oakes, M.W. Clark, J.A. Lake, A.T. Matheson, and W. Zillig, "A new ribosome structure," *Science* 225 (1984): 510–512, at 512.

33. Lake et al., "Eocytes: A new ribosome structure."

34. James A. Lake, Michael W. Clark, Eric Henderson, Shawn P. Fay, Melanie Oakes, Andrew Scheinman, J.P. Thornber, and R.A. Mah, "Eubacteria, halobacteria, and the origin of photosynthesis: The photocytes," *Proceedings of the National Academy of Sciences* 82 (1985): 3716–3720.

35. Ibid., 3716.

36. Ibid., 3720.

37. M. Sarich and A.C. Wilson, "Immunological time scale for hominid evolution," *Science* 158 (1967): 1200–1203.

38. M.C. King and A.C. Wilson, "Evolution at two levels in humans and chimpanzees," *Science* 188 (1975): 107–116 (see also commentary, *Science* 189 [1975]: 446–447); M.C. King and A.C. Wilson, "Our close cousin, the chimpanzee," *New Scientist* 67 (1975): 16–18.

39. W.M. Brown, E.M. Prager, A. Wang, and A.C. Wilson, "Mitochondrial DNA sequences of primates: Tempo and mode of evolution," *Journal of Molecular Evolution* 18 (1982): 225–239; R.L. Cann and A.C. Wilson, "Length mutations in human mitochondrial DNA," *Genetics* 104 (1983): 699–711.

40. A.C. Wilson, "The molecular basis of evolution," *Scientific American* 253 (1985): 164–173, at 170.
41. Ibid.
42. Carl Woese to Wolfram Zillig, September 20, 1985 (Woese Papers, University of Illinois).
43. R. Garrett, "The uniqueness of archaebacteria," *Nature* 318 (1985): 233. See also Zillig and Kandler, eds., *Archaebacteria '85*; W. Zillig and O. Kandler, *International Workshop on Biology and Biochemistry of Archaebacteria* (Stuttgart: Gustav Fischer, 1986).
44. Wolfram Zillig to Carl Woese, May 15, 1984 (Woese Papers, University of Illinois).
45. This was the beginning of what Woese saw to be the inherent bias in *Nature*'s editorship regarding the archaebacteria. Seven years earlier, Newmark had rejected Woese's paper reporting the startling discovery that *Thermoplasma acidophilum*, the first nonmethanogenic archaebacteria, was a member of the archaebacteria. He considered it to be of insufficient interest to readers of *Nature* (see chapter 14).
46. Kjems and Garrett, "An intron in the 23S rRNA gene of the archeabacterium *Desulfurococcus mobile*"; Daniels et al., "Transcription and excision of a large intron."
47. Garrett, "The uniqueness of archaebacteria," 234.
48. Ibid., 235.
49. James Lake to Roger Garrett, December 10, 1985 (Roger Garrett Papers, University of Copenhagen).
50. Ibid.
51. James Lake, "An alternative to archaebacterial dogma," *Nature* 319 (1986): 626 [emphasis original].
52. Wolfram Zillig to Carl Woese, January 22, 1986 (Woese Papers, University of Illinois).
53. Zillig et al., "The phylogeny of archaebacteria"; Woese et al., "The phylogenetic relationships of three sulfur dependent archaebacteria."
54. W. Zillig, "Archaebacterial status quo is defended," *Nature* 320 (1986): 220.
55. Marina Stöffler-Meilicke, Claudia Böhme, Olaf Strobel, August Böck, and Georg Stöffler, "Structure of ribosomal subunits of *M. vannielii*: Ribosomal morphology as a phylogenetic marker," *Science* 231 (1986): 1306–1308, at 1308.
56. Carl Woese, Norman Pace, and Gary Olsen, "Are arguments against archaebacteria valid?" *Nature* 320 (1986): 401–402.
57. Lederer, "Archaebacterial status quo is defended."
58. James Lake, "In defence of bacterial phylogeny," *Nature* 321 (1986): 657–658.
59. James A. Lake, "Origin of the eukaryotic nucleus determined by rate-invariant analysis of rRNA sequences," *Nature* 331 (1988): 184–186, at 185. See also David Penny, "What was the first living cell?" *Nature* 331 (1988): 111.
60. Lake, "Origin of the eukaryotic nucleus," 184; see also Lake, "A rate-independent technique for analysis of nucleic acid sequences: Evolutionary parsimony," *Molecular Biology and Evolution* 4 (1987): 167–191.

61. Maria C. Rivera and James Lake, "Evidence that eukaryotes and eocyte prokaryotes are immediate relatives," *Science* 257 (1992): 74–76. See Lake, Jonathan Moore, Anne Simonson, and Rivera, "Fulfilling Darwin's dream," in J. Sapp, ed., *Microbial phylogeny and evolution* (New York: Oxford University Press 2005), 184–206.

62. Wolfram Zillig to Carl Woese, February 1988 (Woese Papers, University of Illinois).

63. G.J. Olsen and C.R. Woese, "A brief note concerning archaebacterial phylogeny," *Canadian Journal of Microbiology* 35 (1989): 119–123. See also Joe Felsenstein, "Perils of molecular introspection," *Nature* 335 (1988): 118.

64. Manolo Gouy and Wen-Hsiung Li, "Phylogenetic analysis based on rRNA sequences supports the archaebacterial rather than the eocyte tree," *Nature* 339 (1989): 145–147.

65. J.A. Lake, "Archaebacterial or eocyte tree," *Nature* 343 (1990): 418–420.

66. Wen-Hsiung Li to Geoffrey North, assistant editor of *Nature*, June 29, 1989 (Woese Papers, University of Illinois).

67. Manolo Gouy and Wen-Hsiung Li, "Archaebacterial or eocyte tree?" *Nature* 343 (1990): 419.

68. L. Jin and M. Nei, "Limitations of the evolutionary parsimony method of phylogenetic analysis," *Molecular Biology and Evolution* 7 (1990): 82–102.

69. C.R. Woese, "Bacterial evolution," *Microbiological Reviews* 51 (1987): 221–271.

70. Ibid., 227.

71. Ibid., 232.

72. Ibid., 232.

73. Ibid., 262.

74. Robert M. Schwartz and Margaret O. Dayhoff, "Origins of prokaryotes, eukaryotes, mitochondria, and chloroplasts," *Science* 199 (1978): 395–403.

75. Johann Peter Gogarten, Henrik Kibak, Peter Dittrich, Lincoln Taiz, Emma Jean Bowman, Barry J. Bowman, Morris F. Manolson, Ronald Poole, Takayasu Date, Tairo Oshima, Jin Konishi, Kimitoshi Denda, and Masasuke Yoshida, "Evolution of the vacuolar H+-ATPase: Implications for the origin of eukaryotes," *Proceedings of the National Academy of Sciences* 86 (1989) 6661–6665, at 6664.

76. Ibid., 6665.

77. Naoyuki Iwabe, Kei-ichi Kma, Masami Hasegawa, Syozo Osawa, and Takashi Miyata, "Evolutionary relationship of archaeabacteria, eubacteria, and eukaryotes inferred from phylogenetic trees of duplicated genes," *Proceedings of the National Academy of Sciences* 86 (1989): 9355–9359, at 9358.

CHAPTER 19

1. See Lynn Margulis, *Symbiosis in cell evolution* (San Francisco: W.H. Freeman, 1981), 19.

2. For a review, see T. Cavalier-Smith, "The kingdom Chromista: Origin and systematics," *Progress in Phycological Research* 4 (1986): 309–347; Cavalier-Smith, "A revised six-kingdom system of life," *Biological Reviews* 73 (1998): 203–266.

3. T. Cavalier-Smith, "Endosymbiotic origin of the mitochondrial envelope," In W. Schwemmler and H.E.A. Schenk, eds., *Endocytobiology II. Intracellular space as an oligogenetic ecosystem* (Berlin: De Gutyer, 1983), 265–279; Cavalier Smith, "A six-kingdom classified and unified phylogeny," in Schwemmler and Schenk, eds., *Endocytobiology II*, 1027–1034.

4. T. Cavalier-Smith, "The kingdom of organisms," *Nature* 324 (1986): 416–417.

5. Ibid., 416.

6. Carl Woese, Otto Kandler, and Mark Wheelis, "Towards a natural system of organisms: Proposal for the domains Archaea, Bacteria, and Eucarya," *Proceedings of the National Academy of Sciences* 87 (1990): 4576–4579.

7. J. Kjems and R.A. Garrett, "Novel splicing mechanism for the rRNA intron in the archaebacterium *Desulfurococcus mobilis*," *Cell* 54 (1988): 693–703.

8. C. Woese, "A manifesto for microbial genomics," *Current Biology* 8/22 (1999): R781–R783.

9. Carl Woese to the author, November 25, 2006.

10. C. Woese to O. Kandler, October 6, 1989; C. Woese to Wolfram Zillig, October 8, 1989 (Woese Papers, University of Illinois).

11. Otto Kandler to Carl Woese October 13, 1989 (Kandler Papers, Universität Münchin).

12. Ibid.

13. V.B.D. Skerman, V. McGowan, and P.H.A. Sneath, "Approved lists of bacterial names," *International Journal of Systematic Bacteriology* 30 (1980): 225–420.

14. B.J. Tindall, "Misunderstanding the bacteriological code," *International Journal of Systematic Bacteriology* 49 (1999): 1313–1316.

15. Otto Kandler to Carl Woese, October 16, 1989 (Kandler Papers, Universität Münchin).

16. Carl Woese to Otto Kandler, October 23, 1989 (Woese Papers, University of Illinois).

17. R.Y. Stanier, E.A. Adelberg, J.L. Ingraham, and M.L. Wheelis, *Introduction to the microbial world* (Englewood Cliffs: Prentice Hall, 1979); Stanier, Ingraham, Wheelis, and P.R. Painter, *The microbial world*, 5th ed. (Englewood Cliffs: Prentice Hall, 1986).

18. Mark Wheelis to Carl Woese, December 8, 1989 (Woese Papers, University of Illinois).

19. Woese explained the situation in a December 14, 1989, letter to Kandler and included a copy of Wheelis's letter to him to consider. Kandler replied,

> I have no objections if Mark becomes a co-author, although I am not convinced that it is fully justified. Your original "urkingdoms" can only be understood as a new level above kingdoms. Thus, Marks suggestion is not new in principle. His contribution is that he reactivated your original idea. (Otto Kandler to Carl Woese, January 5, 1990 [Kandler Papers, Universität Münchin])

20. Otto Kandler to Carl Woese, January 5, 1990 (Kandler Papers, Universität Münchin).

21. Carl Woese to Otto Kandler, January 5, 1990 (Woese Papers, University of Illinois).

22. Carl Woese to Otto Kandler, January 8, 1990 (Woese Papers, University of Illinois).

23. Otto Kandler to Carl Woese, January 15, 1989 (Kandler Papers, Universität Münchin).

24. Carl R. Woese and George E. Fox, "Phylogenetic structure of the prokaryotic domain: The primary kingdoms," *Proceedings of the National Academy of Sciences* 74 (1977): 5088–5090.

25. Otto Kandler to Carl Woese, November 22, 1989 (Kandler Papers, Universität Münchin).

26. Otto Kandler to Carl Woese, December 7, 1989 (Kandler Papers, Universität Münchin).

27. Carl Woese to Otto Kandler, December 8, 1989 (Kandler Papers, Universität Münchin).

28. Otto Kandler to Carl Woese, February 27, 1990; Otto Kandler to Carl Woese, March 6, 1990 (Kandler Papers, Universität Münchin).

29. Otto Kandler to Carl Woese, November 22, 1989 (Kandler Papers, Universität Münchin).

30. Otto Kandler to Carl Woese, March 9, 1990 (Kandler Papers, Universität Münchin).

31. Woese et al., "Towards a natural system of organisms," 4578.

32. Carl Woese to Otto Kandler, October 23, 1989 (Woese Papers, University of Illinois).

33. Carl Woese to Otto Kandler, November 2, 1989 (Woese Papers, University of Illinois).

34. Otto Kandler to Carl Woese, November 8, 1989 (Kandler Papers, Universität Münchin).

35. Carl Woese to Otto Kandler, December 28, 1989 (Woese Papers, University of Illinois).

36. Carl Woese to Otto Kandler, January 27, 1990 (Woese Papers, University of Illinois).

37. Otto Kandler to Carl Woese, March 14, 1990 (Kandler Papers, Universität Münchin).

38. Woese et al., "Towards a natural system of organisms," 4578.

39. Ibid., 4578–4579.

40. Ibid., 4578.

41. Ibid.

42. R.H. Whittaker, "New concepts of kingdoms of organisms," *Science* 163 (1969): 150–160; Whittaker and L. Margulis, "Protist classification and the kingdoms of organisms," *Biosystems* 10 (1978): 3–18; Margulis and K.V. Schwartz, *Five kingdoms and illustrated guide to the phyla of life on Earth* (San Francisco: W.H. Freeman, 1988).

43. Woese et al., "Towards a natural system of organisms," 4576.

44. Ibid.

45. Ibid., 4577.

46. Ibid.

47. Ibid.

48. See C. Woese, "The Archaea: Their history and significance," in M. Kates, D.J. Kushner, and A.T. Matheson, eds., *The biochemistry of Archaea (Archaebacteria)* (Dordrecht: Elsevier, 1993), vii–xxxix, at vii. See also Roger Garrett and Hans-Peter Klenk, eds., *Archaea: Evolution, physiology and molecular biology* (Maldan, MA: Blackwell, 2007).

49. Carl Woese to Wolfram Zillig, March 20, 1992 (Woese Papers, University of Illinois).

CHAPTER 20

1. See M. Dietrich, "Paradox and persuasion: Negotiating the place of molecular evolution within evolutionary biology," *Journal of the History of Biology* 31 (1998): 87–111.

2. T. Dobzhansky, "Evolutionary and population genetics," *Science* 142 (1963): 1131–1135.

3. E. Mayr, "The new versus the classical in science," *Science* 141 (1963): 763.

4. See M. Kimura, *The neutral theory of molecular evolution* (Cambridge: Cambridge University Press, 1983); G.L. Morgan, "Emile Zuckerkandl, Linus Pauling, and the molecular evolutionary clock, 1959–1965," *Journal of the History of Biology* 31 (1998): 155–178. See also M. Dietrich, "The origins of the neutral theory of molecular evolution," *Journal of the History of Biology* 27 (1994): 21–59; Dietrich, "Paradox and persuasion"; W. Provine, "The neutral theory of molecular evolution in historical perspective," in B. Takahata and J. Crow, eds., *Population biology of genes and molecules* (Tokyo: Baifukan, 1990), 17–31.

5. Carl Woese, "Bacterial evolution," *Microbiological Reviews* 51 (1987): 221–271, at 226.

6. See Dietrich, "Paradox and persuasion"; Provine, "The neutral theory of molecular evolution in historical perspective."

7. G.G. Simpson, *Concession to the improbable: An unconventional autobiography* (New Haven: Yale University Press, 1978), 269. The scope and significance of the neutral theory in molecular evolution remain unsettled. See M. Nei, *Molecular evolutionary genetics* (New York: Columbia University Press, 1987).

8. See E. Mayr, "A natural system of organisms," *Nature* 348 (1990): 491.

9. Ernst Mayr and William Provine, eds., *The evolutionary synthesis* (Cambridge: Harvard University Press, 1980); E. Mayr, *The growth of biological thought* (Cambridge: Harvard University Press, 1982).

10. In 1982, Mayr was not convinced that "numerical methods were superior to the human computer," but he did admit that there were indications that suggest that if sufficiently large numbers of molecules were evaluated simultaneously, they were "more reliable than an indiscriminate morpho-phenetic analysis" of the numerical taxonomists. Mayr, *The growth of biological thought*, 245.

11. Careful expert that he was when it came to multicellular organisms, he seemed to show a careless ennui when dealing with the microbial world. His historical statements about the microbial concepts were error-ridden. He became a great champion of the prokaryote–eukaryote dichotomy and so, too, a promulgator of the great prokaryote myth. He falsely asserted that "the fundamental

difference between eukaryote and prokaryote [was] originally pointed out by Chatton in 1938" (see chapter 22). Mayr, "A natural system of organisms." Mayr stated that Roger Stanier and Cornelis van Niel introduced the terms and concepts to English readers in 1942, 20 years before they actually did so. Mayr, *The growth of biological thought*, 244.

12. Mayr, *The growth of biological thought*, 244.

13. Ibid.

14. See Mitchell L. Sogin, John H. Gunderson, Hille J. Elwood, Rogelio A. Alonso, and Debra A. Peattie, "Phylogenetic meaning of the kingdom concept: An unusual ribosomal RNA from *Giardia lamblia*," *Science* 243 (1989): 75–77.

15. Mayr, "A natural system of organisms."

16. Ibid.

17. Horace Judson, *The eighth day of creation: Makers of the revolution in biology* (New York: Simon and Schuster, 1979).

18. Evelyn Fox Keller, *The century of the gene* (Cambridge: Harvard University Press, 2000).

19. Carl Woese to Ernst Mayr, December 13, 1987 (Woese Papers, University of Illinois); C.R. Woese, "Bacterial evolution."

20. Carl Woese to Ernst Mayr, December 13, 1987 (Woese Papers, University of Illinois).

21. Ibid.

22. See, e.g., Verne Grant, *The origin of adaptations* (New York: Columbia University Press, 1963), 88–89.

23. Carl Woese to Ernst Mayr, January 4, 1988 (Woese Papers, University of Illinois).

24. W. Hennig, *Phylogenetic systematics*, translation by D.D. Davis and R. Zangerl (Urbana: University of Illinois Press, 1966).

25. E. Mayr and P.D. Ashlock, *Principles of systematic zoology* (New York: McGraw Hill, 1991).

26. E. Mayr to Carl Woese, January 14, 1991 (Woese Papers, University of Illinois).

27. Otto Kandler to Carl Woese, November 16, 1982 (Kandler Papers, Universität München).

28. Otto Kandler to Carl Woese, January 23, 1991 (Kandler Papers, Universität München).

29. Otto Kandler to Carl Woese, December 12, 1990 (Kandler Papers, Universität München).

30. Carl Woese, Otto Kandler, and Mark Wheelis, "A natural classification," *Nature* 351 (1991): 528–529; Kandler, Woese, and Wheelis, note to *Nature*, January 1991, unpublished draft (Woese Papers, University of Illinois).

31. Woese et al., "A natural classification," 529.

32. Ernst Mayr to Carl Woese, January 30, 1991 (Woese Papers, University of Illinois).

33. Carl Woese to Mark Wheelis, February 11, 1991 (Woese Papers, University of Illinois).

34. Teresa Scranney, assistant production editor of *Nature*, to Carl Woese, February 20, 1991 (Woese Papers, University of Illinois).

35. Ernst Mayr to Carl Woese, March 1, 1991 (Woese Papers, University of Illinois). Mayr emphasized that he did not lump together the two bacterial taxa. He recognized them as subdomains. Taxonomic rank, he remarked, "is determined not by age, but by degree of difference."
36. Woese et al., "A natural classification."
37. E. Mayr fax to C. Woese, June 6, 1991 (Woese Papers, University of Illinois).
38. E. Mayr, "More natural classification," *Nature* 353 (1991): 122.
39. Ford Doolittle, "Still more natural classification," unpublished ms. submitted to *Nature*, October 1, 1991 (Woese Papers, University of Illinois).
40. Patrick Forterre, "Neutral terms," *Nature* 355 (1992): 305.
41. T. Cavalier-Smith, "Bacteria and eukaryotes," *Nature* 356 (1992): 570.
42. Horishi Hori, Yukio Satow, Isao Inoue, and Mitsuo Chihara, "Archaebacteria vs metabacteria: Phylogenetic tree of organisms indicated by comparisons of 5S ribosomal RNA sequences," in S. Osawa and T. Honjo, eds., *Evolution of life* (New York: Springer, 1991), 325–336.
43. M.L. Wheelis, O. Kandler, and C.R. Woese, "On the nature of global classification," *Proceedings of the National Academy of Sciences* 89 (1992): 2930–2934.
44. Ibid., 2933.
45. L. Margulis and R Guerrero, "Kingdoms in turmoil," *New Scientist* 23 (1991): 46–50.
46. Lynn Margulis, *Symbiosis in cell evolution*, 2nd ed. (New York: W.H. Freeman, 1993), 2, 38.
47. See A. Balows, H.G. Trüper, M. Dworkin, W. Harder, and K.H. Schleifer, eds., *The prokaryotes*, 2nd ed. (New York: Springer, 1992), Vol. 1: vii.
48. J.G. Olsen, C.R., Woese, and R.J. Overbeek, "The winds of (evolutionary) change: Breathing new life into microbiology," *Journal of Bacteriology* 176 (1994): 1–6.
49. C.R. Woese, "There must be a prokaryote somewhere: Microbiology's search for itself," *Microbial Reviews* 58 (1994): 1–9.
50. As two microbiologists quipped in *Nature*, "It is a truth universally acknowledged that there are only two kinds of bacteria. One is *Escherichia coli* and the other is not." J.A. Downie and J.P. Young, "Genome sequencing: The ABC of symbiosis," *Nature* 412 (2001): 597–598.
51. Woese, "There must be a prokaryote somewhere," 6.
52. Carl Woese to the author, January 27, 2006.
53. E. Mayr, "Two empires or three?" *Proceedings of the National Academy of Sciences* 95 (1998): 9720–9723, at 9720.
54. Ibid., 9721.
55. Ibid., 9722.
56. Ibid., 9721.
57. Ibid., 9722.
58. Ibid.
59. See John Corliss, "Toward a nomenclatural protist perspective," in L. Margulis, H.I. McKhann, and L. Olendzenski, eds., *Illustrated glossary of protoctista* (Boston: Johns and Bartlett, 1993), xxvii–xxxii.
60. E. Mayr, "Two empires or three?" 9721.

61. Ibid., 9723.

62. C.R. Woese, "Default Taxonomy; Ernst Mayr's view of the microbial world," *Proceedings of the National Academy of Sciences* 95 (1998): 11043–11046, at 11045.

63. Ibid., 11044.

64. Ibid., 11046.

65. R.S. Breed, E.G.D. Murray, and A. Parker Hitchens, eds., *Bergey's manual of determinative bacteriology*, 6th ed. (Baltimore: Williams and Wilkins, 1957), 9.

66. A.J. Kluyver, "Microbial metabolism and the energetic basis of life," in A.J. Kluyver and C.B. van Niel, *The microbe's contribution to biology* (Cambridge: Harvard University Press, 1956), 1–30, at 2.

67. Ibid., 3.

68. William B. Whitman, David C. Coleman, and William J. Wiebe, "Prokaryotes: The unseen majority," *Proceedings of the National Academy of Sciences* 95 (1998): 6578–6583.

69. Woese, "Default taxonomy," 11045.

70. D.A. Stahl, D.J. Lane, G.J. Olsen, and N.R. Pace, "The analysis of hydrothermal vent-associated symbionts by ribosomal RNA sequences," *Science* 224 (1984): 409–411; Pace, Stahl, Lane, and Olsen, "The analysis of natural microbial populations by ribosomal RNA sequences," *Advances in Microbial Ecology* 9 (1986): 1–55.

71. E.F. DeLong, G.S. Wickham, and N.R. Pace, "Phylogenetic stains: Ribosomal RNA-based probes for the identification of single cells," *Science* 243 (1989): 1360–1363.

72. Norman Pace, "A molecular view of microbial diversity and the biosphere," *Science* 276 (1997): 734–740.

73. N.T. Perna, G. Plunkett 3rd, V. Burland, B. Mau, J.D. Glasner, D.J. Rose, G.F. Mayhew, et al., "Genome sequence of enterohaemorrhagic *Escherichia coli* O157:H7," *Nature* 409 (2001): 529–533.

74. W.F. Doolittle, Y. Boucher, C.L. Nesbø, C.J. Douady, J.O. Andersson, and A.J. Roger, "How big is the iceberg of which organellar genes in nuclear genomes are but the tip?" *Philosophical Transactions: Biological Sciences* 358 (2003): 39–58, at 48.

75. The Crafoord Prize honors basic research in certain fields that are not eligible for Nobel Prize consideration. Mayr was awarded the prize four years jointly with John Maynard Smith and George C. Williams.

76. Ernst Mayr to Carl Woese, February 27, 2003 (Woese Papers, University of Illinois).

77. Carl Woese to Ernst Mayr, June 17, 2004 (Woese Papers, University of Illinois).

CHAPTER 21

1. See Ravi Jain, Maria C. Rivera, and James A. Lake, "Lateral gene transfer among genomes: The complexity hypothesis," *Proceedings of the National Academy of Sciences* 96 (1999): 3801–3806.

2. H.C. Neu, "The crisis in antibiotic resistance," *Science* 257 (1992): 1064–1073.

3. J.G. Lawrence and H. Ochman, "Molecular archaeology of the *Escherichia coli* genome," *Proceedings of the National Academy of Sciences* 95 (1998): 9413–9417. See also B.W. Wren, "Microbial genome analysis: Insight into virulence, host adaptation and evolution," *Nature Reviews Genetics* 1 (2000): 30–39.

4. S. Sonea and P. Panisset, *The new bacteriology* (Boston: Jones and Bartlett, 1983); Sonea and L.G. Mathieu, *Prokaryotology* (Montreal: Les Presses de Université de Montréal, 2000).

5. Norman Anderson, "Evolutionary significance of virus infection," *Nature* 227 (1970): 1346–1347, at 1346.

6. Michael Syvanen, "Conserved regions in mammalian b-globins: Could they arise by cross-species gene exchange?" *Journal of Theoretical Biology* 107 (1984): 685–696; Michael Syvanen, "Cross-species gene transfer: Implications for a new theory of evolution," *Journal of Theoretical Biology* (1985): 112, 333–343.

7. Dorothy Jones and P.H.A. Sneath, "Genetic transfer and bacterial taxonomy," *Bacteriological Reviews* 34 (1970): 40–81.

8. R.Y. Stanier, "Toward an evolutionary taxonomy of the bacteria," *Recent Advances in Microbiology* 7 (1971): 595–604, at 597.

9. Darryl Reanney, "Commentary: Gene transfer as a mechanism of microbial evolution," *Bioscience* 27 (1977): 340–344, at 341. See also Reanney, "Extrachromosomal elements as possible agents of adaptation and development," *Bacteriological Reviews* 40 (1976): 552–590.

10. Carl Woese, "Bacterial evolution," *Microbiological Reviews* 51 (1987): 221–271, at 227. This paper was cited some 300 times in the first two years, and today it has more than 2,000 citations.

11. Ibid.

12. Ibid., 230.

13. Ibid., 231.

14. Ibid.

15. W.M. Fitch and K. Upper, "The phylogeny of tRNA sequences provides evidence for ambiguity reduction in the origin of the genetic code," *Cold Spring Harbor Symposia on Quantitative Biology* 52 (1987): 759–767.

16. Peter Gogarten, "The early evolution of cellular life," *Trends in Ecology and Evolution* 10 (1995): 147–150, at 150.

17. Patrick Forterre and Hervé Philippe, "The last universal common ancestor (LUCA), simple or complex," *Biological Bulletin* 196 (1999): 373–377, at 375.

18. Carl R. Woese, "Archaebacteria and cellular origins: An overview," in Otto Kandler, ed., *Archaebacteria. Proceedings of the 1st International Workshop on Archaebacteria*, Munich, June 27th–July 1st 1981 (Stuttgart: Gustav Fischer, 1982), 1–17, at 13.

19. Ibid.

20. Ibid.

21. Ibid.

22. Woese, "Bacterial evolution," 264.

23. O. Kandler, "The early diversification of life," in S. Bengtson, ed., *Early life on Earth: Nobel symposium no. 84* (New York: Columbia University Press, 1994), 152–160; Kandler, "Cell wall biochemistry in Archaea and its phylogenetic

implications," *Journal of Biological Physics* 20 (1994): 165–169. See also Karl Popper, "Progenote or protogenote," *Science* 250 (1990): 1070.

24. R.D. Fleischmann, M.D. Adams, O. White, R.A. Clayton, E.F. Kirkness, A.R. Kerlavage, et al., "Whole-genome random sequencing and assembly of *Haemophilus influenzae* Rd.," *Science* 269 (1995): 468–470.

25. D.J. Bult, O. White, G.J. Olsen, and C.R. Woese, "Complete genome sequence of the methanogenic archaeon, *Methanococcus jannaschii*," *Science* 273 (1996): 1058–1073; Olsen and Woese, "Lessons from an Archaeal genome: What are we learning from *Methanococcus jannaschii*?" *Trends in Genetics* 12 (1996): 377–379.

26. J. Raymond, O. Zhaxybayeva, J.P. Gogarten, S.Y. Gerdes, and R.E. Blankenship, "Whole-genome analysis of photosynthetic prokaryotes," *Science* 298 (2002): 1616–1620.

27. Berend Snel, Martijn A. Huynen, and Bas E, Dutilh. "Genome trees and the nature of genome evolution," *Annual Review of Microbiology* 59 (2005): 191–209.

28. Lawrence and Ochman, "Molecular archaeology of the *Escherichia coli* genome."

29. Ibid., 9414.

30. Ibid., 9416.

31. M.R. Rose and W.F. Doolittle, "Parasitic DNA—the origin of species and sex," *New Scientist* 98 (1983): 787–789; Doolittle, T.R.L Kirkwood, and M.A.H Dempster, "Selfish DNAs with self-restraint," *Nature* 307 (1984): 501–502; L.C. Schalkwyk, R.L. Charlebois, and W.F. Doolittle, "Insertion sequences on plasmid pHV1 of *Haloferax volcanii*," *Canadian Journal of Microbiology* 39 (1992): 201–206; Doolittle, "What are the archaebacteria and why are they important?" *Biochemical Journal* 58 (1992): 1–6.

32. W.F. Doolittle, "Archaebacteria coming of age," *Trends in Genetics* 1 (1985): 268–26; Doolittle, "The evolutionary significance of the archaebacteria," *Annals of the New York Academy of Sciences* 503 (1986): 72–77; Doolittle, "Bacterial evolution," *Canadian Journal of Microbiology* 34 (1987): 547–551.

33. See James R. Brown and W. Ford Doolittle, "Archaea and the prokaryote-to-eukaryote transition," *Microbiology and Molecular Biology Reviews* 61 (1997): 456–502.

34. See J.R. Brown, Y. Masuchi, F.T. Robb, and W.F. Doolittle, "Evolutionary relationships of bacterial and archaeal glutamine synthetase genes," *Journal of Molecular Evolution* 38 (1994): 566–576.

35. Radhey Gupta and Bhag Singh, "Cloning of the HSP70 gene from *Halobacterium marismortui*: Relatedness of archaebacterial HSP70 to its eubacterial homologs and a model for the evolution of the HSP70 gene," *Journal of Bacteriology* 174 (1992): 4594–4605.

36. O. Tiboni, P. Cammarano, and A.M. Sanangelantoni, "Cloning and sequencing of the gene encoding glutamine synthetase I from the archaeum *Pyrococcus woesei*: Anomalous phylogenies inferred from analysis of archaeal and bacterial glutamine synthetase I sequences," *Journal of Bacteriology* 175 (1993): 2961–2969; Cammarano, P. Palm, R. Creta, E. Ceccarelli, A.M. Sanangerantoni,

and O. Tiboni, "Early evolutionary relationships among known life forms inferred from elongation factor EF-2 (Ef-G) sequences: Phylogenetic coherence and structure of the archael domain," *Journal of Molecular Evolution* 34 (1992): 396–405; Y. Kumada, D.R. Benson D. Hillemann, T.J. Hosted, D.A. Rochefort, C.J. Thompson, W. Wohlleben, and Y. Tateno, "Evolution of the glutamine synthetase gene, one of the oldest existing and functioning genes," *Proceedings of the National Academy of Sciences* 90 (1993): 3009–3013. See also Nadia Benachenhou-Lahfa, Patrick Forterre, and Bernard Labeda, "Evolution of glutamate dehydrogenase genes: Evidence for two paralogous protein families and unusual branching patterns of the archaebacteria in the universal tree of life," *Journal of Molecular Evolution* 36 (1993): 335–346.

37. E. Hilario and J.P. Gogarten, "Lateral transfer of ATPase genes—the tree of life becomes the net of life," *BioSystems* 31 (1993): 111–119.

38. Gogarten, "The early evolution of cellular life," 150.

39. Brown et al., "Evolutionary relationships of bacterial and archaeal glutamine synthetase genes," 576.

40. J.R., Brown and W.F. Doolittle, "Root of the universal tree of life based on ancient aminoacyl-tRNA synthetase gene duplications," *Proceedings of the National Academy of Sciences* 92 (1995): 2441–2445, at 2444.

41. J.R. Brown, F.T. Robb, R. Weiss, and W.F. Doolittle, "Evidence for the early divergence of tryptophanyl- and tyrosyl-tRNA synthetases," *Journal of Molecular Evolution* 45 (1997): 9–16.

42. Brown and Doolittle, "Archaea and the prokaryote-to-eukaryote transition."

43. Norman Pace, "A molecular view of microbial diversity and the biosphere," *Science* 276 (1997): 734–740.

44. Ibid., 735.

45. See Ford Doolittle, "If the tree of life fell, would it make a sound?" in Jan Sapp, ed., *Microbial phylogeny and evolution* (New York: Oxford University Press, 2005), 119–133, at 130.

46. Ford Doolittle, "At the core of the Archaea," *Proceedings of the National Academy of Sciences* 93 (1996): 8797–8799, at 8799.

47. Charles Darwin, *On the origin of species*, facsimile of 1859 edition (Cambridge Harvard University Press, 1964), 489.

48. W. Ford Doolittle and Eric Bapteste, "Pattern pluralism and the tree of life hypothesis," *Proceedings of the National Academy of Sciences* 104 (2007): 2043–2049, at 2047.

49. Ibid., 2048.

50. Doolittle, "At the core of the Archaea," 8799.

51. Ibid.

52. See also E.V. Koonin A.R. Musegian, M.Y. Galperin, and D.R. Walker, "Comparison of archaeal and bacterial genomes: Computer analysis of protein sequences predicts novel functions and suggests a chimeric origin for the Archaea," *Molecular Microbiology* 25 (1997): 619–637.

53. Elizabeth Pennisi, "Genome data shake tree of life," *Science* 280 (1998): 672–674; Pennisi, "Is it time to uproot the tree of life?" *Science* 284 (1999): 1305–1307.

54. The full genome of *Aquifex* was sequenced in 1998: G. Deckert, P.V. Warren, T. Gaasterland, W.G. Young, A.L. Lenox, D.E. Graham, R. Overbeek, M.A. Snead, M. Keller, M. Aujay, R. Huber, R.A. Feldman, J.M. Short, G.J. Olsen, and R.V. Swanson, "The complete genome of the hyperthermophilic bacterium *Aquifex aeolicus*," *Nature* 392 (1998): 353–358.

55. Pennisi, "Genome data shake tree of life," 672.

56. Ibid., 673.

57. Ibid., 542.

58. See E.F. DeLong, "Archaea in coastal marine environments," *Proceedings of the National Academy of Sciences* 89 (1992): 5685–5689.

59. Edward DeLong, "Archael means and extremes," *Science* 280 (1998): 542–543.

60. See also J. Furhman, K. McCallum, and A. Davis "Novel major archaeal group from marine plankton," *Nature* 356 (1992): 148; S.M. Barns, C.F. Delwiche, J.D. Palmer, and N.R. Pace, "Perspectives on archaeal diversity, thermophily and monophyly from environmental rRNA sequences," *Proceedings of the National Academy of Sciences* 93 (1996): 9188–9193.

61. Steven J. Hallam, Konstantinos T. Konstantinidis, Nik Putnam, Christa Schleper, Yoh-ichi Watanabe, Junichi Sugahara, Christina Preston, José de la Torre, Paul M. Richardson, and Edward F. DeLong, "Genomic analysis of the uncultivated marine crenarchaeote *Cenarchaeum symbiosum*," *Proceedings of the National Academy of Sciences* 103 (2006): 18296–18301.

62. Edward DeLong to the author, June 18, 2008.

63. Carl Woese, "The universal ancestor," *Proceedings of the National Academy of Sciences* 95 (1998): 6854–6859, at 6854.

64. Ibid.

65. Ibid., 6857. See also Carl Woese, "Interpreting the universal phylogenetic tree," *Proceedings of the National Academy of Sciences* 97 (2000): 8392–8396.

66. W. Ford Doolittle, "Phylogenetic classification and the universal tree," *Science* 284 (1999): 2124–2128; Doolittle, "Uprooting the tree of life," *Scientific American* (February 2000): 90–95.

67. Doolittle, "Phylogenetic classification and the universal tree," 2126.

68. Tsuneaki Asai, Dmitry Zaporojets, Craig Squires, and Catherine Squires, "An *Escherichia coli* strain with all chromosomal rRNA operons inactivated: Complete exchange of rRNA genes between bacteria," *Proceedings of the National Academy of Sciences* 96 (1999): 1971–1976; see also Wai Ho Yap, Zhenshui Zhang, and Yue Wang, "Distinct types of rRNA operons exist in the genome of the actinomycete *Thermomonospora chromogena* and evidence for lateral transfer of an entire rRNA operon," *Journal of Bacteriology* 181 (1999): 5201–5209; Leo M. Schouls, Corrie S. Schot, and Jan A. Jacobs, "Lateral transfer of segments of the SSU rRNA genes between species of the *Streptococcus anginosus* group," *Journal of Bacteriology* (2003): 7241–7246.

69. Doolittle, "Phylogenetic classification and the universal tree," 2127.

70. J. Peter Gogarten, W. Ford Doolittle, and Jeffrey G. Lawrence, "Prokaryotic evolution in light of gene transfer," *Molecular Biology and Evolution* 19 (2002): 2226–2238.

71. J. Peter Gogarten, W. Ford Doolittle, and Jeffrey G. Lawrence, "Prokaryotic evolution in light of gene transfer," *Molecular Biology and Evolution* 19 (2002): 2235.

72. Ibid.

73. H. Ochman, J.G. Lawrence, and E.A. Groisman, "Lateral gene transfer and the nature of bacterial innovation," *Nature* 405 (2000): 299–304; Gogarten et al., "Prokaryotic evolution in light of gene transfer"; J. Eisen, "Lateral gene transfer among microbial genomes: New insights from complete genome analysis," *Current Opinion in Genetics and Development* 10 (2000): 606–611.

74. W.F. Doolittle, Y. Boucher, C.L. Nesbø, C.J. Douady, J.O. Andersson, and A.J. Roger, "How big is the iceberg of which organellar genes in nuclear genomes are but the tip?" *Philosophical Transactions: Biological Sciences* 358 (2003): 39–58, at 47.

75. Ibid.

76. Ibid., 53.

77. Robert L. Charlebois and W. Ford Doolittle, "Computing prokaryotic gene ubiquity: Rescuing the core from extinction," *Genome Research* 1 (2004): 2469–2477.

78. See, e.g., Jain et al., "Lateral gene transfer among genomes"; C.R. Woese, G.J. Olsen, M. Ibba, and D. Söll, "Aminoacyl-tRNA synthetases, the genetic code, and the evolutionary process," *Microbiology and Molecular Biology Reviews* 64 (2000): 202–236; V. Daubin, M. Gouy, and G. Perrière, "A phylogenomic approach to bacterial phylogeny: Evidence of a core of genes sharing a common history," *Genome Research* 12 (2002): 1080–1090; O. Matte-Taillez, C. Brochier, P. Forterre, and H. Philippe, "Archaeal phylogeny based on ribosomal proteins," *Molecular Biology and Evolution* 19 (2002): 631–639; Wolfgang Ludwig and Karl-Heinz Schleifer, "The molecular phylogeny of bacteria based on conserved genes," in Jan Sapp, ed., *Microbial phylogeny and evolution*, 70–99; Snel et al., "Genome trees and the nature of genome evolution."

79. C.G. Kurland, "Paradigm lost," in J. Sapp ed., *Microbial phylogeny and evolution*, 207–223, at 220.

80. C.G. Kurland, B. Canback, and O.G. Berg, "Lateral gene transfer: A critical review," *Proceedings of the National Academy of Sciences* 19 (2003): 9658–9692.

81. Jeffrey G. Lawrence and Howard Ochman, "Reconciling the many faces of lateral gene transfer," *Trends in Microbiology* 10/1 (2002): 1–4, at 1.

82. Mark A. Ragan, "Detection of lateral gene transfer among microbial genomes," *Current Opinion in Genetics and Development* 11 (2001): 620–626; Lawrence and Ochman, "Reconciling the many faces of lateral gene transfer"; Lisa B. Koski, Richard A. Morton, and G. Brian Golding, "Codon bias and base composition are poor indicators of laterally transferred genes," *Molecular Biology and Evolution* 18 (2001): 404–412.

83. Robert G. Beiko, Timothy J. Harlow, and Mark A. Ragan, "Highways of gene sharing in prokaryotes," *Proceedings of the National Academy of Sciences* 102 (2005): 14332–14337. See also Snel et al., "Genome trees and the nature of genome evolution"; Tal Dagan and William Martin, "Ancestral genome

sizes specify the minimum rate of lateral gene transfer during prokaryote evolution," *Proceedings of the National Academy of Sciences* 104 (2007): 870–875.

84. Doolittle and Bapteste, "Pattern pluralism and the tree of life hypothesis," 2046.
85. Ibid., 2048.
86. Ibid., 2044.
87. Ford Doolittle to the author, May 11, 2008.
88. The scope and significance of hybridization in plants and animal evolution remains unresolved. See L.H. Rieseberg, T.E. Wood, and E. Baack. "The nature of plant species," *Nature* 440 (2006): 524–527; Rieseberg and J.H. Willis, "Plant speciation," *Science* 317 (2007): 910–914; M.L. Arnold, *Natural hybridization and evolution* (Oxford: Oxford University Press, 1997); Thomas E. Dowling and Carol L. Secor, "The role of hybridization and introgression in the diversification of animals," *Annual Revue of Ecology and Systematics* 28 (1997): 593–619; N.H. Barton, "The role of hybridization in evolution," *Molecular Ecology* 10 (2001): 551–568; Aaron O. Richardson and Jeffrey D. Palmer, "Lateral gene transfer in plants," *Journal of Experimental Botany* 58 (2007): 1–9,
89. Lynn Margulis and René Fester, eds., *Symbiosis as a source of evolutionary innovation: Speciation and morphogenesis* (Boston: MIT Press, 1991); N. Kondo, N. Nikoh, N. Ijichi, M. Shimada, and T. Fukatsu, "Genome fragment of *Wolbachia* endosymbiont transferred to X chromosome of host insect," *Proceedings of the National Academy of Sciences* 99 (2002): 14280–14285;. N. Nikohl, K. Tanaka, F. Shibata, N. Kondo, M. Hizumes, M. Shimada, and T. Fukatsu, "Wolbachia genome integrated in an insect chromosome: Evolution and fate of laterally transferred endosymbiont genes," *Genome Research* 18 (2008): 272–280; J.C.D. Hotopp, M.E. Clark, D.C.S.G. Oliveira, J.M. Foster, P. Fischer, M.C.M. Torres, J.D. Giebel, N. Kumar, N. Ishmael, S. Wang, et al., "Widespread lateral gene transfer from intracellular bacteria to multicellular eukaryotes," *Science* 317 (2007): 1753–1756. See, e.g., Richardson and Palmer, "Lateral gene transfer in plants"; Kondo et al., "Genome fragment of *Wolbachia* endosymbiont transferred to X chromosome of host insect."
90. Roswitha Löwer, Johannes Löwer, and Reinherd Kurth, "The viruses in all of us: Characteristics and biological significance of human endogenous retrovirus sequences," *Proceedings of the National Academy of Sciences* 93 (1999): 5177–5184. See J. Robin Harris, "Placental endogenous retrovirus (ERV): Structural, functional, and evolutionary significance," *BioEssays* 20 (1998): 307–316; Frederic Bushman, *Lateral DNA transfer: Mechanisms and consequences* (New York: Cold Spring Harbor Laboratory Press, 2002); F. Ryan, "Genomic creativity and natural selection: A modern synthesis," *Biological Journal of the Linnaean Society* 88 (2006): 655–672.
91. See Jan Sapp, *Genesis: The evolution of biology* (New York: Oxford University Press, 2003), 234–251.
92. V. Daubin and H. Ochman, "Start up entities in the origin of new genes," *Current Opinion in Genetics and Development* 14 (2004): 616–619;

M.G. Weinbauer and F. Rassoulzadegan, "Are viruses driving microbial diversification and diversity?" *Environmental Microbiology* 6 (2004): 1–11; C.A. Suttle, "Viruses in the sea," *Nature* 437 (2005): 356–361.

93. Roger W. Hendrix, "Evolution: The long evolutionary reach of viruses," *Current Biology* 9 (1999): R914–R917; Hendrix, Jeffrey G. Lawrence, Graham F. Hatfull, and Sherwood Casjens, "The origins and ongoing evolution of viruses," *Trends in Microbiology* 8 (2000): 504–507; Lawrence, Graham F. Hatfull, and Roger W. Hendrix, "Imbroglios of viral taxonomy: Genetic exchange and failings of phenetic approaches," *Journal of Bacteriology* 184 (2002): 4891–4905; E.V. Koonin and V.V. Dolja, "Evolution of complexity in the viral world: The dawn of a new vision," *Virus Research*, 117 (2006): 1–4. P. Forterre, "The origin of viruses and their possible roles in major evolutionary transitions," *Virus Research* 117 (2006): 5–16.

CHAPTER 22

1. C.R. Woese, "There must be a prokaryote somewhere: Microbiology's search for itself," *Microbiological Reviews* 58 (1994): 1–9.

2. Ibid., 1.

3. Ibid., 6.

4. See Jan Sapp, "The prokaryote-eukaryote dichotomy: Meanings and mythology," *Microbiology and Molecular Biology Reviews* 69 (2005), 292–305; Sapp, "The nine lives of Gregor Mendel," in H.E. Le Grand, ed., *Experimental inquiries* (Dordrecht: Kluwer, 1990), 137–166.

5. See A.N. Glazer, "Roger Yate Stanier, 1916–1982: A transcendent journey," *International Microbiology* (2001) 4: 59–66, at 63.

6. E. Mayr, "Two empires or three?" *Proceedings of the National Academy of Sciences* 95 (1998): 9720–9723, at 9720.

7. W.F. Doolittle, "At the core of the Archaea," *Proceedings of the National Academy of Science* 93 (1996): 8797–8799, at 8798.

8. Mayr, "Two empires or three?", at 8723.

9. D.A. Walsh and W.F Doolittle, "The real domains of life," *Current Biology* 15 (2005): R237–R240.

10. Norman R. Pace, "Time for a change," *Nature* 441 (2006): 289.

11. William Martin and Eugene Koonin, "A positive definition of prokaryotes," *Nature* 442 (2006): 24.

12. Ibid.

13. Lynn Margulis and Michael Dolan, "Advances in biology reveal truth about prokaryotes," *Nature* 445 (2007): 21.

14. T. Cavalier-Smith, "Concept of a bacterium still valid in prokaryote debate," *Nature* 446 (2007): 257.

15. T. Cavalier-Smith, "The origin of eukaryote and archaebacterial cells," *Annals of the New York Academy of Sciences* 503 (1987): 17–54.

16. T. Cavalier-Smith, "The neomuran origin of Archaebacteria, the negibacterial root of the universal tree and bacterial megaclassification," *International Journal of Systematic and Evolutionary Microbiology* 52 (2002): 7–76.

17. Nigel Goldenfeld and Carl Woese, "Biology's next revolution," *Nature* 445 (2007): 369.

18. C.M. Waters and B.L. Bassler, "Quorum sensing: Cell-to-cell communication in bacteria," *Annual Review of Cell and Developmental Biology* 21 (2005): 319–346.

19. Goldenfeld and Woese, "Biology's next revolution."

20. Cavalier-Smith, "Concept of a bacterium still valid in prokaryote debate."

21. Norman Pace, "More on the definition of the prokaryote," unpublished letter to *Nature*, August 31, 2006 (Pace Papers, University of Colorado).

22. Patrick Forterre and David Pranishvilli, "Should we execute prokaryotes?" (unpublished paper); Patrick Forterre to Carl Woese, January 13, 2007 (Woese Papers, University of Illinois).

23. J.A. Fuerst and R.I. Webb, "Membrane-bounded nucleoid in the eubacterium *Gemmata obscuriglobus,*" *Proceedings of the National Academy of Sciences* 88 (1991): 8184–8188, at 8188.

24. Carl Woese to Lucy Odling-Smee, January 19, 2007 (Woese Papers, University of Illinois).

25. Norman Pace, "The molecular tree of life changes how we see, teach microbial diversity," *Microbe* 3 (2008): 15–20.

26. John Ingraham, "Pro- and anti-'procaryote,'" *Microbe* 3 (2008): 33.

27. Ibid.

28. Da-Fei Feng, Glen Cho, and Russell F. Doolittle, "Determining divergence times with a protein clock: Update and reevaluation," *Proceedings of the National Academy of Sciences* 94 (1997): 13028–13033.

29. Masami Hasagawa and Walter Fitch, "Dating the cenancestor of organisms," *Science* 274 (1996): 1750; Russell F. Doolittle, Da-Fei Feng, Simon Tsang, Glen Cho, and Elizabeth Little, "Determining divergence times of the major kingdoms of living organisms with a protein clock," *Science* 271 (1996): 470–477.

30. G.E. Fox, E. Stackebrandt, R.B. Hespell, J. Gibson, J. Maniloff, T.A. Dyer, R.S. Wolfe, et al., "The phylogeny of prokaryotes," *Science* 209 (1980): 457–463, at 463.

31. M. L Wheelis, O. Kandler, and C.R. Woese, "On the nature of global classification," *Proceedings of the National Academy of Sciences* 89 (1992): 2930–2934, at 2933.

32. Ibid.

33. T. Fenchel and B.J. Finlay, *Ecology and evolution in anoxic worlds* (Oxford: Oxford University Press, 1995).

34. T. Cavalier-Smith, "The origin of cells, a symbiosis between genes, catalysts and membranes," *Cold Spring Harbor Symposia on Quantitative Biology* 52 (1987): 805–824.

35. This way of expressing the different symbiosis theories was depicted by Hyman Hartman and Alexei Fedorov, "The origin of the eukaryotic cell: A genomic investigation," *Proceedings of the National Academy of Sciences* 99 (2002): 1420–1425, at 1420.

36. Wolfram Zillig, "Comparative biochemistry of archaea and bacteria," *Current Opinion in Genetics and Development* 1 (1991): 544–551, at 545.

37. Reinhard Hensel, Peter Zwickl, Stefan Fabry, Jutta Lang, and Peter Palm, "Sequence comparison of glyceraldehyde-3-phosphate dehydrogenases from the three urkingdoms: Evolutionary implication," *Canadian Journal of Microbiology* 35 (1989): 81–85, at 85.

38. Radhey S. Gupta, Karen Aitken, Mizied Falah, and Bhag Singh, "Cloning of *Giardia lamblia* heat shock protein Hsp70 homologs: Implications regarding origin of eukaryotic cells and of endoplasmic reticulum," *Proceedings of the National Academy of Sciences* 92 (1994): 2895–2899, at 2899; Gupta and G.B. Golding, "The origin of the eukaryotic cell," *Trends in Biochemical Sciences* 21 (1996): 166–170; Gupta, "Molecular sequences and the early history of life," in J. Sapp, ed., *Microbial phylogeny and evolution: Concepts and controversies* (New York: Oxford University Press, 2005), 160–183.

39. R.S. Gupta and G.B. Golding, "Evolution of Hsp70 gene and its implications regarding relationships between archaebacteria, eubacteria, and eukaryotes," *Journal of Molecular Evolution* 37 (1993): 573–582.

40. R.S. Gupta, "Protein phylogenies and signature sequences: A reappraisal of evolutionary relationships among archaebacteria, eubacteria, and eukaryotes," *Microbiology and Molecular Biology Reviews* 62 (1998): 1435–1491; Gupta, "The *natural* evolutionary relationships among prokaryotes," *Critical Reviews of Microbiology* 26 (2000): 111–131.

41. Wolfgang Ludwig and Karl-Heinz Schleifer, "The molecular phylogeny of bacteria based on conserved genes," in Sapp, ed., *Microbial phylogeny and evolution*, 70–98.

42. Ibid., 76.

43. J.A. Lake, E. Henderson, M.W. Clark, and A.T. Matheson, "Mapping evolution with ribosome structure: Intralineage constancy and interlineage variation," *Proceedings of the National Academy of Sciences* 79 (1982): 5948–5952, at 5951.

44. J.A. Lake, and M.C. Rivera, "Was the nucleus the first symbiont?" *Proceedings of the National Academy of Sciences* 91 (1994): 2880–2881; Lake, Jonathan Moore, Anne Simonson, and Maria Rivera, "Fulfilling Darwin's dream," in Sapp, *Microbial phylogeny and evolution*, 184–206, at 189.

45. Lynn Margulis, *The origin of eukaryotic cells* (New Haven: Yale University Press, 1970), 183.

46. Lynn Margulis, *Symbiosis in cell evolution*, 2nd ed. (San Francisco: W.H. Freeman, 1981), 5.

47. J.L. Hall and D.J. Luck, "Basal body-associated DNA: *In situ* studies in *Chlamydomonas reinhardtii*," *Proceedings of the National Academy of Sciences* 92 (1995): 5129–5133.

48. Michael Chapman, Michael Dolan, and Lynn Margulis, "Centrioles and kinetosomes: Form, function, and evolution," *Quarterly Review of Biology* 75 (2000): 409–429; Margulis, Chapman, Ricardo Guerrero, and John Hall, "The last eukaryotic common ancestor (LECA): Acquisition of cytoskeletal motility from aerotolerant spirochetes in the Proterozoic eon," *Proceedings of the National Academy of Sciences* 103 (2007): 13080–13085, at 13080.

49. Hannah Melintsky, Frederick Rainey, and Lynn Margulis, "The karyomastigont model of eukaryosis," in Sapp, ed., *Microbial phylogeny and evolution*, 261–280, at 262.

50. Ibid., 276.

51. A.J. Roger, "Reconstructing early events in eukaryotic evolution," *American Naturalist* 154 (1999): S146–S163; P.J. Keeling, "A kingdom's progress: Archezoa and the origin of eukaryotes," *BioEssays* 20 (1998): 87–95; T. Cavalier-Smith, "A revised six-kingdom system of life," *Biological Reviews* 73 (1998): 203–266.

52. R.P. Hirt, J.M. Logsdon, Jr., B. Healy, M.W. Dorey, W.F. Doolittle, and T.M. Embley, "Microsporidia are related to fungi: Evidence from the largest subunit of RNA polymerase II and other proteins," *Proceedings of the National Academy of Sciences* 96 (1999): 580–585.

53. C.G. Clark and A.J. Roger, "Direct evidence for secondary loss of mitochondria in *Entamoeba histolytica*," *Proceedings of the National Academy of Sciences* 92 (1995): 6518–6521; Roger, Clark, and W.F. Doolittle, "A possible mitochondrial gene in the early-branching amitochondriate protist *Trichomonas vaginalis*," *Proceedings of the National Academy of Sciences* 93 (1996): 14618–14622; A. Germot, H. Philippe, and H. Le Guyader, "Evidence for loss of mitochondria in microsporidia from a mitochondrial-type HSP70 in *Nosema locustae*," *Molecular and Biochemical Parasitology* 87 (1997): 159–168.

54. D.G. Searcy, D.B. Stein, and G.R. Green, "Phylogenetic affinities between eukaryotic cells and a thermoplasmic mycoplasm," *BioSystems* 10 (1978): 19–28; Searcy, "Origins of mitochondria and chloroplasts from sulphur-based symbioses," in H. Hartman and K. Matsuno, eds., *The origin and evolution of the cell* (Singapore: World Scientific, 1992), 47–78; Searcy, "Syntrophic models for mitochondrial origin," in J. Seckbach, ed., *Symbiosis: Mechanisms and model systems* (Kluwer: Doordrecht, 2002), 163–183.

55. W. Martin and M. Müller, "The hydrogen hypothesis for the first eukaryote," *Nature* 392 (1998): 37–41.

56. For a historical overview, see Jan Sapp, "Mitochondria and their host: Morphology to molecular phylogeny," in W. Martin and M. Müller, eds., *Mitochondria and hydrogenosomes* (Heidelberg: Springer, 2006), 57–84.

57. While most accounts emphasize the importance of ATP production, some emphasize that the advantage of the proto-mitochondria lay first in detoxifying the host of oxygen. G.E. Andersson and C.G. Kurland, "Origins of mitochondria and hydrogenosomes," *Current Opinion in Microbiology* 2 (1999): 535–541.

58. Margulis, *Symbiosis in cell evolution*, 208.

59. J. Maynard Smith and E. Szathmáry, *The major transitions in evolution* (New York: Freeman, 1995), 23.

60. C.G. Kurland, "Evolution of mitochondrial genomes and the genetic code," *BioEssays* 14 (1992): 709–714; Kurland, "Something for everyone: Horizontal gene transfer in evolution," *EMBO Reports* 1 (2000): 92–95; J.F. Allen, "Control of gene expression by redox potential and the requirement for chloroplast and mitochondrial genomes," *Journal of Theoretical Biology* 165 (1993): 609–631; H.L. Race, R.G. Hermann, and W. Martin, "Why have organelles retained genomes?" *Trends in Genetics* 15 (1999): 364–370; W.F. Doolittle, "You are what you eat: A gene transfer ratchet could account for bacterial

genes in eukaryotic nuclear genomes," *Trends in Genetics* 14 (1998): 307–311; J.N. Timmis, M.A. Ayliffe, C.Y. Huang, and W. Martin, "Endosymbiotic gene transfer: Organelle genomes forge eukaryotic chromosomes," *Nature Review Genetics* 5 (2004): 123–136.

61. D.G. Lindmark and M. Müller, "Hydrogenosome, a cytoplasmic organelle of the anaerobic flagellate *Trichomonas foetus*, and its role in pyruvate metabolism," *Journal of Biological Chemistry* 248 (1973): 7724–7728; M. Müller "The hydogenosome," *Journal of General Microbiology* 139 (1993): 2879–2889; M.T. Embley and W. Martin, "A hydrogen-producing mitochondrion," *Nature* 396 (1998): 517–519; T.M. Embley, B.J. Finlay, P.L. Dyal, R.P. Hirt, M. Wilkinson, and A.G. Williams, "Multiple origins of anaerobic ciliates with hydrogenosomes within the radiation of aerobic ciliates," *Proceedings of the Royal Society of London Series B* 262 (1995): 87–93; A. Akhmanova, F. Voncken, T. van Alen, A. van Hoek, B. Boxma, G. Vogels, M. Veenhuis, and H.P. Hackstein, "A hydrogenosome with a genome," *Nature* 396 (1998): 527–528; M. Müller and W. Martin, "The genome of *Rickettsia prowazekii* and some thoughts on the origins of mitochondria and hydrogenosomes," *BioEssays* 21 (1999): 377–381.

62. W. Martin, M. Hoffmeister, C. Rotte, and K. Henze, "An overview of endosymbiotic models for the origins of eukaryotes, their ATP-producing organelles (mitochondria and hydrogenosomes) and their heterotropic lifestyle," *Biological Chemistry* 382 (2001): 1521–1539, at 1526.

63. M.W. Gray, "Contemporary issues in mitochondrial origins and evolution," in J. Sapp, ed., *Microbial evolution: Concepts and controversies* (New York: Oxford University Press, 2005), 224–237, at 226.

64. W.F. Doolittle, "A paradigm gets shifty," *Nature* 392 (1998): 15–16; Doolittle, "You are what you eat"; Andersson and Kurland, "Origins of mitochondria and hydrogenosomes"; Roger, "Reconstructing early events in eukaryotic evolution."

65. Stephen Jay Gould, *Wonderful life: The Burgess Shale and the nature of history* (London: Hutchison Radius, 1989), 310; W.F. Doolittle, Y. Boucher, C. L Nesbø, C.J. Douady, J.O. Andersson, and A.J. Roger, "How big is the iceberg of which organellar genes in nuclear genomes are but the tip?" *Philosophical Transactions: Biological Sciences* 358 (2003): 39–58.

66. John Maynard Smith and Eors Szathmáry, *The origins of life: From the birth of life to the origin of language* (New York: Oxford University Press, 1999), 107. See also Maynard Smith and Szathmáry, *The great transitions in life* (New York: W.H. Freeman, 1995), 195.

67. See J.H. Werren, "Heritable microorganisms and reproductive parasitism," in Jan Sapp, ed., *Microbial evolution and phylogeny* (New York: Oxford University Press, 2005), 290–316.

68. Julie C. Dunning Hotopp, M.E. Clark, D.C. Oliveira, J.M. Foster, P. Fischer, M.C. Torres, J.D. Giebel, et al., "Widespread lateral gene transfer from intracellular bacteria to multicellular eukaryotes," *Science* 317 (2007): 1753–1755; Eugene A. Gladyshev, Matthew Meselson, and Irina R. Arkhipova, "Massive horizontal gene transfer in bdelloid rotifers," *Science* 320 (2008): 1210–1213.

69. Keeling, "A kingdom's progress," 93. The Archezoa as a primitive subkingdom fell, but the conceptual relic persists.

70. Gray, "Contemporary issues in mitochondrial origins and evolution," 226.

71. See Russell Doolittle, "The origins and evolution of eukaryotic proteins," *Philosophical Transactions of the Royal Society of London Series B* 349 (1995): 235–240.

72. E.A. Minchin, "The evolution of the cell," *Report of the Eighty-Fifth Meeting of the British Association for the Advancement of Science, September 7–11* (1915), 437–464.

73. Ibid., 456.

74. Ibid., 460.

75. R. Stanier, "Some aspects of the biology of cells and their possible evolutionary Significance," *Symposia of the Society for General Microbiology* 20 (1970): 1–38, at 31.

76. Tom Cavalier Smith, "The simultaneous symbiotic origin of mitochondria, chloroplasts and microbodies," *Annals of the New York Academy of Sciences* 503 (1987): 55–71, at 56.

77. See Ricardo Guerrero, Carlos Pedros-Alio, Isabel Esteve, Jordi Mas, David Chase, and Lynn Margulis, "Predatory prokaryotes: Predation and primary consumption evolved in bacteria," *Proceedings of the National Academy of Sciences* 83 (1986): 2138–2142, at 2138.

78. T. Cavalier-Smith, "The simultaneous symbiotic origin of mitochondria, chloroplasts and microbodies," 56.

79. D.E. Wujek, "Intracellular bacteria in the blue-green-algae *Pluerocapsa minor,*" *Transactions of the American Microscopical Society* 98 (1979): 143–145. See also C.D. von Dohlen, S. Kohler, S.T. Alsop, and W.R. McManus, "Mealybug α-proteobacterial ensymbionts contain α-proteobacterial symbionts," *Nature* 412 (2001): 433–435.

80. T. Horiike, K. Hamada, S. Kanaya, and T. Shinozawa, "Origin of eukaryotic cell nuclei by symbiosis of archaea in bacteria is revealed by homology-hit analysis," *Nature Cell Biology* 2 (2001): 210–214.

81. H. Hartman, "The origin of the eukaryotic cell," *Speculations in Science Technology* 7 (1984): 77–81.

82. Hartman and Fedorov, "The origin of the eukaryotic cell," 1424.

83. Mitchell L. Sogin, "Early evolution and the origin of eukaryotes," *Current Opinion in Genetics and Development* 1 (1991): 457–463, at 457.

84. See Mitchell L. Sogin, John H. Gunderson, Hille J. Elwood, Rogelio A. Alonso, and Debra A. Peattie, "Phylogenetic meaning of the kingdom concept: An unusual ribosomal RNA from *Giardia lamblia,*" *Science* 243 (1989): 75–77.

85. Sogin, "Early evolution and the origin of eukaryotes," 458.

86. Ibid., 461.

87. Hartman and Fedorov, "The origin of the eukaryotic cell," 1420.

88. See, e.g., C.G. Kurland, L.J. Collins, and D. Penny, "Genomics and the irreducible nature of eukaryote cells," *Science* 312 (2006): 1011–1014; Joel B. Dacks and Mark C. Field, "Evolution of the eukaryotic membrane-trafficking

system: Origin, tempo and mode," *Journal of Cell Science* 120 (2007): 2977–2985. See also Patrick Forterre, "Three RNA cells for ribosomal lineages and three DNA viruses to replicate their genomes: A hypothesis for the origin of cellular domain," *Proceedings of the National Academy of Sciences* [103] (2006): 3669–3674.

89. For criticisms of this model see, Melintsky et al., "The karyomastigont model of eukaryosis"; Michael Dolan, "The missing piece: The microtubule cytoskeleton and the origin of eukaryotes," in Sapp, ed., *Microbial phylogeny and evolution*, 281–289.

INDEX

Adaptor hypothesis, 151, 152, 153, 154, 356, 357

Adelberg, Edward, 85, 86, 92, 93, 98, 100, 102, 137, 261

Agassiz, Louis, 24, 77, 80

Algae, 14, 26, 39, 74. *See also* Blue-green algae

Alligators and birds, 185

Allopatric speciation, 82

Allsopp, Allan, 86, 125

Alternative RNA splicing, 246, 258

Altmann, Richard, 138

Ambler, Richard, 235, 240

American Society of Microbiology, 233

Amino acid sequencing, x, 208, 235, 260, 268, 276, 315, 317, 318

Amoeba, 13, 20, 74, 77, 96, 105

Amoeboplasm, 117, 118, 307

Amorphozoa, 23

Anacystis nidulans, 233

Analogy, 20
  in taxonomy, 11, 22

Anastomosis, 76, 82. *See also* Evolution; Hybridization; Lateral gene transfer; Symbiosis

Anderson, N., 140, 284

Anderson, E.S., 141

Anemones, xii, 14, 115

Animalia, 27, 73, 106, 107, 108, 198, 261, 265

Animals. *See also* Comparative anatomy; Embryology
  infusoria and, 13, 15, 19, 22, 23, 24
  kingdom, viii, ix, x, xii, 7, 16, 23, 24, 27, 73, 74, 106, 107, 108, 109, 110, 261, 265

origins of, 59, 306 (*see also* Protists; Protoctista; Protozoa)
  reality of, 8, 9
  sex, 19

Annealing, 295

Annelids, 13, 14

Anthrax, 46

Anthropomorphism, vii

Antibiotic resistance, 138, 139, 140, 141, 157, 235, 284, 308, 309, 315

Antibiotics, development of, 138

Apes, 34, 155, 185, 149

Arachnids, 13

Archaea, viii, xi. *See also* Archaebacteria; Eocyte; Metabacteria; Neomura
  differences from Bacteria, 274
  formal naming of, xvi, 258–266
  kingdoms of, 263–264
  in nonextreme habitats, 293
  opposition to, 269–282
  and prokaryote concept, 258, 301, 302
  as proto-eukaryote, 312
  as sister to eukaryotes, 290, 307

Archaebacteria. *See also* Archaea; Eocyte; Metabacteria; Neomura
  groupings within, 245
  introns of, 245
  lipids of, 168, 189, 193–198, 203, 215–217, 237, 239, 265, 274, 285, 294
  the media and, 177–190
  molecular genetics, 210–216, 243, 244, 249
  proto-eukaryote, 236, 245–251
  reception in Germany, 199–216

Archaebacteria (*continued*)
  reception in the U.S., 181–186, 178, 190, 198
  relations to eukaryotes, 216–219
  ribosomal subunit morphology, 247–251
  RNA polymerase of, 207–216, 219, 258, 285
  5S rRNA analysis of, 190, 217–219
  16S rRNA analysis of, 166–168, 180–184, 187–194, 198–204, 226
  transfer RNAs of, 175, 181, 186, 200, 246, 290
  viruses of, 215, 244, 250
  walls, xv, 44, 173, 175, 179, 199, 206, 215–218, 243, 258, 264, 274, 285, 302, 304
  workshops, 243, 244
Archaeon era, 197
Archetista, 95
Archetype, 19
Archezoa, 257, 307, 309
Aristotle, 6, 13, 32
Arrhenius, Svante, 49
Atlas of Protein Sequence and Structure, 151
Atmosphere, primitive, 169, 222, 224, 225, 309–310
Atmospheric change, 309–310
Atomic evolution, 110
Autotroph, as origin of life, 62, 63, 220–225, 314
Avery, Oswald, 113
Aves, 134

Bacillus subtilis, 51, 74, 233
Bacteria. *See also* Archaea; Archaebacteria; Eubacteria; Monera; Prokaryotes; Schizomycetes
  as animals, 5, 8, 29
  autotrophic, 54, 62, 63
  as common name, 47
  conjugation, 81, 93, 98, 136, 137, 141, 298, 315
  Domain-defined, 265
  classification as Chaos, 8, 57, 68
  distinguished from viruses, 95, 96, 109
  early polyphyletic conceptions of, 59, 100, 103, 106
  F factor, 137, 138, 141
  as fission fungi, 46, 50, 58, 62, 88, 101
  Gram negative, 53, 54, 101, 202, 182, 308
  Gram positive, 53, 101, 102, 202, 203, 264, 299, 308, 309
  heterotropic, 88, 96, 192, 220, 221, 224, 228
  as laboratory domesticates, x, 258
  medical conceptions of, xiii, 45, 46, 48, 55, 58, 67, 73, 272
  nucleus and, xiv, 19, 36, 38, 41, 42, 43, 48, 70, 73, 74, 88, 91, 92, 96, 98, 101, 102, 104, 109, 315
  as plants, xiii, 44, 46, 47, 50, 49, 52, 71, 72
  pleomorphic concepts of, 51, 52, 68, 314
  R factor, 141
  sex and, xiv, 79, 81, 93, 94, 96, 98, 137, 138, 301
Bacterial genetics, xiv, 93, 79, 136. *See also* Molecular genetics
Bacterial kingdom, 71, 84, 101, 141. *See also* Monera
Bacterial spores, 47, 48, 53, 59, 92, 101, 146, 190, 240
Bacteriology
  as applied science, xii, xiii, 45, 46, 54, 56, 57, 58, 60, 138
  systematic, 57, 58, 72, 87, 199, 244, 300
Bacteriophage, 66, 95, 109, 119, 150. *See also* Viruses
Bacteriorhodopsin, 189
Bacteriota, 261
Bacterium, as genus, 29
Baer, Karl von, 33
Balch, William, 165–169, 237
Banks, Joseph, 6
Barber, Mary, 138–139, 141
Barghoorn, Elso, 223
Barker, Horace, 164, 169, 174
Bassler, Bonnie, 305
Bather, Francis, 69, 70
Bathybius haeckellii, 41
Bauplan, 19, 21
Bayley, Stanley, 190
Beijerinck, Martins, 54–56, 162
Bellemare, Guy, 190
Benda, Carl, 118
Berg, Paul, 244
Bergey, David, 47
Bergey's Manual of Determinative Bacteriology, 279
Bergey's Manual of Systematic Bacteriology, 300
Bernal, J.D., 221
Big bang theory, 147, 286

Billroth, Theodore, 51
Bioblasts, 118
Biofuels, 195, 201
Bioinformatics, 130, 187
Biological species concept, 81, 336
Biotypes, 81, 91
Birds, vii, 13, 14, 23
　　bacteria and, 277–279
　　bats and, 60, 181, 227
　　dinosaurs and, 33–34, 112, 134, 185,
　　　230, 273
　　species, 278
Bishop, David, 158
Bisset, Kenneth, 113
Blue-green algae. *See also* Chloroplasts;
　　Chromacae; Cyanobacteria;
　　Cyanophyceae
　　chloroplasts and, 41, 92, 116, 117, 121,
　　　122, 125, 171, 231, 232, 233
　　directed panspermia and, 240
　　excluded from bacteria, 73, 84, 88, 92,
　　　94, 101, 102, 103, 107
　　as first organisms, 39, 63, 65, 76, 240
　　fossil record of, 103, 169, 223
　　as intermediary between prokaryote and
　　　eukaryote, 73, 125, 233
　　in kingdom Prokaryotae, 106
　　in kingdom Protista, 36, 39, 41, 94
　　in kingdom Protophyta, 95
　　as monera, 36, 39, 41, 74, 77, 88, 102,
　　　109, 121, 315
　　nucleus and, 42, 43, 86, 94
　　as prokaryotic cells, 96, 98, 102, 104,
　　　107, 202, 315
　　as schizophytae, 50, 72
Body plans, 112, 118
Bolton, E.T., 152
Bonen, Linda, 232–234, 237, 288
Bonnet, Charles, 11
Book of Genesis, 35, 222
Bory de Saint Vincent, Jean-Baptiste,
　　15–16, 21, 23
Botany, 6, 21, 41, 47, 71–74, 80, 201, 202, 206
Boveri, Theodor, 116, 307
Breed, Robert, 65, 67, 69, 86, 87, 89
Brenner, Sydney, 147, 149, 150
Brock, Thomas, 192, 193, 194, 197, 203,
　　212, 220
Brown, James, 289, 290
Brown, Robert, 20
Bryant, Marvin, 164, 195, 200

Buchanan, Robert, 107
Buchner, Hans, 51
Buffon, Georges-Louis Leclerc, 8, 9, 11,
　　15, 21, 80
Bulloch, William, 47
Bu'Lock, John, 197
Burgess Pass, 112
Burrows, Jeanne, 137
Butenandt, Adolf, 209, 371
Bütschli, Otto, 42

Cain, 311
Caldariella, 211, 212
Calman, William, 83
Cambrian explosion, 306
Cambrian fossils, 112
Cassin, John, 17, 26
Cavalier-Smith, Tom, 125, 257, 276, 304,
　　307, 312
Cell theory, xii, 16, 17, 19, 20, 23, 24, 104
Cenancestor, 286, 289
Center for Prokaryote Genome
　　Analysis, 259
Centrioles, 113, 116, 122, 123, 124, 309
Centrosomes, 74, 116
Cesalpino, Andreas, 8
Cetus Corporation, 244
Challenger, 41, 49
Chaos infusorium, 8
Charlebois, Robert, 296
Chatton, Edouard, 96, 97, 98, 102, 301
Chemical evolution, 110, 178
Chemical paleogenetics, 129
Chemosynthetic bacteria, 54, 56, 88
Chimpanzees, 185
Chlamydia, 107
Chlamydomonas, 93, 115
Chloroplasts
　　coined, 34
　　genetics, 125
　　symbiotic origin, 41, 116, 124, 126,
　　　171, 231, 232, 233, 234, 238, 243,
　　　249, 257, 285, 288, 309, 310,
　　　315, 316
Chromacae, 41. *See also* Blue-green algae;
　　Cyanobacteria
Chromatin
　　bacterial, 42, 43, 73, 91, 117, 172,
　　　311, 312
　　coined, 42
Chromista, 257

Chromosomes
  bacterial, 92, 141, 269, 304
  coined, 43
  eukaryotic, 113, 116, 123, 141, 306
  as symbionts, 116, 307
Chromosome theory of inheritance, 66,
  70, 78
Chronocyte, 313
Chronometer universal, 316
Cilia, 19, 116, 118, 122
Ciliates, 19, 39, 264
Clade, definition, 133, 253, 278, 304
Cladism, 273, 275
Cladistics, 133, 134, 231, 249, 254, 272,
  273, 275, 278
Classification. *See also* Cladistics;
  Gene; Genomics; Monophyletic
  classification; Natural kinds;
  Phenetics; Phylogeny, coined;
  Ribosomal RNA; Taxonomy
  hierarchical, xiv, 7, 11, 30, 58, 59,
    82, 86, 91, 119, 142, 292, 296,
    314, 317
  natural, xiv, 10, 11, 21, 23, 26, 28, 50,
    59–62, 68–77, 80–83, 86–91,
    101–104, 182, 188, 196, 236, 244,
    264–275, 283, 301, 314
Clostridium, 153, 289
  as first organisms, 221, 228, 231, 238
Cocci, 47, 89
Code redundancy of, 288
Coding problem. *See* Genetic code
Codon
  assignments, 149, 150
  coined, 167
  usage and LGT, 288
Coenzyme M, 164, 165, 167, 193, 200
Cohen, Seymour, 123
Cohn, Ferdinand
  on bacteria as plants, 45, 49
  bacterial classification of, 51
  on bacteria's place in nature, 47, 48
  on blue-green algae, 70
  discovery of spores, 47, 48
  on Panspermia, 49
  on pure cultures, 47
Cold war, 120
Colloids, 65
Committee on Characterization and
  Classification of Bacterial Types, 60,
  62, 63, 67, 68

Common descent, viii, 29, 61, 76, 77, 104,
  105, 133
  and lateral gene transfer, viii, 136, 142,
    291, 298
Comparative anatomy, 14, 21
Comparative biochemistry, 85, 100,
  244, 399
Comparative embryology, 33, 326
Comparative morphology, 91, 128
Comparative physiology, 119
Complexity
  cellular levels of, 173, 176, 224, 236, 237
  chemical, 110, 224
  hypothesis, 241, 283
  metabolic, 89
  molecular, 152, 153, 156, 175, 214,
    235, 238
  morphological, 59, 60, 73, 89, 101, 110,
    113, 135, 274
  symbiotic, 115
Condillac, Etienne, 6
Conjugata, subkingdom, 26
Conjugation. *See* Bacteria
Conrad, Henry, 84
Convergence. *See* Evolution
Convoluta roscoffensis, 115
Cook, James, 6
Copeland, Edwin, 72
Copeland, Herbert, on kingdoms, 72–79,
  84, 88, 101, 108, 109, 328
Coral, 7, 15. *See also* Polyps
Craford prize, 280
Creator, 8, 35
Crenarchaea, 293
Crenarchaeota, 263, 264
Crick, Francis, 249
  adaptor hypothesis of, 151–154
  on directed Panspermia, 239–240
  on double helix, 146
  frozen accident theory of, 151
  on protein taxonomy, 129
  on triplet code, 147
Crocodiles. *See* Birds
Cronquist, Arthur, 125
Crown galls, 121
Crustaceans, 13
Cunningham, Scott, 254
Cuvier, Georges, 14, 15, 21
Cyanobacteria. *See also* Blue-green algae
  as monera, 36, 41 160, 202, 223, 233,
    257, 264, 289

chloroplast origin, 41, 234, 238, 257, 264, 289

fossil record, 223, 316

missing link between prokaryotes and eukaryotes, 233

most advanced prokaryote, 233

Cyanophyceae, 39, 96. *See also* Blue-green algae; Cyanobacteria

Cyclogeny, 68

Cytochrome *c*, 130, 228, 235, 236, 241, 285

Cytoplasm
as cell body, 42
coined, 20
and division of labor, 38
as symbiont, xii, 116, 117, 118, 220, 307–313

Cytoplasmic heredity, 140

Cytoskeleton and phagocytosis, 322

D'Herelle, Félix
Bacteriophages, 66, 118
Lamarckian concepts of, 136
on origin of life, 66, 299
on symbiosis in evolution, 118

Dalhousie University, 232, 234

Daphnia, 96

Darlington, C.D., 79, 120

Darwin, Charles, vii, 75, 271, 273, 275, 309
agnosticism of, 35
on common descent, 29, 30, 34
Creator, 35
on divergence, 30
on embryological characters, 33
on essential characters, xiii, 11, 32, 43, 182
on evolutionary progress, 11, 32
on genealogical classification, xiii, 30–32, 279, 282
on genealogical tree, 30–32, 317, 327
on Haeckel, 33
on homologous and adaptive or analogous characters, 31, 188
on the inheritance of acquired characteristics, 10
on Lamarck and, 11
myths of, 11
on natural selection, 29, 78
nominalist concept of species, 80
on the origin of life, 34, 35
on the origin of man, 34
on pangenes, 10
recapitulation theory and, 32

species definition
on the struggle for existence, 29
on the unity of life, xiii, 34

Darwinian evolution, vii, ix, x, xi, 10, 25, 34, 35, 78, 79, 81, 110, 134–139, 188, 269, 302, 317, 318

Darwinian transition, 294

Darwinism, transcended, 317

Davaine, Casimer, 47

Dayhoff, Margaret, 127, 130, 131, 222, 228, 229, 230, 231, 235, 255, 260

Dead Sea, 188

De Bary, Anton, 54, 115

De Ley, Jozeph, 132, 135

Delbrück, Max, 95, 96, 178

Delft School, 265

DeLong, Edward, 293

Demerec, Milislav, 136

Demoulin, Vincent, 231

Der Hoeven, Jan van, 22, 24

De Rosa, Mario, 211, 212

Descent of Man, 34

Descent with modification, 29, 30, 238, 283, 291, 317

Desmids, 19, 24

Deutche Gesellshaft für Hygiene und Mikrobiologie, 213

Devonian, 135

Diatoms, 17, 19, 24, 72, 74

Dickerson, Richard, 235, 236, 239, 379

Diderms, 308

Dinosaurs, 34, 134, 278. *See also* Birds

Diseases, 7, 39, 45, 46, 47, 70, 96, 138, 190, 214

Divergence, 30, 91, 92, 98, 110, 128, 135, 169, 180, 185, 188, 228, 267, 273, 285, 306, 317

Diversa Corporation, 293

Division of labor. *See also* Symbiosis
ecological, 30
divergence and, 30
nucleus and cytoplasm, 20, 38, 39
physiological, 20
social sciences and, 325

DNA, x, 15, 93. *See also* Genetic code; Lateral gene transfer; Molecular clock; Progenotes concept; Transcription; Translation
bacterial, 93, 98, 113, 120, 121, 136, 137, 142, 259
centrioles and, 123, 309

DNA (*continued*)
chloroplast, 115, 121, 315
cloning, 214, 244
code, 105, 146, 147, 286
composition, 131, 132
of eucaryotic nucleus, 122, 234, 246, 333
finger printing, 205
homology, 127, 132, 205
hybridization, 131, 132
information, 105, 128, 129
mitochondrial, 121, 122, 234, 249, 315
origin, 312
recombinant, 128, 138
replication, 170, 274, 286, 288, 318
as secret of life, 268
selfish, 287, 289
sequencing, 244 (*see also* Genomics)
structure, 166
and transformations, 93
viral, 95, 107
DNA-dependent RNA polymerase, 207, 209–214
DNA-RNA hybridization, 143, 144, 150, 158, 206, 245
Dobell, Clifford, 5, 73
Dobzhansky, Theodosius, 268
Bacteria, 81
selection, 78
species concept of, 81
Doi, Roy, 143
Dolan, Michael, 304
Domain. *See also* Archaea; Bacteria; Empire; Eucarya; Superkingdom; Urkingdom
controversy over, 266–281, 284, 292, 297, 300, 303, 313, 315, 316
eubacterial, 266
institutional, 71, 207
prokaryotic, 162, 175, 176, 183, 184, 262
taxonomic proposal, 257–266
third, 214
Doolittle, Russel, 306
Doolittle, W.F.
arrival at Dalhousie, 232
on bacterial diversity, 280
concept of universal tree, 289–290, 291, 294, 295, 296, 298
early support for three-domain proposal, 275, 302
on genetic core, 290, 296, 297, 298
on lateral gene transfer, 282, 288–296

on "majority rule" classification, 383, 290, 292, 296
on mitochondrial and chloroplast origins, 233, 234, 288
nominalist views of, 290, 292, 294, 302
Pace and, 232
philosophical views of, 294
on the progenotes, 295
Spiegelman and, 232
Woese and, 232, 294
Double helix, 147, 151. *See also* DNA
Doudoroff, Michael, 85, 86, 92, 93, 98, 100, 102, 137, 261
Dougherty, Ellsworth, 102, 103
Drosophila, 70, 253, 258
Dubnau, David, 143
Dubos, René
on blue-green algae nucleus, 92
comparative physiology, 101
on polyphyletic nature of monera, 101
on symbiosis, 121, 136
Dujardin, Félix, 19
Dysentery, 46, 66, 70, 118, 137, 139

Eck, Richard, 127, 131, 222
E. coli, 148, 280, 284
antibiotic resistance, 139, 284
elephants and, 104, 105
genomics and, 259
LGT and, 288
as model organism, 105, 150, 158, 192, 210, 258–260, 315
prokaryote concept and, 277, 301
salmonella and, 288
viruses of, 209, 215
Ecology, vii, 100, 107, 108
coined, 28
microbial, xiv, 54, 56, 126, 160, 164, 166, 169, 285, 286, 317
Edicaran biota, 112
Ehrenberg, Christian Gottfried, 17–19, 21, 23, 24, 47
Elephants, 18, 104, 105, 157, 301
Elongation factors, 256, 286, 289
Embranchements, 14
Embryology and species transmutation, 53
and genealogical classification, 32, 33, 43, 91, 132
Empire, 7, 15, 123, 262, 263, 264, 266, 270, 276, 277
Enderlein, Günther, 77

Endosymbiosis, 115, 124, 171, 173. *See also*
    Symbiosis
Engler, Adolf, 72
Enterobacteriaceae, 141, 144
Eocyte concept, 245–247, 250, 257
Eörs, Szathmáry, 310
Episome, 139
Essential characters, xii, xv. *See also*
    Genetic core
    Biochemical processes, 165
    Buffon on, 8, 9
    Darwin on, 32, 143, 292
    Lamarck on, xiii, 11, 12, 32, 143, 292
    Minchin on, 69
    Molecular processes, 171, 186, 278, 283,
        285, 317
    Prokaryote concept and, 104
    Proteins as, 125
Essence of the organism, 283, 290
Essentialism, 290, 302, 336
Etherial kingdom, 15
Eubacteria order, 62, 90, 101, 104, 175
Eubacterial urkingdoms, 175, 176, 183,
        189, 190, 204, 257, 214, 216, 237, 257
    introns of, 258
    relations with archaebacteria, 216, 217,
        219, 238, 245, 246, 249, 257, 263,
        275, 276, 278
    relations with eukaryote, 219, 220,
        263, 276, 307
    16S rRNA analysis of, 205–206,
        226, 238
Eucarya, xi, 262, 264, 295, 300, 306,
        316. *See also* Eukaryote
    origin, 306, 308
    sister to Archaea, 289
Eucaryota, 106, 257, 261, 264, 315
Eukaryote (or eucaryote). *See also*
        Archezoa; Eocyte concept; Eucarya;
        Eukaryota; Prokaryote;Protists;
        Protozoa; Urkaryote
    genomic analysis of, 312
    number of species, 278
    origin of, 176, 258, 261, 289, 307–313
    origin of term, 96
    prokaryote and, 286–87
Eukaryotic, defined, 96, 98, 102
Euryarchaeota, 263, 264
Evolution. *See also* Darwinian evolution;
        Lamarckian evolution; Lateral gene
        transfer; Natural selection; Symbiosis

chemical, 110, 178, 376
convergent, 69, 101, 102, 130, 134, 193,
        194, 197, 198, 203, 227, 236, 303
cosmic, 110, 111, 131, 222
cultural, 111
gradual, xi, 10, 14, 78, 115, 119, 125,
        126, 241, 317
Lamarckian, 78, 79, 119, 136
molecular, 129, 130, 271, 274, 311
neutral theory of, 130, 278, 182, 268
non-Darwinian, 130, 131, 137, 176, 268,
        271, 317
physiological, 89
progression, 32, 33, 112
reticulated, xv, 82, 83, 96, 119, 127, 141,
        142, 144, 184, 238, 286
saltational, 11, 78, 115, 124, 126, 317
tempo and mode, 241, 294, 296
Evolutionary direction, 29, 89, 107 108,
        118, 124, 222
Evolutionary parsimony method, 251,
        252, 253
Evolutionary progression, 32, 33, 112
Evolutionary synthesis, ix, 78–82, 119,
        269, 272
Evolutionary trends, 11, 12, 78, 89, 94,
        112, 203
Exobiology, 177
Exons, 246
Extremophiles, 194, 215. *See also* Archaea,
        Archaebacteria

Famintsyn, Andrei, 54
Fedorov, Alexei, 313
Feldman, Robert, 293
Felsenstein, Joseph, 242, 248
Ferredoxin, 228, 230, 255
First cause, 25, 35
Fisher, R.A., 78
Fitch, Walter, 130, 186, 228, 239, 286
Fitt, Peter, 210
Five kingdom model, 107, 198, 257, 265, 276
Flagellata, 39, 74
Flemming, Walther, 42, 43
Foraminifera, 8, 23
Forterre, Patrick, 276, 286
Fossil record
    of blue-green algae, 169, 223, 316
    Cambrian, 112
    classification and, 50, 83, 182, 184, 227, 230
    of infusoria, 17, 21

Fossil record (*continued*)
  living, 135
  of microbes, 17, 21, 59, 83, 91, 112,
      113, 124, 156, 169, 179, 223, 227,
      282, 284, 316
  molecules and, 156
  Postcambrian, 182
  progressive nature of, 112
  rates of evolutionary change and, 241, 284
Fox, George, 300
  on Archaea, 266
  on "big tree," 226, 236–241
  on cell walls, 168, 169, 201, 203, 204
  collaboration with Balch, 166–167
  collaboration with Woese, 161–205,
      220, 217
  development of dendograms, 160,
      204, 258
  on the discovery of the archaebacteria,
      161, 165, 166–169, 262
  on eukaryote origin, 176, 289, 307
  on halophile and methanogen lipids,
      168, 186, 195, 196
  on Hori and Osawa, 219
  on lateral gene transfer, 235, 236
  on progenotes, 168, 170–171, 237
  prokaryote-eukaryote dichotomy, 162,
      170, 175, 300
  secondary structure of 5S rRNA, 218
  third kingdom of life, 175
  at the University of Houston, 185
  Watson and, 160, 186
Frameshift experiments, 147
Frozen accident theory, 150–151
Fuerst, John, 305
Functionism, 294
Fungi. *See also* Schizomycetes
  algae and, 108
  as animals, 24
  Bacteria as, xiii, 8, 46, 50, 51, 58, 68,
      84, 88, 101
  as kingdom, xiv, 107, 108, 109, 111, 198,
      265, 269, 276
  as laboratory domesticates, 70
  lichens and, 54
  as plants, 24, 72
  as protist, 74, 98, 108
  molecular phylogenetics of, 289
  mycorrhizal, 54
  nuclei of, 109
  as reducers, 108

sex in, 17, 18
  as subkingdom, 26

Gamow, George, 167
Garden of Eden, 222
Garrett, Roger, 249, 250
GC percent, 132, 135, 147, 212, 238, 288
Gene. *See also* DNA; Genetic code;
      Genetics; Genomics; Lateral gene
      transfer
  bacteriophages and, 66, 299
  classification and, 69, 127
  clusters, xi, 285, 296
  doubling, 230, 254, 255
  informational, 283, 296
  as master, 129
  medicine, 128
  modern synthesis and, xi, 78
  operational, 283
  sentence, 128
  split, 245
Genetic code. *See also* DNA; Transcription;
      Translation
  cracking of, 146–148
  diamond, 147
  dictionary, 151
  evolution of, xi, 287, 145–162, 170–175,
      222, 258
  lateral gene transfer and, 140, 284
  redundancy of, 288
  table, 148
  universality of, 125, 131, 140, 143, 156,
      217, 286
Genetic Code, The, 153–155
Genetic core, xv, 143–145, 158, 182, 278,
      283, 290, 294–318. *See also* Essential
      characters
Genetics. *See also* Molecular genetics
  biochemical, 93
  classical, xi, 66, 70, 74, 78, 120
  cytoplasmic, 119, 120
  population, xi, 78, 91, 110, 137, 272, 302
Genomics, xiv, 259, 282, 287, 293, 296, 306
  environmental, 316
  lateral gene transfer and, 287–299, 336
Genus, reality of, 7
Germ theory, 314
Giardia lamblia, 308, 309
Gibbons, N.E., 183
Gibor, Aharon, 125
Gibson, Jane, 236, 237, 359

Gilbert, Walter, 244
Gilmour, John, 84
God, 7, 35, 37, 80, 234
Goethe, Wolfgang, 19, 37
Gogarten, Peter, 255, 286, 289, 296
Goksøyr, Jostein, 122
Goldenfeld, Nigel, 304, 305
Goldfuss, George, 21
Golgi bodies, 118
Gorilla, 21
Gould, Stephen, 187
Gouy, Manolo, 253
Grades, 26, 69, 77, 276, 302
Gram, Christian, 53
Grant, Verne, 171
Gray, Michael, 232, 234
Great chain of being, 7, 11
Gunsalus, I.C., 163, 261
Gupta, Radhey, 308

Haeckel, Ernst, 28, 29
    chloroplast origin, 41, 116
    Christian God, 29, 37
    criticism of Darwin's theory, 35
    criticisms of, 33, 50
    Cyanophyceae, 39, 41
    Darwin and, 29, 34
    Goethe and, 29, 37
    Huxley and, 34
    Lamarck and, 29
    monera, 36–40, 77, 120
    monist philosophy, 29, 36, 37
    Naturphilosophie and, 29
    origins of life, 35, 36, 37, 41, 76, 104
    phylogeny, 28, 33
    protista, 36, 39, 40, 48, 74
    protophyta, 39
    protozoa, 39
    recapitulation theory, 33, 36
    spontaneous generation, 36, 48
Haldane, J.B.S., 78, 88, 149, 228, 299
Halophiles, extreme, xv, 162. *See also*
        Archaea; Archaebacteria: Eocyte;
        Metabacteria
    ancient organisms, 190, 196, 197, 200
    archaebacteria, 191, 204, 216, 219,
        257, 258
    eubacteria and, 247–248, 249, 250
    eukaryote and, 245
    introns of, 246, 260
    lipids, 189, 194, 195, 216

    methanogens and, 251
    photosynthesis, 189, 248
    5S rRNA analysis of, 190, 218, 219
    16S rRNA analysis of, 163, 189, 191, 202
    transcription enzymes, 211, 212, 216
    tRNAs of, 193
    walls, 178, 189, 202–204, 216
Halvorson, H. Orin, 163
Hartman, Hyman, 312–313
Hayes, William, 93, 137
Hedges, Robert, 141–142
Hegel, G.W.F., 29
Hemoglobin, 129, 228
Hemprich, Wilhelm, 18
Hennig, Willi, 133, 134, 292. *See also*
        Cladistics
Henrici, Arthur, 104
Heredity. *See also* DNA; Gene; Genetic
        code; Genetics; Lateral gene transfer;
        Plasmids; Symbiosis
    cytoplasmic, 120, 136, 140
    infective, 140, 139
    Mendelian, 120
Hershey, Alfred, 95, 178
Hesse, Fanny Angelina, 51
Hesse, Walther, 51
Heterotrophs, xiv, 65, 88, 89, 135, 146,
        149, 221, 224, 238, 315
Hilario, Elena, 289
Hippos, 104
Hiroshima University, 218
Hoagland, Mahon, 152
Hogg, John, 23–26
Holism, 294
Hominids, 34, 185
Homology, 21, 31, 32, 68, 86
    molecular, 132, 170, 215, 218, 219, 227,
        233, 234
    gene duplication and, 228
Hooke, Robert, 4, 5, 202
Hooker, Joseph, 10, 32, 35
Hopkins, Frederick, 105
Hoppe-Seyler, Felix, 192
Hori, Horishi, 217–219, 276
Horizontal gene transfer. *See* Lateral gene
        transfer
Horowitz, Norman, 221–222
Hsp60, 308
Hsp70, 289, 308
Human genome project, 259
Humans and apes, 34, 185, 249

Humboldt, Alexander von, 17, 18
Huxley, Julian, 78, 79, 83
Huxley, Thomas Henry
  Darwin and, 31
  Haeckel and, 41
  on phylogeny and classification, 34, 40
  on protoplasm, 37
  on the unity of life, 105
Hybridization. *See also* Lateral gene
    transfer; Symbiosis
  animal, 82, 298, 338
  hybridization, 298
  molecular, 131, 132, 137, 143, 144, 145,
    150, 158, 206, 245, 285, 193
  plant, 7, 82, 83, 119, 142, 298, 338
Hydra, 4, 41, 68, 115
Hydrogen hypothesis, 330

Idealism, 11
Idealists, 86–87
Igarashi, Richard, 143
Infusoria, 8, 13–21, 23, 25, 37, 39, 43,
    54, 104
Ingraham, John, 306
Inheritance of acquired characteristics, 10,
    12, 78, 119, 120
Insects
  classification, 10, 13, 104
  evolution, 181, 227
  fossils, 112
  hereditary symbionts in, 298, 311
Institute for Genomic Research, The, 287
International Code of Nomenclature, 261
International Society for Cell Biology, 122
Introns, 246, 289
  archaebacteria, 246, 250, 258, 285
  eubacteria, 258
  spliceosomal, 246, 302, 304
Invertebrates, 10, 13, 15, 21, 112, 276

Jacob, François, 95
  episomes coined, 139
  on gene action, 129
  messenger RNA, 149, 150
  operon theory, 149
  on origin of genetic code, 155, 156, 185
  on the universality of the code, 100,
    105, 156
Jahn, Frances, 95, 109
Jahn, Theodore, 95, 109
Jannasch, Holger, 180–200

Jin, Lin, 253
Jones, Dorothy, 127, 141, 284
Journal of Bacteriology, 57, 58, 60
Journal of Molecular Evolution, 130, 169,
    191, 218
Jukes, Thomas, 130, 217, 268
Juni, Elliot, 183
Just-so stories, 126, 357

Kaiser-Wilhelm Institute for
    Biochemistry, 209
Kamen, Martin, 235, 236, 240
Kandler, Otto, 189
  on ancient relics, 200, 216
  on archaebacteria relations with
    eukaryotes, 244–245
  on "big tree," 226
  Bryant's visit, 200
  complementarity with Woese's program,
    204–206, 213, 214
  defense of three domains, 267, 268,
    273–276, 287
  on definition of the eukaryotic cell, 307
  dispute with Mayr, 273–276
  education, 201–202
  fosters archaebacterial research in
    Germany, 210–213, 243
  on genealogy and classification, 267, 273
  influence in microbiology in Germany,
    199, 206–208
  meets Woese and Fox, 188, 201
  on the metabacteria, 217–219
  on naming the three domains,
    257–266, 301
  on progenotes concept, 287
  recommends Stackebrandt, 205
  on walls of archaebacteria,199, 207
  on walls of gram-positive bacteria,
    201–203
  on walls of halophiles, 200, 204
  on walls of methanogens, 175, 183, 200,
    202, 204, 210
  on Woese, 243
  Wolfe and, 199, 200
Kant, Immanuel, 19
Kappa particles, 40
Karyote, superkingdom, 257
Kates, Morris, 194, 195
Kilmer, Joyce, 236
Kimura, Motoo, 130, 268
King, Jack, 268

King, Mary-Claire, 249
Kingdoms. *See* Domain; Empire;
    Superkingdom; Urkingdom;
    individual kingdoms
Kipling, Rudyard, 126
Kiriakoff, Sergius, 133
Klein, Richard, 125
Kligler, I.J., 63
Kluyver, Albert Jan, 75, 85
    on microbial biomass, 280
    on microbial classification of, 86, 89, 90,
        92, 135, 265
    on microbial diversity, 162, 279
    on unity of biochemistry, 105
Knoll, Andrew, 223
Koch, Robert, 45, 47, 51, 52
Kölliker, Rudolf, 20
Koonin, Eugene, 303
Kurland, Charles, 297

Lactobacillus, 46, 208, 209
Lake, James, 246, 257
    electron microscopy of, 247
    eocyte urkingdom of, 246
    eukaryote origin of, 308
    evolutionary parsimony method of,
        249–252
    opposition to archaebacteria concept of,
        246–256, 277
    Photocyta urkingdom of, 246, 251
    superkingdoms parkaryotes and
        karyotes, 246, 251, 257
Lamarck, Jean Baptiste, xiii, 14
    against reticulated classification, 11
    Darwin and, 10, 25, 29, 32, 90, 112, 143
    genealogy and classification, xiii, 11, 12,
        14, 15, 299, 317
    great chain of being and, 10
    historical myths, 9, 10, 32
    mechanisms of evolution, 12
    on spontaneous generation, 14, 104
    vertebrate and invertebrate
        classification, 13
Lamarckian evolution, xiv, 61, 78, 79, 139,
    141, 305, 314, 315
Lambda phage, 141
Lander, Eric, 259
Langworthy, Thomas, 194, 196, 197
Lateral gene transfer. *See also*
    Hybridization; Symbiosis
    antibiotics crises and, 138, 284

bacterial genome structure and, 142,
    238, 285
common descent and, viii, 136, 142,
    291, 298
conjugation and, 137, 141, 298, 315
detection of, 283, 284, 288–289
ideology and, 297
Lamarckian principles and, 78, 141,
    139, 305
numerical taxonomy and, 143, 144, 283,
    290, 292, 296, 316
origin of the genetic code and, 140, 284
as paradigm change, 296
plants and animals, 284
progenotes concept, 173–174, 287, 294
rooting tree of life and, 289, 290, 302
transduction, 95, 120, 121, 136, 137,
    138, 141, 284, 288
transformation, 93, 121, 136, 315
Lavosier, Anton, 6
Lawrence, Jeffrey, 288, 296
Leder, Philip, 149
Lederberg, Esther, 136
Lederberg, Joshua
    coins exobiology, 177
    on conjugation, 93, 137
    demonstrates bacterial recombination, 93
    on infective heredity, 120
    on lysogeny, 120
    on the origin of mitochondria, 121
    on plasmids, 121
    on replicating plating, 136
    studies of phage, 95
    on "third form of life," 180
    on transduction, 95, 120, 136
Lederer, Hermann, 251
Leeuwenhoek, Antony van, 3, 4, 5, 8, 17,
    18, 55
Leeuwenhoek, Medal, 56
Leursen, Kenneth, 166
Lewontin, Richard, 197
Li, Wen-Hsiung, 253
Lichens, 9, 26, 54, 115
Lindley, John, 21
Linnaeus Carolus, 6, 7, 8, 15, 22, 44, 58,
    74, 172, 292
Lister, Joseph, 45, 46, 137
Lithophyta, 15
Lithozoa, 15, 21
Loeffer, Friedrich, 52
Logic of Life, The, 155–156

Louis, Gregory, 210
Lovejoy, Arthur, 21
LUCA, 276, 286, 287
Ludwig, Wolfgang, 308
Luehrsen, Kenneth, 190, 191
Luria, Salvador, 95, 136, 163
    bacterial hybridization, 137
    bacterial species, 137
    reception to third form of life, 178
Lutman, Benjamin Franklin, 73
Lwoff, André
    on lysogeny, 95
    Stanier and, 95, 96
    on virus concept, 95
Lysenko, T.D., 120
Lysogeny, 95, 120

MacLeod, Colin, 93
Magrum, Linda, 176, 191
Mammals, 13, 14, 33, 112,
    273, 280
Manganese effect, 208
Maniloff, Jack, 163, 168
Manwaring, W.H., 79
Margoliash, Emmanel, 130
Margulis, Lynn, 257, 265
    antimolecular sentiment of, 233
    on archaebacteria, 198, 276
    on five kingdoms, 255, 276, 277
    on origins of eukaryote, 122, 123, 309
    on origins of mitochondria, 122,
        124, 309
    on origins of plastids, 122
    on prokaryote-eukaryote dichotomy,
        277, 304
    on three-domain concept, 276
Marine Biological Laboratory, 160, 166
Mars, 178, 195, 222
Martin, William
    on eukaryote origins, 309–310
    on prokaryote, 303–305
Matheson, Alistar, 190, 219, 248
Matthaei, Heinrich, 148
Maugh, Thomas, 186
Max Planck Institute for Biochemistry,
    209, 210, 244
Mayr, Ernst
    against three domains, 269–281
    animal hybridization, 82
    archaebacteria, 270
    changing views of kingdoms, 269, 270

cladistics, 134, 273
criticisms of cladism, 272, 273, 275
debate with Woese, 269–281
defense of prokaryote-eukaryote
    dichotomy, 267–281, 302
empires of, 277
evolutionary synthesis and, 269
on evolutionary taxonomy, 134, 195, 267
historical accounts, 269
on kingdoms, 269, 270
on molecular phylogenetics, 268, 270, 272
on speciation, 82
species concept of, 80, 81
McBride, Barry, 164
McCarthy, Brian, 132, 144
McCarty, Maclyn, 93
McMaster University, 190
Medusinae, 21
Meiosis, 81, 113, 278
Mendel, 301
Mendelian genetics, ix, 78
Merezhkowsky, Constantin, 116, 117, 307
Meselson, Mathew, 149, 150
Messenger RNA, 147, 149, 150, 209, 246,
    304, 312
Metabacteria, 217, 219, 276
Metaphyta, 109, 265
Metazoa, 109, 265
Methanogenesis, 272
Methanogens. See also Archaea;
    Archaebacteria; Euryarchaeota
    adaptations and, 186, 188, 197, 198
    ancient organisms, 169, 180, 186, 190,
        197, 198, 225, 239, 245, 263
    biochemistry of, 163, 164, 165, 183, 200
    biofuels, 169, 200
    convergent evolution and, 197, 198
    halophiles and, 245, 247, 249, 251,
        253, 254
    histone gene in, 260
    lipids of, 194–197, 200, 239
    mesophilic, 163, 196–200, 293, 313
    ribosome structure of, 247, 248, 249, 250
    5S rRNA of, 190
    16S rRNA classification of, xv, 162,
        165–167, 174, 175, 182–184, 187, 190,
        199, 204
    sewage and, 200
    thermophilic, 163
    as third form of life, 168, 175, 178, 186
    transcription enzymes of, 210, 211, 245

tRNA of, 175, 189, 193, 238
walls of, 168, 173, 175, 200, 202, 204
Meyer, Terrance, 235, 236, 250
Microbial World, The, 85, 86, 91–95,
    98–103, 137, 182, 184, 261
Microbiology, 73, 86, 155, 146, 163. *See also*
    Bacteriology; *Microbial World, The*
    as applied science, xii, 60, 85, 102, 119,
        138, 199, 233, 234, 279
    classical, 55,163, 181–184, 277, 294, 301
Micrococcaceae, 205
Micrococcus, 50, 51, 205
Microscope
    compound, xiii, 4, 18, 19, 41, 50, 58
    electron, xiv, 55, 92, 95, 98, 121, 144
Migula, Walter, 60, 62
Mill, John Stuart, 81
Miller, Stanley, 195
Minchin, Edward, 42, 43, 69, 311
Mineral kingdom, 7, 15, 22, 26, 34
Mitochondria
    absent in some protists, 176, 257, 308,
        309, 310, 311, 312
    anthropomorphism and, 123, 310
    coined, 118
    cytochrome *c* in, 123, 130, 228, 231
    DNA in, 122, 315
    electron microscopy of, 98
    endogenous conception of, 124, 125,
        233, 234
    lacking in bacteria, 98, 113, 157, 304, 307
    pleomorphism, 118, 122
    purple bacteria and, 231, 238, 206, 310
    alpha-proteobacteria and, 310
    respiration and, xi, 98
    16S rRNA analysis of, xv, 233, 234,
        288, 316
    symbiotic origin of, xi, 120–125, 116,
        118, 171, 176, 203, 228, 229, 230,
        234, 285, 288, 309, 310, 311, 315
Mitochondrial Eve, 249
Mitosis, xi, 43, 70, 73, 74, 79, 92, 98, 113,
    122, 176, 278, 312
Mixotricha paradoxa, 122
Model organism. See *E. coli*
Modern Synthesis, The, ix, 78, 79, 269. *See*
    also Evolutionary synthesis
Molds, 9, 72, 73, 111
Molecular clock, 130, 241, 258, 268, 361
Molecular genetics, 113, 243, 244, 249,
    268, 289, 296

Mollusk, 33
Monas, 13, 14
Monera. *See also* Prokaryotes
    Haeckel's denied, 43
    kingdom, xiii, xiv, 36, 73–77, 84, 86,
        88, 91, 94, 107, 111, 115, 127, 198,
        257, 265, 269, 270, 276, 315
    monophyletic views of, 75, 101, 103, 110
    negative definition of, 74, 88, 102, 265
    nucleus and, 28, 38–42, 73
    polyphyletic views of, 36, 39, 74–75,
        101, 102, 103, 275
    precellular entities, 28, 34, 36–42,
        116, 171
    prokaryotes and, xiv, 102
    proteins, phylogenies and, 127
    as protist, 28, 36–39, 94, 98, 103, 104
    viruses and, 111
Monist philosophy, 36–37
Monod, Jacques, 95, 105, 149
Monoderms, 308
Monophyletic classification, 108, 133, 134
    Archaebacteria and, 253, 255, 273
    Eocyte and, 249, 252
    monera and, 39, 101, 105, 110
    prokaryote and, 106, 258, 265, 273,
        301, 306
Moore, Benjamin, 65, 88
Moore, R.L. 144
Mother. *See also* Origin of life
    of eukaryotes, 246, 307–311
    universal, 105, 286, 289, 295
Muller, H.J.
    on genes and viruses, 66, 95, 299
    on species, 91
Müller, Miklós, 330
Munich Polytechnic Institute, 201, 207
Murien, 189. *See also* Peptidoglycan
Murray, R.G.E., 106, 107, 183, 184
Mushrooms, 8, 9, 22
Mycetozoa, 39
Mychota kingdom, 77, 335
Mycobacteria, 89
Mycoplasm, 117, 118, 307
Mycoplasma, 68, 192, 193, 196, 218
Myxomyctes, 72

Nägeli, Carl von, 46, 47, 51, 58
NASA, 177, 180, 195, 362
Nass, Margit, 122
Nass, Sylvan, 122

National Institutes of Health, 148, 213, 259, 362
National Jewish Hospital and Research Center, 232, 371
National Research Council of Canada, 190, 194
National Science Foundation, 177, 362
Natural classification. *See* Classification
Natural history, ix, xv, 6, 8, 17, 18, 19, 29, 34, 63, 69, 77, 126, 199, 207, 243, 258, 280, 317
Natural kinds, 8, 22, 77, 81, 282, 290, 302
Natural law, 123
Natural selection, vii, xviii, 10, 29, 35, 78, 115, 118, 119, 130, 136, 187, 188, 195, 221, 268, 288
Nature letters, 249–253, 275, 303–306
Naturphilosophie, 19, 29
Nazar, Ross, 190
Nei, Masatoshi, 253
Nematodes, 9, 311
Neomura, 304
Neurath, Hans, 227, 228
Neurospora, 113
New Systematics, The, 79–84
New York Times, 178, 179
Nirenberg, Marshall, 148, 149
Nitrifying bacteria, 55, 61, 63, 74, 76
Nitrogen-fixing bacteria, 54, 55, 121
Nominalism, 8, 9, 80, 294
Nucleic acid, coined, 42
Nuclein, 42
Nucleus. *See also* Eukaryote; Mitosis
    bacteria and (*see* Monera; Prokaryote)
    coined, 20
    definitions, 38, 41–43, 73
    origin of, xii, 172, 220, 238, 252, 307–313
Numerical taxonomy. *See* Phenetics; Taxonomy

Oceans, 41, 49, 280, 293. *See also* Origin of life
Ochman, Howard, 288
Ockham's razor, 125
Oken, Lorenz, 19
Oligonucleotide cataloguing, 159, 160, 166, 181, 184, 200, 204, 241
Olsen, Gary, 251, 252, 259, 288, 290
Ontogeny
    coined, 28

phylogeny and, 33, 43
Oparin, Aleksandr, 66, 88, 149, 220–225, 228
Operon, 149
Orangutans, 185
Origin of life. *See also* Monera; Progenotes concept
    Autotrophs first, 59, 62, 63, 65, 88, 96, 135, 146, 163, 169, 220, 222, 224, 238, 315, 316
    Heterotrophs first, xiv, 65, 66, 88–90, 220–225, 228, 238, 315, 339
    ocean and, 41, 49, 119, 135, 169, 221, 222, 223, 225, 286
Origin of species, ix, x, 10, 29–35, 78, 80, 81, 118, 119, 124, 269. *See also* Evolution; Evolutionary synthesis; Lateral gene transfer; Natural selection; Symbiosis
Origin of Species, The, xi, 10, 25, 29–35, 275
Orla-Jensen, Sigurd, 59–63, 88, 135
Oró, John, 185, 186, 239
Orthologous genes, defined, 228, 231
Osawa, Syozo, 217–219
Osborn, Henry Fairfield, 53
Outgroup, defined, 230, 248, 254, 255
Owen, Richard, 20, 21
    on intelligent design, 25
    kingdoms of, 22, 23, 25, 77
    on origin of life, 21

Pace, Norman, 232, 251
    on genetic core, 290
    microbial diversity, 232, 280, 290, 293
    on prokaryote-eukaryote dichotomy, 277, 303–306
    universal tree, 291
Paleogenetics, 129
Paleontology, 14, 20, 23, 25, 90. *See also* Fossil record
pan-adaptationism, 187, 188
Pangenes, 10
Pangloss, Dr., 188
Panlateralism, 292, 296, 297, 297, 298, 317
Panspermia, 49, 89, 239, 240
Pansporella perplex, 96, 97
Paralogous genes, defined, 228
Paramecium, 93, 120
Parkaryote superkingdom, 246, 251, 257
Pascher, Adolf, 101
Pasteur Institute, 45, 66, 84, 96, 149, 166

Pasteur, Louis, 45, 48, 51, 85, 86, 190, 279, 314, 329
Pauling, Linus, 129, 130, 296
Penicillin, 39, 92, 136, 138
Pennisi, Elizabeth, 293
Peptidoglycan
  amino acid composition of, 202–205
  evolution of, 196, 200, 204, 216, 305
  lacking in archaebacteria, 168, 173, 189, 200, 204, 206, 305
  prokaryote definition, xv, 173, 175, 189, 301
Phagocytosis, 311, 312
Phenetics, 133, 382. See also Numerical taxonomy
Philippe, Herve, 286
Philosophie Zoologique, 10, 13
Photocyta, 246, 249, 251, 257, 271
Phylogeny, coined, 28. See also Cladistics; Classification; Ribosomal RNA; Taxonomy
Phytomonera, 59. See also Blue-green algae
Phytozoa, 21
Pithecanthropus alalus, 34
Planctomycetes, 305
Plants. See also Bacteria; Blue-green algae; Chloroplasts
  animal-plants, xii, 7, 8, 9, 15
  diversity of, 270
  fission plants, xiii
  fossil record of, 81
  genealogies, 14
  higher, 57, 65, 72, 74
  hybridization, 7, 81, 82, 119, 137, 142, 298
  kingdom, viii, 7, 16, 23, 73, 74, 106, 107, 108, 111, 261, 265, 276
  lower, 71
  origins of, x, xii, 14, 25, 35, 39, 40, 74, 94, 108, 110, 112, 116, 182, 223, 270, 271, 307, 315 (see also Symbiosis)
  polyploidy, 81, 82
  producers, 108
  reality of, 8, 9, 21–26
  reticulated descent, 83, 119
  soul, 32
  species concept and, 91
Plasmagenes, 150
Plasmids, 121, 136, 138, 141, 125, 287
Plato, 19, 336
Platonic idealism, 11

Plaut, Walter, 115, 121
Plentitude, principle of, 11
Pleomorphism, 51, 52, 68, 314
Pliny, 7
Pneumococcus, 93
Pollard, Ernest, 146
Polygastria, 19, 22, 23, 24
polymerase chain reaction, 214, 244
Polyps, 8, 9, 13, 15, 16, 68
Ponnamperuma, Cyril, 178, 186
Population genetics. See Genetics
Portier, Paul, 118, 119
Pouchet, Félix, 36
Precambrian, 112, 113, 135, 182
Predators, origin of, 311
Predatory bacteria, 313
Premonera, 102
Prévot, André, 71, 84, 86, 101
Primalia, xii, 26, 27
Primordial soup, xiv, 135, 146, 149, 220, 221, 222, 223, 244, 315
Principles of Numerical Taxonomy, 133, 143
Principles of Systematic Zoology, 273, 274
Pringsheim, Ernst, 101, 102, 127
Proceedings of the National Academy of Sciences, 174, 177, 264, 276, 277
Progenotes concept, 146, 170–175, 185, 216, 223, 237, 245, 261
Prokaryota. See also Monera
  Domain, 279
  Empire, 277
  kingdom, 266, 271
  superkingdom, 106, 276, 315
Prokaryotes, xv, xvi, 85–99, 104, 106, 162. See also Monera
  dispute over, 273–277, 303–306, 316, 317 (see also Archaea; Archaebacteria)
  essentialism and, 302
  Monophyletic views of, 102, 103, 316
  myth, 301
  negative definition, 104, 107, 113, 157, 257, 258, 265, 301, 302, 303, 304
  Polyphetic views of, 103, 173, 176, 258, 326, 273, 303
  progenotes and, 171–174, 176
Prophage, 95
Protists. See also Archezoa; Eukaryote
  anaerobic, 307, 310, 311
  diversity, 108, 270, 278, 312
  first observed, 4, 8
  higher and lower, 94, 104, 106

Protists (*continued*)
   kingdom, viii, xii, xiv, 28, 36, 39–44,
     71, 72, 74, 77, 94, 106–109, 111, 131,
     198, 265, 268–269, 276, 315
   paraphyletic, 265, 110
   phylogeny, 135, 271, 307, 313
   polyphyletic group, 109
   subdomain, 270
   symbionts in, xi, 115, 122, 123, 232,
     298, 310
Protoamoeba, 77
Protoctista, 24–26, 77, 108, 324
*Protomastigote*s, 96
Protophyta, 24, 29
   kingdom, 95
Protoplasm, 37, 42, 92, 96, 117, 280,
   286, 312
Protozoa, 96. *See also* Eukaryote
   as class, 21
   as kingdom, xii, 20, 23, 24, 25, 27, 39,
     79, 77
   as lower animals and animals, 71, 72
   nucleus, 96
   phylogeny, 96, 135
   within protista, 109
   sex, 93–94, 98
Psychodes, 15
Psychodiaire kingdom, 15–16, 21, 23
Pure cultures, 51, 55, 164
Purple bacteria, 62, 88, 92, 160, 206,
   231, 235, 238, 255, 264, 285, 286,
   309, 368

Quorum sensing, 305

Radiolarians, 13, 17, 29
Raven, Peter, 107, 125
Realists, 106–108
Reanney, Darryl, 284
Recapitulation theory, 33, 36, 326
Reductionism, 153, 272, 294
Reptiles, 13, 14, 112, 134, 185
Reticulated evolution. *See* Evolution
Ribosomal RNA, phylogenies, x, 144, 145,
   154–159, 166–170, 174, 175, 180–184,
   204, 215, 226, 231–238, 249, 253, 254,
   274, 285, 289–298, 303, 308, 315, 316
Ribosomes
   chloroplast, 121, 122
   function, x, 143, 147, 149, 151, 153
   mitochondria, 121, 122

origin, 155
prokaryotic, 124
Rickettsia, 95, 107
Rifampicin, 210, 212, 214, 245
Ris, Hans, 115, 121
RNA polymerase. *See* Archaebacteria
RNA splicing, 246, 258
RNA world, 155
Roberts Richard, 245
Robinow, C.F., 92, 98

Sager, Ruth, 125
Saint-Hilaire, Geoffroy, 14
Salmonella, 284, 288
Saltational evolution. *See* Evolution; Lateral
   gene transfer; Symbiosis
Sanger Frederick, 129, 158, 159, 166,
   244, 271
Sarich, Vincent, 249
Sauerkraut, 201, 208
Schelling, Friedrich, 19, 29
Schimper, Andreas, 116
Schizomycetes, 46, 47, 50, 58, 62, 88, 101.
   See also Bacteria
Schizophyta, 31. *See also* Blue-green algae
   kingdom, 72
Schleiden, Matthias, 20
Schleifer, Karl-Heinz, 202, 203, 206, 308
Schramm, Gerhard, 209
Schwartz, Robert, 228–231, 235, 255
Science letters, 231, 240, 241
Searcy, Denis, 330
Sedgwick, T.W., 57
Semantides, 130
Sex
   bacterial, xiv, 93, 94, 96, 98, 137, 302
   cells, 11, 19, 298
   control, 93
   Eukaryotes, 304
   hormones, 219
Sharp, Phillip, 246
Shigella, 66, 137, 139
Siebold, Karl von, 41, 44
Signature analysis, 160, 161, 166, 167,
   313, 400
Simpson, George Gaylord, 241, 268, 337
Smith, John Maynard, 310
Sneath, Peter
   bacterial phylogeny, 135, 141, 144
   on lateral gene transfer, 127, 141, 284
   numerical taxonomy, 12, 134

Society of American Bacteriologists, 57, 58, 60, 62, 67, 86
Sogin, Mitchell, 251, 312
Sokal, Robert, 158, 159, 358
Soluble RNA, 143, 147, 152. *See also* Transfer RNA
Sonneborn, Tracy, 120, 175, 181
Speciation, 91, 110. *See also* Hybridization; Lateral gene transfer; Symbiosis
   Allopatric, 82
   in animals, 82
   Bacteria and, 284, 288
   and gene history, 228, 272
   in plants, 81, 82
   symbiosis and, 118, 119
Species
   animals and, 68, 81
   bacteria, ix, x, xiv, 7, 47, 50, 51, 54, 58, 59, 67, 68, 69, 70, 81, 91, 137, 285, 290, 292, 295, 297, 298, 305, 314, 316, 317, 318
   concepts of, 33, 67, 68, 69, 81, 82, 314
   numbers, 103, 226, 278
   physiological, 70
   plants and, 68, 81–82
   reality of, 9, 77, 81, 338
Spencer, Herbert, 20
Spiegelman, Sol, 149, 150, 158, 163, 232
Spirillum, 19, 47, 51
Spirochaeta, 19, 122, 309
Spirodiscus, 19
Spliceosomes, 246, 304
Split genes, 246
Sponges, 33, 294
   as animal, 22
   as invertebrates, 15
   in kingdom Protozoa, 23
   as plant-animals, xii, 7, 8
   as plants, 24
   as subkingdom *Spongiae*
Spontaneous generation, 13, 14, 36, 39, 40, 45, 48, 49, 63, 79
Spores. *See* Bacterial spores
Sprague, Thomas, 83
Stackebrandt, Erko, 205, 206, 217, 226, 236
Stanier, Roger. See also *Microbial World, The*; Van Niel, C.B.
   anatomy of bacteria, 86, 88, 91–98, 202
   on bacterial sex, 88, 93, 137
   on blue-green algae, 107, 108, 102
   adopts kingdom Monera, 88
   renounces kingdom Monera, 91, 94

   adopts kingdom Protista, 94
   lateral gene transfer, 138, 284
   adopts monophytic conception of bacteria, 100, 103–104
   attempts a natural classification, 87–90, 135
   renewed hope in natural classification, 132
   renounces a natural classification, 85, 90, 91, 106, 126, 235
   on phagocytosis, 311
   on polyphyletic conception of bacteria, 103
   on prokaryote and eukaryote dichotomy, 96–99, 102, 301
   on prokaryote before eukaryote, 114
   on realists and idealists, 87–88
Staphylococci, 51, 138, 139, 352
Stebbins, George, 82
Stent, Gunther, 123
Stetter, Karl, 199
   education, 207
   grouping the archaebacteria, 214, 217, 251
   manganese effect, 208–209
   partnership with Zillig, 208–217
   professor at Regensburg, 215
   relations with Kandler, 207, 210
   RNA polymerases of archaebacteria, 210–212, 214
   student protests of the 1960s, 207, 371
   thermoacidophiles, 211–214
Stocker, Bruce, 137
Stöffler Georg, 251
Streptolydigin, 212, 214, 245
Student protests, 29, 116, 120, 222
Subkingdoms, 26, 109, 257, 261
Sulfolobus, 245. *See also* Archaebacteria; Archaea
   creanarchaeota, 263, 264
   discovery of, 192
   eukaryotes and, 217 235, 246–250
   halophiles and, 196, 197, 203, 210, 215
   introns of, 246
   lipids of, 194–197
   methanogens and, 194–197, 214
   RNA polymerase of, 211–214
   16S rRNA analysis of, 168, 191, 193, 196, 211
   Thermoplasma and, 193, 194, 203, 210
   viruses of, 215, 250
   walls of, 203

Superkingdoms. See also Domain; Empire;
    Urkingdom
    Eucaryota, 106, 107, 183, 257, 266,
        276, 315
    karyotes, 244, 251, 252
    parkaryotes, 244, 251, 252
    Procaryota, 106, 107, 183, 257, 266,
        276, 315
Symbiogenesis, 116, 312. See also Symbiosis
symbiome, 299
Symbiosis
    aboriginal property of life, 171, 173,
        174, 217
    anthropomorphisms, 123, 310, 311
    biological philosophy, 121, 299
    centrioles and, 116, 122
    chloroplasts origins, xi, 54, 116
    Chromista and, 257
    coined, 115
    Convoluta roscoffensis and, 115
    crown galls and, 121
    DNA screening and, 311
    evolutionary synthesis and, 119, 121,
        269, 311
    germ theory and, 14, 119, 121
    insects and, 311
    Lamarckism and, 78
    lichens, 54
    medicine and, 119, 121
    Mendelian genetics and, 119
    mitochondria origins, xi, 116, 121, 122,
        123, 171, 309–310
    mutualism and, 123, 164
    mycorrhiza, 54
    nematodes and, 311
    nitrogen-fixing bacteria, 54, 121
    opposition to, 119–120
    origin of eukaryote, xii, 116, 122, 299,
        307–312
    origin of new genes, 118, 122, 123,
        125, 171
    origin of species and, 118, 119
    phagocytosis and, 311–312
    protists and, xi, 298, 304, 307, 317
    16S rRNA analysis and, 232–234
    saltationism and, 115, 122
    speculation and, 120, 125, 126
    struggle for existence and, 120
    viruses and, 119, 121
    Wolbachia and, 311
Systema Naturae, 272

Systematics. 100, 134. See also
    Classification; Taxonomy
    bacterial, 87, 92, 181, 244
    cladism and, 134, 273, 292
    modern synthesis and, 79–84
    16S rRNA and, 181, 306
Syvanen, Michael, 284

Tachytelic species, 241
Tatum, Edward, 93
Taxonomy. See also Cladistics;
    Classification; Phenetics
    aims of, 20, 60, 67, 270, 278, 279
    anthropomorphism and, 262–263
    botanical, 72, 73, 83, 84, 90
    conventions of, 6, 77
    grades and, 26, 69, 77, 276, 302
    numerical, 133, 134, 143–145, 306
    phylogeny and, xiii, xiv, 21, 40, 63, 68,
        69, 80, 82–94, 108, 133, 188, 227,
        278, 279
    revolution in, ix, 60, 186, 188, 205, 226,
        266, 289, 314
    statistical, 61, 134
    zoological, 82, 83, 90, 273, 274
Teleology, 19, 29, 29
Tempo and Mode of Evolution, 241
Thermoacidophiles, 263
Thermodynamics and evolution, 271
Thermoplasma
    DNA-dependent RNA polymerase of,
        210, 214, 235
    eukaryotes and, 309
    discovery of, 192
    lipids of, 194–197
    mycoplasma and, 192
    RNA polymerase of
    16S rRNA analysis, xv, 163, 187,
        191, 193, 196, 198
    sulfolobus and 193, 194, 203,
        210, 235
    walls and, 168, 192, 193
Thermoproteales, 214, 245, 250
Thermus aquaticus, 214
Thomas, Lewis, 123
Thomson, Wyville, 49
Tilney-Basset, R.A.E., 126
Tobacco mosaic virus. See Viruses
Tomakomai, 218–219
Tornabene, Thomas, 194–200, 216
Transcendantal anatomists, 14, 33

Transcription, 168, 172, 207, 207–216, 219, 258, 265, 274, 283–290, 295, 303–306, 313, 318. *See also* DNA-dependent RNA polymerase; Genetic code; Genetic core
Transduction, 95, 120, 121, 136–138, 141, 284, 298
Transfer RNA, 152, 168, 175, 181, 186, 206, 246, 290
  of archaebacteria, 189, 217, 138, 235, 246, 247, 289
  concepts of, 151–153
  of *E. coli* and yeast, 158
  evolution of, 154–157
Transformation, 78, 93, 136, 138, 315, 333. *See also* Lateral gene transfer
Transgenic organisms, 128, 215, 284
Translation, 143, 149–153, 155
  evolution of, 145, 146, 149, 153–157, 168, 170–175, 185, 216, 223, 237, 245, 261
Tree of life, vii, viii, x, xi, xii, xv, 18, 30, 37, 41, 69, 76, 80–82, 102, 127, 130, 131, 132, 156, 169, 193, 198, 216, 218, 219, 226–242, 249, 252, 254, 264, 281, 283, 291, 316, 317
  images, 12, 31, 38, 64, 75, 90, 97, 103, 108, 109, 110, 111, 117, 229, 230, 257, 252, 254, 255, 264, 291, 295, 296, 304
  lateral gene transfer and, 76, 82, 119, 142, 282, 284, 286, 288, 289, 290, 293–300, 302, 304, 317, 318
  rooting, 228–230, 253–255, 260, 275, 278, 290, 302
"Tree of Porphyry," 7
Triceratops, 134
Trilobites, 112
Tubercle bacilli, 46, 52
Turrill, William, 81
Twort, Frederick W., 65, 79
Typhoid, 46, 52, 53, 109
Tyrannosaurus, 134

Unity of biochemistry, 105, 165
Unity of life, xiv, xix, 14, 34, 75, 100, 101, 103, 105, 107, 109, 111, 113, 165, 315
Universal phylogenetic tree. *See* Tree of life
University of Illinois, 137, 145, 149, 150, 163, 164, 195, 232, 261. *See also* Fox, George; Woese, Carl; Wolfe, Ralph

University of Munich, 47, 199, 201, 202, 206, 210, 244, 369, 371. *See also* Kandler, Otto
University of Regensburg, 215
Uralgae, 125
Urkaryote, 176, 171, 245, 247
Urkingdom, 175, 176, 199, 187, 191, 194, 204, 210, 215, 218, 236, 245, 246, 251

Van Niel, C.B., 301, 265
  summer course, 86, 164
  adopts kingdom monera, 88
  renounces kingdom monera, 91, 94
  attempts a natural classification of bacteria, 87–90, 135, 239
  renounces a natural classification of bacteria, 85, 90, 91, 106, 235, 265
  on prokaryote-eukaryote dichotomy, 96–99, 102, 103, 301
  on realists and idealists, 87–88
Venter, Craig, 288
Venus, 224
Vermeer, Johannes, 3
Vermes, 8
Vertebrates, 21, 22, 41, 140, 182, 250
Vibrio, 13, 19, 47, 51
Vietnam war, 207
Viking program, 178, 195
Virchow, Rudolf, 20
Virologists, 121
Virulence, 96
Viruses. *See also* Lateral gene transfer
  Archaebacterial, 215, 244, 250
  Archetista kingdom 115
  cancer and, 158
  as disease or poison, 18, 46, 55, 95
  as distinct from a cell, 95, 96
  diversity of, 299
  as fluid, 55
  as genes, 66, 95, 110
  gene therapy and, 138
  as genetic particles, 93, 95, 299
  human genome and, 298
  as mini bacteria, 65, 115
  molecular biology of the gene and, 146, 148, 150, 205, 209
  as monera, 121
  origin of, 66, 110, 299
  origin of the genetic code and, 140, 284, 287
  as protists 298

Viruses (*continued*)
  as protophyta, 95
  saltational changes, 150
  as symbionts, 119, 120, 121
  tobacco mosaic, 55, 331, 341
Volvox, 8, 13

Wächtershäuser, Günter, 245
Wallace, Alfred Russel, 29
Wallin, Ivan, 118, 119
Watanabe, Tsutomu, 139
Watasé, Shôsaburô, 116, 307
Watson, James, 146, 160, 186, 209, 249
Weber, T.A., 65
Wheelis, Mark, 261, 262, 264, 265, 267,
    273, 274, 275, 296, 301, 307
Whitehead, Alfred North, 153
Whitman, William, 279
Whittaker, Robert, 265
Wilkins, Maurice, 249
Wilkinson, J.F., 186, 216
Wilson, Allan, 249
Wilson, Edmund, 42, 43
Wilson, Thomas, 17, 26
Winogradsky, Sergie, 54, 86, 91
Winslow, C.-E.A., 57, 60, 61, 86
Woese, Carl. *See also* Archaea;
    Archaebacteria; Tree of life
  archaebacteria discovery, 162–167, 175
  on archaebacteria-eukaryote
      relationship, 244–246, 253
  on archaebacteria as primitive, 169–170,
      191, 216, 219
  arrival at University of Illinois, 150
  code cracking and, 146–148
  Crafoord Prize, 280
  Crick and, 146, 147, 151, 152, 156,
      239, 240
  on Darwinian transition, 294
  education of, 146
  eocyte concept and, 247, 248, 251
  on eukaryotic origins, 170, 171, 175,
      176, 237, 245, 246, 258, 307
  on evolution of the genetic code, 145,
      149–156, 176
  Fox and, 160–170, 175, 176, 181–186,
      201–205, 210, 217–219, 226, 235–
      241, 257, 262, 289, 300, 307, 358
  on frozen accident theory, 151–152
  at General Electric Knowles
      Laboratory, 146

on the genetic core, 241, 285
on genomics, 259
holism of, 153, 272, 294
on Hori and Osawa, 217–220
Kandler and, 201–205, 212–213,
    257–266
Lake and, 247–249
Lamarckism and, 305
on lateral gene transfer, 235, 236, 241,
    285, 286, 294, 295, 305
Margulis and, 173, 174, 198
Mayr and, 269–282
noncladistic views of, 273
oligonucleotides cataloging, 158–171
origins of life, 220–225
at Pasteur Institute, 149
progenotes concept, 170–174,
    286–287, 300
on prokaryote-eukaryote dichotomy,
    157, 170, 176, 236, 258, 277–281, 301
reciprocating ratchet mechanism, 154, 160
on rooting the universal tree, 254–255
Sogin and, 159
Spiegelman and, 150, 158, 178
Stanier and, 183, 239, 265, 301
on symbiosis, 171, 173, 176, 232, 233
on tempo and mode, 241, 242, 267, 268,
    271, 274, 287
on three domains, 257–266
on tRNA as enzyme, 153
turn to microbial phylogenetics, 176
on universal tree, 236–243
Wolfe and, 163–167, 178–180, 236–239
at Yale, 146
Zillig and, 199–214, 219, 243–251,
    260, 262
Zuckerkandl and, 172–174, 187, 191, 239
Wöhler, Friedrich, 22
Wolbachia, 311
Wolfe, Ralph
  on archaebacteria, 166, 167, 169, 205, 214
  on coenzyme M, 164, 165, 167
  culturing methanogens, 163–165, 195
  Kandler and, 199–201, 204, 210
  on media and third form of life, 178
  methanogen lipids and, 195–197
  on unity of biochemistry, 165
  universal phylogenetic tree and, 236–239
  Woese and, 163–167, 178–180, 236–239
Wollman Elie, 95, 139
Woolhouse, W.H., 126

World War I, 69, 96
World War II, 10, 68, 70, 78, 85, 92, 120, 139, 201, 361
Worms, 10, 13, 14, 104, 112
Wright, Sewall, 78

Yanofsky, Charles, 232
Yčas, Martynas, 148
Yeast, 4, 47, 55, 73, 93, 130, 150, 158, 168, 211, 217
Yellowstone Park, 192
Yogurt, 46, 208
Young, Richard, 177, 178

Zamecnik, Paul, 152
Zillig, Wolfram, 260, 266
  Butenandt and, 209
  on archaebacteria as ancient, 245
  on archaebacterial viruses, 215, 244, 250
  on archaebacteria relations with eukaryotes, 219, 244–246
  on eocyte concept, 247, 248, 251
  grouping the archaebacteria, 214, 217, 245
  on Hori and Osawa, 217–219
  Lake and, 247–249
  RNA polymerases of archaebacteria, 210–212, 214
  Schramm and, 206
  Stetter and, 199, 207–212
  on symbiotic origin of eukaryote, 307–308
  on thermoacidophiles, 211–214
  Woese and, 199, 210–214, 219, 243–251, 260, 262
Zinder, Norton, 95, 120
Zoology, 22, 29, 71, 73, 80, 103, 269, 273, 274
Zoomonera, 39
Zoophyta, xii, 7, 8, 9
Zuckerkandl, Emile, 296
  editor of *Journal of Molecular Evolution*, 130, 172, 191
  on lateral gene transfer, 174
  on the molecular clock concept, 129–130
  on paleogenetics, 129
  Pauling and, 129
  on the progenotes concept, 172–174
  Woese and, 172–174, 187, 191, 239